PREFACE

Credit for the longevity of this work belongs to the original two authors, Lawrence Kinsler and Austin Frey, both of whom have now passed away. When Austin entrusted us with the preparation of the third edition, our goal was to update the text while maintaining the spirit of the first two editions. The continued acceptance of this book in advanced undergraduate and introductory graduate courses suggests that this goal was met. For this fourth edition, we have continued this updating and have added new material.

Considerable effort has been made to provide more homework problems. The total number has been increased from about 300 in the previous editions to over 700 in this edition. The availability of desktop computers now makes it possible for students to investigate many acoustic problems that were previously too tedious and time consuming for classroom use. Included in this category are investigations of the limits of validity of approximate solutions and numerically based studies of the effects of varying the various parameters in a problem. To take advantage of this new tool, we have added a great number of problems (usually marked with a suffix "C") where the student may be expected to use or write computer programs. Any convenient programming language should work, but one with good graphing software will make things easier. Doing these problems should develop a greater appreciation of acoustics and its applications while also enhancing computer skills.

The following additional changes have been made in the fourth edition: (1) As an organizational aid to the student, and to save instructors some time, equations, figures, tables, and homework problems are all now numbered by chapter and section. Although appearing somewhat more cumbersome, we believe the organizational advantages far outweigh the disadvantages. (2) The discussion of transmitter and receiver sensitivity has been moved to Chapter 5 to facilitate early incorporation of microphones in any accompanying laboratory. (3) The chapters on absorption and sources have been interchanged so that the discussion of beam patterns precedes the more sophisticated discussion of absorption effects. (4) Derivations from the diffusion equation of the effects of thermal conductivity on the attenuation of waves in the free field and in pipes have been added to the chapter on absorption. (5) The discussions of normal modes and waveguides

iii

have been collected into a single chapter and have been expanded to include normal modes in cylindrical and spherical cavities and propagation in layers. (6) Considerations of transient excitations and orthonormality have been enhanced. (7) Two new chapters have been added to illustrate how the principles of acoustics can be applied to topics that are not normally covered in an undergraduate course. These chapters, on finite-amplitude acoustics and shock waves, are not meant to survey developments in these fields. They are intended to introduce the relevant underlying acoustic principles and to demonstrate how the fundamentals of acoustics can be extended to certain more complicated problems. We have selected these examples from our own areas of teaching and research. (8) The appendixes have been enhanced to provide more information on physical constants, elementary transcendental functions (equations, tables, and figures), elements of thermodynamics, and elasticity and viscosity.

New materials are frequently at a somewhat more advanced level. As in the third edition, we have indicated with asterisks in the Contents those sections in each chapter that can be eliminated in a lower-level introductory course. Such a course can be based on the first five or six chapters with selected topics from the seventh and eighth. Beyond these, the remaining chapters are independent of each other (with only a couple of exceptions that can be dealt with quite easily), so that topics of interest can be chosen at will.

With the advent of the handheld calculator, it was no longer necessary for textbooks to include tables for trigonometric, exponential, and logarithmic functions. While the availability of desktop calculators and current mathematical software makes it unnecessary to include tables of more complicated functions (Bessel functions, etc.), until handheld calculators have these functions programmed into them, tables are still useful. However, students are encouraged to use their desktop calculators to make fine-grained tables for the functions found in the appendixes. In addition, they will find it useful to create tables for such things as the shock parameters in Chapter 17.

From time to time we will be posting updated information on our web site: www.wiley.com/college/kinsler. At this site you will also be able to send us messages. We welcome you to do so.

We would like to express our appreciation to those who have educated us, corrected many of our misconceptions, and aided us: our coauthors Austin R. Frey and Lawrence E. Kinsler; our mentors James Mcgrath, Edwin Ressler, Robert T. Beyer, and A. O. Williams; our colleagues O. B. Wilson, Anthony Atchley, Steve Baker, and Wayne M. Wright; and our many students, including Lt. Thomas Green (who programmed many of the computer problems in Chapters 1–15) and L. Miles.

Finally, we offer out heartfelt thanks for their help, cooperation, advice, and guidance to those at John Wiley & Sons who were instrumental in preparing this edition of the book: physics editor Stuart Johnson, production editor Barbara Russiello, designer Kevin Murphy, editorial program assistants Cathy Donovan and Tom Hempstead, as well as to Christina della Bartolomea who copy edited the manuscript and Gloria Hamilton who proofread the galleys.

Alan B. Coppens   **James V. Sanders**
Black Mountain, NC   *Monterey, CA*

CONTENTS

CHAPTER 3
VIBRATIONS OF BARS

CHAPTER 4
THE TWO-DIMENSIONAL WAVE EQUATION:
VIBRATIONS OF MEMBRANES AND PLATES

CHAPTER 8
ABSORPTION AND ATTENUATION OF SOUND

CHAPTER 9
CAVITIES AND WAVEGUIDES

CHAPTER 10
PIPES, RESONATORS, AND FILTERS

CHAPTER 11
NOISE, SIGNAL DETECTION, HEARING, AND SPEECH

CHAPTER 12
ARCHITECTURAL ACOUSTICS

CHAPTER 13
ENVIRONMENTAL ACOUSTICS

CHAPTER 14
TRANSDUCTION

CHAPTER 15
UNDERWATER ACOUSTICS

CHAPTER   16
SELECTED NONLINEAR ACOUSTIC EFFECTS

CHAPTER   17
SHOCK WAVES AND EXPLOSIONS

APPENDIXES

FUNDAMENTALS OF VIBRATION

1.1 INTRODUCTION

Before beginning a discussion of acoustics, we should settle on a system of units. Acoustics encompasses such a wide range of scientific and engineering disciplines that the choice is not easy. A survey of the literature reveals a great lack of uniformity: writers use units common to their particular fields of interest. Most early work has been reported in the CGS (centimeter–gram–second) system, but considerable engineering work has been reported in a mixture of metric and English units. Work in electroacoustics and underwater acoustics has commonly been reported in the MKS (meter–kilogram–second) system. A codification of the MKS system, the SI (Le Système International d'Unités), has been established as the standard. This is the system generally used in this book. CGS and SI units are equated and compared in Appendix A1.

Throughout this text, "log" will represent logarithm to the base 10 and "ln" (the "natural logarithm") will represent logarithm to the base e.

Acoustics as a science may be defined as the generation, transmission, and reception of energy as vibrational waves in matter. When the molecules of a fluid or solid are displaced from their normal configurations, an internal elastic restoring force arises. It is this elastic restoring force, coupled with the inertia of the system, that enables matter to participate in oscillatory vibrations and thereby generate and transmit acoustic waves. Examples include the tensile force produced when a spring is stretched, the increase in pressure produced when a fluid is compressed, and the restoring force produced when a point on a stretched wire is displaced transverse to its length.

The most familiar acoustic phenomenon is that associated with the sensation of sound. For the average young person, a vibrational disturbance is interpreted as sound if its frequency lies in the interval from about 20 Hz to 20,000 Hz (1 Hz = 1 hertz = 1 cycle per second). However, in a broader sense acoustics also includes the *ultrasonic* frequencies above 20,000 Hz and the *infrasonic* frequencies below 20 Hz. The natures of the vibrations associated with acoustics are many, including the simple sinusoidal vibrations produced by a tuning fork, the complex vibrations generated by a bowed violin string, and the nonperiodic motions associated with an explosion, to mention but a few. In studying vibrations it is advisable to begin with the simplest type, a one-dimensional sinusoidal vibration that has only a single frequency component (a pure tone).

1.2 THE SIMPLE OSCILLATOR

If a mass m, fastened to a spring and constrained to move parallel to the spring, is displaced slightly from its rest position and released, the mass will vibrate. Measurement shows that the displacement of the mass from its rest position is a sinusoidal function of time. Sinusoidal vibrations of this type are called *simple harmonic vibrations*. A large number of vibrators used in acoustics can be modeled as simple oscillators. Loaded tuning forks and loudspeaker diaphragms, constructed so that at low frequencies their masses move as units, are but two examples. Even more complex vibrating systems have many of the characteristics of the simple systems and may often be modeled, to a first approximation, by simple oscillators.

The only physical restrictions placed on the equations for the motion of a simple oscillator are that the restoring force be directly proportional to the displacement (Hooke's law), the mass be constant, and there be no losses to attenuate the motion. When these restrictions apply, the frequency of vibration is independent of amplitude and the motion is simple harmonic.

A similar restriction applies to more complex types of vibration, such as the transmission of an acoustic wave through a fluid. If the acoustic pressures are so large that they no longer are proportional to the displacements of the particles of fluid, it becomes necessary to replace the normal acoustic equations with more general equations that are much more complicated. With sounds of ordinary intensity this is not necessary, for even the noise generated by a large crowd at a football game rarely causes the amplitude of motion of the air molecules to exceed one-tenth of a millimeter, which is within the limit given above. The amplitude of the shock wave generated by a large explosion is, however, well above this limit, and hence the normal acoustic equations are not applicable.

Returning to the simple oscillator shown in Fig. 1.2.1, let us assume that the restoring force f in newtons (N) can be expressed by the equation

$$f = -sx \tag{1.2.1}$$

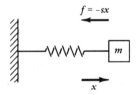

Figure 1.2.1   Schematic representation of a simple oscillator consisting of a mass m attached to one end of a spring of spring constant s. The other end of the spring is fixed.

where x is the displacement in meters (m) of the mass m in kilograms (kg) from its rest position, s is the *stiffness* or *spring constant* in N/m, and the minus sign indicates that the force is opposed to the displacement. Substituting this expression for force into the general equation of linear motion

$$f = m\frac{d^2x}{dt^2} \tag{1.2.2}$$

where d^2x/dt^2 is the acceleration of the mass, we obtain

$$\frac{d^2x}{dt^2} + \frac{s}{m}x = 0 \tag{1.2.3}$$

Both s and m are positive, so that we can define a constant

$$\boxed{\omega_0^2 = s/m} \tag{1.2.4}$$

which casts our equation into the form

$$\boxed{\frac{d^2x}{dt^2} + \omega_0^2 x = 0} \tag{1.2.5}$$

This is an important linear differential equation whose general solution is well known and may be obtained by several methods.

One method is to assume a trial solution of the form

$$x = A_1 \cos \gamma t \tag{1.2.6}$$

Differentiation and substitution into (1.2.5) shows that this is a solution if $\gamma = \omega_0$. It may similarly be shown that

$$x = A_2 \sin \omega_0 t \tag{1.2.7}$$

is also a solution. The complete general solution is the sum of these two,

$$x = A_1 \cos \omega_0 t + A_2 \sin \omega_0 t \tag{1.2.8}$$

where A_1 and A_2 are arbitrary constants and the parameter ω_0 is the *natural angular frequency* in radians per second (rad/s). Since there are 2π radians in one cycle, the *natural frequency* f_0 in hertz (Hz) is related to the natural angular frequency by

$$f_0 = \omega_0/2\pi \tag{1.2.9}$$

Note that either decreasing the stiffness or increasing the mass lowers the frequency. The *period T* of one complete vibration is given by

$$T = 1/f_0 \tag{1.2.10}$$

1.3 INITIAL CONDITIONS

If at time $t = 0$ the mass has an initial displacement x_0 and an initial speed u_0, then the arbitrary constants A_1 and A_2 are fixed by these initial conditions and the subsequent motion of the mass is completely determined. Direct substitution into (1.2.8) of $x = x_0$ at $t = 0$ will show that A_1 equals the initial displacement x_0. Differentiation of (1.2.8) and substitution of the initial speed at $t = 0$ gives $u_0 = \omega_0 A_2$, and (1.2.8) becomes

$$x = x_0 \cos \omega_0 t + (u_0/\omega_0) \sin \omega_0 t \qquad (1.3.1)$$

Another form of (1.2.8) may be obtained by letting $A_1 = A \cos \phi$ and $A_2 = -A \sin \phi$, where A and ϕ are two new arbitrary constants. Substitution and simplification then gives

$$x = A \cos(\omega_0 t + \phi) \qquad (1.3.2)$$

where A is the *amplitude* of the motion and ϕ is the *initial phase angle* of the motion. The values of A and ϕ are determined by the initial conditions and are

$$A = [x_0^2 + (u_0/\omega_0)^2]^{1/2} \qquad \text{and} \qquad \phi = \tan^{-1}(-u_0/\omega_0 x_0) \qquad (1.3.3)$$

Successive differentiation of (1.3.2) shows that the speed of the mass is

$$u = -U \sin(\omega_0 t + \phi) \qquad (1.3.4)$$

where $U = \omega_0 A$ is the *speed amplitude,* and the acceleration of the mass is

$$a = -\omega_0 U \cos(\omega_0 t + \phi) \qquad (1.3.5)$$

In these forms it is seen that the displacement lags 90° ($\pi/2$ rad) behind the speed and that the acceleration is 180° (π rad) out of phase with the displacement, as shown in Fig. 1.3.1.

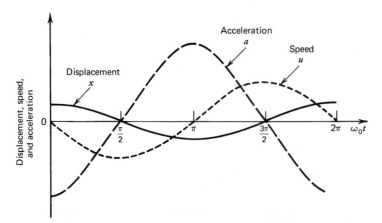

Figure 1.3.1   The speed u of a simple oscillator always leads to the displacement x by 90°. Acceleration a and displacement x are always 180° out of phase with each other. Plotted curves correspond to $\phi = 0°$.

1.4 ENERGY OF VIBRATION

The mechanical energy E of a system is the sum of the system's potential energy E_p and kinetic energy E_k. The potential energy is the work done in distorting the spring as the mass moves from its position of static equilibrium. Since the force exerted by the mass on the spring is in the direction of the displacement and equals $+sx$, the potential energy E_p stored in the spring is

$$E_p = \int_0^x sx\,dx = \tfrac{1}{2}sx^2 \tag{1.4.1}$$

Expression of x by (1.3.2) gives

$$E_p = \tfrac{1}{2}sA^2\cos^2(\omega_0 t + \phi) \tag{1.4.2}$$

The kinetic energy possessed by the mass is

$$E_k = \tfrac{1}{2}mu^2 \tag{1.4.3}$$

Expression of u by (1.3.4) gives

$$E_k = \tfrac{1}{2}mU^2\sin^2(\omega_0 t + \phi) \tag{1.4.4}$$

The total energy of the system is

$$E = E_p + E_k = \tfrac{1}{2}m\omega_0^2 A^2 \tag{1.4.5}$$

where use has been made of $s = m\omega_0^2$, $U = \omega_0 A$, and the identity $\sin^2\sigma + \cos^2\sigma = 1$. The total energy can be rewritten in alternate forms,

$$E = \tfrac{1}{2}sA^2 = \tfrac{1}{2}mU^2 \tag{1.4.6}$$

The total energy is a constant (independent of time) and is equal either to the maximum potential energy (when the mass is at its greatest displacement and is instantaneously at rest) or to the maximum kinetic energy (when the mass passes through its equilibrium position with maximum speed). Since the system was assumed to be free of external forces and not subject to any frictional forces, it is not surprising that the total energy does not change with time.

   If all other quantities in the above equations are expressed in MKS units, then E_p, E_k, and E will be in joules (J).

1.5 COMPLEX EXPONENTIAL METHOD OF SOLUTION

Throughout this book, complex quantities will often, but not always, be represented by **boldface** type. One exception is the definition $j \equiv \sqrt{-1}$. We will use the engineering convention of representing the time dependence of oscillatory functions by $\exp(j\omega t)$, rather than the physics convention of $\exp(-i\omega t)$, because of the many close analogies between acoustics and engineering applications. In many cases, consonance between apparently disparate sources can be resolved by making the transformation of j to $-i$. This may in some cases result in an exchange of complex functions from one type to another, but the textual context will usually resolve any ambiguities. Readers unacquainted with complex numbers should refer to Appendixes A2 and A3.

A more general and flexible approach to solving linear differential equations of the form (1.2.5) is to postulate

$$\mathbf{x} = \mathbf{A}e^{\gamma t} \tag{1.5.1}$$

Substitution gives $\gamma^2 = -\omega_0^2$ or $\gamma = \pm j\omega_0$. Thus, the general solution is

$$\mathbf{x} = \mathbf{A}_1 e^{j\omega_0 t} + \mathbf{A}_2 e^{-j\omega_0 t} \tag{1.5.2}$$

where $\mathbf{A}_1$ and $\mathbf{A}_2$ are to be determined by initial conditions, $\mathbf{x}(0) = x_0$ and $d\mathbf{x}(0)/dt = u_0$. This results in two equations

$$\mathbf{A}_1 + \mathbf{A}_2 = x_0 \quad \text{and} \quad \mathbf{A}_1 - \mathbf{A}_2 = u_0/j\omega_0 = -ju_0/\omega_0 \tag{1.5.3}$$

from which

$$\mathbf{A}_1 = \tfrac{1}{2}(x_0 - ju_0/\omega_0) \quad \text{and} \quad \mathbf{A}_2 = \tfrac{1}{2}(x_0 + ju_0/\omega_0) \tag{1.5.4}$$

Note that $\mathbf{A}_1$ and $\mathbf{A}_2$ are complex conjugates, so there are really only two constants a and b, where $\mathbf{A}_1 = a - jb$ and $\mathbf{A}_2 = a + jb$. This must be the case since the differential equation is of second order with two independent solutions and, therefore, with two arbitrary constants to be determined by two initial conditions. Substitution of $\mathbf{A}_1$ and $\mathbf{A}_2$ into (1.5.2) yields

$$\mathbf{x} = x_0 \cos \omega_0 t + (u_0/\omega_0) \sin \omega_0 t \tag{1.5.5}$$

which is identical with (1.3.1). Satisfying the initial conditions, which are both real, caused the imaginary part of $\mathbf{x}$ to vanish as an automatic consequence.

In practice it is unnecessary to go through the mathematical steps required to make the imaginary part of the general solution vanish, for *the real part of the complex solution is by itself a complete general solution of the original real differential equation.* Thus, for example, if we express $\mathbf{A}_1 = a_1 + jb_1$ and $\mathbf{A}_2 = a_2 + jb_2$ in (1.5.2) and, before applying initial conditions, take the real part, we have

$$\text{Re}\{\mathbf{x}\} = (a_1 + a_2) \cos \omega_0 t - (b_1 - b_2) \sin \omega_0 t \tag{1.5.6}$$

Now, application of the initial conditions yields $a_1 + a_2 = x_0$ and $b_1 - b_2 = u_0/\omega_0$ so that $\text{Re}\{\mathbf{x}\}$ is identical with (1.3.1). Similarly, a complete solution is obtained if the displacement is written in the complex form

$$\mathbf{x} = \mathbf{A}e^{j\omega_0 t} \tag{1.5.7}$$

where $\mathbf{A} = a + jb$, and only the real part is considered,

$$\text{Re}\{\mathbf{x}\} = a \cos \omega_0 t - b \sin \omega_0 t \tag{1.5.8}$$

From the form (1.5.7), which will be used frequently throughout this book, it is particularly easy to obtain the complex speed $\mathbf{u} = dx/dt$ and the complex acceleration $\mathbf{a} = d\mathbf{u}/dt$ of the mass. The complex speed is

$$\mathbf{u} = j\omega_0 \mathbf{A} e^{j\omega_0 t} = j\omega_0 \mathbf{x} \tag{1.5.9}$$

and the complex acceleration is

$$\mathbf{a} = -\omega_0^2 \mathbf{A} e^{j\omega_0 t} = -\omega_0^2 \mathbf{x} \tag{1.5.10}$$

The expression $\exp(j\omega_0 t)$ may be thought of as a *phasor* of unit length rotating counterclockwise in the complex plane with an angular speed ω_0. Similarly, any complex quantity $\mathbf{A} = a + jb$ may be represented by a phasor of length $A = \sqrt{a^2 + b^2}$, making an angle $\phi = \tan^{-1}(b/a)$ counterclockwise from the positive real axis. Consequently, the product $\mathbf{A} \exp(j\omega_0 t)$ represents a phasor of length A and initial phase angle ϕ rotating in the complex plane with angular speed ω_0 (Fig. 1.5.1). The real part of this rotating phasor (its projection on the real axis) is

$$A \cos(\omega_0 t + \phi) \tag{1.5.11}$$

and varies harmonically with time.

From (1.5.9) we see that differentiation of $\mathbf{x}$ with respect to time gives $\mathbf{u} = j\omega_0 \mathbf{x}$, and hence the phasor representing speed leads that representing displacement by a phase angle of 90°. The projection of this phasor onto the real axis gives the instantaneous speed, the speed amplitude being $\omega_0 A$. Equation (1.5.10) shows that the phasor $\mathbf{a}$ representing the acceleration is out of phase with the displacement phasor by π rad, or 180°. The projection of this phasor onto the real axis gives the instantaneous acceleration, the acceleration amplitude being $\omega_0^2 A$.

It will be the general practice in this textbook to analyze problems by the complex exponential method. The chief advantages of the procedure, as compared with the trigonometric method of solution, are its greater mathematical simplicity and the relative ease with which the phase relationships among the various mechanical and acoustic variables can be determined. However, care must be taken to obtain the *real* part of the complex solution to arrive at the correct physical equation.

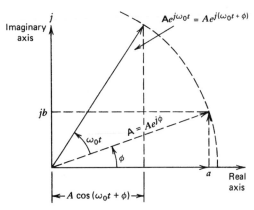

Figure 1.5.1 Physical representation of a phasor $A \exp[j(\omega_0 t + \phi)]$.

1.6 DAMPED OSCILLATIONS

Whenever a real body is set into oscillation, dissipative (frictional) forces arise. These forces are of many types, depending on the particular oscillating system, but they will always result in a *damping* of the oscillations—a decrease in the amplitude of the free oscillations with time. Let us first consider the effect of a *viscous* frictional force f_r on a simple oscillator. Such a force is assumed proportional to the speed of the mass and directed to oppose the motion. It can be expressed as

$$f_r = -R_m \frac{dx}{dt} \tag{1.6.1}$$

where R_m is a positive constant called the *mechanical resistance* of the system. It is evident that mechanical resistance has the units of newton-second per meter (N·s/m) or kilogram per second (kg/s).

A device that generates such a frictional force can be represented by a dashpot (shock absorber). This system is suggested in Fig. 1.6.1a. A simple harmonic oscillator subject to such a frictional force is usually diagrammed as in Fig. 1.6.1b.

If the effect of resistance is included, the equation of motion of an oscillator constrained by a stiffness force $-sx$ becomes

$$m\frac{d^2x}{dt^2} + R_m\frac{dx}{dt} + sx = 0 \tag{1.6.2}$$

Dividing through by m and recalling that $\omega_0 = \sqrt{s/m}$ we have

$$\boxed{\frac{d^2x}{dt^2} + \frac{R_m}{m}\frac{dx}{dt} + \omega_0^2 x = 0} \tag{1.6.3}$$

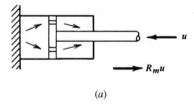

(a)

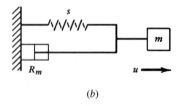

(b)

Figure 1.6.1   *(a)* Representative sketch of a dashpot with mechanical resistance R_m.
(b) Schematic representation of a damped, free oscillator consisting of a mass m attached to a spring of spring constant s and a dashpot with mechanical resistance R_m.

This equation may be solved by the complex exponential method. Assume a solution of the form

$$\mathbf{x} = \mathbf{A}e^{j\gamma t} \tag{1.6.4}$$

and substitute into (1.6.3) to obtain

$$[\gamma^2 + (R_m/m)\gamma + \omega_0^2]\mathbf{A}e^{j\gamma t} = 0 \tag{1.6.5}$$

Since this must be true for all time,

$$\gamma^2 + (R_m/m)\gamma + \omega_0^2 = 0 \tag{1.6.6}$$

or

$$\gamma = -\beta \pm (\beta^2 - \omega_0^2)^{1/2} \tag{1.6.7}$$

$$\beta = R_m/2m \tag{1.6.8}$$

In most cases of importance in acoustics, the mechanical resistance R_m is small enough so that $\omega_0 > \beta$ and γ is complex. Also, notice that if $R_m = 0$ then

$$\gamma = \pm(-\omega_0^2)^{1/2} = \pm j\omega_0 \tag{1.6.9}$$

and the problem has been reduced to that of the undamped oscillator. This suggests defining a new constant ω_d by

$$\omega_d = (\omega_0^2 - \beta^2)^{1/2} \tag{1.6.10}$$

Now, γ is given by

$$\gamma = -\beta \pm j\omega_d \tag{1.6.11}$$

and ω_d is seen to be the *natural angular frequency* of the damped oscillator. Note that ω_d is always less than the natural angular frequency ω_0 of the same oscillator without damping.

The complete solution is the sum of the two solutions obtained above,

$$\mathbf{x} = e^{-\beta t}(\mathbf{A}_1 e^{j\omega_d t} + \mathbf{A}_2 e^{-j\omega_d t}) \tag{1.6.12}$$

As in the nondissipative case, the constants $\mathbf{A}_1$ and $\mathbf{A}_2$ are in general complex. As noted earlier, the real part of this complex solution is the complete general solution. One convenient form of this general solution is

$$x = Ae^{-\beta t}\cos(\omega_d t + \phi) \tag{1.6.13}$$

where A and ϕ are real constants determined by the initial conditions. Figure 1.6.2 displays the time history of the displacement of a damped harmonic oscillator for various values of β.

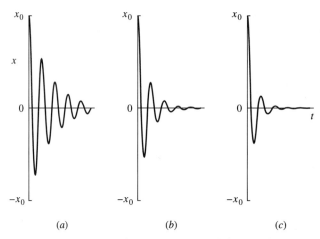

Figure 1.6.2   Decay of an underdamped, free oscillator.
Initial conditions: $x_0 = 1$ and $u_0 = 0$. (a) $\beta/\omega_0 = 0.1$. (b) $\beta/\omega_0 = 0.2$. (c) $\beta/\omega_0 = 0.3$.

The amplitude of the damped oscillator, defined as $A\exp(-\beta t)$, is no longer constant but decreases exponentially with time. As with the undamped oscillator, the frequency is independent of the amplitude of oscillation.

One measure of the rapidity with which the oscillations are damped by friction is the time required for the amplitude to decrease to $1/e$ of its initial value. This time τ is the *relaxation time* (other names include *decay modulus, decay time, time constant*, and *characteristic time*) and is given by

$$\tau = 1/\beta = 2m/R_m \tag{1.6.14}$$

The quantity β is the *temporal absorption coefficient*. (As with τ there are a variety of names for β; we mention only one.) The smaller R_m, the larger τ is and the longer it takes for the oscillations to damp out.

If the mechanical resistance R_m is large enough, then $\omega_0 \leq \beta$ and the system is no longer oscillatory; a displaced mass returns asymptotically to its rest position. If $\beta = \omega_0$, the system is known as *critically damped*.

The solution (1.6.13) is the real part of the complex solution

$$\mathbf{x} = \mathbf{A}e^{-\beta t}e^{j\omega_d t} \tag{1.6.15}$$

where $\mathbf{A} = A\exp(j\phi)$. If we rearrange the exponents,

$$\mathbf{x} = \mathbf{A}e^{j(\omega_d + j\beta)t} \tag{1.6.16}$$

we can define a *complex angular frequency*

$$\boldsymbol{\omega}_d = \omega_d + j\beta \tag{1.6.17}$$

whose real part is the angular frequency ω_d of the damped motion and whose imaginary part is the temporal absorption coefficient β. This convention of assimilating the angular frequency and the absorption coefficient into a single complex quantity often proves useful in investigating damped vibrations, as we will see in subsequent chapters.

1.7 FORCED OSCILLATIONS

A simple oscillator, or some equivalent system, is often driven by an *externally applied force* $f(t)$. The differential equation for the motion becomes

$$m\frac{d^2x}{dt^2} + R_m\frac{dx}{dt} + sx = f(t) \qquad (1.7.1)$$

Such a system is suggested in Fig 1.7.1.

For the case of a sinusoidal driving force $f(t) = F\cos\omega t$ applied to the oscillator at some initial time, the solution of (1.7.1) is the sum of two parts—a *transient* term containing two arbitrary constants and a *steady-state* term that depends on F and ω but does not contain any arbitrary constants. The transient (homogeneous) term is obtained by setting F equal to zero. Since the resulting equation is identical with (1.6.3), the transient term is given by (1.6.13). Its angular frequency is ω_d. The arbitrary constants are determined by applying the initial conditions to the total solution. After a sufficient time interval $t \gg 1/\beta$, the damping term $\exp(-\beta t)$ makes this portion of the solution negligible, leaving only the steady-state term whose angular frequency ω is that of the driving force.

To obtain the steady-state (particular) solution, it will be advantageous to replace the real driving force $F\cos\omega t$ by its equivalent complex driving force $\mathbf{f} = F\exp(j\omega t)$. The equation then becomes

$$m\frac{d^2\mathbf{x}}{dt^2} + R_m\frac{d\mathbf{x}}{dt} + s\mathbf{x} = Fe^{j\omega t} \qquad (1.7.2)$$

The solution of this equation gives the complex displacement $\mathbf{x}$. Since the real part of the complex driving force $\mathbf{f}$ represents the actual driving force $F\cos\omega t$, the real part of the complex displacement will represent the actual displacement.

Because $\mathbf{f} = F\exp(j\omega t)$ is periodic with angular frequency ω, it is plausible to assume that $\mathbf{x}$ must be also. Then, $\mathbf{x} = \mathbf{A}\exp(j\omega t)$, where $\mathbf{A}$ is in general complex. Equation (1.7.2) becomes

$$(-\mathbf{A}\omega^2 m + j\mathbf{A}\omega R_m + \mathbf{A}s)e^{j\omega t} = Fe^{j\omega t} \qquad (1.7.3)$$

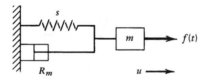

Figure 1.7.1   Schematic representation of a damped, forced oscillator consisting of a mass m driven by a force $f(t)$ attached to a spring of spring constant s and a dashpot with mechanical resistance R_m.

Solving for **A** yields the complex displacement

$$\mathbf{x} = \frac{1}{j\omega} \frac{Fe^{j\omega t}}{R_m + j(\omega m - s/\omega)} \tag{1.7.4}$$

and differentiation gives the complex speed

$$\mathbf{u} = \frac{Fe^{j\omega t}}{R_m + j(\omega m - s/\omega)} \tag{1.7.5}$$

These last two equations can be cast into somewhat simpler form if we define the *complex mechanical input impedance* $\mathbf{Z}_m$ of the system

$$\mathbf{Z}_m = R_m + jX_m \tag{1.7.6}$$

where the *mechanical reactance* X_m is

$$X_m = \omega m - s/\omega \tag{1.7.7}$$

The mechanical impedance $\mathbf{Z}_m = Z_m \exp(j\Theta)$ has magnitude

$$Z_m = [R_m^2 + (\omega m - s/\omega)^2]^{1/2} \tag{1.7.8}$$

and phase angle

$$\Theta = \tan^{-1}(X_m/R_m) = \tan^{-1}[(\omega m - s/\omega)/R_m] \tag{1.7.9}$$

The dimensions of mechanical impedance are the same as those of mechanical resistance and are expressed in the same units, $N \cdot s/m$, often defined as *mechanical ohms*. It is to be emphasized that, although the mechanical ohm is analogous to the electrical ohm, these two quantities do not have the same units. The electrical ohm has the dimensions of voltage divided by current; the mechanical ohm has the dimensions of force divided by speed.

Using the definition of $\mathbf{Z}_m$ we may write (1.7.5) in the simplified form

$$\boxed{\mathbf{Z}_m = \mathbf{f}/\mathbf{u}} \tag{1.7.10}$$

which gives a most important physical meaning to the complex mechanical impedance: $\mathbf{Z}_m$ is *the ratio of the complex driving force* $\mathbf{f} = F\exp(j\omega t)$ *to the resultant complex speed* $\mathbf{u}$ *of the system at the point where the force is applied.* If, for the driving frequency of interest, the complex impedance $\mathbf{Z}_m$ is known, then we can immediately obtain the complex speed

$$\mathbf{u} = \mathbf{f}/\mathbf{Z}_m \tag{1.7.11}$$

and make use of $\mathbf{u} = j\omega\mathbf{x}$ to obtain the complex displacement

$$\mathbf{x} = \mathbf{f}/j\omega\mathbf{Z}_m \tag{1.7.12}$$

Thus, knowledge of $\mathbf{Z}_m$ is equivalent to solving the differential equation.

The actual displacement is given by the real part of (1.7.4),

$$x = (F/\omega Z_m)\sin(\omega t - \Theta) \qquad (1.7.13)$$

and the actual speed is given by the real part of (1.7.5),

$$u = (F/Z_m)\cos(\omega t - \Theta) \qquad (1.7.14)$$

[both with the help of (1.7.8) and (1.7.9)]. The ratio F/Z_m gives the maximum speed of the driven oscillator and is the speed amplitude. Equation (1.7.14) shows that Θ is the phase angle between the speed and the driving force. When this angle is positive, it indicates that the speed lags the driving force by Θ. When this angle is negative, it indicates that the speed leads the driving force.

1.8   TRANSIENT RESPONSE OF AN OSCILLATOR

Before continuing the discussion of the simple oscillator it will be well to consider the effect of superimposing the transient response on the steady-state condition. The complete general solution of (1.7.2) is

$$x = Ae^{-\beta t}\cos(\omega_d t + \phi) + (F/\omega Z_m)\sin(\omega t - \Theta) \qquad (1.8.1)$$

where A and ϕ are two arbitrary constants whose values are determined by the initial conditions.

As a special case, let us assume that $x_0 = 0$ and $u_0 = 0$ at time $t = 0$ when the driving force is first applied, and that β is small compared to ω_0. Application of these conditions to (1.8.1) gives

$$A = (F/Z_m^2)[(X_m/\omega)^2 + (R_m/\omega_d)^2]^{1/2}$$
$$\tan\phi = (\omega/\omega_d)(R_m/X_m) \qquad (1.8.2)$$

Representative curves showing the relative importance of the steady-state and transient terms in producing a combined motion are plotted in Fig. 1.8.1. The effect of the transient is apparent in the left portion of these curves, but near the right end the transient has been so damped that the final steady state is nearly reached. Curves for other initial conditions are analogous, in that the wave form is always somewhat irregular immediately after the application of the driving force, but soon settles into the steady state.

Another important transient is the *decay transient*, which results when the driving force is abruptly removed. The equation of this motion is that of the damped oscillator, (1.6.13), and its angular frequency of oscillation is ω_d not ω. The constants giving the amplitude and phase angle of this motion depend on the part of its cycle in which the driving force is removed. It is impossible to remove the driving force without the appearance of a decay transient, although the effect will be negligible if the amplitude of the driving force is very slowly reduced to zero or the damping is very strong. The decay transient characteristics of mechanical vibrator elements are of particular importance when considering the fidelity of response of sound reproduction components such as loudspeakers and microphones. An example of an overly slow decay is a noticeable

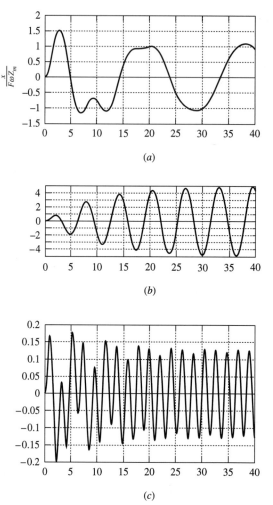

Figure 1.8.1 Transient response of a damped, forced oscillator with $\beta/\omega_d = 0.1$, $x_0 = 0$, and $u_0 = 0$. (*a*) $\omega/\omega_d = \frac{1}{3}$. (*b*) $\omega/\omega_d = 1$. (*c*) $\omega/\omega_d = 3$.

"hangover" at the natural frequency produced by some poorly designed loud-speaker systems.

1.9 POWER RELATIONS

The *instantaneous power* Π_i in watts (W) supplied to the system is equal to the product of the instantaneous driving force and the resulting instantaneous speed. Substituting the appropriate real expressions for the steady-state force and speed,

$$\Pi_i = (F^2/Z_m) \cos \omega t \cos(\omega t - \Theta) \qquad (1.9.1)$$

It should be noted that the instantaneous power Π_i is *not* equal to the real part of the product of the complex driving force $\mathbf{f}$ and the complex speed $\mathbf{u}$.

In most situations the *average power* Π being supplied to the system is of more significance than the instantaneous power. This average power is equal to the total work done per complete vibration divided by the time of one vibration,

$$\Pi = \frac{1}{T}\int_0^T \Pi_i \, dt = \langle \Pi_i \rangle_T \tag{1.9.2}$$

Substitution of Π_i in this equation gives

$$\begin{aligned}
\Pi &= \frac{F^2}{Z_m T}\int_0^T \cos \omega t \cos(\omega t - \Theta)\, dt \\
&= \frac{F^2}{Z_m T}\int_0^T (\cos^2 \omega t \cos \Theta + \cos \omega t \sin \omega t \sin \Theta)\, dt \\
&= \frac{F^2}{2Z_m}\cos \Theta
\end{aligned} \tag{1.9.3}$$

This average power supplied to the system by the driving force is not permanently stored in the system but is dissipated in the work expended in moving the system against the frictional force $R_m u$. Since $\cos \Theta = R_m/Z_m$, then (1.9.3) may be written as

$$\Pi = F^2 R_m / 2Z_m^2 \tag{1.9.4}$$

The average power delivered to the oscillator is a maximum when the mechanical reactance X_m vanishes, which from (1.7.7) occurs when $\omega = \omega_0$. At this frequency $\cos \Theta$ has its maximum value of unity ($\Theta = 0$) and Z_m its minimum value R_m.

1.10 MECHANICAL RESONANCE

The *resonance angular frequency* ω_0 is defined as that at which the mechanical reactance X_m vanishes and the mechanical impedance is pure real with its minimum value, $Z_m = R_m$. As has just been noted, at this angular frequency a driving force will supply maximum power to the oscillator. In Section 1.2, ω_0 was found to be the natural angular frequency of a similar undamped oscillator and also the angular frequency of maximum speed amplitude. At $\omega = \omega_0$, (1.7.14) reduces to

$$u_{res} = (F/R_m)\cos \omega_0 t \tag{1.10.1}$$

and the displacement (1.7.13) reduces to

$$x_{res} = (F/\omega_0 R_m)\sin \omega_0 t \tag{1.10.2}$$

(Note that ω_0 does not give the maximum displacement amplitude, which occurs at the angular frequency minimizing the product ωZ_m. It can be shown that this occurs when $\omega = \sqrt{\omega_0^2 - 2\beta^2}$.)

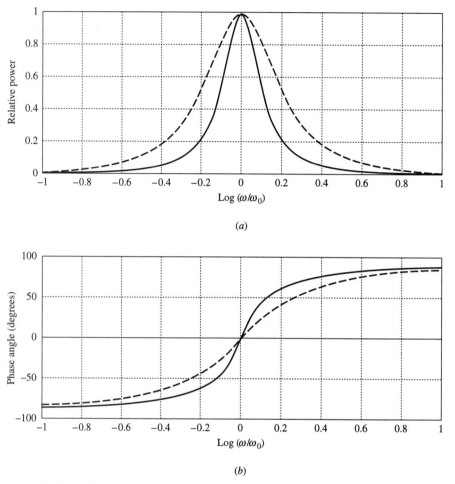

Figure 1.10.1   Response of a simple driven mechanical oscillator. (*a*) Input power relative to its value at resonance. (*b*) Phase angle Θ. Solid lines correspond to $Q = 2$. Dashed lines correspond to $Q = 1$.

If the average power (1.9.4) is plotted as a function of the frequency of a driving force of constant amplitude, a curve similar to Fig. 1.10.1*a* is obtained. It has a maximum value of $F^2/2R_m$ at the resonance frequency and falls at lower and higher frequencies. The sharpness of the peak of the power curve is primarily determined by R_m/m. If this ratio is small, the curve falls off very rapidly—a *sharp resonance*. If, on the other hand, R_m/m is large, the curve falls off more slowly and the system has a *broad resonance*. A more precise definition of the sharpness of resonance can be given in terms of the *quality factor* Q of the system, defined by

$$Q = \omega_0/(\omega_u - \omega_l) \tag{1.10.3}$$

where ω_u and ω_l are the two angular frequencies, above and below resonance, respectively, at which the average power has dropped to one-half its resonance value.

It is also possible to express Q in terms of the mechanical constants of the system. From (1.9.4) it is evident that the average power will be one-half of its resonance value whenever $Z_m^2 = 2R_m^2$. This corresponds to

$$R_m^2 + X_m^2 = 2R_m^2 \quad \text{or} \quad X_m = \pm R_m \tag{1.10.4}$$

Since $X_m = \omega m - s/\omega$, the two values of ω that satisfy this requirement are

$$\omega_u m - s/\omega_u = R_m \quad \text{and} \quad \omega_l m - s/\omega_l = -R_m \tag{1.10.5}$$

The elimination of s between these equations yields

$$\omega_u - \omega_l = R_m/m \tag{1.10.6}$$

so that

$$Q = \omega_0 m/R_m = \omega_0/2\beta \tag{1.10.7}$$

with the help of (1.6.8). Use of (1.6.14) for the relaxation time τ of this oscillator gives

$$Q = \tfrac{1}{2}\omega_0\tau \tag{1.10.8}$$

The sharpness of the resonance of the driven oscillator is directly related to the length of time it takes for the free oscillator to decay to $1/e$ of its initial amplitude. Furthermore, the number of oscillations taken for this decay is $(\omega_d/\omega_0)Q/\pi$ or about Q/π for weak damping. Thus, if an oscillator has a Q of 100 and a natural frequency 1000 Hz, it will take $(100/\pi)$ cycles or 32 ms to decay to $1/e$ of its initial amplitude. It should also be noted that $Q/2\pi$ is the ratio of the mechanical energy of the oscillator driven at its resonance frequency to the energy dissipated per cycle of vibration. Proof of this is left as an exercise (Problem 1.10.3).

When the oscillator is driven at resonance the phase angle Θ is zero and the speed u is in phase with the driving force f. When ω is greater than ω_0 the phase angle is positive, and when ω approaches infinity u lags f by an angle that approaches 90°. When ω is less than ω_0 the phase angle is negative, and as ω approaches zero u leads f by 90°. Figure 1.10.1b shows the dependence of Θ on frequency for a typical oscillator. In systems having relatively small mechanical resistance, the phase angles of both speed and displacement vary rapidly in the vicinity of resonance.

1.11 MECHANICAL RESONANCE AND FREQUENCY

Mechanical systems driven by periodic forces can be grouped into three different classes. (1) Sometimes it is desired that the system respond strongly to only *one* particular frequency. If the mechanical resistance of a simple oscillator is small, its impedance will be relatively large at all frequencies except those in the immediate vicinity of resonance, and such an oscillator will consequently respond strongly only in the vicinity of resonance. Some common examples are tuning forks, the resonators below the bars of a xylophone, and magnetostrictive sonar transducers. (2) In other applications it is desired that the system respond strongly to a series of discrete frequencies. The simple oscillator does not have this property, but mechanical systems that do behave in this manner can be designed. These will be considered in subsequent chapters. (3) A third type of use requires that the system respond more or less uniformly to a wide range of frequencies. Examples include the vibrator elements of many electroacoustic and mechanoacoustic transducers: microphones, loudspeakers, hydrophones, many sonar transducers, and the sounding board of a piano.

In different applications, the quantity whose amplitude is supposed to be independent of frequency may be different. In some cases the displacement amplitude is to be independent of frequency; in others it is the speed amplitude or the amplitude of the acceleration that is to be invariant. By a suitable choice of the stiffness, mass, and mechanical resistance, a simple oscillator can be made to satisfy any of these requirements over a limited frequency range. These three special cases of frequency-independent driven oscillators are known as *stiffness-*, *resistance-*, and *mass-controlled* systems, respectively.

A *stiffness-controlled* system is characterized by a large value of s/ω for the frequency range over which the response is to be flat. In this range both ωm and R_m are negligible in comparison with s/m and $\mathbf{Z}_m$ is very nearly equal to $-js/\omega$, so that

$$x \approx (F/s)\cos \omega t \qquad (1.11.1)$$

It should be noted that, although the displacement amplitude is independent of frequency, the speed amplitude is not, nor is the acceleration amplitude.

A *resistance-controlled* system is one for which R_m is large in comparison with X_m. This will be true when an oscillator of relatively high mechanical resistance is operated in the vicinity of resonance. Then

$$u \approx (F/R_m)\cos \omega t \qquad (1.11.2)$$

so that the speed amplitude is essentially independent of frequency, although both the displacement amplitude and the acceleration are not.

A *mass-controlled* system is characterized by a large value of ωm over the desired frequency range. Then s/ω and R_m are negligible and $\mathbf{Z}_m$ is approximately equal to $j\omega m$. Neither displacement nor speed amplitudes are independent of frequency, but

$$a \approx (F/m)\cos \omega t \qquad (1.11.3)$$

so the acceleration amplitude is independent of frequency.

All driven mechanical vibrator elements are resistance-controlled for frequencies nearly equal to their resonant frequency, but for vibrators of low mechanical resistance the range of relatively flat response is extremely narrow. Similarly, all driven vibrators are stiffness-controlled for frequencies well below f_0, and mass-controlled for frequencies well above f_0. A suitable choice of mechanical constants will place any of these systems in the desired part of the frequency range, but the computed values are sometimes very difficult to attain in practice.

*1.12 EQUIVALENT ELECTRICAL CIRCUITS FOR OSCILLATORS

Many vibrating systems are mathematically equivalent to corresponding electrical systems. For example, consider a simple series electrical circuit containing inductance L, resistance R, and capacitance C, driven by an impressed sinusoidal voltage $V \cos \omega t$, as suggested in Fig. 1.12.1a. The differential equation for the current $\mathbf{I} = d\mathbf{q}/dt$, where $\mathbf{q}$ is the complex charge, is

$$L\frac{d\mathbf{I}}{dt} + R\mathbf{I} + \frac{\mathbf{q}}{C} = \mathbf{V} \tag{1.12.1}$$

with $\mathbf{V} = V\exp(j\omega t)$. This equation may be written

$$L\frac{d^2\mathbf{q}}{dt^2} + R\frac{d\mathbf{q}}{dt} + \frac{\mathbf{q}}{C} = \mathbf{V} \tag{1.12.2}$$

which has the same form as (1.7.2). Thus, the steady-state solution for $\mathbf{q}$ is

$$\mathbf{q} = \frac{1}{j\omega}\frac{\mathbf{V}}{R + j(\omega L - 1/\omega C)} \tag{1.12.3}$$

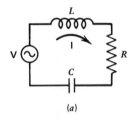

(a)

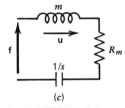

(b)

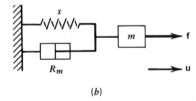

(c)

Figure 1.12.1 Equivalent series systems.
(a) Series electrical circuit driven with voltage $\mathbf{V}$. All elements experience the same current $\mathbf{I}$.
(b) Mechanical system with mass m driven by force $\mathbf{f}$ and attached to a spring of spring constant s and dashpot of mechanical resistance R_m. All elements move with the same speed $\mathbf{u}$. (c) The electrical equivalent of the mechanical system in (b).

and the current is $I = V/Z$, where

$$Z = R + j(\omega L - 1/\omega C) \tag{1.12.4}$$

We see that the electrical circuit of Fig. 1.12.1a is the mathematical analog of the damped harmonic oscillator of Fig. 1.12.1b. The current I in the electrical system is equivalent to the speed u in the mechanical system, the charge q is equivalent to the displacement x, and the applied voltage V is equivalent to the applied force f. Furthermore, the impedances for these two systems have similar forms, with the mechanical resistance R_m analogous to the electrical resistance R, the mass m analogous to the electrical inductance L, and the mechanical stiffness s analogous to the reciprocal of the electrical capacitance C. By direct comparison of (1.12.1) with (1.7.1), it can be seen that the resonance angular frequency of the electrical circuit is

$$\omega_0 = 1/\sqrt{LC} \tag{1.12.5}$$

and the average power dissipated is

$$\Pi = (V^2/2Z)\cos\Theta \tag{1.12.6}$$

The elements in the electrical system (Fig. 1.12.1a) are said to be in $series$ because they experience the same current. Similarly the elements in the mechanical system (Fig. 1.12.1b) can be represented by the series circuit of Fig. 1.12.1c: they experience the same displacement and, therefore, the same speed.

If a simple mechanical oscillator is driven by a sinusoidal force applied to the normally fixed end of the spring as suggested by Fig. 1.12.2a, then the mass and the spring experience the same force and this combination is represented by a $parallel$ circuit, as shown in Fig. 1.12.2b. The speed of the driven end of the spring is equivalent to the current entering the parallel circuit, and the speed u_m of the mass is equivalent to the current flowing through the inductor.

Other equivalent systems are shown in Figs. 1.12.3 and 1.12.4.

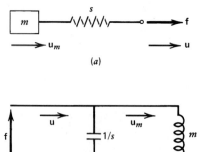

(a)

(b)

Figure 1.12.2   Equivalent parallel systems. (a) Mechanical system with mass attached to a spring and with the other end of the spring driven by a force f. The elements feel the same force but have different speeds. (b) The equivalent electrical circuit with inductance m and capacitance $1/s$. All elements experience the same voltage but carry different currents.

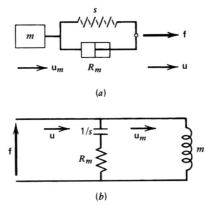

(a)

(b)

Figure 1.12.3   Equivalent series–parallel systems. (a) Mechanical system with mass attached to a combination of spring and dashpot with the other end of the spring/dashpot driven. The dashpot and spring both move with the same speed. They experience different forces, but the sum of forces is equal to the force on the mass. (b)The equivalent electrical circuit with inductance, resistance, and capacitance. The capacitance and the resistance share the same current and the sum of the voltages across them equals the voltage across the inductance.

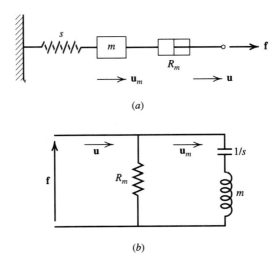

(a)

(b)

Figure 1.12.4   Equivalent series–parallel systems. (a) Mechanical system with mass attached between a spring and a dashpot. One end of the spring is fixed and the dashpot is driven. The mass and spring share the same speed while the sum of forces on them equals the force on the dashpot. (b) The equivalent mechanical circuit. The capacitance and the inductance carry the same current and the sum of the voltages across them equals the voltage across the resistance.

1.13 LINEAR COMBINATIONS OF SIMPLE HARMONIC VIBRATIONS

In many important situations that arise in acoustics, the motion of a body is a linear combination of the vibrations induced separately by two or more simple harmonic excitations. It is easy to show that the displacement of the body is then the sum of the individual displacements resulting from each of the harmonic excitations. Combining the effects of individual vibrations by linear addition is valid for the majority of cases encountered in acoustics. In general, the presence of one vibration does not alter the medium to such an extent that the characteristics of other vibrations are disturbed. Consequently, the total vibration is obtained by a *linear superposition* of the individual vibrations.

One case is the combination of two excitations that have the same angular frequency ω. If the two individual displacements are given by

$$\mathbf{x}_1 = A_1 e^{j(\omega t + \phi_1)} \quad \text{and} \quad \mathbf{x}_2 = A_2 e^{j(\omega t + \phi_2)} \tag{1.13.1}$$

their linear combination $\mathbf{x} = \mathbf{x}_1 + \mathbf{x}_2$ results in a motion $A \exp[j(\omega t + \phi)]$, where

$$A e^{j(\omega t + \phi)} = (A_1 e^{j\phi_1} + A_2 e^{j\phi_2}) e^{j\omega t} \tag{1.13.2}$$

Solution for A and ϕ can be accomplished easily if the addition of the phasors $A_1 \exp(j\omega t)$ and $A_2 \exp(j\omega t)$ is represented graphically, as in Fig. 1.13.1. From the projections of each phasor on the real and imaginary axes,

$$A = [(A_1 \cos\phi_1 + A_2 \cos\phi_2)^2 + (A_1 \sin\phi_1 + A_2 \sin\phi_2)^2]^{1/2}$$

$$\tan\phi = \frac{A_1 \sin\phi_1 + A_2 \sin\phi_2}{A_1 \cos\phi_1 + A_2 \cos\phi_2} \tag{1.13.3}$$

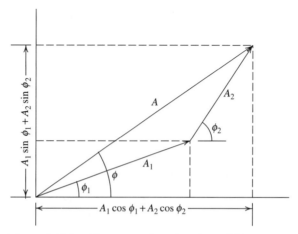

Figure 1.13.1   Phasor combination $A \exp(j\phi) = A_1 \exp(j\phi_1) + A_2 \exp(j\phi_2)$ of two simple harmonic motions having identical frequencies.

The real displacement is

$$x = x_1 + x_2 = A\cos(\omega t + \phi) \tag{1.13.4}$$

where A and ϕ are given by (1.13.3). The linear combination of two simple harmonic vibrations of identical frequency yields another simple harmonic vibration of this same frequency, having a different phase angle and an amplitude in the range $|A_1 - A_2| \leq A \leq (A_1 + A_2)$.

With the help of Fig. 1.13.1, it is clear that the addition of more than two phasors can be accomplished by drawing them in a chain, head to tail, and then taking their components on the real and imaginary axes. Thus, it may readily be shown that the vibration resulting from the addition of any number n of simple harmonic vibrations of identical frequency has amplitude A and phase angle ϕ given by

$$A = \left[\left(\sum A_n \cos\phi_n\right)^2 + \left(\sum A_n \sin\phi_n\right)^2\right]^{1/2} \tag{1.13.5}$$

$$\tan\phi = \sum A_n \sin\phi_n \Big/ \sum A_n \cos\phi_n$$

Thus, any linear combination of simple harmonic vibrations of identical frequency produces a new simple harmonic vibration of this same frequency. For example, when two or more sound waves overlap in a fluid medium, at each point in the fluid the periodic sound pressures of the individual waves combine as described above.

The expression for the linear combination of two simple harmonic vibrations of *different* angular frequencies ω_1 and ω_2 is

$$\mathbf{x} = A_1 e^{j(\omega_1 t + \phi_1)} + A_2 e^{j(\omega_2 t + \phi_2)} \tag{1.13.6}$$

The resulting motion is not simple harmonic, so that it cannot be represented by a simple sine or cosine function. However, if the ratio of the larger to the smaller frequency is a rational number (commensurate), the motion is periodic with angular frequency given by the greatest common divisor of ω_1 and ω_2. Otherwise, the resulting motion is a nonperiodic oscillation that never repeats itself. The linear combination of three or more simple harmonic vibrations that have different frequencies has characteristics similar to those discussed for two.

The linear combination of two simple harmonic vibrations of nearly the same frequency is easy to interpret. If the angular frequency ω_2 is written as

$$\omega_2 = \omega_1 + \Delta\omega \tag{1.13.7}$$

then the combination is

$$\mathbf{x} = A_1 e^{j(\omega_1 t + \phi_1)} + A_2 e^{j(\omega_1 t + \Delta\omega t + \phi_2)} \tag{1.13.8}$$

This can be reexpressed as

$$\mathbf{x} = (A_1 e^{j\phi_1} + A_2 e^{j(\phi_2 + \Delta\omega t)})e^{j\omega_1 t} \tag{1.13.9}$$

and then cast into the form

$$\mathbf{x} = Ae^{j(\omega_1 t + \phi)} \tag{1.13.10}$$

where

$$A = [A_1^2 + A_2^2 + 2A_1A_2\cos(\phi_1 - \phi_2 - \Delta\omega t)]^{1/2}$$

$$\tan\phi = \frac{A_1\sin\phi_1 + A_2\sin(\phi_2 + \Delta\omega t)}{A_1\cos\phi_1 + A_2\cos(\phi_2 + \Delta\omega t)} \tag{1.13.11}$$

The resulting vibration may be regarded as *approximately* simple harmonic, with angular frequency ω_1, but with both amplitude A and phase ϕ varying slowly at a frequency of $\Delta\omega/2\pi$. It can be shown that the amplitude of the vibration waxes and wanes between the limits $(A_1 + A_2)$ and $|A_1 - A_2|$. The effect of the variation in phase angle is somewhat more complicated. It modifies the vibration in such a manner that its frequency is not strictly constant, but the average angular frequency may be shown to lie somewhere between ω_1 and ω_2, depending on the relative magnitudes of A_1 and A_2. In the sounding of two pure tones of slightly different frequencies, this variation in amplitude results in a rhythmic pulsing of the loudness of the sound known as *beating.* As an example let us consider the special case $A_1 = A_2$ and $\phi_1 = \phi_2 = 0$. The equations (1.13.11) become

$$A = A_1[2 + 2\cos(\Delta\omega t)]^{1/2}$$

$$\tan\phi = \frac{\sin(\Delta\omega t)}{1 + \cos(\Delta\omega t)} \tag{1.13.12}$$

The amplitude ranges between $2A_1$ and zero, and the beating is very pronounced. Audible beats and other associated phenomena will be discussed in more detail in Chapter 11.

1.14 ANALYSIS OF COMPLEX VIBRATIONS BY FOURIER'S THEOREM

In the preceding section we noted that the linear combination of two or more simple harmonic vibrations with commensurate frequencies leads to a complex vibration that has a frequency determined by the greatest common divisor. Conversely, by means of a powerful mathematical theorem originated by Fourier, it is possible to analyze any complex periodic vibration into a harmonic array of component frequencies.

Stated briefly, this theorem asserts that any single-valued periodic function may be expressed as a summation of simple harmonic terms whose frequencies are integral multiples of the repetition rate of the given function. Since the above restrictions are normally satisfied in the case of the vibrations of material bodies, the theorem is widely used in acoustics.

If a certain vibration of period T is represented by the function $f(t)$, then Fourier's theorem states that $f(t)$ may be represented by the harmonic series

$$f(t) = \tfrac{1}{2}A_0 + A_1 \cos \omega t + A_2 \cos 2\omega t + \cdots + A_n \cos n\omega t + \cdots$$
$$+ B_1 \sin \omega t + B_2 \sin 2\omega t + \cdots + B_n \sin n\omega t + \cdots$$

(1.14.1)

where $\omega = 2\pi/T$ and the A's and B's are constants to be determined.

The formulas for evaluating these constants (derived in standard mathematical texts) are

$$A_n = \frac{2}{T} \int_0^T f(t) \cos n\omega t \, dt$$

$$B_n = \frac{2}{T} \int_0^T f(t) \sin n\omega t \, dt$$

(1.14.2)

Whether or not these integrations are feasible will depend on the nature and complexity of the function $f(t)$. If this function exactly represents the combination of a finite number of pure sine and cosine vibrations, the series obtained by computing the above constants will contain only these terms. Analysis, for instance, of simple beats will yield only the two frequencies present. Similarly, the complex vibration constituting the sum of three pure musical tones will analyze into those frequencies alone. On the other hand, if the vibration is characterized by abrupt changes in slope, like sawtooth waves or square waves, then the entire infinite series must be considered for a complete equivalence of motion. If $f(t)$ and df/dt are piecewise continuous over the interval $0 \le t \le T$, it is possible to show that the harmonic series is always convergent. However, jagged functions will require the inclusion of a large number of terms merely to achieve a reasonably good approximation to the original function, and there may be difficulties close to discontinuities. Fortunately, the majority of vibrations encountered in acoustics are relatively smooth functions of time. In such cases, the convergence is rather rapid and only a few terms must be computed.

Depending on the nature of the function being expanded, some terms in the series may be absent. If the function $f(t)$ is symmetrical with respect to $f = 0$, the constant term A_0 will be absent. If the function is *even*, $f(t) = f(-t)$, then all sine terms will be missing. An *odd* function, $f(t) = -f(-t)$, will cause all cosine terms to be absent.

In analyzing the perception of sound, a factor enabling us to reduce the number of higher frequency terms to be computed is that the subjective interpretation of a complex sound vibration is often only slightly altered if the higher frequencies are removed or ignored.

Let us apply the above analysis to a square wave of unit amplitude and period T, defined as

$$f(t) = \begin{cases} +1 & 0 \le t < T/2 \\ -1 & T/2 \le t < T \end{cases}$$

(1.14.3)

and repeating every period. Substitution into (1.14.2) yields all $A_n = 0$, $B_n = 0$ for n even, and

$$B_n = 4/n\pi \qquad n = 1, 3, 5, \ldots$$

(1.14.4)

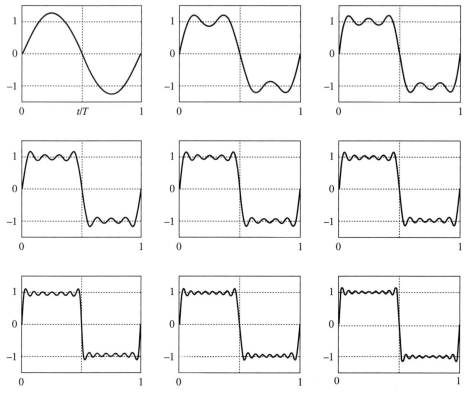

Figure 1.14.1   The Fourier series representation of a square wave vibration of unit amplitude and period T showing the results of including the lowest nonzero harmonics one at a time.

Note that A_0 is zero because of the symmetry of the motion about $f = 0$. All A_n are zero since the function is odd. The B_n are zero for even n because of the symmetry of $f(t)$ within each half-period. The complete harmonic series equivalent to the square wave vibration is

$$f(t) = \frac{4}{\pi}\left(\sin\omega t + \frac{1}{3}\sin 3\omega t + \frac{1}{5}\sin 5\omega t + \cdots + \frac{1}{n}\sin n\omega t + \cdots\right) \quad (1.14.5)$$

Plotted in Fig. 1.14.1 are results obtained by retaining various numbers of terms of the series. Differences among the plots are quite apparent. Because of the discontinuities, the Fourier series develops visible overshoot near these times if a large enough number of terms are retained.

*1.15 THE FOURIER TRANSFORM

Two fundamental methods are available for the analysis of pulses and other signals of finite duration: the Laplace transform and the Fourier transform. While the Laplace transform is a common approach, the underlying physics is somewhat hidden and there must be no motion before some specified time. We will follow most acoustics texts and use the second approach, Fourier analysis. (Actually, these two methods are closely related, the principal differences being the temporal restriction and the mathematical nomenclature.)

It has been demonstrated in Section 1.14 that a repeating waveform of period T can be considered as a sum of sinusoidal components whose frequencies are integral multiples of the fundamental frequency $f = 1/T$. If we now consider a nonrepeating waveform as being one of a family whose sequential members are identical in shape and uniformly spaced a large time T apart, and then allow T to become infinite, the fundamental frequency of the motion must approach zero, and the summation over all harmonics must be replaced by an integration over all frequencies.

Thus, if $\mathbf{f}(t)$ is a transient disturbance, we can write the general expression

$$\mathbf{f}(t) \int_{-\infty}^{\infty} \mathbf{g}(w)e^{jwt}\,dw \tag{1.15.1}$$

where w is the angular frequency. (We have chosen w rather than ω for notational reasons that will appear later, and because it is the "dummy" variable of integration.) The quantity $\mathbf{g}(w)$ is the *spectral density* of $\mathbf{f}(t)$. The integration region $-\infty < w < 0$ introduces the concept of "negative" frequency, but from

$$e^{j(\pm wt)} = \cos wt \pm j\sin wt \tag{1.15.2}$$

this is no more than a means of generating complex conjugates.

Given $\mathbf{f}(t)$, inversion of the integral to obtain the spectral density $\mathbf{g}(w)$ of the transient function yields

$$\mathbf{g}(w) = \frac{1}{2\pi}\int_{-\infty}^{\infty} \mathbf{f}(t)e^{-jwt}\,dt \tag{1.15.3}$$

(A proof, being rather mathematical and peripheral to our interests, will not be offered. Consult any standard text on Fourier transforms.) The pair (1.15.1) and (1.15.3) constitute one form of the *Fourier integral transforms*. Examination of the pair reveals that if $\mathbf{f}$ has some dimension $\lceil$ (such as m, N, Pa, or J), then $\mathbf{g}$ has dimension $\lceil$ ·s (m·s, N·s, Pa·s, J·s).

As an example, assume that $\mathbf{f}(t)$ represents a single extremely short but strong force such as striking an oscillator with a hammer or a drumhead with a drumstick. Such impulses can be approximated by the *Dirac delta function*, defined by

$$\delta(t) = 0 \qquad t \neq 0$$
$$\int_{-\infty}^{\infty} \delta(t)\,dt = 1 \tag{1.15.4}$$

The integral is dimensionless, so in general $\delta(v)$ has the dimension of $1/v$ where v is the variable of integration. One representation of $\delta(t)$ is

$$\delta(t) = \begin{cases} 0 & |t| > \varepsilon/2 \\ 1/\varepsilon & |t| \leq \varepsilon/2 \end{cases} \tag{1.15.5}$$

in the limit $\varepsilon \to 0$.

Substitution of $\mathbf{f}(t) = \delta(t)$ into (1.15.3) yields

$$\mathbf{g}(w) = \frac{1}{2\pi}\int_{-\infty}^{\infty} \delta(t)e^{-jwt}\,dt \tag{1.15.6}$$

Now, use of (1.15.5) shows that, since $\delta(t)$ is nonzero only where $|t| \leq \varepsilon/2$, the limits can be replaced by $\pm \varepsilon/2$. As $\varepsilon \to 0$, $\exp(-jwt)$ can be replaced by its value at $w = 0$, which leaves the result

$$\mathbf{g}(w) = \frac{1}{2\pi} \int_{-\varepsilon/2}^{\varepsilon/2} \delta(t)\,dt = \frac{1}{2\pi} \tag{1.15.7}$$

Thus, all frequencies are equally present in $\delta(t)$. (In this case, $\mathbf{f}$ has dimension $1/s$ and $\mathbf{g}$ is dimensionless.)

Conversely, if we write $\mathbf{g}(w)$ as consisting only of a single frequency,

$$\mathbf{g}(w) = \delta(w - \omega) \tag{1.15.8}$$

then

$$\mathbf{f}(t) \int_{-\infty}^{\infty} \delta(w - \omega)e^{jwt}\,dw = e^{j\omega t} \tag{1.15.9}$$

and the spectral density of a monofrequency signal is a delta function centered on that frequency. [In this second case, $\delta(w - \omega)$ has dimension s, as does $\mathbf{g}(w)$, and $\mathbf{f}(t)$ is dimensionless.]

The utility of this approach can be demonstrated by a simple exercise. Let $\mathbf{F}(t)$ be an impulsive force applied to an oscillator and express $\mathbf{F}(t)$ in terms of its Fourier components

$$\mathbf{F}(t) = \int_{-\infty}^{\infty} \mathbf{G}(w)e^{jwt}\,dw \tag{1.15.10}$$

where the spectral density $\mathbf{G}(w)$ is found from (1.15.3). Each of these monofrequency force components

$$\mathbf{f}(w, t) = \mathbf{G}(w)e^{jwt} \tag{1.15.11}$$

will generate a monofrequency complex speed component $\mathbf{u}(w, t)$ given from (1.7.10) by

$$\mathbf{u}(w, t) = \mathbf{f}(w, t)/\mathbf{Z}(w) = [\mathbf{G}(w)/\mathbf{Z}(w)]e^{jwt} \tag{1.15.12}$$

where $\mathbf{Z}(w)$ is the input mechanical impedance of the oscillator at the angular frequency w. Now, $\mathbf{G}(w)/\mathbf{Z}(w)$ is the spectral density of the speed and the resultant transient speed $\mathbf{U}(t)$ of the oscillator is, therefore,

$$\mathbf{U}(t) = \int_{-\infty}^{\infty} \mathbf{u}(w, t)\,dw = \int_{-\infty}^{\infty} \frac{\mathbf{G}(w)}{\mathbf{Z}(w)} e^{jwt}\,dw \tag{1.15.13}$$

It can be verified by direct substitution of (1.15.13) into (1.7.1) that $\mathbf{U}(t)$ is the solution for the applied force $f(t) = \mathbf{F}(t)$.

The physical interpretation of this approach is quite important and straightforward. If an arbitrary force is applied to a mechanical system, the resultant motion can be found by resolving the force into its individual frequency components, obtaining the motion resulting from each of these monofrequency components, and then assembling the resulting motion by combining the individual monofrequency motions. This is the very same case we encountered in periodic, nonharmonic forces, except that integrals must replace summations because the individual frequency components are not discrete but are continuously distributed over a range of frequencies.

While evaluation of these integrals may be difficult and involve special techniques (such as calculus of residues) or approximations (such as the method of stationary phase), tables

Table 1.15.1   Fourier integral transforms for a few simple functions when all relevant integrals are proper

$f(t)$	$g(w)$	$f(t)$	$g(w)$
$\dfrac{d^n f(t)}{dt^n}$	$(jw)^n g(w)$	$e^{j\omega t}$	$\delta(w - \omega)$
$(-jt)^n f(t)$	$\dfrac{d^n g(w)}{dw^n}$	$\delta(t - \tau)$	$\dfrac{1}{2\pi} e^{-jw\tau}$
$f(t)e^{j\omega t}$	$g(w - \omega)$	$1(t + \tau) - 1(t - \tau)$	$\dfrac{1}{\pi} \dfrac{\sin w\tau}{w}$
$f(t - \tau)$	$g(w)e^{-jw\tau}$	$e^{-bt} \cdot 1(t)$	$\dfrac{1}{2\pi} \dfrac{1}{jw + b}$
$\delta(t)$	$\dfrac{1}{2\pi}$	$e^{j\omega t} \cdot 1(t)$	$\dfrac{1}{2\pi} \dfrac{j}{\omega - w}$
1	$\delta(w)$	$(\cos \omega t) \cdot 1(t)$	$\dfrac{1}{2\pi} \dfrac{jw}{\omega^2 - w^2}$
$1(t)$	$\dfrac{1}{2\pi} \dfrac{1}{jw}$	$(\sin \omega t) \cdot 1(t)$	$\dfrac{1}{2\pi} \dfrac{\omega}{\omega^2 - w^2}$

of transformation pairs $f(t)$ and $g(w)$ are easily accessible, although the lack of a generally followed convention often entails a fair amount of calculation to cast the tabulations into the desired form.

It is useful to define the unit step function (Heaviside unit function) $1(t)$ as

$$1(t) = \int_{-\infty}^{t} \delta(t)\,dt = \begin{cases} 0 & t < 0 \\ 1 & t > 0 \end{cases} \tag{1.15.14}$$

This dimensionless function [sometimes designated as $u(t)$, which we avoid to prevent confusion with the particle speed] is used as a multiplier to designate functions $f(t) \cdot 1(t)$ that are zero for $t < 0$ and then assume their indicated behavior for $t > 0$. Table 1.15.1 presents transform pairs for a few simple cases consistent with (1.15.1) and (1.15.3).

An interesting and valuable relationship between the effective duration Δt of a signal and the effective bandwidth Δw of its spectral density is

$$\Delta w\, \Delta t \sim 2\pi \tag{1.15.15}$$

We will not prove (1.15.15), but Problem 1.15.9 demonstrates it for a pulse. This relationship, well known in quantum mechanics and signal processing, says in effect that the broader the frequency spectrum of a transient signal, the more concentrated in time it will be, and vice versa. Thus, it is expected, and plausible, that the greater the duration of a gated sinusoidal wave, the narrower its frequency spectrum.

Note that this is consistent with the limiting cases of (1) a delta function in time, which has an infinitely wide spectrum, and (2) a monofrequency oscillation $\cos \omega t$, for which the spectral density is a pair of delta functions centered at $w = \pm\omega$. Other examples are (1) a square pulse of unit amplitude and duration Δt with a spectrum $(\pi w)^{-1} \sin(w\Delta t/2)$ as shown in Fig. 1.15.1, and (2) a cosinusoidal pulse of four cycles of period T and constant amplitude turned on for a time interval $\Delta t = 4T$, which has a spectrum containing two principal peaks at $\pm\omega$ as shown in Fig. 1.15.2.

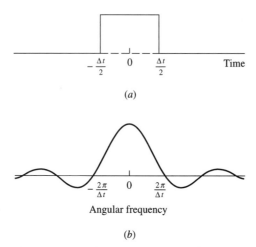

(a)

(b)

Figure 1.15.1 (a) A square pulse of duration Δt. (b) The spectrum of this pulse.

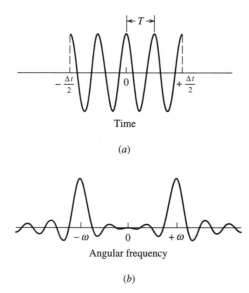

(a)

(b)

Figure 1.15.2 (a) A cosinusoidal pulse of angular frequency ω and period T turned on for a time interval $\Delta t = 4T$. (b) The spectrum of this pulse.

Modern signal processing systems are digital in that they sample the signal at discrete times and then analyze the resulting set of discrete numbers instead of a continuous function. This analysis is carried out using the *discrete Fourier transform* (DFT). The DFT is computationally intensive, requiring N^2 complex multiplications and additions, where N is the number of desired terms. To reduce computation time, an algorithm called the *fast Fourier transform* (FFT) has been developed. For the discussion of the DFT and FFT, see any book on signal processing, such as Burdic, *Underwater Acoustic System Analysis*, Prentice Hall (1991).

Problems

1.2.1. Given two springs of stiffness s and two bodies of mass M, find the natural frequencies of the systems sketched below.

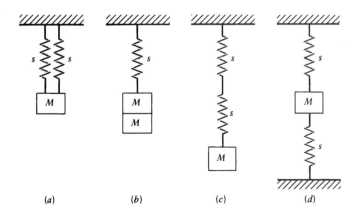

(a)   (b)   (c)   (d)

1.3.1. At time $t = T/2$ the speed of a simple oscillator of angular frequency ω_0 has maximum amplitude U and positive value. Find $x(t)$.

1.3.2. A simple oscillator whose natural frequency is 5 rad/s is displaced a distance 0.03 m from its equilibrium position and released. Find (a) the initial acceleration, (b) the amplitude of the resulting motion, and (c) the maximum speed attained.

1.3.3C. For a simple oscillator, plot the displacement as a function of t/T for the following initial conditions: (a) $u_0 = 0$ and $x_0/A = -1, 0, 1$; (b) $x_0/A = 1$ and $u_0/\omega_0 A = -1$, 0, 1.

1.4.1. Show for any (undamped) simple oscillator that $E_k(max) = E_p(max)$.

1.4.2. If the mass m_s of a spring is not negligible compared with the mass m attached to the spring, the additional inertia of the spring will result in a reduced frequency of vibration. Assume the speed of any element of the spring is proportional to its distance y from the fixed end of the spring. (a) Calculate the total energy of the system. (b) From this, derive the differential equation for the displacement of the mass m and show that the mass oscillates with a frequency $\omega_0 = \sqrt{s/m_e}$, where $m_e = m + m_s/3$. Problem 1.4.1 may be helpful.

1.5.1. Given that the real part of $\mathbf{x} = \mathbf{A}\exp(j\omega t)$ is $x = A\cos(\omega t + \phi)$, show that the real part of $\mathbf{x}^2$ does not equal x^2.

1.5.2. Find the real part, magnitude, and phase of (a) $\sqrt{x + jy}$, (b) $A\exp[j(\omega t + \phi)]$, and (c) $[1 + \exp(-2j\theta)]\exp(j\theta)$.

1.5.3. Given the two complex numbers $\mathbf{A} = A\exp[j(\omega t + \theta)]$ and $\mathbf{B} = B\exp[j(\omega t + \phi)]$, find (a) the real part of $\mathbf{AB}$, (b) the real part of $\mathbf{A/B}$, (c) the real part of $\mathbf{A}$ times the real part of $\mathbf{B}$, (d) the phase of $\mathbf{AB}$, and (e) the phase of $\mathbf{A/B}$.

1.5.4. Given the complex numbers $\mathbf{A} = x + jy$ and $\mathbf{B} = X + jY$, find (a) the magnitude of $\mathbf{A}$, (b) the magnitude of $\mathbf{B}$, (c) the magnitude of $\mathbf{AB}$, (d) the real part of $\mathbf{AB}$, (e) the phase of $\mathbf{AB}$, and (f) the real part of $\mathbf{A/B}$.

1.6.1. A mass of 0.5 kg hangs on a spring. When an additional mass of 0.2 kg is attached to the spring, the spring stretches an additional 0.04 m. When the 0.2 kg mass is abruptly removed, the amplitude of the ensuing oscillations of the 0.5 kg mass is observed to decrease to $1/e$ of its initial value in 1.0 s. Compute values for R_m, ω_d, A, and ϕ.

1.6.2. Verify that for a critically damped oscillator $x = (A + Bt)\exp(-\beta t)$ satisfies the equation of motion.

1.6.3. Show that if $\beta \ll \omega_0$ then $\omega_d \approx \omega_0[1 - \frac{1}{2}(\beta/\omega_0)^2]$.

1.6.4. A damped oscillator whose general solution is $x = A\exp(-\beta t)\cos(\omega_d + \phi)$ starts at rest with a positive speed u_0. Find A.

1.6.5. For the damped simple oscillators of Fig. 1.6.2, find A and ϕ for each case.

1.7.1. (a) What is the general expression for the acceleration of a damped oscillator driven by a force $F\cos\omega t$? (b) Derive an expression for the angular frequency maximizing the acceleration.

1.7.2. From (1.7.9) find the angular frequencies for which Θ goes to (a) 0, (b) $\pi/2$, (c) $-\pi/2$, and (d) has magnitude $\pi/4$.

1.7.3. A mass M is connected to a rigid foundation by a spring and dashpot (spring constant s and mechanical resistance R_m) and is constrained to move perpendicular to the foundation. A second mass m is attached to M by an arm of length L and rotates with angular frequency ω about an axis perpendicular to the motion of M. Find the resulting steady-state speed amplitude of M.

1.7.4. The inertial switch that activates an airbag in an automobile can be modeled as a spring of stiffness s with one end attached to a case fixed to the vehicle and the other end attached to a mass m free to move within the case. When the case is decelerated, the mass compresses the spring to activate a switch that releases the air into the airbag. The case is decelerated at a constant rate a. (a) Find the equation of motion for the mass. (b) By direct substitution, show that the motion of the mass is $x = \frac{1}{2}at^2 + (a/\omega_0^2)(\cos\omega_0 t - 1)$, where $\omega_0^2 = s/m$. (c) Find the minimum deceleration required to compress the spring a distance X, and express the answer in terms of the force necessary to statically compress the spring the same distance X.

1.7.5. A mass m is attached to a spring of stiffness s. The motion of the other end of the spring is attached to a table whose acceleration is $A\exp(j\omega t)$ with A a constant. (a) Show that the ratio of the acceleration of the mass to that of the table is $[1 - (\omega/\omega_0)^2]^{-1}$, where $\omega_0^2 = s/m$. (b) Plot this ratio as a function of ω/ω_0 for $0 < \omega/\omega_0 < 5$. (c) Comment on the applicability of this system as a vibration isolator.

1.7.6C. An oscillator with mass 0.5 kg, stiffness 100 N/m, and mechanical resistance 1.4 kg/s is driven by a sinusoidal force of amplitude 2 N. Plot the speed amplitude and the phase angle between the displacement and speed as a function of the driving frequency and find the frequencies for which the phase angle is 45°.

1.8.1. An oscillator at rest experiences a force $F\sin\omega_0 t$ beginning at $t = 0$. If $\beta \ll \omega_0$, show that $x(t) \approx -(F/\omega_0 R_m)[1 - \exp(-\beta t)]\cos\omega_0 t$.

1.8.2. An undamped oscillator is driven beginning at $t = 0$ with a force $F\sin\omega t$, where $\omega \neq \omega_0$. (a) Find the resultant speed of the mass if it is at rest at $t = 0$. (b) Sketch the waveform of the speed if $\omega = 2\omega_0$. (c) If a small amount of damping is introduced and the driving frequency is far below resonance, show that the steady-state solution is approximated by $u(t) \approx (\omega F/s)\cos\omega t$.

1.8.3. The displacement of a damped oscillator driven by a square wave forcing function with fundamental angular frequency ω is shown below for (a) $\omega/\omega_d \gg 1$ and (b) $\omega/\omega_d \ll 1$. Explain the behaviors of these curves on physical grounds.

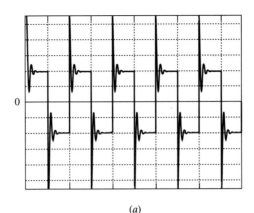

(a)

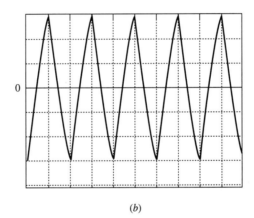

(b)

1.9.1. Show that the instantaneous power Π_i is given by $\mathrm{Re}\{\mathbf{f}\} \cdot \mathrm{Re}\{\mathbf{u}\}$ but not by $\mathrm{Re}\{\mathbf{fu}\}$.

1.9.2. A damped oscillator is driven with a force $\mathbf{f} = F\exp(j\omega t)$ and has speed $\mathbf{u} = U\exp[j(\omega t - \Theta)]$ with $U = F/Z_m$. (a) Show that the power Π consumed is $\frac{1}{2}\mathrm{Re}\{\mathbf{fu}^*\}$, where $\mathbf{u}^*$ is the complex conjugate of $\mathbf{u}$. (b) Show that $\mathrm{Re}\{\mathbf{fu}^*\} = \mathrm{Re}\{\mathbf{f}^*\mathbf{u}\}$.

1.9.3. Verify that in the steady state the power dissipated by the frictional force in the damped driven oscillator is equal to that being supplied by the driving force.

1.9.4. Use the average power supplied to an oscillator by a driving force and the total energy E stored in the oscillator to obtain the relationship $dE/dt = 2\beta E$. Explain the physical meaning of this result. Does this relationship mandate exponential decay?

1.10.1. Show that $Z_m = \omega_0 m[(\omega/\omega_0 - \omega_0/\omega)^2 + 1/Q^2]^{1/2}$.

1.10.2. A mass of 0.5 kg hangs on a spring. The stiffness of the spring is 100 N/m, and the mechanical resistance is 1.4 kg/s. The force (N) driving the system is $f = 2\cos 5t$. (a) What will be the steady-state values of the displacement amplitude, speed amplitude, and average power dissipation? (b) What is the phase angle between speed and force? (c) What is the resonance frequency and what would be the displacement amplitude, speed amplitude, and average power dissipation at this frequency and for the same force magnitude as in (a)? (d) What is the Q of the system, and over what range of frequencies will the power loss be at least 50% of its resonance value?

1.10.3. When a simple oscillator is driven at its resonance frequency, show that the ratio of the energy dissipated per cycle to the total mechanical energy present is $2\pi/Q$.

1.10.4. Derive equations that give the two angular frequencies corresponding to the half-power points of a driven oscillator. Show that they are given approximately by $\omega_0 \pm R_m/2m$.

1.10.5. Derive an equation for Q from $d\Theta/df$ evaluated at $f = f_0$.

1.10.6. For a lightly damped ($\beta/\omega_0 \ll 1$) driven oscillator, show that, to second order in β/ω_0, $\frac{1}{2}(\omega_l + \omega_u) = \omega_0 + \frac{1}{2}(\beta/\omega_0)^2$.

1.10.7. The resonance curve of an oscillator can be obtained experimentally with the use of a *wave analyzer* (an electronic instrument that automatically sweeps the driving frequency of the applied voltage while plotting the output current as a function of frequency). These instruments have a sweep rate that can be changed from "slow" to "fast." Plotted below are resonance curves for the same oscillator obtained at (I) a fast sweep rate and (II) a slow sweep rate. The horizontal scale is 1 Hz/division. Give a qualitative explanation why curve (II) differs so radically from the expected response.

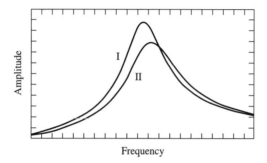

Frequency

1.12.1. A mass m is fastened to one end of a horizontal spring of stiffness s, and a horizontal driving force $F\sin\omega t$ is applied to the other end of the spring. (*a*) Assuming no damping, determine the equation giving the motion of the driven end of the spring as a function of time. (*b*) Show that the expression for the speed of this end of the spring is analogous to that giving the current into a parallel LC electrical circuit. (*c*) If the constants of the above system are $F = 3\,\mathrm{N}$, $s = 200\,\mathrm{N/m}$, and $m = 0.5\,\mathrm{kg}$, compute and plot curves showing how the displacement and speed amplitudes of the driven end of the spring vary with frequency in the range $0 < \omega < 100\,\mathrm{rad/s}$.

1.12.2. Find the mechanical impedances, the resonance frequencies, and the equivalent electrical circuits for the following systems.

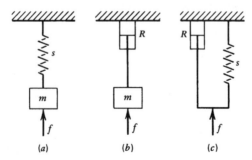

1.12.3. Masses m and M are connected by a spring of stiffness s and the smaller mass m is driven with an external force. (*a*) Obtain the equivalent electrical circuit. (*b*) Obtain the resonance angular frequency ω_0. (*c*) How does ω_0 change if the mass M is the one that is driven?

1.12.4C. For each of the three equivalent circuits of Figs. 1.12.2–1.12.4, plot the amplitudes of the displacement and speed for a frequency span showing all significant aspects of the motion. Assume $m = 0.5$ kg, $s = 100$ N/m, $R_m = 1.4$ kg/s, and $F = 2$ N.

1.13.1. Show that the amplitude A_n of the displacement resulting from the linear addition of n harmonic vibrations all of the same amplitude A and frequency but having different initial phase angles of $\phi_1 = \varepsilon$, $\phi_2 = 2\varepsilon$, $\phi_3 = 3\varepsilon, \ldots, \phi_n = n\varepsilon, \ldots$ is given by

$$A_n = \frac{A\sin(n\varepsilon/2)}{\sin(\varepsilon/2)}$$

1.13.2. Assume a damped oscillator is driven with a force composed of two terms of angular frequencies ω_1 and ω_2. (a) Evaluate Π_i. (b) Show that the power Π dissipated is the sum of the powers dissipated by each of the terms acting alone.

1.13.3. Show that the sum of two simple harmonic vibrations with the same amplitudes A and initial phase angles $\phi = 0$, but with different angular frequencies ω_2 and ω_1, is $x = 2A\sin(\Delta\omega/2)\exp[j(\omega_1 + \Delta\omega/2)t]$, where $\omega_2 = \omega_1 + \Delta\omega$.

1.13.4C. Plot the coherent sum of two sine waves of equal displacement amplitudes and frequencies but with relative phases differing from $0°$ to $360°$ in steps of $45°$.

1.13.5C. Make plots of the sum of two signals with equal amplitudes but with different frequencies f and $f + \Delta f$ for different values of $(f + \Delta f)/f$. Observe the behavior of the resulting wave as $f + \Delta f$ is increased from f to $2f$ (an octave). If you have sound output on your computer, correlate the waveform and the quality of the sound.

1.13.6C. An amplitude modulated (AM) signal is $x = [1 + m\cos(2\pi ft)]\cos(2\pi Ft)$, where F is the carrier frequency, f is the signal frequency, and m is the modulation index. (a) Use trigonometric identities to show that this signal is composed of three components: the carrier F and two sidebands $F \pm f$. (b) Plot x for $F = 20$ kHz and $f = 1$ kHz for $m = 0.5, 1$, and 2.

1.14.1. Show by direct calculation that the square wave of unit amplitude is represented by (1.14.5).

1.14.2. Show that the Fourier components of the fixed, fixed string of length L pulled aside a distance h at its midpoint are $A_n = (1/n^2)(8h/\pi^2)\sin(n\pi/2)$. *Hint:* Note this has the same profile in space as a triangular wave in time over half its period.

1.14.3. (a) Show that a sawtooth wave falling linearly from $+1$ to -1 over each period T of motion can be expressed as $f(t) = (1 - 2t/T)$ for $0 \le t < T$ and repeats for each period. (b) Determine that the Fourier coefficients for the waveform are $A_n = 0$ and $B_n = 2/n\pi$ for $n = 1, 2, 3, \ldots$.

1.15.1. Using the spectral density $\mathbf{g}(w)$ of the force $F\exp(jwt)$ solve (1.15.13) for the resultant speed of an oscillator with mechanical impedance $\mathbf{Z}_m$.

1.15.2. One end of a dashpot of mechanical resistance R_m is attached to a wall and the other end is struck at $t = 0$ by a force $F(t) = \mathscr{F}\delta(t)$ where $\mathscr{F} = 1$ N·s is the impulse. (a) Obtain the speed and displacement of the struck end by the Fourier transform technique. (b) Confirm the results of (a) by direct solution of (1.7.1).

1.15.3. (a) Derive $\mathbf{g}(w)$ for $\mathbf{f}(t) = \exp(-bt) \cdot 1(t)$ from (1.15.3). Note the role b plays in reducing the upper limit of the integration to zero. (b) Derive $\mathbf{g}(w)$ for $\mathbf{f}(t) = 1(t)$ from (1.15.3) by allowing $b \to 0$. Explain physically why this result is acceptable even though in the limit the integration becomes improper.

1.15.4. A simple oscillator at rest is struck with a force $\mathbf{f}(t) = \mathcal{F}\delta(t)$ where $\mathcal{F} = 1 \text{ N·s}$ is the impulse. Find the displacement and speed of the mass using Fourier transforms.

1.15.5. Obtain the spectral density for $\mathbf{f}(t) = \exp(j\omega t) \cdot 1(t)$ from (1.15.3). *Hint:* Introduce a little attenuation by multiplying $\mathbf{f}(t)$ by $\exp(-bt)$, evaluate the integral, and then let $b \to 0$.

1.15.6. A simple oscillator at rest is struck with a force $F(t) = F \cdot 1(t)$ where $F = 1 \text{ N}$. (*a*) Find the displacement and speed of the oscillator using Section 1.15. (*b*) Confirm the result by direct solution of (1.7.1).

1.15.7. A simple oscillator is suspended vertically with the spring unstretched. At time $t = 0$ the mass is released so that it is suddenly subjected to the force of gravity. Solve for the resulting displacement as a function of time.

1.15.8. Using Table 1.15.1, obtain the spectral density of $\mathbf{f}(t) = [\exp(-bt) \sin \omega t] \cdot 1(t)$.

1.15.9. Assume that a rectangular wave pulse is turned on at $t = 0$ and is turned off at a later time Δt. (*a*) Show that the wave can be represented by $\mathbf{f}(t) = [1(t) - 1(t - \Delta t)]$. (*b*) Show that the frequency spectrum of this wave is $\mathbf{g}(w) = -(j/2\pi w)[1 - \exp(-jw\,\Delta t)]$. (*c*) Determine the interval Δw between the first two zeros of $\mathbf{g}(w)$ along the $\pm w$ axes and explain why this contains the most important part of the spectral density. (*d*) Show that $\Delta w\,\Delta t = 4\pi$.

1.15.10C. Make plots of the frequency spectrum of a sawtooth waveform defined by

$$f(t) = a[1 - 2(t/T - n)] \qquad n \le t/T \le (n+1)$$

for all integer n. Retain the same number of terms as in each of the plots for the square wave in Fig. 1.14.1. Comment on the relative importance of overshoot for the two waveforms.

1.15.11. (*a*) Show that the spectral density of $\cos(\omega t + \phi)$ is

$$\mathbf{g}(w) = \tfrac{1}{2}[\delta(w - \omega)e^{j\phi} + \delta(w + \omega)e^{-j\phi}]$$

(*b*) From (*a*) find the spectral densities of $\cos \omega t$ and $\sin \omega t$.

1.15.12. (*a*) Derive the spectrum of the pulse of Fig. 1.15.1. (*b*) Find the maximum value of the spectral density. (*c*) Find the width Δw of the central peak between the first nulls and show that this gives $\Delta w\,\Delta t \approx 4\pi$. (*d*) Find the width $\Delta w'$ of the central peak between the points where the curve has half its maximum value and show that this gives $\Delta w'\,\Delta t \approx 2.4\pi$. *Hint:* Use trial and error to obtain the relevant value of $w\,\Delta t/2$.

1.15.13. If the pulse in Fig. 1.15.2 has unit amplitude, show that

$$\mathbf{g}(w) = -\frac{1}{\pi}\frac{w}{\omega^2 - w^2}\sin\left(4\pi\frac{w}{\omega}\right)$$

C h a p t e r 2

TRANSVERSE MOTION: THE VIBRATING STRING

2.1 VIBRATIONS OF EXTENDED SYSTEMS

In the previous chapter it was assumed that the mass moves as a rigid body with no rotation so that it could be considered concentrated at a single point. However, most vibrating bodies are not so simple. For example, the diaphragm of a loudspeaker has its mass distributed over its surface and the cone does not move as a unit. The same occurs for a piano string and for the surface of a cymbal. Rather than beginning with the study of such complicated vibrations, we consider first the ideal vibrating string, the most readily visualized physical system involving the propagation of waves. Even this simple system is a hypothetical one; certain simplifying assumptions must be made that cannot be completely realized in practice. Nevertheless, the results obtained are extremely important because they yield a fundamental understanding of wave phenomena.

2.2 TRANSVERSE WAVES ON A STRING

If a portion of a stretched string is displaced from its equilibrium position and released, it is observed that the displacement does not remain fixed in its initial position, but breaks up into two separate disturbances that propagate along the string, one moving to the right and the other to the left with equal speed, as suggested by Fig. 2.2.1. Furthermore, it is observed that the speed of propagation of a *small disturbance* is independent of the shape and amplitude of the initial displacement and depends only on the mass per unit length of the string and its tension. Experiment and theory show that this speed is given by $c = \sqrt{T/\rho_L}$, where c is in m/s, T is the tension in N, and ρ_L is the *linear density* (mass per unit length) of the string in kg/m. A propagating transverse disturbance is referred to as a *transverse traveling wave*.

37

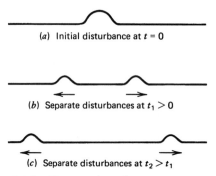

(a) Initial disturbance at $t = 0$

(b) Separate disturbances at $t_1 > 0$

(c) Separate disturbances at $t_2 > t_1$

Figure 2.2.1   Propagation of a transverse disturbance along a stretched string. (a) Initial stationary disturbance at $t = 0$. (b) Disturbances moving to right and left at $t_1 > 0$. (c) Disturbances moving to right and left at $t_2 > t_1$.

2.3 THE ONE-DIMENSIONAL WAVE EQUATION

By considering the forces that tend to return the string to its equilibrium position, it is possible to derive a *wave equation*. Solutions of this wave equation satisfying the appropriate initial and boundary conditions will completely define the motion of the string.

Assume a string of uniform linear density ρ_L and negligible stiffness, stretched to a tension T great enough that the effects of gravity can be neglected. Also assume that there are no dissipative forces (such as those associated with friction or with the radiation of acoustic energy). Figure 2.3.1 isolates an infinitesimal element of the string with equilibrium position x and equilibrium length dx. If the transverse displacement of this element from its equilibrium position y is small, the tension T remains constant along the string and the difference between the y components of the tension at the two ends of the element is

$$df_y = (T \sin \theta)_{x+dx} - (T \sin \theta)_x \qquad (2.3.1)$$

where θ is the angle between the tangent to the string and the x axis, $(T \sin \theta)_{x+dx}$ is the value of $T \sin \theta$ at $x + dx$, and $(T \sin \theta)_x$ is its value at x. Applying the Taylor's

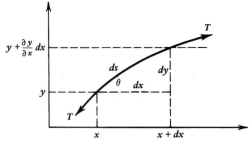

Figure 2.3.1   Forces acting on a string element of length ds.

series expansion

$$f(x + dx) = f(x) + \left(\frac{\partial f}{\partial x}\right)_x dx + \frac{1}{2}\left(\frac{\partial^2 f}{\partial x^2}\right)_x dx^2 + \cdots \tag{2.3.2}$$

to (2.3.1) gives

$$df_y = \left[(T\sin\theta)_x + \frac{\partial(T\sin\theta)}{\partial x} dx + \cdots\right] - (T\sin\theta)_x = \frac{\partial(T\sin\theta)}{\partial x} dx \tag{2.3.3}$$

where we have retained only the lowest-order nonvanishing terms. If θ is small, $\sin\theta$ may be replaced by $\partial y/\partial x$ and the net transverse force on the element becomes

$$df_y = \frac{\partial}{\partial x}\left(T\frac{\partial y}{\partial x}\right) dx = T\frac{\partial^2 y}{\partial x^2} dx \tag{2.3.4}$$

Since the mass of the element is $\rho_L\, dx$ and its acceleration in the y direction is $\partial^2 y/\partial t^2$, Newton's law gives

$$df_y = \rho_L\, dx \frac{\partial^2 y}{\partial t^2} \tag{2.3.5}$$

Combination of (2.3.4) and (2.3.5) then yields a *one-dimensional wave equation*,

$$\boxed{\frac{\partial^2 y}{\partial x^2} = \frac{1}{c^2}\frac{\partial^2 y}{\partial t^2}} \tag{2.3.6}$$

where the constant c^2 is defined by

$$\boxed{c^2 = T/\rho_L} \tag{2.3.7}$$

2.4 GENERAL SOLUTION OF THE WAVE EQUATION

Equation (2.3.6) is a second-order, partial differential equation. Its complete solution contains two arbitrary, but twice differentiable, functions

$$y(x, t) = y_1(ct - x) + y_2(ct + x) \tag{2.4.1}$$

one of argument $(ct - x)$ and the other of argument $(ct + x)$. Direct substitution of y_1 and y_2 into the wave equation and successive applications of the chain rule with the arguments $w = (ct \pm x)$,

$$\frac{\partial f}{\partial t} = \frac{df}{dw}\frac{\partial w}{\partial t} = c\frac{df}{dw} \quad \text{and} \quad \frac{\partial f}{\partial x} = \frac{df}{dw}\frac{\partial w}{\partial x} = \pm\frac{df}{dw} \tag{2.4.2}$$

will verify that they are solutions. Examples of such functions include $\log(ct \pm x)$, $(t \pm x/c)^2$, $\sin[\omega(t \pm x/c)]$, $\exp[j\omega(t \pm x/c)]$, and $\cosh(ct \pm x)$.

2.5 WAVE NATURE OF THE GENERAL SOLUTION

Consider the solution $y_1(ct - x)$. At time t_1 the transverse displacement of the string is given by $y_1(ct_1 - x_1)$, as suggested by Fig. 2.5.1a. At a later time t_2 the same displacement will be given by $y_1(ct_2 - x_2)$, as suggested by Fig. 2.5.1b. A particular value of the transverse displacement found at x_1 when $t = t_1$ will be found at position x_2 when $t = t_2$ if

$$ct_1 - x_1 = ct_2 - x_2 \tag{2.5.1}$$

Thus, this point of the waveform has moved a distance

$$x_2 - x_1 = c(t_2 - t_1) \tag{2.5.2}$$

to the right. The shape of the disturbance remains unchanged and travels along the string to the right at a constant propagation speed c. The function $y_1(ct - x)$ thus represents a *wave* traveling in the $+x$ direction. The speed c with which a particular value of y_1 propagates along the string is called the *phase speed*. It is important to note that, while the waveform propagates with phase speed c, the material elements of the string move transversely about their equilibrium positions with *particle speeds* given by $u(x, t) = \partial y_1/\partial t$. This same argument can be applied to a wave traveling in the $-x$ direction.

The wave shape remains constant as the initial disturbance progresses along the string. This mathematical conclusion is never exactly realized in practice, since the assumptions made in deriving the wave equation are never completely fulfilled for real strings, which always have some bending stiffness and are acted on by dissipative forces. Waves traveling along real strings become distorted. For the relatively flexible strings and the low damping normally encountered in musical instruments, the rate of distortion is quite slight if the amplitude of the disturbance is small. For large amplitudes, on the other hand, the change of wave shape may be pronounced.

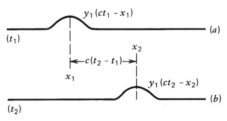

Figure 2.5.1   A transverse wave traveling to the right. (a) Waveform at t_1. (b) Waveform at $t_2 > t_1$. The wave propagates with no distortion and phase speed $c = (x_2 - x_1)/(t_2 - t_1)$.

2.6 INITIAL VALUES AND BOUNDARY CONDITIONS

The functions $y_1(ct - x)$ and $y_2(ct + x)$ are determined by the *initial values* and the *boundary conditions*. For the freely vibrating string, *initial values* at $t = 0$ are determined by the type and point of application of the exciting force applied to the string. For example, the initial wave shape set up by *striking* a string (when a piano is played) is quite different from that established by *plucking* a string (a harp or guitar) or in *bowing* a string (a violin); the functions representing the wave shape are consequently different. They are further determined by the boundary conditions at the ends of the string. Actual strings are always finite in length and must be held in some manner at their ends. For example, if the supports of the strings are rigid, the sum $y_1 + y_2$ is constrained to have zero value at all times at the points of support. When a string is driven to *steady-state* conditions by a periodic external driving force, the functions y_1 and y_2 are periodic with the same frequency, but their other characteristics (such as amplitude of vibration) are determined by the point of application of the force and by the boundary conditions.

2.7 REFLECTION AT A BOUNDARY

Assume that a string is rigidly supported (*fixed*) at $x = 0$. Then $y_1(ct - x)$ and $y_2(ct + x)$ are no longer completely arbitrary since their sum must be zero at all times at $x = 0$. A moment's thought reveals that this boundary condition can be satisfied if

$$y(0, t) = y_1(ct - 0) + y_2(ct + 0) = 0 \qquad (2.7.1)$$

As may be seen in Fig. 2.7.1, the process of reflection at a rigid boundary can be considered as one in which the wave moving to the left is transformed into a wave of opposite displacement traveling to the right,

$$y(x, t) = y_1(ct - x) - y_1(ct + x) \qquad (2.7.2)$$

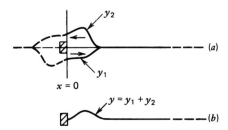

Figure 2.7.1 Reflection of a transverse wave traveling to the left from a rigid end. In (*a*), the dashed-line segment of wave y_2 is shown being reflected to become the solid-line segment of wave y_1. The resultant wave, shown in (*b*), travels to the right. Note that the displacement at $x = 0$ is always zero.

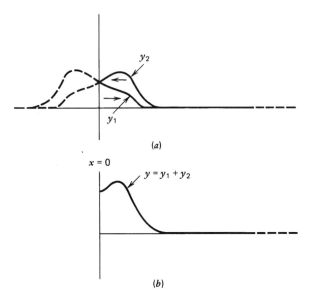

Figure 2.7.2   Reflection from a free end. In (*a*), the
dashed-line segment of wave y_2 traveling to the left is
shown being reflected to become the solid-line segment
of wave y_1. The resultant wave, shown in (*b*), travels to
the right. Note that the slope at $x = 0$ is always zero.

Another example of a simple boundary condition is an end supported so that, although the string is held taut, there is no transverse force on the string. Such an end is termed a *free* end. The absence of a transverse force requires $T \sin \theta$ to vanish. This means that the incident and reflected waveforms must have equal and opposite slopes with respect to x, and this in turn means that the waveforms have identical profiles (viewed from the directions into which they are propagating). Thus,

$$y(x,t) = y_1(ct - x) + y_1(ct + x) \qquad (2.7.3)$$

The process of reflection at a free boundary may be considered as one in which the wave moving to the left is reflected into an identically shaped wave of the same displacement traveling to the right (Fig. 2.7.2) so that the slope $\partial y / \partial x$ is zero at the boundary.

2.8 FORCED VIBRATION OF AN INFINITE STRING

The simplest type of vibration that can be set up on a string results from the application of a *transverse sinusoidal driving force* to one end of an ideal string of infinite length. Since all real strings are of finite length, this particular problem may seem to be of purely academic interest, but its analysis is justified. (1) It furnishes a simple introduction to the study of vibrations of strings of finite length that aids in the understanding of the transmission of acoustic waves. (2) With carefully chosen termination at one end, a string can act as if it were infinitely long.

Consider an ideal string of infinite length extending to the right from $x = 0$, stretched to a tension T, with a transverse driving force $F \cos \omega t$ applied at the end $x = 0$. Assume that the end does not move in the x direction but is free to move in the y direction. As in the previous chapter, let us replace $F \cos \omega t$ with the complex force $\mathbf{f}(t) = F \exp(j\omega t)$. Since the string extends infinitely far in the positive x direction and is excited into motion by the force at its left end, the solution must contain only waves moving to the right,

$$\mathbf{y}(x, t) = \mathbf{y}_1(ct - x) \tag{2.8.1}$$

The boundary condition at $x = 0$ requires

$$\mathbf{y}(0, t) = \mathbf{A}e^{j\omega t} \tag{2.8.2}$$

where $\mathbf{A}$ is a complex constant (whose amplitude and phase eventually will be related to the driving force). Combination gives

$$\mathbf{y}_1(ct) = \mathbf{A}e^{jk(ct)} \tag{2.8.3}$$

where the *wave number* k is defined by

$$\boxed{k = \omega/c} \tag{2.8.4}$$

The solution for all x must be $\mathbf{A} \exp[jk(ct - x)]$ or

$$\mathbf{y}(x, t) = \mathbf{A}e^{j(\omega t - kx)} \tag{2.8.5}$$

Figure 2.8.1a shows the shape of the string at two instants of time and Fig. 2.8.1b the time histories of two points on the string. The elements of the string execute simple harmonic motion about their equilibrium positions with frequency $f = \omega/2\pi$ and period $T = 1/f$. The shape of the string at any instant is a sinusoid of amplitude $A = |\mathbf{A}|$. At fixed time, the shape is a function of x, and when x changes by an amount λ so that $k\lambda = 2\pi$, the displacement and slope of the string are as before. The distance λ between these corresponding points is called the *wavelength* and we see that

$$\boxed{\lambda = 2\pi/k} \tag{2.8.6}$$

This waveform moves to the right with a phase speed $c = \sqrt{T/\rho_L}$ and is called a *harmonic traveling wave*. Because the waveform moves one wavelength in a time equal to one period, frequency and wavelength are related to the phase speed by

$$\boxed{c = \lambda f} \tag{2.8.7}$$

a relationship fundamental to all wave motion. Note that (2.8.7) can also be obtained from (2.8.4) by expressing k as $2\pi/\lambda$ and ω as $2\pi f$.

To relate the amplitude of the wave to the driving force, consider the forces applied to the left end of the string, as shown in Fig. 2.8.2. Since there is no mass

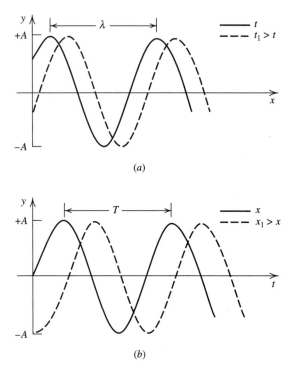

(a)

(b)

Figure 2.8.1   A harmonic wave traveling to the right: (a) spatial behavior at two closely occurring times, with λ the wavelength; and (b) temporal behavior at two closely spaced positions, with T the period.

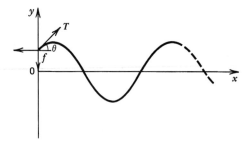

Figure 2.8.2   Forces acting on the end of a driven string. The vertical component of the tension $T\sin\theta$ acting on the driver is balanced by f, the vertical component of the force of the driver on the string.

concentrated at the end of the string, the driver must provide the force necessary to exactly balance the tension: opposite to $T\cos\theta$ horizontally and opposing $T\sin\theta$ vertically, as suggested in the figure. Therefore, the total transverse force $(f + T\sin\theta)$ on the string at the left end must vanish, so that for small values of θ

$$\mathbf{f} = -T\left(\frac{\partial \mathbf{y}}{\partial x}\right)_{x=0} \tag{2.8.8}$$

where the minus sign denotes that the applied force must be directed downward when $(\partial \mathbf{y}/\partial x)_{x=0}$ is positive. Thus, we see that the slope of the string at the forced end is determined by the applied force and the tension in the string, $(\partial \mathbf{y}/\partial x)_{x=0} = -\mathbf{f}/T$. [For example, if $\mathbf{f} = 0$ so that the end of the string is free to move transversely, then $(\partial \mathbf{y}/\partial x)_{x=0} = 0$. This is the boundary condition for a free end, as stated in Section 2.7.] Substitution of $\mathbf{f} = F\exp(j\omega t)$ and (2.8.5) into (2.8.8) gives

$$Fe^{j\omega t} = -T(-jk)\mathbf{A}e^{j\omega t} \tag{2.8.9}$$

so that

$$\mathbf{y}(x, t) = (F/jkT)e^{j(\omega t - kx)} \tag{2.8.10}$$

and the particle speed $\mathbf{u} = \partial \mathbf{y}/\partial t$ becomes

$$\mathbf{u}(x, t) = (F/\rho_L c)e^{j(\omega t - kx)} \tag{2.8.11}$$

Now, let us define the *input mechanical impedance* $\mathbf{Z}_{m0}$ of the string as the ratio of the driving force to the transverse speed of the string at the driving point ($x = 0$),

$$\boxed{\mathbf{Z}_{m0} = \mathbf{f}/\mathbf{u}(0, t)} \tag{2.8.12}$$

Then for the case of the infinite string

$$\mathbf{Z}_{m0} = \rho_L c \tag{2.8.13}$$

The input impedance of an infinite string is a real quantity so that the mechanical load offered by the string is purely resistive. This is to be expected, since an infinite string can propagate energy only away from the driver. The input impedance of an infinite string is a function only of the tension of the string and its mass per unit length. Independent of the applied driving force, it is thus a characteristic property of the *string* and not of the *wave*. For this reason it is called the *characteristic mechanical impedance* of the string. It is analogous to the characteristic electrical impedance of an infinitely long electrical transmission line.

The instantaneous power input to the string is $\Pi_i = fu$ with u evaluated at $x = 0$, or

$$\Pi_i = (F\cos \omega t)\left[(F/\rho_L c)\cos \omega t\right] \tag{2.8.14}$$

The time average over one cycle gives the average power input,

$$\Pi = F^2/2\rho_L c = \tfrac{1}{2}\rho_L c U_0^2 \tag{2.8.15}$$

where

$$U_0 = |\mathbf{u}(0, t)| = F/\rho_L c \tag{2.8.16}$$

is the speed amplitude of the string at $x = 0$.

2.9 FORCED VIBRATION
OF A STRING OF FINITE LENGTH

The behavior of a string of *finite* length forced at one end is considerably more complicated than that of the infinite string. The wave reflected from the support at the far end of the string coexists with the wave traveling toward the support and in turn is reflected from the driven end. However, when the steady state is attained, the solution must be expressible in terms of two harmonic waves traveling in opposite directions:

$$\mathbf{y}(x, t) = \mathbf{A}e^{j(\omega t - kx)} + \mathbf{B}e^{j(\omega t + kx)} \tag{2.9.1}$$

where the complex amplitudes **A** and **B** are determined by the boundary conditions. Let us consider several classes of termination.

(a) The Forced, Fixed String

Assume that a string is driven at one end and fixed at the other. At the left end, the boundary condition is (2.8.8)

$$Fe^{j\omega t} + T\left(\frac{\partial \mathbf{y}}{\partial x}\right)_{x=0} = 0 \tag{2.9.2}$$

at all times. Substitution of (2.9.1) into this boundary condition gives

$$F + T(-jk\mathbf{A} + jk\mathbf{B}) = 0 \tag{2.9.3}$$

Since the string is rigidly supported at $x = L$, the displacement at this point is always zero so that

$$\mathbf{A}e^{-jkL} + \mathbf{B}e^{jkL} = 0 \tag{2.9.4}$$

Solving (2.9.3) and (2.9.4) simultaneously for **A** and **B**, we have

$$\mathbf{A} = \frac{Fe^{jkL}}{2jkT \cos kL}$$

$$\mathbf{B} = -\frac{Fe^{-jkL}}{2jkT \cos kL} \tag{2.9.5}$$

Substitution of these constants into (2.9.1) gives

$$\mathbf{y}(x, t) = \frac{F}{2jkT \cos kL}\left(e^{j[\omega t + k(L-x)]} - e^{j[\omega t - k(L-x)]}\right) \tag{2.9.6}$$

or factoring the exp($j\omega t$) and simplifying,

$$\mathbf{y}(x, t) = \frac{F}{kT}\frac{\sin[k(L-x)]}{\cos kL}e^{j\omega t} \tag{2.9.7}$$

Thus, we have two different but equivalent ways of looking at the solution: (2.9.6) can be interpreted as two waves of equal amplitudes and wavelengths traveling in opposite directions on the string. On the other hand, (2.9.7) describes a waveform that does not propagate along the string; instead, the string oscillates while the waveform remains stationary. Such a wave is called a *standing wave* and is characterized mathematically by an amplitude that depends on the position along the string. These two descriptions reveal that a combination of waves of equal amplitudes traveling in opposite directions gives rise to a stationary vibration with a spatially dependent amplitude. This ability to view standing waves as combinations of traveling waves, and vice versa, will often be utilized in dealing with wave motion.

Consideration of the term $\sin[k(L - x)]$ in (2.9.7) shows that there are positions, called *nodes*, where the displacement is zero at all times. These locations are given by $k(L - x) = q\pi$ for $q = 0, 1, 2, \ldots, \leq kL/\pi$. The positions x_q of the nodes are then

$$x_q = L - q\lambda/2 \qquad q = 0, 1, 2, \ldots, \leq 2L/\lambda \qquad (2.9.8)$$

A representative standing wave is shown in Fig. 2.9.1, where the instantaneous displacements of the string at various times have been sketched. The distance between nodes is $\lambda/2$. The moving portions of the string between the nodes are called *loops*, and the positions of *maximum* displacement are called *antinodes*.

Note that the position of the driver with respect to the nodes is a function of frequency. If L is an integral multiple of $\lambda/2$, a node will occur at the position of the driver. If the frequency is then increased, the wavelength decreases, causing the node to migrate away from the driver. An antinode will exist at the driver for driving frequencies such that L is an odd multiple of $\lambda/4$.

The migration of the nodes with varying driving frequency is accompanied by some startling changes in the amplitude at the antinodes. The denominator of (2.9.7) becomes zero at driving frequencies such that $\cos kL = 0$,

$$kL = (2n - 1)\pi/2 \qquad n = 1, 2, 3, \ldots \qquad (2.9.9)$$

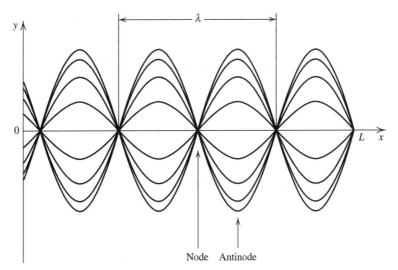

Figure 2.9.1 The shape of a string of length L at several different times for a standing wave. The nodes are separated by $\lambda/2$.

Since $\omega/k = c$, this gives

$$f_{rn} = [(2n - 1)/4](c/L) \tag{2.9.10}$$

The string has its strongest vibrations when the driving frequency has one of the values f_{rn}. These are the *resonance frequencies*. The infinite amplitudes of vibration predicted by (2.9.7) at resonance do not occur in actual strings because the assumptions of small θ, constant T, and no losses are violated. However, the amplitude will be a maximum at these frequencies. Note that at resonance there is an antinode at the driven end, so that $u(0, t)$ is as large as possible.

Similarly, the frequencies for which the amplitude is a minimum are determined by the condition $\cos kL = \pm 1$,

$$kL = n\pi \qquad n = 1, 2, 3, \ldots \tag{2.9.11}$$

or

$$f_{an} = (n/2)(c/L) \tag{2.9.12}$$

[It will be observed from (2.9.7) that these minimum amplitudes decrease progressively with increasing frequency.] These frequencies are the *antiresonance frequencies*. At antiresonance, there is a node at the driven end, so that $u(0, t) = 0$. [In reality, the presence of dissipation causes $u(0, t)$ to be finite, but small; this will be studied in more detail later.]

The input mechanical impedance $\mathbf{Z}_{m0}$ is given by (2.8.12),

$$\mathbf{Z}_{m0} = Fe^{j\omega t}/\mathbf{u}(0, t) \tag{2.9.13}$$

which for the case of the forced, fixed string yields

$$\mathbf{Z}_{m0} = -j\rho_L c \cot kL \tag{2.9.14}$$

This impedance is a pure reactance so that no power is absorbed by the string. (For a lossless string with a fixed end, there is no way for energy to leave the system.)

Consideration of the input impedance leads to the same conclusions: whenever $\cot kL = 0$, the input impedance is zero and the amplitude of vibration is consequently a maximum. *The resonance frequencies of any mechanical system are defined in general as those frequencies for which the input mechanical reactance goes to zero.* For the forced, fixed string this yields the resonance frequencies given by (2.9.10). At the antiresonance frequencies given by (2.9.12), $\mathbf{Z}_{m0}$ is infinite and the motion of the driven end of the string is infinitesimally small, although the remainder of the string is in motion. When $\mathbf{Z}_{m0}$ is not purely reactive, the specification of antiresonance becomes more complicated. This will be investigated in Chapter 3.

For very low frequencies the input impedance has the limiting value

$$\mathbf{Z}_{m0} \rightarrow -j\rho_L c/kL = -jT/\omega L \tag{2.9.15}$$

which is identical with the input impedance of a spring having stiffness $s = T/L$.

Caution should be exercised in applying the concepts of resonance and antiresonance as developed above to a real driven string. In any physically realizable system, the driving force (usually originating from an electrical voltage) is transferred to the string through a transducer. This transducer has a mechanical

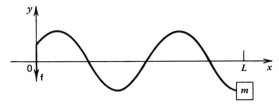

Fig. 2.9.2 The forced, mass-loaded string. The string is driven at $x = 0$ and the mass is constrained to move transversely at $x = L$.

impedance of its own, which can significantly affect the behavior of the system. The full implications of this will be left until the discussion of the driven pipe in Chapter 10.

*(b) The Forced, Mass-Loaded String

If the string is terminated at $x = L$ not with a rigid support but with one possessing inertance so that it behaves like a mass, as sketched in Fig. 2.9.2, then analysis of the motion becomes more complicated. As before, the solution must still be of the form (2.9.1), and the boundary condition at $x = 0$ is still (2.9.2),

$$Fe^{j\omega t} + \rho_L c^2 \left(\frac{\partial \mathbf{y}}{\partial x} \right)_{x=0} = 0 \tag{2.9.16}$$

where we have replaced T with $\rho_L c^2$.

The condition at $x = L$ is now different: the force applied to the mass must be $-T(\partial y / \partial x)_{x=L}$ since a negative slope at $x = L$ results in an upward force in the $+y$ direction. By Newton's second law this becomes

$$-\rho_L c^2 \left(\frac{\partial \mathbf{y}}{\partial x} \right)_{x=L} = m \left(\frac{\partial^2 \mathbf{y}}{\partial t^2} \right)_{x=L} \tag{2.9.17}$$

Substitution of (2.9.1) into (2.9.16) yields

$$F = -\rho_L c^2 (-jk\mathbf{A} + jk\mathbf{B}) \tag{2.9.18}$$

as before, but substitution into (2.9.17) gives a new equation

$$-\rho_L c^2 (-jk\mathbf{A}e^{-jkL} + jk\mathbf{B}e^{jkL}) = m(j\omega)^2 (\mathbf{A}e^{-jkL} + \mathbf{B}e^{jkL}) \tag{2.9.19}$$

Solution for **A** and **B** gives

$$\mathbf{A} = -\frac{Fe^{jkL}}{2\omega \rho_L c} \frac{1 + (j\omega m / \rho_L c)}{(\omega m / \rho_L c) \cos kL + \sin kL}$$

$$\mathbf{B} = -\frac{Fe^{-jkL}}{2\omega \rho_L c} \frac{1 - (j\omega m / \rho_L c)}{(\omega m / \rho_L c) \cos kL + \sin kL} \tag{2.9.20}$$

Note that **A** and **B** are complex conjugates. The wave traveling to the left has the same amplitude as that traveling to the right. The complex speed of the string, $\mathbf{u} = \partial \mathbf{y} / \partial t = j\omega \mathbf{y}$, is

$$\mathbf{u}(x, t) = -j \frac{F}{\rho_L c} \frac{\cos[k(L - x)] - (\omega m / \rho_L c) \sin[k(L - x)]}{(\omega m / \rho_L c) \cos kL + \sin kL} e^{j\omega t} \tag{2.9.21}$$

and the input mechanical impedance is

$$\mathbf{Z}_{m0} = j\rho_L c \frac{(\omega m/\rho_L c) + \tan kL}{1 - (\omega m/\rho_L c)\tan kL} \tag{2.9.22}$$

Again, $\mathbf{Z}_{m0}$ is purely reactive.

Resonance frequencies occur when the input reactance vanishes, which is equivalent to equating the numerator of $\mathbf{Z}_{m0}$ to zero. This results in

$$\tan kL = -(m/m_s)kL \tag{2.9.23}$$

where $m_s = \rho_L L$ is the mass of the string. There is no explicit solution of this transcendental equation. For very small mass loading, $m \ll m_s$, and for values of kL that are not too large, $\tan kL \approx 0$ or $kL \approx n\pi$, the condition of resonance for a forced, free string. Such a result is to be expected, since for very light loading the string is essentially free at the end $x = L$. Similarly, for heavy mass loading ($m \gg m_s$) the mass acts very much like a rigid support, and the resonance frequencies approach those of a forced, fixed string. The general case of intermediate mass loading can be solved readily with a hand calculator or by graphical means. If we plot both $\tan kL$ and $-(m/m_s)/kL$ on the same axes as functions of kL, the resonance frequencies will correspond to the values of kL for which the curves intersect.

For example, in the special case $m = m_s$, the values of kL satisfying (2.9.23) are

$$kL = 2.03, 4.91, 7.98, \ldots \tag{2.9.24}$$

as shown in Fig. 2.9.3. The lowest resonance frequency, given by $k_1 L = 2.03$, is $f_1 = (2.03/2\pi)(c/L)$. This lies between the lowest resonance frequency of a forced, free string and the lowest resonance frequency of a forced, fixed string. The higher resonance frequencies are not integral multiples of the lowest resonance frequency. For example, the ratio of the frequency of the second resonance to that of the lowest resonance is $4.91/2.03 = 2.42$.

The locations of the nodes on the string are altered by mass loading. The nodes fall where $\mathbf{u}(x, t) = 0$, found from the vanishing of the numerator of (2.9.21):

$$\tan[k(L - x_q)] = \rho_L c/\omega m \qquad q = 0, 1, 2, \ldots, \leq 2L/\lambda \tag{2.9.25}$$

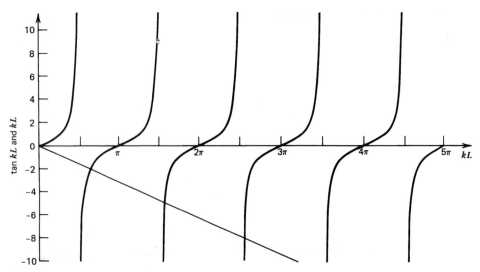

Figure 2.9.3   Graphical solution of $\tan kL = -kL$ for the resonance frequencies of a forced, mass-loaded string with $m = m_s$. The roots are $kL = 2.03, 4.91, 7.98, \ldots$.

Since the right side of this equation gets larger as the frequency decreases, the node found at $x_0 = L$ for very high frequencies moves to smaller values of x as the frequency is lowered, until at very low frequencies it is one-quarter wavelength from the end and there is an antinode at $x = L$. This means that the end at $x = L$ appears rigid at high frequencies and free at low frequencies.

(c) The Forced, Resistance-Loaded String

For a final example of the behavior of a forced string of finite length, let the end at $x = L$ be attached to a dashpot constrained to move transversely. The trial solution is (2.9.1) and the boundary condition at $x = 0$ is (2.8.8). Now, however, at $y = L$ we must have the force $R_m(\partial \mathbf{y}/\partial t)_{x=L}$ balancing the force $-T(\partial \mathbf{y}/\partial t)_{x=L}$, so

$$-\rho_L c^2 \left(\frac{\partial \mathbf{y}}{\partial x}\right)_{x=L} = R_m \left(\frac{\partial \mathbf{y}}{\partial t}\right)_{x=L} \tag{2.9.26}$$

We could continue toward a solution as before, but a little subtlety will save a lot of work. The solution must behave as $\exp(j\omega t)$, so that (2.9.26) can be rewritten as

$$-\rho_L c^2 \left(\frac{\partial \mathbf{y}}{\partial x}\right)_{x=L} = \frac{R_m}{j\omega} \left(\frac{\partial^2 \mathbf{y}}{\partial t^2}\right)_{x=L} \tag{2.9.27}$$

Note that if we replace m in (2.9.17) with $R_m/j\omega$, then (2.9.17) and (2.9.27) become the same: we can use the formulas of the preceding example if we substitute R_m for $j\omega m$ everywhere. This yields new expressions for $\mathbf{A}$ and $\mathbf{B}$,

$$\mathbf{A} = -\frac{Fe^{jkL}}{2\omega\rho_L c}\frac{1 + (R_m/\rho_L c)}{(R_m/j\rho_L c)\cos kL + \sin kL}$$
$$\tag{2.9.28}$$
$$\mathbf{B} = -\frac{Fe^{-jkL}}{2\omega\rho_L c}\frac{1 - (R_m/\rho_L c)}{(R_m/j\rho_L c)\cos kL + \sin kL}$$

Thus $\mathbf{A}$ and $\mathbf{B}$ are no longer equal in amplitude. Indeed, $|\mathbf{B}|/|\mathbf{A}| = |\rho_L c - R_m|/(\rho_L c + R_m) \leq 1$ so that the wave traveling to the left has smaller amplitude than that traveling to the right. This is physically plausible: since the dashpot dissipates energy, more must flow into it than out of it. This new result will have significant effects on the wave pattern of the string. The complex speed is found by substituting R_m for $j\omega m$ in (2.9.21),

$$\mathbf{u}(x,t) = \frac{F}{\rho_L c}\frac{\cos[k(L-x)] + j(R_m/\rho_L c)\sin[k(L-x)]}{(R_m/\rho_L c)\cos kL + j\sin kL}e^{j\omega t} \tag{2.9.29}$$

and the mechanical input impedance likewise is found from (2.9.22),

$$\mathbf{Z}_{m0} = \rho_L c\frac{(R_m/\rho_L c) + j\tan kL}{1 + j(R_m/\rho_L c)\tan kL} \tag{2.9.30}$$

Detailed analysis of this and similar forced, resonant systems will be undertaken in Chapter 3, where it will be seen that in general R_m or $j\omega m$ is replaced by $\mathbf{Z}_m$, the mechanical impedance of the termination. Here we content ourselves with two observations:

1. The speed amplitude $U(x) = |\mathbf{u}(x,t)|$ is found from (2.9.29):

$$U(x) = \frac{F}{\rho_L c}\left(\frac{\cos^2[k(L-x)] + (R_m/\rho_L c)^2\sin^2[k(L-x)]}{(R_m/\rho_L c)^2\cos^2 kL + \sin^2 kL}\right)^{1/2} = \frac{F}{\rho_L c}\frac{numerator}{denominator} \tag{2.9.31}$$

The *numerator* varies between 1 and $R_m / \rho_L c$ as x decreases from L to 0, and the *denominator* has fixed finite value that depends on the driving angular frequency $\omega = kc$. Thus, $U(x)$ has relative maxima and minima, but no exact nulls.

2. In the special case $R_m = \rho_L c$ the string behaves exactly like one of infinite length. There is no reflection from the terminated end at $x = L$ when the impedance of the termination is matched to the characteristic impedance of the string.

2.10 NORMAL MODES OF THE FIXED, FIXED STRING

Let us now turn our attention to a different class of solutions to the wave equation for finite strings. Rather than forcing the string into motion by driving one end, let the string be fixed at both ends and excited into motion by some initial displacement (or impact) along its length, much like a plucked guitar string or struck piano string.

Since the string is fixed at both ends, the boundary conditions are $\mathbf{y} = 0$ at $x = 0$ and $x = L$. A trial solution satisfying the wave equation is

$$\mathbf{y}(x,t) = \mathbf{A}e^{j(\omega t - kx)} + \mathbf{B}e^{j(\omega t + kx)} \tag{2.10.1}$$

and application of the boundary conditions gives

$$\mathbf{A} + \mathbf{B} = 0$$
$$\mathbf{A}e^{-jkL} + \mathbf{B}e^{jkL} = 0 \tag{2.10.2}$$

The first of these requires $\mathbf{B} = -\mathbf{A}$ and this, substituted into the second, gives

$$2j\mathbf{A}\sin kL = 0 \tag{2.10.3}$$

This second boundary condition can be satisfied two ways. (1) Let $\mathbf{A} = 0$. This gives $\mathbf{y} = 0$, the trivial solution of no motion. (2) Let $\sin kL = 0$. This choice requires

$$kL = n\pi \qquad n = 1, 2, 3, \ldots \tag{2.10.4}$$

(The value $n = 0$ is not allowed, since for the string fixed at both ends this also corresponds to no motion.) This equation shows that only the discrete values $k = k_n = n\pi/L$ lead to solutions. Furthermore, since $\omega/k = c$, only certain frequencies are allowed,

$$f_n = \omega_n/2\pi = (n/2)(c/L) \tag{2.10.5}$$

Thus, there is a family of solutions, each of the form

$$\mathbf{y}_n(x,t) = \mathbf{A}_n(\sin k_n x)e^{j\omega_n t} \tag{2.10.6}$$

where $\mathbf{A}_n$ is the complex amplitude of the nth solution.

If we replace $\mathbf{A}_n$ with $A_n - jB_n$, then the real transverse displacement of the nth solution is

$$\mathbf{y}_n(x, t) = (A_n \cos \omega_n t + B_n \sin \omega_n t) \sin k_n x \qquad (2.10.7)$$

The constants A_n and B_n must be determined from the initial conditions.

Application of the boundary conditions has limited the viable solutions of the wave equation to a series of discrete functions (2.10.7). These functions are called *eigenfunctions* or *normal modes*. [Strictly speaking, the spatial function $\sin k_n x$ is the normal mode and $y_n(x, t)$ is the product of the normal mode with the related oscillatory function of time. However, y_n is often referred to as the normal mode, and we will sometimes adopt that usage.] Associated with each of these solutions is a unique frequency known as the *eigenfrequency, natural frequency,* or *normal mode frequency.* For the fixed, fixed string of our example, the eigenfrequencies are given by (2.10.5), and use of $\lambda_n f_n = c$ shows that $L = n\lambda_n/2$. An integral number of half-wavelengths encompasses the length of the fixed, fixed string.

The normal mode with the lowest eigenfrequency has $n = 1$ and is called the *fundamental mode.* Its eigenfrequency $f_1 = 2(c/L)$ is called the *fundamental* or *first harmonic.* The eigenfrequencies with $n = 2, 3, \ldots$ are called *overtones.* In the case of the fixed, fixed strings, $f_n = nf_1$, and the overtones are *harmonics,* integral multiples of the fundamental. Thus, the second harmonic is the *first* overtone, and so forth. (Another less confusing terminology, which should be more widely accepted, names the overtones as *partials;* by convention, the fundamental is the *first partial,* the second harmonic or first overtone is the *second partial,* and so forth.) As we will see in the next section, for more realistic boundary conditions the overtones of a freely vibrating string are not necessarily integral multiples of the fundamental. (Equivalently, the second and higher partials are not harmonic with the first.)

The complete solution for a rigidly supported, freely vibrating string is the sum of all the individual modes of vibration represented by (2.10.7):

$$\mathbf{y}(x, t) = \sum_{n=1}^{\infty} (A_n \cos \omega_n t + B_n \sin \omega_n t) \sin k_n x \qquad (2.10.8)$$

Assume that at time $t = 0$ the string is distorted from its normal linear configuration, the displacement at each point being

$$y(x, 0) \qquad (2.10.9)$$

and the corresponding speed being

$$u(x, 0) = \left(\frac{\partial y}{\partial t}\right)_{t=0} \qquad (2.10.10)$$

Now, if (2.10.8) is to describe the position of the string at all times, it must represent it at $t = 0$ so that

$$y(x, 0) = \sum_{n=1}^{\infty} A_n \sin k_n x \qquad (2.10.11)$$

Its derivative with respect to time must also represent the speed at $t = 0$ and hence

$$u(x, 0) = \sum_{n=1}^{\infty} \omega_n B_n \sin k_n x \qquad (2.10.12)$$

Applying the Fourier theorem (Section 1.14) to (2.10.11) and (2.10.12) gives

$$A_n = \frac{2}{L} \int_0^L y(x, 0) \sin k_n x \, dx$$

$$\qquad (2.10.13)$$

$$B_n = \frac{2}{\omega_n L} \int_0^L u(x, 0) \sin k_n x \, dx$$

(a) A Plucked String

Assume that a string is initially pulled aside a distance h at its center and then released. Here $u(x, 0)$ is zero and all coefficients B_n will vanish. Applying the results of Problem 1.14.2 gives

$$A_n = [8h/(n\pi)^2] \sin(n\pi/2) \qquad (2.10.14)$$

so that $A_2 = A_4 = A_6 = \cdots = 0$ and $A_1 = 8h/\pi^2, A_3/A_1 = -1/9, A_5/A_1 = 1/25$, and so forth. The values of the coefficients A_n determine the amplitudes of the various harmonic modes of vibration of the string. All vibrations corresponding to the even harmonics are absent. Each of these absent modes has a node where the string was initially pulled aside. Harmonics having a node at the point where the string is plucked cannot be excited.

(b) A Struck String

If the string is struck a blow at its midpoint, then the initial distribution of transverse speed can be approximated as $u(x, 0) = \mathcal{U}\delta(x - L/2)$ where $\mathcal{U}$ has dimension m^2/s. Since there is no initial displacement, all the coefficients A_n are zero in (2.10.8). The coefficients B_n are given by (2.10.13),

$$B_n = \frac{2\mathcal{U}}{\omega_n L} \int_0^L \delta(x - L/2) \sin k_n x \, dx = \frac{2}{n} \frac{\mathcal{U}}{\pi c} \sin \frac{n\pi}{2} \qquad (2.10.15)$$

As with the string plucked at its midpoint, harmonics (the even ones) having a node where the string is struck are absent. The odd harmonics are present with relative amplitudes $B_n/B_1 = 1, -1/3, 1/5, -1/7, \ldots$ for odd n.

*2.11 EFFECTS OF MORE REALISTIC BOUNDARY CONDITIONS ON THE FREELY VIBRATING STRING

Any yielding of the supports modifies the motion of the string, for the boundary conditions are no longer $y = 0$ at its ends. Instead, at these points the impedance of the string must equal the transverse mechanical impedance of the support.

Assume that the left end of the string is attached to a support whose mechanical impedance is $\mathbf{Z}_{m0}$. For example, let the string be attached at $x = 0$ to an undamped harmonic oscillator. The mechanical impedance of the oscillator presented to the string at $x = 0$ is $\mathbf{Z}_{m0} = j(\omega m - s/\omega)$. This assumption, that the support can be replaced by the elements of a simple harmonic oscillator constrained to move transversely to the string, is representative of many real interactions wherein the support exhibits inertia and resilience. The transverse force $\mathbf{f}$ exerted *by the string on the mass* is given by

$$\mathbf{f} = T\left(\frac{\partial \mathbf{y}}{\partial x}\right)_{x=0} \tag{2.11.1}$$

The boundary condition at $x = 0$ is $\mathbf{u}(0, t) = \mathbf{f}/\mathbf{Z}_{m0}$. Using (2.11.1) gives

$$\mathbf{u}(0, t) = \frac{1}{\mathbf{Z}_{m0}} T\left(\frac{\partial \mathbf{y}}{\partial x}\right)_{x=0} \tag{2.11.2}$$

Similarly, the condition at $x = L$ is

$$\mathbf{u}(L, t) = -\frac{1}{\mathbf{Z}_m} T\left(\frac{\partial \mathbf{y}}{\partial x}\right)_{x=L} \tag{2.11.3}$$

where $\mathbf{Z}_m$ is the mechanical impedance of the support at $x = L$.

If the mechanical impedance at the support is infinite, then the above requires $\mathbf{u}(L, t) = 0$ and therefore $\mathbf{y}(L, t) = 0$, the condition for a fixed end. If the support offers no restraint to the transverse motion of the string, its mechanical impedance is zero and the boundary condition must be $(\partial \mathbf{y}/\partial x)_{x=L} = 0$, the condition for a free end.

(a) The Fixed, Mass-Loaded String

Assume that the string is fixed at $x = 0$ and that the support at $x = L$ can be characterized by a mass m. The boundary conditions are

$$\mathbf{u}(0, t) = 0$$
$$\mathbf{u}(L, t) = -\frac{T}{j\omega m}\left(\frac{\partial \mathbf{y}}{\partial x}\right)_{x=L} \tag{2.11.4}$$

Application of the first boundary condition to the general harmonic solution (2.9.1) yields

$$\mathbf{y}(x, t) = -2j\mathbf{A}(\sin kx)e^{j\omega t} \tag{2.11.5}$$

Substitution of this into the second boundary condition then gives

$$j\omega \sin kL = -(T/j\omega m)\cos kL \tag{2.11.6}$$

This can be rearranged,

$$\cot kL = (m/m_s)kL \tag{2.11.7}$$

where $m_s = \rho_L L$ is the mass of the string. Figure 2.11.1 illustrates the graphical solution of (2.11.7) for a few values of m/m_s. If m/m_s is large, the solutions approach $kL = n\pi$ and the string behaves as if it were fixed at both ends. As m/m_s is reduced, the allowed values of kL increase, thereby *raising* the normal mode frequencies. Furthermore, these frequencies are no longer related by integers: *the overtones are not harmonics of the fundamental.* Since the frequencies are raised and there must be a node at $x = 0$, the last node at the other end of

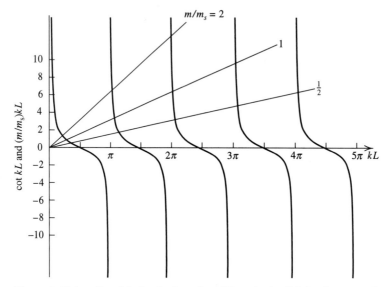

Figure 2.11.1   Graphical solution of $\cot kL = (m/m_s)kL$ for the normal modes of a fixed, mass-loaded string with $m/m_s = 0.5, 1.0, 2.0$.

the string lies within $x = L$. As the mass gets smaller, this last node moves toward $L - \lambda/4$ and in the limit $m/m_s = 0$ there is an antinode at $x = L$.

(b) The Fixed, Resistance-Loaded String

As a second (quite different) case, consider the effects on the standing wave of a support having finite resistance and no reactance. Assume that the string is fixed at $x = 0$ and attached at $x = L$ to a dashpot constrained to move transversely. The boundary conditions are

$$\mathbf{u}(0, t) = 0 \quad \text{and} \quad \mathbf{u}(L, t) = -\frac{T}{R_m}\left(\frac{\partial \mathbf{y}}{\partial x}\right)_{x=L} \tag{2.11.8}$$

Because of the damping provided by the dashpot, the standing wave will decay with time. As in Section 1.6, we introduce the complex angular frequency $\boldsymbol{\omega} = \omega + j\beta$, where ω is the angular frequency and β is the temporal absorption coefficient. Since there are no losses on the string except at the boundary, our solution must still satisfy the lossless wave equation (2.3.6) and, since $\partial^2 \mathbf{y}/\partial t^2 = -\boldsymbol{\omega}^2 \mathbf{y}$, we are led to the result $\partial^2 \mathbf{y}/\partial x^2 = -(\boldsymbol{\omega}/c)^2 \mathbf{y}$. This suggests solutions of the form

$$\mathbf{y}(x, t) = e^{j(\boldsymbol{\omega}t \pm \mathbf{k}x)} \tag{2.11.9}$$

where the spatial factor has a *complex wave number* $\mathbf{k} = k + j\alpha$ given by

$$\boldsymbol{\omega}^2 = c^2 \mathbf{k}^2 \quad \text{or} \quad \boldsymbol{\omega} = c\mathbf{k} \tag{2.11.10}$$

Now relate k and α to ω and β. Substitution of $\boldsymbol{\omega} = \omega + j\beta$ and $\mathbf{k} = k + j\alpha$ into (2.11.10) and collection of real and imaginary parts gives

$$(\omega - ck) + j(\beta - c\alpha) = 0 \tag{2.11.11}$$

Both real and imaginary parts must vanish. This yields the pair of relations

$$\omega/k = c \quad \text{and} \quad \beta = \alpha c \tag{2.11.12}$$

Substitution of the trial solution

$$\mathbf{y}(x, t) = \mathbf{A}e^{j(\omega t - kx)} + \mathbf{B}e^{j(\omega t + kx)} \tag{2.11.13}$$

into the boundary conditions yields

$$\mathbf{y}(x, t) = -2j\mathbf{A}(\sin \mathbf{k}x)e^{j\omega t} \tag{2.11.14}$$
$$\sin \mathbf{k}L = j(\rho_L c / R_m) \cos \mathbf{k}L$$

This last equation must be satisfied by both its imaginary and real parts,

$$\cos kL \sinh \alpha L = (\rho_L c / R_m) \cos kL \cosh \alpha L \tag{2.11.15}$$
$$\sin kL \cosh \alpha L = (\rho_L c / R_m) \sin kL \sinh \alpha L$$

(Readers unfamiliar with trigonometric functions of complex angles should consult Appendix A3.) Solving simultaneously, we get two possible solutions:

$$\sin kL = 0 \quad \text{and} \quad \tanh \alpha L = \rho_L c / R_m \tag{2.11.16}$$

or

$$\cos kL = 0 \quad \text{and} \quad \tanh \alpha L = R_m / \rho_L c \tag{2.11.17}$$

For weak damping, $R_m \ll \rho_L c$ rules out the first possible solution (2.11.16) because $\tanh x$ must always be less than unity. Then, since $\tanh x \approx x$ for $x \ll 1$, the second possibility (2.11.17) gives

$$\alpha L \approx R_m / \rho_L c \tag{2.11.18}$$

and $kL = (n - \frac{1}{2})\pi$ so that, since the end at $x = 0$ is fixed, the end at $x = L$ is an antinode. In this limit of small damping we can use the approximations $\sinh \alpha x \approx \alpha x$ and $\cosh \alpha x \approx 1$ in (2.11.14) with the result that the particle displacement has magnitude

$$|\mathbf{y}(x, t)| \approx 2Ae^{-\alpha ct}[\sin^2 kx + (\alpha x)^2 \cos^2 kx]^{1/2} \tag{2.11.19}$$

The wave pattern resembles that of the fixed, free string, but the motion decays exponentially with a temporal absorption coefficient $\beta = \alpha c$ and the nodes (occurring where $\sin kx_q = 0$) are not exactly zero but have amplitude $2\alpha x_q A \exp(-\alpha ct)$. Analogously, the antinodes have amplitude $2A \exp(-\alpha ct)$.

(c) The Fixed, Fixed Damped String

Up to this point we have neglected the effects of the surrounding medium on the motion of the string. One effect of the medium provides a resistive force that opposes the motion. As with a simple oscillator, this frictional force damps the free vibrations and reduces their frequency slightly. Part of the energy being dissipated by the string heats the surrounding medium and part goes into the radiation of sound.

Losses resulting from the motion of the string through the surrounding medium can be accounted for by introducing a dissipative term into the wave equation for the string. We can be guided by the form of the differential equation for the damped harmonic oscillator. The loss term is proportional to the particle speed and is additive to the acceleration. Thus, an appropriate wave equation including losses is

$$\frac{\partial^2 y}{\partial t^2} + 2\beta \frac{\partial y}{\partial t} - c^2 \frac{\partial^2 y}{\partial x^2} = 0 \qquad (2.11.20)$$

In assuming a trial solution, we can be guided by the previous example. We expect damping, spatial and/or temporal, so a plausible starting point is (2.11.13),

$$\mathbf{y}(x, t) = \mathbf{A}e^{j(\omega t - \mathbf{k}x)} + \mathbf{B}e^{j(\omega t + \mathbf{k}x)} \qquad (2.11.21)$$

where both $\boldsymbol{\omega}$ and $\mathbf{k}$ can possibly be complex. Let the string be fixed at $x = 0$ and $x = L$. Application of the boundary conditions gives

$$\mathbf{y}(x, t) = -2j\mathbf{A}(\sin \mathbf{k}x)e^{j\omega t}$$
$$\sin \mathbf{k}L = 0 \qquad (2.11.22)$$

The only way the boundary condition at $x = L$ can be satisfied is for the wave number to be pure real, $\mathbf{k} = k$. It is easy to see that this must be true for any combination of boundary conditions that does not allow energy to be taken from the string by a support. Now, direct substitution of (2.11.22) into the lossy wave equation (2.11.20) and recognition that $\mathbf{k}$ must be replaced with k gives the equation that $\boldsymbol{\omega}$ must satisfy,

$$\boldsymbol{\omega}^2 - 2j\beta\boldsymbol{\omega} = c^2 k^2 \qquad (2.11.23)$$

Solution is immediate,

$$\boldsymbol{\omega} = [(ck)^2 - \beta^2]^{1/2} + j\beta \qquad (2.11.24)$$

and the physical interpretation of the result is straightforward. The real part is the natural angular frequency ω_d of the damped string. The product ck is the natural angular frequency ω_0 of the identical string without damping. The relationship is $\omega_d = \sqrt{\omega_0^2 - \beta^2}$, exactly as for the damped oscillator [see (1.6.10)]. The temporal absorption coefficient is β, also just as before. The slightly lower natural frequency for the specified wave number means that the phase speed for the standing wave on the damped string is decreased slightly. It is easy to show (see Problem 2.11.8) for the cases of motion described by standing waves that the phase speed c_d with damping is related to the phase speed c of the undamped string by $c_d/c = \sqrt{1 - (\beta/\omega)^2}$.

Similar results can be obtained for waves traveling on an infinitely long damped string. In this case, it is the wave number that ends up being generalized to a complex quantity whose imaginary part is the spatial attenuation factor α. Analysis is left to Problem 2.11.9.

Another effect of the medium is to add an effective mass per unit length to the string. This may not be negligible in a liquid medium, or at low frequencies in a gaseous medium. An analytical treatment of this additional inertance on a vibrating system is not considered here but is dealt with in Chapter 8.

2.12 ENERGY OF VIBRATION OF A STRING

A vibrating string contains kinetic energy because of the speeds with which its various portions are moving, and potential energy because the string must be stretched as it is deformed from its rest position. (Refer to Fig. 2.3.1.) The element of string between x and $x + dx$ has a mass $\rho_L\, dx$, and if it moves with speed $\partial y/\partial t$ its kinetic energy dE_k is

$$dE_k = \tfrac{1}{2}\rho_L(\partial y/\partial t)^2\, dx \tag{2.12.1}$$

This element is also stretched to a length ds because of the deformation. Since the left end of the element has transverse displacement $y(x, t)$ and the right end is displaced $y(x, t) + (\partial y/\partial x)\, dx$, the increase in length is

$$ds - dx = [(dx)^2 + (\partial y/\partial x)^2(dx)^2]^{1/2} - dx = \{[1 + (\partial y/\partial x)^2]^{1/2} - 1\}\, dx \tag{2.12.2}$$

For the small displacements assumed, this can be simplified with the approximation $\sqrt{1 + \varepsilon} \approx 1 + \varepsilon/2$,

$$ds - dx \approx \tfrac{1}{2}(\partial y/\partial x)^2\, dx \tag{2.12.3}$$

The product of the tension $T = \rho_L c^2$ of the string and the extension of the element gives the potential energy of deformation,

$$dE_p = \tfrac{1}{2}\rho_L c^2(\partial y/\partial x)^2\, dx \tag{2.12.4}$$

The energy per unit length dE/dx is the sum $(dE_k/dx + dE_p/dx)$,

$$\frac{dE}{dx} = \frac{1}{2}\rho_L c^2\left[\left(\frac{\partial y}{\partial x}\right)^2 + \left(\frac{1}{c}\frac{\partial y}{\partial t}\right)^2\right] \tag{2.12.5}$$

and the total energy of the string is the integration of the energy per unit length over the entire length of the string

$$E = \frac{1}{2}\rho_L c^2\int_{string}\left[\left(\frac{\partial y}{\partial x}\right)^2 + \left(\frac{1}{c}\frac{\partial y}{\partial t}\right)^2\right] dx \tag{2.12.6}$$

As an example, let us apply this to the case of the freely vibrating string fixed at $x = 0$ and at $x = L$. The real displacement for each mode is given by

$$y_n(x, t) = A_n \sin k_n x \cos(\omega_n t + \theta_n) \tag{2.12.7}$$

where $k_n L = n\pi$ and $n = 1, 2, 3, \ldots$. The total displacement of the string is the sum of all allowed modes

$$y(x, t) = \sum_{n=1}^{\infty} y_n(x, t) \tag{2.12.8}$$

When the total displacement is squared and substituted into (2.12.6), there will be products $\sin k_n x \sin k_m x$ and $\cos k_n x \cos k_m x$. It is straightforward (and left as an exercise) to show that when these individual products are integrated over the length of the string, the result is

$$\int_0^L (\sin k_n x \sin k_m x)\, dx \;=\; \int_0^L (\cos k_n x \cos k_m x)\, dx \;=\; \frac{L}{2}\delta_{nm} \qquad (n, m \neq 0) \quad (2.12.9)$$

where the *Kronecker delta* δ_{nm} is unity for $n = m$ and is zero otherwise. Thus, all the cross terms vanish, and the only remaining products are those for which $n = m$ and whose integrations are each $L/2$. The result is that the total energy is simply the sum of the energies of the individual modes,

$$E \;=\; \sum_n E_n \qquad (2.12.10)$$

and the energy contained in each mode is

$$E_n \;=\; \tfrac{1}{4}\rho_L L(\omega_n A_n)^2 \;=\; \tfrac{1}{4} m_s U_n^2 \qquad (2.12.11)$$

where $U_n = \omega_n A_n$ is the maximum amplitude of the particle speed of each mode and $m_s = \rho_L L$ is the mass of the string.

For this fixed, fixed string plucked in the middle (pulled aside a distance h and released), the modal amplitudes are given by (2.10.14). Substitution gives

$$E \;=\; m_s c^2 \left(\frac{4}{\pi}\frac{h}{L}\right)^2 \sum_{odd} \frac{1}{n^2} \;=\; 2 m_s c^2 \left(\frac{h}{L}\right)^2 \qquad (2.12.12)$$

with the help of $1 + 1/3^2 + 1/5^2 + \cdots = \pi^2/8$. (This summation is verified in Problem 2.13.6.) The energy of the fundamental is 9 times that of the third harmonic, 25 times that of the fifth, and so on.

It is quite evident that variations in the position at which the string is plucked will alter the harmonic content and therefore the quality of the sound.

*2.13 NORMAL MODES, FOURIER'S THEOREM, AND ORTHOGONALITY

We have seen that in some standing waves the total energy of the system is simply the sum of the energies present in each vibrating mode. Furthermore, we have found that expressing the collective motion of the string as a Fourier superposition of the normal modes of the system is often a relatively straightforward process. Those cases presenting the easiest possibilities for analysis of motion and the distribution of energy satisfy an *orthogonality* condition. It is appropriate at this point to discuss some aspects of orthogonality and the advantages in dealing with systems whose normal modes exhibit this condition.

First, consider a lossless system excited into motion by some initial condition. We found that such systems could be represented as a summation of appropriate normal modes, each with a wave number k_n and multiplied by an oscillating function of time with natural angular frequency $\omega_n = ck_n$.

Given the wave equation (2.3.6), assume that the vibration of the system it describes can be represented as a composite of oscillatory functions of time. Try solutions of the form

$$\mathbf{y}(x, t) \;=\; Y(x) e^{j(\omega t + \phi)} \qquad (2.13.1)$$

with the Y's yet to be determined. Substitution into the wave equation results in a second-order differential equation for $Y(x)$,

$$\frac{d^2Y}{dx^2} + k^2Y = 0 \tag{2.13.2}$$

where $k = \omega/c$. This is a one-dimensional *Helmholtz equation* (sometimes called a "time-independent wave equation"). Its general solution can be written as

$$Y(x) = A\cos kx + B\sin kx \tag{2.13.3}$$

The boundary conditions at $x = 0$ and $x = L$ determine the allowed wave numbers k_n (and thereby the allowed natural angular frequencies ω_n). Thus, there will be solutions

$$y_n(x, t) = Y_n(x)\cos(\omega_n t + \phi_n) \tag{2.13.4}$$

where the initial phase angles are determined by the initial conditions. The functions $Y_n(x)$ are the normal modes of the system.

An important property of the solutions Y_n can now be established. Let us multiply the Helmholtz equation for Y_n by Y_m and vice versa:

$$Y_m\frac{d^2Y_n}{dx^2} + Y_m k_n^2 Y_n = 0$$
$$Y_n\frac{d^2Y_m}{dx^2} + Y_n k_m^2 Y_m = 0 \tag{2.13.5}$$

Now, subtract one from the other, rearrange, and integrate the result over the distance between the boundaries of the string:

$$(k_n^2 - k_m^2)\int_0^L Y_nY_m\,dx = \int_0^L \left(Y_n\frac{d^2Y_m}{dx^2} - Y_m\frac{d^2Y_n}{dx^2}\right)dx \tag{2.13.6}$$

The integrand on the right is a perfect differential, so (2.13.6) becomes

$$(k_n^2 - k_m^2)\int_0^L Y_nY_m\,dx = \left(Y_n\frac{dY_m}{dx} - Y_m\frac{dY_n}{dx}\right)\Bigg|_0^L \tag{2.13.7}$$

A little examination shows that for any combination of fixed and free ends, the right side of the above equation vanishes for all combinations of n and m. On the left, therefore, the integral must vanish unless $(k_n^2 - k_m^2) = 0$. In the special case $n = m$ both sides are zero regardless of the value of the integral. These conditions define *orthogonality*. A set of functions Y_n is orthogonal over a specified interval L if the integration of all products Y_nY_m of the functions over L vanish except for the case $n = m$. In our example, the Y_n form an orthogonal set if

$$\int_0^L Y_nY_m\,dx = C_n\delta_{nm} \tag{2.13.8}$$

where δ_{nm} is the Kronecker delta defined in (2.12.9). If the amplitudes of the Y_n's are so chosen that $C_n = 1$, then the functions are normalized and the set is termed *orthonormal*.

Note that there is nothing sacred about the choice of x as the independent variable in the Helmholtz equation. An equally valid Helmholtz equation is

$$\frac{d^2f}{dt^2} + \omega^2f = 0 \tag{2.13.9}$$

If "boundary" conditions are imposed at $t = 0$ and $t = T$ that are consistent with the vanishing of the right side of

$$(\omega_n^2 - \omega_m^2) \int_0^T f_n f_m \, dt = \left(f_n \frac{df_m}{dt} - f_m \frac{df_n}{dt} \right) \Bigg|_0^T \tag{2.13.10}$$

then the set of functions f_n is an orthogonal set.

An example of an orthogonal set of functions was presented in Section 1.14. It can be shown by direct integration that the Fourier components, $\sin n\omega t$ and $\cos n\omega t$ for $n = 0, 1, 2, \ldots$, form an orthogonal set over the interval $0 \le t \le T$, where $\omega T = 2\pi$.

Given orthogonality, a very useful relationship can be verified quite easily. If $f(t)$ is expanded in a set of *orthonormal* functions $f_n(t)$,

$$f(t) = \sum_{n=1}^{\infty} a_n f_n(t) \tag{2.13.11}$$

then the a_n^2 can be related to an integration of $f^2(t)$ by *Parseval's identity,*

$$\int_0^T f^2(t) \, dt = \sum_{n=1}^{\infty} a_n^2 \tag{2.13.12}$$

(See Problem 2.13.5.) Specifically, if this is applied to the Fourier series written in (1.14.1), the identity becomes

$$\frac{2}{T} \int_0^T f^2(t) \, dt = \frac{1}{2} A_0^2 + \sum_{n=1}^{\infty} (A_n^2 + B_n^2) \tag{2.13.13}$$

Whether we are dealing with a periodic function of time $f(t)$ or a periodic function of space $Y(x)$, the use of orthogonal functions can simplify description of the system considerably. For example, in calculating the energy of the freely vibrating fixed, fixed string, the manipulations could have been simplified with the use of orthogonality. From the above discussion, it is clear that the functions $\sin k_n x$ with $k_n L = n\pi$ for $n = 1, 2, \ldots$ are an orthogonal set over the interval $0 \le x \le L$, as are the functions $\cos k_n x$. Consequently, in the integration of the energy density over the length of the string, the only terms that will not integrate to zero are those for which $n = m$.

2.14 OVERTONES AND HARMONICS

As noted previously, the lowest natural frequency of a vibrating system is termed the *fundamental*, and the higher natural frequencies are termed *overtones*. We have also seen that, if the supports of a string are perfectly rigid, the overtones are harmonics; but if otherwise, the overtones are in general not harmonic.

Nonharmonicity of the overtones is often encountered in musical instruments. For example, the vibration of a violin string is coupled to the sounding board by a bridge for more efficient radiation of energy. The bridge then acts as a reactive termination since it must flex somewhat to communicate motion to the sounding board. As a result, the natural frequencies resulting from the free vibration of the plucked string will be slightly nonharmonic.

Another effect that is sometimes important is that the free vibrations of many real strings differ from those of an idealized string: if the string has innate bending stiffness, as is true for a piano string, then the observed overtones will be higher than predicted on the basis of an ideal string. Since the effect of stiffness increases with increasing frequency, the higher overtones of a real piano string become increasingly sharp with respect to the fundamental. It is just this effect, plus the fact that the piano string is not rigidly fixed at both ends, that gives the piano some of its distinctive tonal quality. This also contributes to the fact that a well-tuned piano has a "stretched" scale—the high notes tend to be a little sharp and the low notes a little flat. Pianos are commonly designed so that the point of impact of the hammer is about one-seventh to one-eighth of the way from one end of the strings. This would suggest that these seventh or eighth partials should be nearly absent. In reality, the finite width of the hammer, its mass, and its "rest" on the strings allows these partials to be excited. (When the guitar or violin string is plucked, however, partials are suppressed as expected from the placement of the finger.) If the displacements of the piano strings are studied after they are struck, it will be seen that the motions tend to decay (because of the loss of energy), the different overtones decaying at different rates. Equally significant, since the overtones are nonintegral multiples of the fundamental, the waveform is not stationary but shows considerable change in shape as the relative phases of the fundamental and overtones change with time. This is true for most percussive instruments, such as timpani, cymbals, *plucked* violin, piano, xylophone, woodblocks, and so forth. The sound they generate is a superposition of slowly decaying fundamental and more rapidly decaying nonharmonic overtones.

Instruments that are blown or bowed, on the other hand, like the oboe, *bowed* violin, organ, trumpet, and so forth, are *forced* vibrating systems. In this case, the forcing function is usually made up of harmonics. (The violin string is pulled to one side by the rosined hair of the bow until it snaps back and is reacquired by the bow; the motion is something like a sawtooth wave with a well-defined period and identical cycle-to-cycle behavior.) The motion of air within the clarinet and the associated periodic clapping of its reed are identically repeated cycle by cycle. Thus, the steady-state notes produced by these *driven* instruments consist of frequencies that are integral multiples of the lowest driving frequency. The relative amounts of these higher overtones control the tone color or *timbre* of the instrument.

It must be remembered, however, that these forced vibrations are initiated at some definite time so that there is an initial interval during which the transient vibrations are also strong enough to be heard. These affect the sound in much the same way as the transient solution affects the initial behavior of the forced, damped oscillator. This transient is often of considerable importance in identifying a particular instrument: the ear seems to exhibit memory in that these initial effects aid in keeping track of each instrument while others are also sounding. Even though an individual instrument contributes negligible power to the output of a full orchestra, the mind is often able to make use of these transient "fingerprints" as one aid in identifying a particular instrument in the general uproar.

Problems

2.3.1. What forms do the equations of motion for an idealized string take if either (*a*) the linear density varies with position, or (*b*) the string hangs vertically supported only at the upper end?

2.4.1. By direct substitution show that each of the following is a solution of the wave equation: (*a*) $f_1(x - ct)$, (*b*) $\ln[a(ct - x)]$, (*c*) $a(ct - x)^2$, and (*d*) $\cos[a(ct - x)]$. Similarly, show that each of the following is not a solution of the wave equation: (e) $a(ct - x^2)$ and (f) $at(ct - x)$.

2.5.1. Sketch $y = A\exp(-a|ct - x|)$ for $t = 0, 1$, and 2 s. Let $c = 5$ cm/s, $a = 3$ cm^{-1}, and $A = 1$ cm. What is the significance of the displacement of these curves?

2.7.1C. Construct a waveform of finite spatial extent of your choice (right triangle, isosceles triangle, semicircle, etc.) that moves to the right with constant speed without changing shape. Then construct a waveform moving to the left with the shape required to simulate (a) a fixed end and (b) a free end.

2.7.2C. Construct a waveform of finite spatial extent of your choice (right triangle, isosceles triangle, semicircle, etc.) that simulates an initially stationary displacement in the middle of a string with fixed ends. Simulate the subsequent motion of the string as the initial displacement separates into two oppositely moving waves that reflect twice from the fixed ends and recombine into the initial shape.

2.8.1. Consider the waveform $y = 4\cos(3t - 2x)$ propagating on a string of linear density 0.1 g/cm, where y and x are in centimeters and t is in seconds. (a) What are the amplitude, phase speed, frequency, wavelength, and wave number? (b) What is the particle speed of the element at $x = 0$ at $t = 0$?

2.8.2. An infinite string ($-\infty < x \le 0$) of linear density ρ_L and under tension T is attached at $x = 0$ to a second infinite string ($0 < x < \infty$) under the same tension but of linear density $2\rho_L$. If a wave of angular frequency ω and amplitude A is traveling in the $+x$ direction on the first string, find the amplitude of the wave traveling on the second string.

2.8.3. An infinite string ($-\infty < x \le 0$) under tension T is attached at $x = 0$ to two parallel infinite strings ($0 < x < \infty$) under tensions $T/3$ and $2T/3$, respectively. The linear density ρ_L is the same for all strings. If a wave of angular frequency ω and amplitude A is traveling in the $+x$ direction on the first string, find the amplitudes of the waves traveling on each of the other two strings.

2.9.1. Evaluate the mechanical impedance seen by the applied force driving an infinite string at a distance L from a fixed end. Interpret the individual terms in the mechanical impedance.

2.9.2. Evaluate the mechanical impedance seen by the applied force driving a simple harmonic oscillator with an infinite string extending transversely from the mass.

2.9.3. A string is stretched between rigid supports a distance L apart. It is driven by a force $F\cos\omega t$ located at its midpoint. (a) What is the mechanical impedance at the midpoint? (b) Show that the displacement amplitude of the midpoint is $(F/2kT)\tan(kL/2)$. (c) What is the amplitude of the displacement of the point $x = L/4$?

2.9.4. A string of density 0.01 kg/m is stretched with a tension of 5 N from a rigid support at one end to a device producing transverse periodic vibrations at the other end. The length of the string is 0.44 m, and it is observed that, when the driving frequency has a given value, the nodes are spaced 0.1 m apart and the maximum amplitude is 0.02 m. What are the (a) frequency and (b) amplitude of the driving force?

2.9.5. (a) Assume that a forced, fixed string is driven by a source that has constant speed amplitude $\mathbf{u}(0, t) = U_0\exp(j\omega t)$, where U_0 is independent of frequency. Find the frequencies of maximum displacement amplitude of the standing wave. (b) Repeat for a source that has a constant displacement amplitude $\mathbf{y}(0, t) = Y_0\exp(j\omega t)$. (c) Contrast the results of (a) and (b) with the frequencies of mechanical resonance for the forced, fixed string. (d) Does mechanical resonance always coincide with maximum amplitude of the motion?

2.9.6. Note that the input reactance of the forced, mass-loaded string also vanishes when the denominator of $\mathbf{Z}_{m0}$ becomes infinite. This occurs when $\tan kL = \rho_L c / \omega m$ or $\tan kL = (m_s/m)/kL$, which is the reciprocal relationship of that of (2.9.23). Show that for large kL this means that the two possible solutions occur for values of kL spaced about $\pi/2$ apart. These alternate solutions correspond to the condition of *antiresonance*, which will be investigated in a later chapter.

2.9.7. A string of length L, tension T, and linear density ρ_L is fixed at one end and free at the other. Find the resonance frequencies if the string is driven in the middle.

2.9.8C. Plot the envelope of the standing wave on a forced, fixed string for a range of frequencies and constant driving force amplitude. (*a*) Comment on the amplitude of the maximum displacement and the position of the node closest to the driver as a function of frequency. (*b*) As functions of frequency, plot the displacement amplitude of the driver, the maximum displacement amplitude of the string, and the magnitude of the input mechanical impedance. Comment on significant aspects of these plots.

2.9.9C. (*a*) Find the lowest three resonance values of kL for the driven, mass-loaded string with $m/m_s = 1$. (*b*) In terms of the resonance number n, write an expression for kL valid for large values of kL. (*c*) For a constant amplitude driving force, plot the shape of the string for several frequencies above and below the lowest resonance frequency.

2.9.10C. A driven, resistance-loaded string is terminated in a load resistance $R_m = 0.1\rho_L c$. (*a*) Plot the real and imaginary parts of $\mathbf{Z}_{m0}$ for $0 < kL < 3\pi$. (*b*) Plot the amplitude of the particle speed as a function of position for several frequencies on each side of those for which the mechanical reactance in (*a*) equals zero.

2.9.11. (*a*) Show that the expression $(m/m_s)kL$ in (2.9.23) can be expressed as the ratio of the mass at $x = L$ divided by the mass of the string per loop of wave. *Hint:* Use $kL = n\pi$. (*b*) Express the restriction "$m \ll m_s$ and kL not too large" in terms of the mass of the support and the mass per loop. (*c*) In these terms, characterize a free end and a fixed end at $x = L$.

2.10.1. A stretched string of length L is plucked at a position $L/3$ by producing an initial displacement h and then releasing the string. Determine the resulting amplitudes for the fundamental and the first three harmonic overtones. Sketch the wave shapes of these individual waves and the shape of the string resulting from the linear combination of these waves at $t = 0$. Repeat for $t = L/c$, where c is the transverse wave speed on the string.

2.10.2. Given a string, fixed at both ends, with ρ_L, L, and T specified so that the phase speed c and the fundamental resonance frequency f are known numbers, obtain the phase speed c' in terms of c and the fundamental resonance f' in terms of f if another string of the same material is used but: (*a*) the length is doubled, (*b*) the mass per unit length is quadrupled, (*c*) the cross-sectional area is doubled, (*d*) the tension is reduced to half, and (*e*) the diameter of the string is doubled.

2.10.3. Show that the work done in displacing the center of a fixed, fixed string by an amount h equals the sum of the energies present in the various modes of vibration when the string is released.

2.11.1. Find kL for the normal modes of a fixed, spring-loaded string when $T = sL$. Sketch the waveforms for the fundamental and the first overtone.

2.11.2. A mass of 0.2 kg is hung from a string of 0.05 kg mass and 1.0 m length. (*a*) What is the speed of transverse waves on the string? (Neglect the weight of the string in computing the tension.) (*b*) What are the frequencies of the fundamental and first overtone modes of transverse vibration of the string? (*c*) When the string is vibrating at its first overtone, what is the ratio of its displacement amplitude at the antinode to that of the mass?

2.11.3. Show for the fixed, resistance-loaded string that if the resistance R_m is much greater than $\rho_L c$ then the string has nodes at $x = L$ and the absorption is given by $\alpha L \approx \rho_L c / R_m$. Does the motion resemble that of a fixed, fixed string?

2.11.4. A string of linear density 0.01 kg/m and of 0.2 m length is stretched between rigid supports to a tension of 10 N. It is loaded at its center with a mass of 0.001 kg. (*a*) What is the fundamental frequency of the system? (*b*) What is the first overtone frequency of the system? (*c*) What is the frequency of the second overtone? *Hint:* Note that the fundamental and even overtones are symmetric about the center of the system, whereas the odd overtones are antisymmetric.

2.11.5C. A fixed, mass-loaded string has $m/m_s = 1$. (*a*) Find kL for the first three normal modes. (*b*) In terms of the mode number n, write an expression for kL valid for large values of kL. (*c*) Compare these latter frequencies to the resonance frequencies of a driven, mass-loaded string with the same m/m_s as found in Problem 2.9.9C.

2.11.6C. A fixed, resistance-loaded string has $R_m = 0.4\rho_L c$. (*a*) Use the weak damping approximation to plot the shape of the third harmonic at times equal to the first five periods of the motion. (*b*) Show that the nodes and antinodes decay as predicted.

2.11.7C. For the fundamental of a fixed, fixed string with fluid damping ($\beta = 0.1$), plot (*a*) the shape of the string for five consecutive periods and (*b*) the maximum displacement of the middle of the string as a function of time. (*c*) From this last graph, find the time for the displacement to decay to $1/e$ of its initial value and compare to the value predicted from β.

2.11.8. Derive the ratio of the phase speed c_d of a string immersed in a damping medium to the phase speed c for the same string when there is no damping.

2.11.9. An infinitely long damped string is excited at its end at $x = 0$ so that the displacement of that end is given by $y(0, t) = \exp(j\omega t)$. Write the complex wave number as $\mathbf{k} = k_d - j\alpha$, where α is the spatial attenuation factor. (*a*) Verify that the complex angular frequency $\boldsymbol{\omega}$ is pure real, $\boldsymbol{\omega} = \omega$. (*b*) Show that $k_d/k = \sqrt{1 + (\alpha/k)^2}$ where k is the wave number in the absence of damping. (*c*) Show that the ratio of the phase speed without damping to that with is $c/c_d = \sqrt{1 + (\alpha/k)^2}$. (*d*) In the case of small damping, $\alpha/k \ll 1$, find a good approximate relationship between α/k and β/ω for traveling and standing waves on strings of the same composition and tension. (*e*) How do the phase speeds for damped standing and damped traveling waves compare?

2.11.10. Following the derivation of Section 2.11(c), obtain the behavior of the fixed, free damped string.

2.12.1. Assume that there are two excited modes on a fixed, fixed string. Demonstrate that (2.12.10) is true.

2.12.2. A standing wave on a fixed, fixed string of length $L = 31.4$ cm and linear density 0.1 g/cm is given by $y = 2\sin(x/5)\cos(3t)$, where y and x are in centimeters and t is in seconds. (*a*) Find the phase speed, frequency, and wave number. (*b*) What is the amplitude of the particle displacement and speed at $x = L/2$ and $x = L/4$? (*c*) Find the energy density at these points. (*d*) How much energy is in the entire length of the string?

2.13.1. By direct evaluation of (2.13.7) show that the normal modes of the freely vibrating string form an orthogonal set if the string is (*a*) fixed, fixed, (*b*) free, fixed, (*c*) free, free.

·

2.13.2. (*a*) Do the normal modes of the freely vibrating fixed, mass-loaded string form an orthogonal set? Show why your answer is true. (*b*) Repeat for a fixed, spring-loaded string.

2.13.3. (*a*) Show that the collection of functions

$$f_0 = (1/T)^{1/2}$$

$$f_n = (2/T)^{1/2} \cos n\omega t \qquad \text{for } n \neq 0$$

$$g_n = (2/T)^{1/2} \sin n\omega t$$

form an orthonormal set over $0 \leq t \leq T$, where $\omega T = 2\pi$. (*b*) Show that if the $f(t)$ of (1.14.1) is expanded in this set of functions in the form

$$f(t) = \sum_{n=0}^{\infty} (a_n f_n + b_n g_n)$$

then the respective coefficients are related by

$$a_0 = (T^{1/2}/2)A_0$$

$$a_n = (T/2)^{1/2} A_n \qquad \text{for } n \neq 0$$

$$b_n = (T/2)^{1/2} B_n$$

2.13.4. Use the results of Problem 2.13.3 to verify (2.13.13).

2.13.5. Verify Parseval's identity by direct substitution of (2.13.11) into (2.13.12) and integration.

2.13.6. Use Parseval's identity to verify that $1 + 1/3^2 + 1/5^2 + \cdots = \pi^2/8$ as asserted in the text following (2.12.12). *Hint:* Start with the square wave and its Fourier series and substitute into the identity.

C h a p t e r 3

VIBRATIONS OF BARS

3.1 LONGITUDINAL VIBRATIONS OF A BAR

Another important wave motion is the propagation of *longitudinal (compressional) waves,* often encountered in solid bars (and, at low frequencies, in gas-filled tubes and ducts with rigid walls). As a longitudinal disturbance moves along a bar, the displacements of particles of the bar are essentially parallel to its axis. When the lateral dimensions of the bar are small compared with its length, each cross-sectional plane of the bar may be considered to move as a unit. (Actually the bar shrinks somewhat laterally as it expands longitudinally, but for thin bars this lateral motion may be neglected.)

A number of acoustic devices utilize longitudinal vibrations in bars. Frequency standards used for producing sounds of definite pitches can be constructed from rods of various lengths. When longitudinal vibrations are excited in such rods, the frequency of vibration is observed to be inversely proportional to the length of the rod (if all are of the same composition). Longitudinal vibrations in nickel tubes are often used to drive the vibrating diaphragm of a sonar transducer. Piezoelectric crystals may be cut so that the frequency of longitudinal vibration in a selected direction in the crystal is used either to control the frequency of an oscillatory electric current or to drive an electroacoustic transducer.

Studying longitudinal vibrations of bars also aids in understanding acoustic waves. The mathematical expressions for the transmission of acoustic plane waves through fluid media are very similar to those for the transmission of compressional waves along a bar. If the fluid is confined to a rigid pipe, there is also a close analogy between the boundary conditions.

3.2 LONGITUDINAL STRAIN

Consider a bar of length L and uniform cross-sectional area S subjected to longitudinal forces. The application of these forces will produce a longitudinal displacement ξ of each of the particles in the bar, and for thin bars this displacement

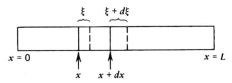

Figure 3.2.1   Longitudinal strain $d\xi/dx$
of an element of length dx in a bar.

will be essentially the same at all points in any particular cross section. Thus, ξ is
assumed to be a function only of distance x along the bar and time t,

$$\xi = \xi(x, t) \tag{3.2.1}$$

Let the coordinates of the left and right ends of the bar be $x = 0$ and $x = L$, and
consider a short segment dx of the unstrained bar lying between x and $x + dx$.
Assume that the forces cause the plane originally located at x to move a distance ξ
to the right, and the plane originally located at $x + dx$ to move a distance $\xi + d\xi$ to
the right (Fig. 3.2.1). The convention adopted in this book is that a *positive* value of
ξ signifies a displacement to the *right* (and a *negative* value to the *left*).

At any time t for small dx the displacement at $x + dx$ can be represented by the
first two terms of a Taylor's series expansion of ξ about x,

$$\xi + d\xi = \xi + (\partial\xi/\partial x)\, dx \tag{3.2.2}$$

Since the left end of the segment has been displaced a distance ξ and the right end
a distance $\xi + d\xi$, the increase in length of the segment is given by

$$(\xi + d\xi) - \xi = d\xi = (\partial\xi/\partial x)\, dx \tag{3.2.3}$$

The *strain* in the segment is defined as the ratio of its change in length $d\xi$ to its
original length dx, or

$$\boxed{\text{strain} = \partial\xi/\partial x} \tag{3.2.4}$$

3.3 LONGITUDINAL WAVE EQUATION

Whenever a bar is strained, elastic forces are produced. These forces act across each
cross-sectional plane in the bar and hold the bar together. Let $f = f(x, t)$ represent
these longitudinal forces, where the convention is adopted of choosing a *positive*
value of f to represent a force of *compression,* as indicated in Fig. 3.3.1, and a *negative*

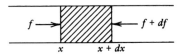

Figure 3.3.1   Compressional forces on an element
of length dx in a bar.

value to represent a force of *tension*. This choice of sign is the opposite of that conventionally taken by many material scientists but has the distinct advantage for us of making the compression of a solid by a positive increment of force analogous to the compression of a fluid by a positive increment in pressure.

The stress in the bar of cross-sectional area S is defined as

$$\boxed{\text{stress} = f/S} \tag{3.3.1}$$

For most materials, if the strain is small the stress is proportional to it. This relationship is known as Hooke's law,

$$f/S = -Y(d\xi/dx) \tag{3.3.2}$$

where Y, the *Young's modulus* or *modulus of elasticity,* is a characteristic property of the material. Since a positive stress results in a negative strain, the minus sign in (3.3.2) ensures a positive value for Y. Values of Y for a number of common solids are given in Appendix A10. Rewriting (3.3.2), we obtain

$$f = -SY(d\xi/dx) \tag{3.3.3}$$

as an expression for the internal longitudinal forces in the bar.

If f represents the internal force at x, then $f + (\partial f/\partial x)\,dx$ represents the force at $x + dx$, and the net force to the right is

$$df = f - [f + (\partial f/\partial x)\,dx] = -(\partial f/\partial x)\,dx \tag{3.3.4}$$

Use of (3.3.3) yields

$$df = SY(\partial^2\xi/\partial x^2)\,dx \tag{3.3.5}$$

The mass of the segment dx is $\rho S\,dx$, where ρ is the *density* (mass per unit volume) of the bar. Therefore, the equation of motion of the segment is

$$(\rho S\,dx)(\partial^2\xi/\partial t^2) = SY(\partial^2\xi/\partial x^2)\,dx \tag{3.3.6}$$

or

$$\boxed{\dfrac{\partial^2\xi}{\partial x^2} = \dfrac{1}{c^2}\dfrac{\partial^2\xi}{\partial t^2}} \tag{3.3.7}$$

A comparison of (3.3.7) with the corresponding (2.3.6) for the transverse motion of a string shows that they have identical form, with longitudinal displacement ξ replacing the transverse displacement y and the phase speed c now given by

$$\boxed{c^2 = Y/\rho} \tag{3.3.8}$$

Thus, the general solution has the same form as that for the transverse wave equation,

$$\xi(x, t) = \xi_1(ct - x) + \xi_2(ct + x) \tag{3.3.9}$$

The complex harmonic solution of (3.3.7) is

$$\xi(x, t) = \mathbf{A}e^{j(\omega t - kx)} + \mathbf{B}e^{j(\omega t + kx)} \tag{3.3.10}$$

where $\mathbf{A}$ and $\mathbf{B}$ are complex amplitude constants and $k = \omega/c$ is the wave number.

Since Young's modulus Y is measured under conditions allowing the strained rod to alter its transverse dimensions, (3.3.8) gives the phase speed only when the solid is a thin bar. When the transverse dimensions of the solid are large compared to a wavelength, a combination of the *bulk modulus* $\mathcal{B}$ and the *shear modulus* $\mathcal{G}$ must be used in place of the Young's modulus to calculate the phase speed. (See Appendix A11.)

3.4 SIMPLE BOUNDARY CONDITIONS

Let the bar be rigidly fixed at both ends, so that $\xi = 0$ at $x = 0$ and at $x = L$ for all times t. (The analysis that follows will be seen to be identical with that of Section 2.10 for a rigidly supported vibrating string.)

Application of $\xi = 0$ at $x = 0$ gives $\mathbf{A} + \mathbf{B} = 0$, so that $\mathbf{B} = -\mathbf{A}$ and (3.3.10) becomes

$$\xi(x, t) = \mathbf{A}e^{j\omega t}(e^{-jkx} - e^{jkx}) = -2j\mathbf{A}e^{j\omega t}\sin kx \tag{3.4.1}$$

The condition $\xi = 0$ at $x = L$ gives $\sin kL = 0$, which requires

$$k_n L = n\pi \qquad n = 1, 2, 3, \ldots \tag{3.4.2}$$

(the same as for a fixed, fixed string). The angular frequencies of the natural modes of vibration are

$$\omega_n = n\pi c/L \quad \text{or} \quad f_n = (n/2)(c/L) \tag{3.4.3}$$

[identical with (2.10.5)]. The complex displacement ξ_n corresponding to the nth mode of vibration is

$$\xi_n(x, t) = -2j\mathbf{A}_n e^{j\omega_n t}\sin k_n x \tag{3.4.4}$$

and the real part is

$$\xi_n(x, t) = (A_n \cos \omega_n t + B_n \sin \omega_n t)\sin k_n x \tag{3.4.5}$$

where the real amplitude constants A_n and B_n are defined by $2\mathbf{A}_n = B_n + jA_n$. The complete solution is the sum of all separate harmonic solutions,

$$\xi(x, t) = \sum_{n=1}^{\infty}(A_n \cos \omega_n t + B_n \sin \omega_n t)\sin k_n x \tag{3.4.6}$$

If the initial conditions of displacement and speed of the bar are known, Fourier's theorem can be used, as in Section 2.10, to evaluate A_n and B_n.

Since a solid bar is very rigid, it is difficult to provide supports of greater rigidity, and the assumed boundary condition is difficult to realize in practice. By contrast, a free end may be achieved readily by placing the bar on soft supports.

When a bar is free to move at an end, there can be no internal elastic force at the end, and hence $f = 0$ at this point. Since $f = -SY(\partial \xi / \partial x)$, this condition is equivalent to

$$\frac{\partial \xi}{\partial x} = 0 \tag{3.4.7}$$

at a free end.

Consider a free, free bar. The condition $\partial \xi / \partial x = 0$ applied to (3.3.10) at $x = 0$ gives

$$-\mathbf{A} + \mathbf{B} = 0 \quad \text{or} \quad \mathbf{B} = \mathbf{A} \tag{3.4.8}$$

so that

$$\xi(x, t) = \mathbf{A}e^{j\omega t}(e^{-jkx} + e^{jkx}) = 2\mathbf{A}e^{j\omega t}\cos kx \tag{3.4.9}$$

Application of $\partial \xi / \partial x = 0$ at $x = L$ gives $\sin kL = 0$ or

$$\omega_n = n\pi c / L \qquad n = 1, 2, 3, \ldots \tag{3.4.10}$$

The natural frequencies of a free, free bar are identical with those of (3.4.3) for a fixed, fixed bar of the same shape and composition. The complex displacement of the nth mode of vibration is

$$\xi_n(x, t) = 2\mathbf{A}_n e^{j\omega_n t}\cos k_n x \tag{3.4.11}$$

and the real displacement is

$$\xi_n(x, t) = (A_n \cos \omega_n t + B_n \sin \omega_n t)\cos k_n x \tag{3.4.12}$$

where now $2\mathbf{A}_n = A_n - jB_n$. In contrast with the fixed, fixed bar, which has nodes at either end, the free, free bar has antinodes at either end as shown by a $\cos k_n x$ term in the above equation, instead of a $\sin k_n x$ as in (3.4.4). A comparison of the nodal patterns for these two types of support is given in Fig. 3.4.1. It should be observed that whenever an *antinode* occurs at the center of the bar the vibrations are *symmetric* with respect to the center: when a segment of the bar left of center is displaced to the left, the corresponding segment to the right of center is also displaced the same distance to the left. Similarly, whenever there is a *node* at the center, the vibrations are *antisymmetric*.

A bar may be rigidly clamped at a point without interfering with any mode of vibration that has a node at this point. However, a mode not having a node at this point will be suppressed. It is impossible to find a position for clamping a free, free bar that will not eliminate some of the normal modes.

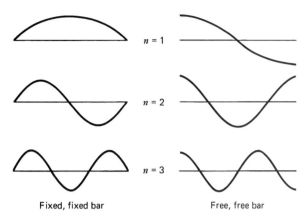

Figure 3.4.1   The lowest three longitudinal standing waves in fixed, fixed and free, free bars.

Next consider a free, fixed bar. Application of $\partial\xi/\partial x = 0$ at $x = 0$ to (3.3.10) gives (3.4.9), and application of $\xi = 0$ at $x = L$ yields $\cos kL = 0$. This requires

$$k_n L = (2n - 1)\pi/2 \qquad n = 1, 2, 3, \ldots \qquad (3.4.13)$$

and the natural frequencies are

$$f_n = [(2n - 1)/4](c/L) \qquad (3.4.14)$$

The frequency of the fundamental is half that of a similar free, free bar, and only the odd-numbered harmonics are present; the frequency of the first overtone of a free, fixed bar is three times that of its fundamental. Because of the absence of even harmonics, the quality of the sound produced by a vibrating free, fixed bar differs markedly from that produced by a free, free bar.

3.5 THE FREE, MASS-LOADED BAR

In many practical applications, a vibrating bar is neither rigidly fixed nor completely free to move at its ends. Instead, it may be loaded with some kind of mechanical impedance, most commonly of the mass-controlled type.

To analyze this type of constraint, consider a bar that is free at $x = 0$ and is loaded with a concentrated mass m at $x = L$. (Ideally, this mass should be a point mass; otherwise it will not move as a unit but will have waves propagated through it.) The boundary condition $\partial\xi/\partial x = 0$ at $x = 0$ applied to (3.3.10) leads again to

$$\xi(x, t) = 2Ae^{j\omega t} \cos kx \qquad (3.5.1)$$

The boundary condition at $x = L$ is obtained by the following argument. Since a positive value for f was chosen to indicate compression of the bar, the reaction to such a force will accelerate the mass attached to the right end of the bar toward the right. Since the mass is attached to the bar, the end of the bar and the mass must experience the same acceleration. Thus, the boundary condition must be

$$\mathbf{f}_L = m \left(\frac{\partial^2 \boldsymbol{\xi}}{\partial t^2} \right)_{x=L} \tag{3.5.2}$$

or, with the help of (3.3.3),

$$-SY \left(\frac{\partial \boldsymbol{\xi}}{\partial x} \right)_{x=L} = m \left(\frac{\partial^2 \boldsymbol{\xi}}{\partial t^2} \right)_{x=L} \tag{3.5.3}$$

Applying the above boundary conditions to $\boldsymbol{\xi}$ gives $kSY \sin kL = -m\omega^2 \cos kL$, or

$$\tan kL = -m\omega c / SY \tag{3.5.4}$$

There is no explicit solution of this transcendental equation. For very small mass loading, however, $m \approx 0$ so that $\tan kL \approx 0$ or $kL \approx n\pi$, which is the condition for the natural frequencies of a free, free bar. Such a result is obviously to be expected, since for very light loadings the bar is essentially free at both ends. Similarly, for heavy mass loadings the mass acts very much like a rigid support, and the natural frequencies approach those of a free, fixed bar.

It should be noted that in practice "fixing" the end of a bar amounts to loading it with a large mass, the mass of the support. For light bars a heavy support will act essentially as an infinite mass, and hence like a rigid restraint, but for heavy bars it may be very difficult, if not impossible, to approximate the fixed condition.

The general case of mass loading can be solved most readily by graphical or numerical means. It will be convenient to replace Y by ρc^2 and to let $m_b = \rho SL$ represent the mass of the bar. Then (3.5.4) becomes

$$\tan kL = -(m/m_b)kL \tag{3.5.5}$$

This transcendental equation is identical with (2.9.23), developed for the forced, mass-loaded string, except that m_b (the mass of the bar) replaces m_s (the mass of the string). Analysis proceeds exactly as before. If we choose the special case $m_b = m$, then the allowed values of kL solving (3.5.5) are $kL = 2.03, 4.91, 7.98, \ldots$. The nodes of the vibrations must occur where

$$\cos kx = 0 \tag{3.5.6}$$

and the fundamental mode, for which $kL = 2.03$, yields a node at

$$2.03\,x/L = \pi/2 \quad \text{or} \quad x = 0.774\,L \tag{3.5.7}$$

In contrast with the free, free bar, the node is no longer at the center but has shifted toward the loading mass, as suggested by Fig. 3.5.1. The bar could be supported at this nodal position without interfering with the fundamental mode of vibration.

Clearly, as the value of m changes from $m \ll m_b$ to $m \gg m_b$, the position of the node of the fundamental shifts from $x \approx L/2$ to $x \approx L$. Thus, the larger the mass

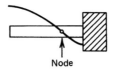

Figure 3.5.1 Fundamental mode of longitudinal vibration of a free, mass-loaded bar.

attached to a free, mass-loaded bar, the more the nodes of each normal mode of vibration are shifted toward the mass-loaded end.

Note that the overtones of the free, mass-loaded bar are not harmonics. The presence of nonharmonic overtones is sometimes advantageous in practical applications. As an illustration, consider a mass-loaded nickel tube that is intended to generate a pure tone and is driven magnetostrictively by alternating currents in a coil mounted on the tube. Harmonic frequency components other than the desired fundamental will be present in the output unless the current produced by the oscillator–amplifier unit driving the tube is well filtered. However, since the overtones of the mass-loaded tube are not harmonics of the fundamental, they will not be resonant at the harmonics of the driving current, and hence will be weakly excited, if at all.

*3.6 THE FREELY VIBRATING BAR: GENERAL BOUNDARY CONDITIONS

For a freely vibrating bar with arbitrary loading on each end, the normal modes of vibration can be determined in terms of the mechanical impedance at each end of the bar. If the mechanical impedance of the support at $x = 0$ is $\mathbf{Z}_{m0}$, the force acting on this support due to the bar is

$$\mathbf{f}_0 = -\mathbf{Z}_{m0}\mathbf{u}(0, t) \tag{3.6.1}$$

where the minus sign arises because a positive compressive force in the bar leads to an acceleration of the support to the left. On the other hand, a positive compressive force at the end $x = L$ leads to an acceleration of the adjacent support to the right so that the force acting on this support is

$$\mathbf{f}_L = +\mathbf{Z}_m\mathbf{u}(L, t) \tag{3.6.2}$$

where $\mathbf{Z}_m$ is the mechanical impedance of the support at the right end of the bar.

These equations can be expressed in terms of the particle displacement by using (3.3.3) to replace the compressive force and writing $\mathbf{u} = \partial\boldsymbol{\xi}/\partial t$,

$$\left(\frac{\partial\boldsymbol{\xi}}{\partial x}\right)_{x=0} = \frac{\mathbf{Z}_{m0}}{\rho_L c^2}\left(\frac{\partial\boldsymbol{\xi}}{\partial t}\right)_{x=0}$$

$$\left(\frac{\partial\boldsymbol{\xi}}{\partial x}\right)_{x=L} = -\frac{\mathbf{Z}_m}{\rho_L c^2}\left(\frac{\partial\boldsymbol{\xi}}{\partial t}\right)_{x=L} \tag{3.6.3}$$

where $\rho_L = \rho S$ is the density per unit length of the bar.

The choice of a trial solution satisfying the lossless wave equation for the bar and the boundary conditions (3.6.3) depends on the natures of the impedances $\mathbf{Z}_{m0}$ and $\mathbf{Z}_m$. If these loads are purely reactive, there can be no loss of acoustic energy so that there is no temporal or spatial damping. An appropriate trial solution would then be (3.3.10). We may go one step further and notice that, since there are no losses, the wave traveling to the right must possess the same energy as that going to the left. The amplitudes must therefore be equal, $|\mathbf{A}| = |\mathbf{B}|$. Application of the boundary conditions (3.6.3) then amounts to determining the phase angles of these complex amplitudes.

On the other hand, if either or both of $\mathbf{Z}_{m0}$ and $\mathbf{Z}_m$ have resistive components, a more general trial solution must be assumed. As was noted in the earlier discussion, in Section 2.11(b) on the freely vibrating string terminated by a resistive support, the presence of resistance requires that there be temporal damping. This means that the temporal behavior of the vibrating bar must be described by a complex angular frequency $\boldsymbol{\omega} = \omega + j\beta$ whose real part is the angular frequency of vibration and whose imaginary part is the temporal absorption coefficient β. Since there are no internal losses in the bar, the wave equation is still (3.3.7). Thus, we postulate

$$\boldsymbol{\xi}(x, t) = (\mathbf{A}e^{-j\mathbf{k}x} + \mathbf{B}e^{j\mathbf{k}x})e^{j\boldsymbol{\omega}t} \qquad (3.6.4)$$

where $\mathbf{k}$ is determined by $\boldsymbol{\omega} = c\mathbf{k}$. Application of the boundary conditions (3.6.3) to the generalized trial solution (3.6.4) yields the pair of equations

$$\mathbf{A} - \mathbf{B} = -(\mathbf{Z}_{m0}/\rho_L c)(\mathbf{A} + \mathbf{B})$$
$$\mathbf{A}e^{-j\mathbf{k}L} - \mathbf{B}e^{j\mathbf{k}L} = (\mathbf{Z}_m/\rho_L c)(\mathbf{A}e^{-j\mathbf{k}L} + \mathbf{B}e^{j\mathbf{k}L}) \qquad (3.6.5)$$

The first equation is solved for $\mathbf{B}$ in terms of $\mathbf{A}$ and this is substituted into the second equation. The results are

$$\mathbf{B} = \frac{1 + (\mathbf{Z}_{m0}/\rho_L c)}{1 - (\mathbf{Z}_{m0}/\rho_L c)}\,\mathbf{A}$$

$$\tan \mathbf{k}L = j\frac{(\mathbf{Z}_{m0}/\rho_L c) + (\mathbf{Z}_m/\rho_L c)}{1 + (\mathbf{Z}_{m0}/\rho_L c)(\mathbf{Z}_m/\rho_L c)} \qquad (3.6.6)$$

Given the impedances $\mathbf{Z}_{m0}$ and $\mathbf{Z}_m$, the properties of the vibration have been obtained, although explicit solution is not in general easy. Any resistive component in $\mathbf{Z}_{m0}$ or $\mathbf{Z}_m$ causes the argument of the tangent to be complex, introducing calculational difficulties in solving this transcendental equation.

*3.7 FORCED VIBRATIONS OF A BAR: RESONANCE AND ANTIRESONANCE REVISITED

In discussing the behavior of a forced, loaded string (Section 2.9), we defined resonance to occur when the speed amplitude was as large as possible and antiresonance to occur when the speed amplitude was as small as possible. It was seen there that resonance corresponded to the vanishing of the input mechanical reactance and that antiresonance corresponded, for purely *reactive* loads, to the reactance becoming infinite. We will now investigate these concepts in more detail and show that they must be modified for loads with a nonzero resistive component. We will find that at resonance the speed amplitude is maximized and at antiresonance it is minimized, but both resonance and antiresonance correspond to the vanishing of the input reactance.

Assume that a bar of length L is driven at $x = 0$ with a force $\mathbf{f}_0 = F_0 \exp(j\omega t)$ and is terminated at $x = L$ by a support possessing a mechanical impedance $\mathbf{Z}_m$. We assume the trial solution (3.3.10). The boundary condition at the forced end is (3.3.3):

$$F_0 e^{j\omega t} = -\rho_L c^2 \left(\frac{\partial \xi}{\partial x}\right)_{x=0} \tag{3.7.1}$$

where $\rho_L = \rho S$ and $Y = \rho c^2$. At the loaded end, the boundary condition is $\mathbf{f}_L = \mathbf{Z}_m \mathbf{u}(L, t)$:

$$\left(\frac{\partial \xi}{\partial x}\right)_{x=L} = -\frac{\mathbf{Z}_m}{\rho_L c^2}\left(\frac{\partial \xi}{\partial t}\right)_{x=L} \tag{3.7.2}$$

Either direct application of these boundary conditions to (3.6.4) or argument by analogy will determine $\mathbf{A}$ and $\mathbf{B}$ and the input mechanical impedance.

Let us argue by analogy. Direct comparison of (3.7.1) and (3.7.2) with (2.9.16) and (2.9.17) reveals that the boundary conditions are identical if we substitute $\mathbf{Z}_m$ for $j\omega m$. Since the trial solution remains unchanged, the same substitution into (2.9.22) gives a generalized form of the input impedance,

$$\mathbf{Z}_{m0} = \rho_L c \frac{(\mathbf{Z}_m/\rho_L c) + j \tan kL}{1 + (\mathbf{Z}_m/\rho_L c)j \tan kL} \tag{3.7.3}$$

If we define a scaled mechanical impedance by

$$\mathbf{Z}_m/\rho_L c = R/\rho_L c + jX/\rho_L c = r + jx \tag{3.7.4}$$

then (3.7.3) can be rewritten as

$$\frac{\mathbf{Z}_{m0}}{\rho_L c} = \frac{r + j(x + \tan kL)}{(1 - x \tan kL) + jr \tan kL} \tag{3.7.5}$$

(Here and to the end of this section, x represents the scaled mechanical reactance.)

It is left as an exercise to show that for $r = 0$ the input impedance is purely reactive and vanishes for frequencies such that $\tan kL = -x$ and becomes infinite when $\tan kL = 1/x$. Since the driving force amplitude is assumed constant, the vanishing of the input impedance $\mathbf{Z}_{m0} = \mathbf{f}_0/\mathbf{u}(0, t)$ means that the speed amplitude at the point of application of the force is infinite, the condition for mechanical *resonance* ($\tan kL = -x$). On the other hand, when the input impedance becomes infinite, the speed amplitude at the driver goes to zero, the condition for mechanical *antiresonance* ($\tan kL = 1/x$).

When the load resistance is not zero, the input impedance (3.7.5) has vanishing reactance if the phases of numerator and denominator are the same. This provides the condition

$$\frac{x + \tan kL}{r} = \frac{r \tan kL}{1 - x \tan kL} \tag{3.7.6}$$

which can be rewritten as a quadratic

$$x \tan^2 kL + (r^2 + x^2 - 1) \tan kL - x = 0 \tag{3.7.7}$$

Under the condition $|r^2 + x^2 - 1| \gg 2|x|$ and with the help of $\sqrt{1 + \varepsilon} \approx 1 + \varepsilon/2$ for small ε, the roots are approximately

$$\tan kL \approx \begin{cases} x/(r^2 + x^2 - 1) \\ -(r^2 + x^2 - 1)/x \end{cases} \tag{3.7.8}$$

Assume that the mechanical support at the right end of the bar has small losses so that $r \ll 1$. Then x must be either large ($x \gg 1$) or small ($x \ll 1$). In either case, the pair of roots simplifies to

$$\tan kL \approx \begin{cases} -x \\ 1/x \end{cases} \tag{3.7.9}$$

Substitution of these roots into (3.7.5) gives (1) for $\tan kL \approx -x$

$$\mathbf{Z}_{m0}/\rho_L c \sim r \tag{3.7.10}$$

regardless of whether $x \gg 1$ or $x \ll 1$, and (2) for $\tan kL \approx 1/x$

$$\mathbf{Z}_{m0}/\rho_L c \sim 1/r \tag{3.7.11}$$

whether x is large or small. (These results are approximate, but simple. Higher accuracy would require much more mathematical manipulation and would result in expressions more complicated than warranted for our discussion.) For small load resistance and both large and small load reactance, the first root $\tan kL \approx -x$ corresponds to resonance, since the input impedance is real and small so that the velocity at the driving point has large amplitude. The second root $\tan kL \approx 1/x$ corresponds to antiresonance since the input impedance is real and large so that the velocity amplitude at the driving point is small. Both resonance and antiresonance frequencies occur when the input mechanical reactance vanishes. The input resistance is small at resonance and large at antiresonance.

These observations are consistent with the standing wave having large amplitude at resonance and small amplitude at antiresonance. For example, for the forced, nearly free bar there must be an antinode close to the end at L so that the maximum particle speed within the bar is nearly $U_L = |\mathbf{u}(L, t)|$. The power transmitted from the bar into the load at L is approximated by

$$\Pi \approx \tfrac{1}{2} U_L^2 R \tag{3.7.12}$$

while the power sent into the bar both at resonance and at antiresonance is

$$\Pi = \tfrac{1}{2} F_0^2 / R_0 \tag{3.7.13}$$

where R_0 is the input mechanical resistance found from (3.7.5). Since the bar itself is assumed to be lossless, these powers must be equal and we can solve for the approximate antinodal speed amplitude, $U_L \approx F_0/\sqrt{R_0 R}$. Substitution of the appropriate values of R_0 gives $U_L \approx F_0/R$ at resonance and $U_L \approx F_0/\rho_L c$ at antiresonance, so that

$$\frac{U_L(\text{antiresonance})}{U_L(\text{resonance})} \approx \frac{R}{\rho_L c} \tag{3.7.14}$$

Since we have assumed $R \ll \rho_L c$, it is clear that the standing wave has much greater amplitude at resonance than at antiresonance.

Examination of the case $r \gg 1$ leads to analogous results and the same conclusions about resonance and antiresonance.

*3.8 TRANSVERSE VIBRATIONS OF A BAR

A bar is capable of vibrating transversely as well as longitudinally, and the internal coupling between strains makes it difficult to produce one motion without the other. For example, if

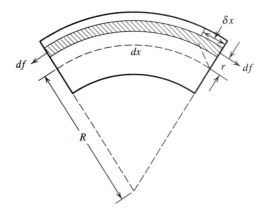

Figure 3.8.1 Bending strains and stresses set
up by the transverse displacements at the ends
of an element of a bar of length dx with radius
of curvature R.

a long thin bar is supported at its center and set into vibration by a hammer blow directed
along the axis of the bar, any slight eccentricity of the blow results in predominantly
transverse vibrations rather than the desired longitudinal vibrations.

Consider a straight bar of length L, having a uniform cross section S with bilateral
symmetry. Let the x coordinate measure positions along the bar, and the y coordinate the
transverse displacements of the bar from its normal configuration. When the bar is bent
as indicated in Fig. 3.8.1, the lower part is compressed and the upper part is stretched.
Somewhere between the top and the bottom of the bar there will be a *neutral axis* whose
length remains unchanged. (If the cross section of the bar is symmetric about a horizontal
plane, this neutral axis will coincide with the central axis of the bar.)

Now consider a segment of the bar of length dx, and assume that the bending of the bar
is measured by the radius of curvature R of the neutral axis. Let $\delta x = (\partial \xi / \partial x)\, dx$ be the
increment of length, due to bending, of a filament of the bar located at a distance r from the
neutral axis. Then the longitudinal force df is given by

$$df = -Y\, dS\, (\delta x / dx) = -Y\, dS\, (\partial \xi / \partial x) \tag{3.8.1}$$

where dS is the cross-sectional area of the filament. The value of δx for the particular filament
considered in Fig. 3.8.1 is positive, so that df is a tension, and consequently negative. For
filaments below the neutral axis δx is negative, giving a positive force of compression.

Now from the geometry $(dx + \delta x)/(R + r) = dx/R$ and hence $\delta x / dx = r/R$. Substitution
into (3.8.1) yields

$$df = -(Y/R)r\, dS \tag{3.8.2}$$

The total longitudinal force $f = \int df$ is zero, negative forces above the neutral axis being
canceled by positive forces below it. However, a bending moment M is present in the bar,

$$M = \int r\, df = -\frac{Y}{R} \int r^2\, dS \tag{3.8.3}$$

If we define a constant κ by

$$\kappa^2 = \frac{1}{S} \int r^2\, dS \tag{3.8.4}$$

then

$$M = -YS\kappa^2/R \tag{3.8.5}$$

The constant κ can be thought of as the *radius of gyration* of the cross-sectional area S, by analogy with the definition of the radius of gyration of a solid. The value of κ for a bar of rectangular cross section and thickness t (measured in the y direction) is $\kappa = t/\sqrt{12}$. For a circular rod of radius a, $\kappa = a/2$.

The radius of curvature R is not in general a constant but is rather a function of position along the neutral axis. If the displacements y of the bar are limited to small values, $\partial y/\partial x \ll 1$, then we may use the approximate relation

$$R = \frac{[1 + (\partial y/\partial x)^2]^{3/2}}{\partial^2 y/\partial x^2} \approx \frac{1}{\partial^2 y/\partial x^2} \tag{3.8.6}$$

Substitution of (3.8.6) into (3.8.5) yields

$$M = -YS\kappa^2(\partial^2 y/\partial x^2) \tag{3.8.7}$$

In the situation illustrated in Fig. 3.8.1, the curvature makes $\partial^2 y/\partial x^2$ negative, and the bending moment M is consequently positive. It is apparent that to obtain the curvature illustrated, the torque applied to the left end of the segment dx must act in a counterclockwise or positive angular direction, so that (3.8.7) gives the torque acting on the left end of the segment both as to magnitude and as to direction. Similarly, the torque acting on the right end of the segment must be clockwise, with the result that it is negative and is therefore represented both in direction and in magnitude by $-M$.

*3.9 TRANSVERSE WAVE EQUATION

The effect of distorting the bar is to produce not only bending moments but also shear forces. Consider an upward shear force F_y acting on the left end of the segment dx as positive (Fig. 3.9.1). Then the associated shear force acting on the right end of the segment must be downward, and is consequently negative. When a bent bar is in a condition of static equilibrium, the torques and shear forces acting on any segment must produce no net turning moment. Taking moments about the left end of the segment of Fig. 3.9.1, we have

$$M(x) - M(x + dx) = F_y(x + dx)\, dx \tag{3.9.1}$$

For segments of small length dx, $M(x + dx)$ and $F_y(x + dx)$ can be expanded in Taylor's expansions about x, and this yields

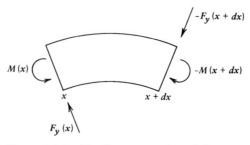

Figure 3.9.1   Bending moments and shear forces set up by the transverse displacements at the ends of an element of a bar of length dx.

$$F_y = -(\partial M/\partial x) = YS\kappa^2 (\partial^3 y/\partial x^3) \tag{3.9.2}$$

where second-order terms in dx have been dropped.

This relation between the shear force F_y and the bending moment M has been derived for a condition of static equilibrium. For transverse vibrations of a bar the equilibrium is dynamic, rather than static, and the right side of (3.9.1) must equal the rate of increase of angular momentum of the segment. However, if the displacement and slope of the bar are limited to small values, the variations in angular momentum may be neglected, and (3.9.2) serves as an adequate approximation for the relation between F_y and y.

The net upward force dF_y acting on the segment dx is then given by

$$dF_y = F_y(x) - F_y(x + dx) = -\left(\frac{\partial F_y}{\partial x}\right)dx = -YS\kappa^2 \left(\frac{\partial^4 y}{\partial x^4}\right)dx \tag{3.9.3}$$

By Newton's second law, this force will give the mass ($\rho S\,dx$) of the segment an upward acceleration $\partial^2 y/\partial t^2$ so that the equation of motion is

$$\frac{\partial^2 y}{\partial t^2} = -(\kappa c)^2 \frac{\partial^4 y}{\partial x^4} \tag{3.9.4}$$

where $c^2 = Y/\rho$. One significant difference between this differential equation and the simpler equation for the transverse waves on a string is the presence of a fourth partial derivative with respect to x, rather than a second partial. As a result, direct substitution shows that functions of the form $f(ct - x)$ are not solutions of (3.9.4). Transverse waves do not travel along the bar with a constant speed c and unchanging shape.

Assume that (3.9.4) may be solved by separation of variables, and write the complex transverse displacement as

$$y(x, t) = \mathbf{\Psi}(x)e^{j\omega t} \tag{3.9.5}$$

Upon substitution, the exponential function of time cancels, leaving a new *total* differential equation involving $\mathbf{\Psi}$,

$$\boxed{\begin{array}{c} \dfrac{d^4 \mathbf{\Psi}}{dx^4} = \left(\dfrac{\omega}{v}\right)^4 \mathbf{\Psi} \\[2mm] v^2 = \omega(\kappa c) \end{array}} \tag{3.9.6}$$

where v has the dimensions of a speed. If we substitute a trial function $\mathbf{\Psi} = \exp(\gamma x)$ into (3.9.6), it is valid for $\gamma = \pm(\omega/v)$ and $\pm j(\omega/v)$. Thus, if we define a quantity g by

$$g = \omega/v \tag{3.9.7}$$

then a complete monofrequency solution can be written as

$$\mathbf{\Psi}(x) = \mathbf{A}e^{gx} + \mathbf{B}e^{-gx} + \mathbf{C}e^{jgx} + \mathbf{D}e^{-jgx}$$
$$y(x, t) = (\mathbf{A}e^{gx} + \mathbf{B}e^{-gx})e^{j\omega t} + \mathbf{C}e^{j(\omega t + gx)} + \mathbf{D}e^{j(\omega t - gx)} \tag{3.9.8}$$

where $\mathbf{A}$, $\mathbf{B}$, $\mathbf{C}$, and $\mathbf{D}$ are arbitrary constants and g is both a *wave number* and also a *spatial attenuation coefficient*. Note that g is proportional to $\sqrt{\omega}$. The solution represents *flexural* disturbances of two kinds: (1) two traveling waves each propagating with a *phase speed* v proportional to $\sqrt{\omega}$ and (2) two standing oscillations that are spatially damped, each with a spatial attenuation coefficient g depending on $\sqrt{\omega}$. Waves of different frequencies travel with different phase speeds, an effect known as *dispersion*. The high-frequency components outrun the low-frequency components, altering the shape of the wave. This is analogous to the transmission of light through glass, wherein the different component frequencies of a light beam travel with different speeds. A vibrating bar is a *dispersive medium* for transverse waves.

The actual solution of (3.9.4) is the real part of (3.9.8). It may conveniently be expressed using hyperbolic and trigonometric identities (see Appendix A3),

$$y(x, t) = [A \cosh gx + B \sinh gx + C \cos gx + D \sin gx] \cos(\omega t + \phi) \qquad (3.9.9)$$

where A, B, C, and D are new real constants. Although these constants are related to the complex constants $\mathbf{A}$, $\mathbf{B}$, $\mathbf{C}$, and $\mathbf{D}$, the relationships are unimportant, since in practice A, B, C, and D are evaluated directly through the application of initial and boundary conditions.

*3.10 BOUNDARY CONDITIONS

Since (3.9.9) contains twice as many arbitrary constants as the corresponding equation for the transverse vibrations of a string, the determination of these constants requires twice as many boundary conditions. This need is fulfilled by the existence of pairs of boundary conditions at the ends of the bar. The particular forms of these conditions depend on the nature of the support and include the following.

(a) Clamped End

If an end of the bar is rigidly clamped, both the displacement and the slope must be zero at that end for all times t. The boundary conditions are therefore

$$y = 0 \quad \text{and} \quad \frac{\partial y}{\partial x} = 0 \qquad (3.10.1)$$

(b) Free End

At a free end there can be neither an externally applied torque nor a shearing force, and hence both M and F_y are zero at the end. However, the displacement and slope are not constrained, except that their values must be small. Then from (3.8.7) and (3.9.2), the boundary conditions are

$$\frac{\partial^2 y}{\partial x^2} = 0 \quad \text{and} \quad \frac{\partial^3 y}{\partial x^3} = 0 \qquad (3.10.2)$$

(c) Simply Supported End

A simply supported end is obtained by constraining that end of the bar between a pair of knife edges mounted perpendicular to the plane of the transverse motion and centered on the neutral axis of the bar (or a pair of needle points, similarly placed on the neutral axis) so that both the transverse displacement and the torque are zero with no constraint on the slope.

$$y = 0 \quad \text{and} \quad \frac{\partial^2 y}{\partial x^2} = 0 \qquad (3.10.3)$$

*3.11 BAR CLAMPED AT ONE END

Assume that a bar of length L is rigidly clamped at $x = 0$ and is free at $x = L$. Then applying the two conditions of (3.10.1) at $x = 0$ to the general solution of (3.9.9) we obtain $A + C = 0$ and $B + D = 0$ so that the general solution reduces to

$$y(x, t) = [A(\cosh gx - \cos gx) + B(\sinh gx - \sin gx)]\cos(\omega t + \phi) \qquad (3.11.1)$$

A further application of the two conditions of (3.10.2) at $x = L$ gives

$$A(\cosh gL + \cos gL) = -B(\sinh gL + \sin gL)$$

$$A(\sinh gL - \sin gL) = -B(\cosh gL + \cos gL) \qquad (3.11.2)$$

While it is impossible for both of these equations to be true for all frequencies, at certain frequencies they become equivalent. To determine these allowed frequencies, divide one equation by the other, thus canceling out the constants A and B. Then cross-multiply and simplify by using the identities $\cos^2 \theta + \sin^2 \theta = 1$ and $\cosh^2 \theta - \sinh^2 \theta = 1$. This gives

$$\cosh gL \cos gL = -1 \qquad (3.11.3)$$

It is easy to obtain the allowed values of gL by numerical techniques, particularly since the hyperbolic cosine grows as $\exp(gL)$ so that the cosine must approach zero very closely for arguments greater than about π. Solutions are found to be

$$gL = \omega L/v = (1.194, 2.988, 5, 7, \ldots)\,\pi/2 \qquad (3.11.4)$$

Substituting $v = \sqrt{\omega \kappa c}$ into (3.11.4) and squaring both sides, we have for the natural frequencies of a transversely vibrating clamped, free bar

$$f = (1.194^2, 2.988^2, 5^2, 7^2, \ldots)\pi\kappa c/8L^2 \qquad (3.11.5)$$

The application of boundary conditions limits the natural modes to a discrete set, just as it does for a vibrating string. However, in contrast with the string, the overtone frequencies are not harmonics of the fundamental. As shown in Table 3.11.1, the first overtone has a frequency more than six times that of the fundamental. If a bar is struck so that the amplitudes of vibration of its overtones are appreciable, the sound produced has a metallic quality. However, the overtones are rapidly damped so that the initial sound mellows into a nearly pure tone at the fundamental. Vibrating reeds in music boxes give good examples of this behavior. Note that the fundamental frequency can be adjusted by varying either the thickness or the length; doubling the length lowers the frequency by a factor of four.

The distribution of nodal points along the bar is more complex than in the examples previously considered, for the nodes are not evenly placed at intervals of $\lambda/2$ but are irregularly spaced. See Fig. 3.11.1. There are *three types* of nodal points where $y = 0$: (1) the node where the bar is clamped, characterized by the additional condition $\partial y/\partial x = 0$; (2) the so-called *true* nodes, lying nearly $\lambda/2$ apart and very close to the points of inflection where $\partial^2 y/\partial x^2 \approx 0$; and (3) the node adjacent to the free end of the bar. (A point of inflection does not lie near this last nodal position but instead is shifted out to the end.) It is also to be noted that the vibrational amplitudes at the various antinodal positions are not the same and that the antinode at the free end has the greatest amplitude of motion.

Table 3.11.1 gives the nodal positions for transverse vibrations of a bar 100 cm in length (clamped at $x = 0$ and free at $x = 100$ cm), the ratios of the frequencies and phase speeds of the overtones to those for the fundamental, and the wavelengths $\lambda = v/f$ for each natural frequency. The increase in phase speed with frequency is quite apparent. As discussed

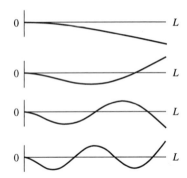

Figure 3.11.1   The four lowest modes of
transverse vibration of a clamped, free bar. Note
the boundary conditions at each end and the
different classes of the nodes.

Table 3.11.1   Transverse vibration characteristics of a clamped,
free bar with $L = 100$ cm

Frequency	Phase Speed	Wavelength (cm)	Nodal Positions (cm from clamped end)
f_1	v_1	335.0	0
$6.267f_1$	$2.50v_1$	133.4	0, 78.3
$17.55f_1$	$4.18v_1$	80.0	0, 50.4, 86.8
$34.39f_1$	$5.87v_1$	57.2	0, 35.8, 64.4, 90.6

earlier, the wavelengths are not in general equal to twice the distance between adjacent
nodes. However, the nodal spacing between true nodes for the third overtone is $\lambda/2$ within
the accuracy of the data $[64.4 - 35.8 = 28.6 = (57.2)/2$ cm$]$.

*3.12 BAR FREE AT BOTH ENDS

Another important kind of transverse vibration is that of a free, free bar. The boundary
conditions are satisfied at $x = 0$ if $A - C = 0$ and $B - D = 0$. Application of the same
conditions at $x = L$ and the same kind of trigonometric and hyperbolic reductions yields
the transcendental equation

$$\cosh gL \cos gL = 1 \tag{3.12.1}$$

As before, numerical solution is relatively simple, and we obtain the natural frequencies for
the transversely vibrating free, free bar,

$$f = (3.011^2, 5^2, 7^2, 9^2, \ldots)\pi\kappa c/8L^2 \tag{3.12.2}$$

Again the overtones are not harmonics of the fundamental.

Table 3.12.1 gives information concerning the frequencies, phase speeds, and nodal
positions of a free, free bar 100 cm long. An inspection of Fig. 3.12.1 shows that the modes

Table 3.12.1   Transverse vibration characteristics of a free, free
bar with $L = 100$ cm

Frequency	Phase Speed	Wavelength (cm)	Nodal Positions (cm from end)
f_1	v_1	133.0	22.4, 77.6
$2.756f_1$	$1.66v_1$	80.0	13.2, 50.0, 86.8
$5.404f_1$	$2.32v_1$	57.2	9.4, 35.6, 64.4, 90.6
$8.933f_1$	$2.99v_1$	44.5	7.3, 27.7, 50.0, 72.3, 92.7

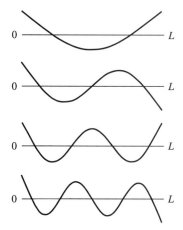

Figure 3.12.1   The four lowest modes of
transverse vibration of a free, free bar. Note the
boundary conditions at each end and the different
classes of the nodes.

of vibration corresponding to the fundamental and all *even* overtones (corresponding to $f_1, f_3, f_5, \ldots$ in the figure) are symmetric about the center. There is a *true antinode* at the center where $\partial y / \partial x = 0$. In contrast, the *odd* overtones ($f_2, f_4, f_6, \ldots$) correspond to antisymmetric modes with respect to the center. In all modes, the nodes are symmetrically distributed about the center. The bar may be supported on a knife edge, or held by a knife-edge clamp, at any nodal point without interfering with the mode of vibration having a node at this point. A knife-edge clamp (or needle-point support) is required, since it must merely restrict the displacement to zero and must not restrict the changes in slope that occur at a node.

Each bar of a xylophone is supported at points corresponding to the nodes of its fundamental. Since the nodes of the accompanying overtones will not in general be located at these same two points, the overtones will rapidly be damped out, leaving the fundamental. This is one of a number of factors that contribute to the mellow sound of a xylophone or marimba.

The free, free bar may be used qualitatively to describe a tuning fork. This is basically a ßU-shaped bar with a stem attached to the center. The bend and the mass-loading of the stem reduce the separation of the two nodes present in the fundamental mode. Compare Fig. 3.12.2 with Fig. 3.12.1. As above, when a tuning fork is struck the overtones rapidly dampen, leaving the fundamental frequency. The stem, at an antinode, vibrates and couples the motion to any surface it touches. The radiation efficiency is enhanced if the surface has large area, or forms a side of a resonator box tuned to the fundamental.

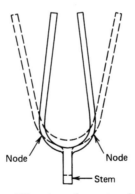

Figure 3.12.2   Vibration of a tuning fork.

If a bar is rigidly clamped at both ends, the boundary conditions $y = 0$ and $\partial y/\partial x = 0$ at the ends $x = 0$ and $x = L$ lead to the same set of natural frequencies as for a free, free bar. However, as is to be expected, the locations of the nodes are different.

*3.13 TORSIONAL WAVES ON A BAR

A bar is capable of vibrating torsionally as well as longitudinally and transversely. For example, if a long, thin bar (or a fiber used as an activator for a torsional-pendulum clock) is fixed at one end and the other end is twisted about the long axis of the bar, the restoring torque will increase as the angle of twist is increased. If the twisted end is then released, a torsional wave will travel down the bar.

For simplicity of discussion, let the bar have a circular cross section of radius a. Isolate an element of the bar of length dx (Fig. 3.13.1a). Break this element into a series of concentric

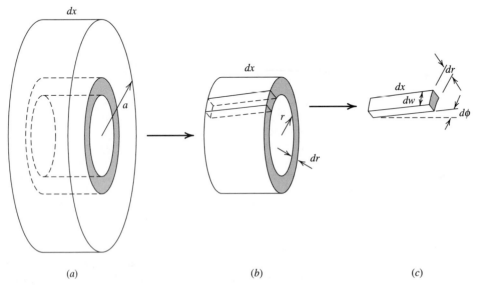

Figure 3.13.1   The element of a circular bar and its subelements used to derive the wave equation for shear waves. (a) A cylinder of radius a and length dx. (b) A cylindrical shell of radius r, thickness dr, and length dx. (c) A plate of width dw, thickness dr, and length dx strained by an angle $d\phi$.

hollow tubes of radius r and thickness dr (Fig. 3.13.1b), and further divide each hollow cylinder into side-by-side rectangular plates of length dx, thickness dr, and (curved) width dw (Fig. 3.13.1c). When the tube is twisted away from equilibrium through a small angle $d\phi$, this rectangle is distorted by an angle $r(d\phi/dx)$ which is the *shearing strain*. The *shearing stress* required to produce that shearing strain is proportional to it (Hooke's law) and the constant of proportionality is the *shear modulus* (the *modulus of rigidity*) $\mathcal{G}$ (see Appendix A11),

$$\text{stress} = \mathcal{G}r(d\phi/dx) \tag{3.13.1}$$

This is the torsional equivalent to (3.3.2). The force df required to produce this distortion is the shearing stress multiplied by the area over which that stress acts,

$$df = \mathcal{G}(dw\,dr)\,r(d\phi/dx) \tag{3.13.2}$$

The torque dM required to produce the strain in the hollow tube of height dx is found by multiplying df by its moment arm r and integrating around the circumference of the tube. Because of the circumferential symmetry, $\int dw = 2\pi r$ so the torque on a tube is

$$d\tau = \mathcal{G}(2\pi r^3)(d\phi/dx)\,dr \tag{3.13.3}$$

The total torque τ twisting this end of the elemental solid cylinder is found by integrating over the concentric tubes from $r = 0$ to $r = a$,

$$\tau = \mathcal{G}\tfrac{1}{2}\pi a^4(d\phi/dx) \tag{3.13.4}$$

The net torque on the elemental cylinder of height dx is the difference of the torques on each end, which by a Taylor's expansion is $\tau(x+dx) - \tau(x) = (d\tau/dx)\,dx = \mathcal{G}(\pi a^4/2)(\partial^2\phi/\partial x^2)\,dx$. This net torque is equal to the moment of inertia of the cylinder $(a^2/2)\,dm$, where $dm = \rho\pi a^2\,dx$, times its angular acceleration $\partial^2\phi/\partial t^2$. This gives the familiar one-dimensional wave equation

$$\frac{\partial^2\phi}{\partial x^2} = \frac{1}{c^2}\frac{\partial^2\phi}{\partial t^2} \tag{3.13.5}$$

with the phase speed c obtained from

$$\boxed{c^2 = \mathcal{G}/\rho} \tag{3.13.6}$$

All the techniques for finding the solutions for waves on strings and longitudinal waves in bars apply here. Examples of boundary conditions applicable to torsional oscillations include (1) fixed end, $\phi = 0$; (2) free end, $\tau = 0$ so that $\partial\phi/\partial x = 0$; and (3) mass-loaded end, $\partial\phi/\partial x = (\partial\phi^2/\partial x^2)I$, where I is the moment of inertia of the load about the axis of the bar.

Problems

3.4.1. A bar of length L is rigidly fixed at $x = 0$ and free to move at $x = L$. (a) Show that only odd integral harmonic overtones are allowed. (b) Determine the fundamental frequency of the bar, if it is composed of steel and has a length of 0.5 m. (c) If a static force F is applied to the free end of the bar so as to displace this end by h, show that, when the bar vibrates longitudinally subsequent to the sudden release of this force, the amplitudes of the various harmonic vibrations are given by $A_n = [8h/(n\pi)^2] \sin(n\pi/2)$. (d) Determine these amplitudes for the above steel bar, if the force is 5000 N and the cross-sectional area of the bar is 0.00005 m^2.

3.4.2. Verify whether or not the normal modes of a longitudinally vibrating bar form an orthogonal set if the boundary conditions are (a) fixed, fixed; (b) free, free; (c) fixed, free.

3.5.1. A steel bar of 0.0001 m^2 cross-sectional area and 0.25 m length is free to move at $x = 0$ and is loaded with 0.15 kg at $x = 0.25$ m. (a) Compute the fundamental frequency of longitudinal vibrations of the above mass-loaded bar. (b) Determine the position at which the bar may be clamped to cause the least interference with its fundamental mode of vibration. (c) When this bar is vibrating in its fundamental mode, what is the ratio of the displacement amplitude of the free end to that of the mass-loaded end? (d) What is the frequency of the first overtone of this bar?

3.5.2. A 2 kg mass is hanging on a steel wire of 0.00001 m^2 cross-sectional area and 1.0 m length. (a) Compute the fundamental frequency of vertical oscillation of the mass by considering it to be a simple oscillator. (b) Compute the fundamental frequency of vertical oscillation of the mass by considering the system to be that of a longitudinally vibrating bar fixed at one end and mass-loaded at the other. (c) Show that for $kL < 0.2$, the equation derived in (b) reduces to (1.2.4).

3.5.3 Are the normal modes orthogonal for (a) a fixed, mass-loaded bar, (b) a free, mass-loaded bar?

3.6.1. A thin bar of length L and mass M is fixed at one end and free at the other. What mass m must be attached to the free end to decrease the fundamental frequency of longitudinal vibration by 25% from its fixed, free value?

3.6.2. A steel bar of 0.2 m length and 0.04 kg mass is loaded at one end with 0.027 kg and at the other end with 0.054 kg. (a) Calculate the fundamental frequency of longitudinal vibration of this system. (b) Calculate the position of the node in the bar. (c) Calculate the ratio of the displacement amplitudes at the two ends of the bar.

3.6.3. Assuming very small losses, find a condition relating the mechanical impedances of the supports at the ends of a bar of length L and longitudinal wave speed c if the bar is to have an integral number of wavelengths between its ends when it is vibrating longitudinally.

3.6.4. Determine an expression giving the fundamental frequency of longitudinal vibrations of a fixed, free bar of length L and mass m, if the reaction of the fixture corresponds to a mechanical reactance of $-js/\omega$ (stiffness).

3.6.5. A bar of length L has circular cross-sectional area S. The material of the bar has a linear density ρ_L and a Young's modulus Y. The bar is terminated at $x = 0$ with a mass m and at $x = L$ by a longitudinal spring of spring constant s. (a) In terms of L, S, ρ_L, s, and Y, find the transcendental equation for kL that must be solved to find the normal modes for longitudinal motion. (b) The bar is aluminum with length 1 m and radius 1 cm. The mass m is 0.848 kg and the spring has $s = 2.23 \times 10^7$ N/m. Find the lowest eigenfrequency. (c) Find the nodal locations.

3.7.1. Show that for $r = 0$ the input impedance $\mathbf{Z}_{m0}$ is purely reactive and has magnitude zero when $\tan kL = -x$ and becomes infinitely large when $\tan kL = 1/x$.

3.7.2. Assume that the load mechanical impedance is $\mathbf{Z}_m/\rho_L c = 1 + jx$. (a) If the load reactance is large ($x \gg 1$) in the frequency range of interest, show that the equations determining resonance and antiresonance are $\tan kL = -x$ and $\tan kL = 1/x$, respectively. (b) If in the frequency range of interest the load reactance is small ($x \ll 1$) show that $\tan kL \approx \pm 1$ and the input impedance becomes $\mathbf{Z}_{m0} \approx \rho_L c$.

3.7.3. A long thin bar of length L is driven by a longitudinal force $F \cos \omega t$ at $x = 0$ and is free at $x = L$. (a) Derive the equation that gives the amplitude of the standing waves set up in the bar. (b) What is the input mechanical impedance? (c) What is the input mechanical impedance of a similar bar of infinite length? (d) If the material of the bar is aluminum, the length is 1.0 m, the cross-sectional area is 0.0001 m², and the amplitude of the driving force is 10 N, plot the amplitude of the driven end of the bar of part (a) as a function of frequency over the range from 200 to 2000 Hz.

3.7.4C. For the longitudinal vibrations of a bar driven at one end by a force of constant amplitude F and loaded at the other with an impedance $R + j(\omega m - s/\omega)$, plot the input power for frequencies covering the first three resonances for (a) three values of R keeping m and s constant, (b) three values of m keeping R and s constant, and (c) three values of s keeping R and m constant.

3.7.5. A thin bar of length L with longitudinal phase speed c is driven at $x = 0$ by a force with adjustable frequency. (a) Find the frequencies for which the driving force experiences an input mechanical impedance equal to the mechanical impedance of the support at $x = L$. (b) At the frequencies of (a), compare the amplitudes of the velocities of the two ends of the bar.

3.7.6. Show that for $r \gg 1$ the conclusions pertaining to resonance and antiresonance following (3.7.11) are also reached.

3.8.1. Find the radius of gyration for a bar of circular cross section of radius a.

3.8.2. Calculate the radius of gyration for a bar of rectangular cross section with thickness t and width w for bending in the direction of (a) the thickness, (b) the width, and (c) transverse to a diagonal.

3.9.1. Show by direct substitution that (3.9.9) is a solution of (3.9.4).

3.9.2. Show that $v = \sqrt{\omega \kappa c}$ has the dimensions of a speed. For what frequency will the transverse vibrations of an aluminum rod of 0.01 m diameter have the same phase speed as that of longitudinal vibrations in the rod?

3.10.1. An aluminum bar of 100 cm length with circular cross section of 1 cm diameter is simply supported at both ends. For transverse vibrations, (a) show that the normal modes are the same as for the fixed, fixed string. (b) Find the frequencies of the normal modes. (c) Are the overtones harmonics as they are for the fixed, fixed string?

3.11.1. For a bar of rectangular cross section with $w = 2t$, one end clamped and the other free, find the ratio of the fundamental frequencies of free vibration for bending in the direction of the thickness to that for bending in the direction of the width.

3.11.2. For a bar of length 100 cm clamped at both ends, find, in terms of the fundamental frequency and phase speed, the frequencies, phase speeds, wavelengths, and nodal positions of the first three normal modes of transverse vibration.

3.11.3C. For a clamped, free bar vibrating in its third transverse mode, (a) plot the displacement amplitude and first three derivatives of the displacement as a function of length. (b) Use these calculated values to show that the boundary conditions are satisfied at both ends and discuss the nature of each node.

3.11.4C. If the same bar used in creating Table 3.11.1 is clamped at both ends, (a) create a table similar to Table 3.11.1, and (b) plot the shape of the first four normal modes.

3.11.5. An aluminum bar 100 cm long with a 1.0 cm radius is clamped at one end and free at the other. (a) Find the frequency of the lowest mode of transverse vibration. (b) If the free end has a displacement amplitude of 5.0 cm, determine all the constants in the equation for the transverse displacement of the bar. (c) Plot the displacement amplitude of the bar.

3.12.1. A steel rod of 0.005 m radius has a length of 0.5 m. (*a*) What is its fundamental frequency of free, free transverse vibrations? (*b*) If the displacement amplitude at the center of the rod is 2 cm when vibrating in its fundamental mode, what is the displacement amplitude at the ends?

3.12.2. Calculate A/B for (*a*) the clamped, free bar and (*b*) the free, free bar.

3.13.1. A 100 cm long aluminum bar has a diameter of 1.0 cm. (*a*) Find the torque required to give one end of the bar a static twist of 360° relative to the other end. (*b*) Find the phase speed of torsional waves on this bar. (*c*) If the bar is rigidly supported in the middle and free at the ends, find the lowest frequency at which it will support a torsional normal mode.

3.13.2. For an aluminum bar free at both ends, find the ratio of the lowest frequencies of a normal mode of longitudinal vibration to the lowest frequency of a normal mode of torsional vibration.

THE TWO-DIMENSIONAL WAVE EQUATION: VIBRATIONS OF MEMBRANES AND PLATES

4.1 VIBRATIONS OF A PLANE SURFACE

Consider transverse vibrations of two-dimensional systems, such as a drumhead or the diaphragm of a microphone. While analysis may seem more complicated because two spatial coordinates are needed to locate a point on the surface and a third to specify its displacement, the equation of motion (subject to the same simplifying assumptions invoked in the previous two chapters) will be merely the two-dimensional generalization of that for a string.

Generalization to two dimensions requires selecting a coordinate system. Choice of a coordinate system matching the boundary conditions (cartesian coordinates for a rectangular boundary and polar coordinates for a circular boundary) will greatly simplify obtaining and interpreting solutions. Unfortunately, the number of useful coordinate systems is strictly limited and, consequently, the number of easily solved membrane problems is similarly restricted.

4.2 THE WAVE EQUATION FOR A STRETCHED MEMBRANE

Assume a membrane is thin, is stretched uniformly in all directions, and vibrates transversely with small displacement amplitudes. Let ρ_S be the *surface density* (kg/m^2) of the membrane, and let $\mathcal{T}$ be the *membrane tension per unit length* (N/m); the material on opposite sides of a line segment of length dl will be pulled apart with a force $\mathcal{T}\, dl$.

In cartesian coordinates the transverse displacement of a point is expressed as $y(x, z, t)$. The force acting on a displaced surface element of area $dS = dx\, dz$ is the sum of the transverse forces acting on the edges parallel to the x and z axes. For the element shown in Fig. 4.2.1 the net vertical force arising from the pair of opposing tensions $\mathcal{T}\, dz$ is

$$\mathcal{T}\, dz \left[\left(\frac{\partial y}{\partial x} \right)_{x+dx} - \left(\frac{\partial y}{\partial x} \right)_x \right] = \mathcal{T} \frac{\partial^2 y}{\partial x^2}\, dx\, dz \tag{4.2.1}$$

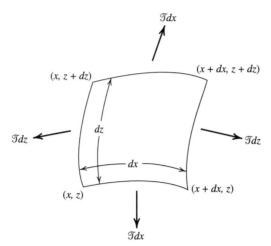

Figure 4.2.1   Elemental area of a membrane
showing the forces acting when the membrane
is displaced transversely.

and that from the pair of tensions $\mathfrak{T}\,dx$ is $\mathfrak{T}(\partial^2 y/\partial z^2)\,dx\,dz$. Equating the sum of
these two to the product of the mass $\rho_S\,dx\,dz$ of the element and its acceleration
$\partial^2 y/\partial t^2$ gives

$$\frac{\partial^2 y}{\partial x^2} + \frac{\partial^2 y}{\partial z^2} = \frac{1}{c^2}\frac{\partial^2 y}{\partial t^2} \tag{4.2.2}$$

with

$$\boxed{c^2 = \mathfrak{T}/\rho_S} \tag{4.2.3}$$

Equation (4.2.2) may be expressed more generally in the form

$$\nabla^2 y = \frac{1}{c^2}\frac{\partial^2 y}{\partial t^2} \tag{4.2.4}$$

where ∇^2 is the *Laplacian operator* (in this case two-dimensional) and (4.2.4) is the
two-dimensional wave equation.
    The form of the Laplacian depends on the choice of the coordinate system. The
Laplacian in two-dimensional cartesian coordinates,

$$\nabla^2 = \frac{\partial^2}{\partial x^2} + \frac{\partial^2}{\partial z^2} \tag{4.2.5}$$

is appropriate for rectangular membranes. For a circular membrane, polar coordinates (r, θ) are preferable and use of

$$\nabla^2 = \frac{\partial^2}{\partial r^2} + \frac{1}{r}\frac{\partial}{\partial r} + \frac{1}{r^2}\frac{\partial^2}{\partial \theta^2} \tag{4.2.6}$$

gives the appropriate wave equation,

$$\frac{\partial^2 y}{\partial r^2} + \frac{1}{r}\frac{\partial y}{\partial r} + \frac{1}{r^2}\frac{\partial^2 y}{\partial \theta^2} = \frac{1}{c^2}\frac{\partial^2 y}{\partial t^2} \tag{4.2.7}$$

Solutions to (4.2.4) will have all the properties of the waves studied previously, generalized to two dimensions. For calculating normal modes on membranes it is conventional to assume the solutions have the form

$$\mathbf{y} = \mathbf{\Psi} e^{j\omega t} \tag{4.2.8}$$

where $\mathbf{\Psi}$ is a function only of position. Substitution and identification of $k = \omega/c$ yields the *Helmholtz equation*,

$$\nabla^2 \mathbf{\Psi} + k^2 \mathbf{\Psi} = 0 \tag{4.2.9}$$

The solutions of (4.2.9) for a membrane with specified shape and boundary conditions are the normal modes of the problem.

4.3 FREE VIBRATIONS OF A RECTANGULAR MEMBRANE WITH FIXED RIM

If a stretched rectangular membrane is fixed at $x = 0, x = L_x, z = 0$, and $z = L_z$, the boundary conditions are

$$y(0, z, t) = y(L_x, z, t) = y(x, 0, t) = y(x, L_z, t) = 0 \tag{4.3.1}$$

Assuming a solution

$$\mathbf{y}(x, z, t) = \mathbf{\Psi}(x, z)e^{j\omega t} \tag{4.3.2}$$

to (4.2.4) gives

$$\frac{\partial^2 \mathbf{\Psi}}{\partial x^2} + \frac{\partial^2 \mathbf{\Psi}}{\partial z^2} + k^2 \mathbf{\Psi} = 0 \tag{4.3.3}$$

Now, apply the method of *separation of variables* by assuming that $\mathbf{\Psi}$ is the product of two functions, each dependent on only one of the dimensions,

$$\mathbf{\Psi}(x, z) = \mathbf{X}(x)\mathbf{Z}(z) \tag{4.3.4}$$

Substitution and division by $\mathbf{X}(x)\,\mathbf{Z}(z)$ gives

$$\frac{1}{X}\frac{d^2 X}{dx^2} + \frac{1}{Z}\frac{d^2 Z}{dz^2} + k^2 = 0 \tag{4.3.5}$$

Since the first term is a function only of x and the second only of z, both must be constants; otherwise the three terms cannot sum to zero for all x and z. This provides the pair of equations

$$\frac{d^2\mathbf{X}}{dx^2} + k_x^2\mathbf{X} = 0 \qquad \frac{d^2\mathbf{Z}}{dz^2} + k_z^2\mathbf{Z} = 0 \tag{4.3.6}$$

where the constants k_x and k_z are related by

$$k_x^2 + k_z^2 = k^2 \tag{4.3.7}$$

Solutions of (4.3.6) are sinusoids, so that

$$\mathbf{y}(x,z,t) = \mathbf{A}\sin\left(k_x x + \phi_x\right)\sin\left(k_z z + \phi_z\right)e^{j\omega t} \tag{4.3.8}$$

where $k_x, k_z, \phi_x,$ and ϕ_z are determined by the boundary conditions. The conditions $y(0,z,t) = 0$ and $y(x,0,t) = 0$ require $\phi_x = 0$ and $\phi_z = 0$, and the conditions $y(L_x,z,t) = 0$ and $y(x,L_z,t) = 0$ require the arguments $k_x L_x$ and $k_z L_z$ to be integral multiples of π. Thus, the standing waves on the membrane are given by

$$\mathbf{y}(x,z,t) = \mathbf{A}\sin k_x x \sin k_z z\, e^{j\omega t}$$

$$k_x = n\pi/L_x \qquad n = 1,2,3,\ldots \tag{4.3.9}$$

$$k_z = m\pi/L_z \qquad m = 1,2,3,\ldots$$

where $|\mathbf{A}|$ is the maximum displacement amplitude. These equations limit the wave numbers k_x and k_z to discrete sets of values, which in turn restrict the natural frequencies for the allowed modes to

$$f_{nm} = \omega_{nm}/2\pi = (c/2)[(n/L_x)^2 + (m/L_z)^2]^{1/2} \tag{4.3.10}$$

This is the two-dimensional extension of the comparable results for the freely vibrating fixed, fixed string. The fundamental frequency is obtained by substitution

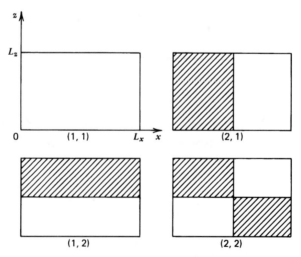

Figure 4.3.1   Schematic representation of four typical normal modes of a rectangular membrane with fixed rim. The modes are designated by the pair of integers (n, m). The hatched areas denote sections of the membrane that vibrate 180° out of phase with the unhatched areas. These areas are separated by nodal lines.

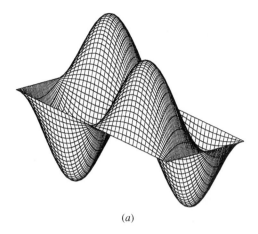

(a)

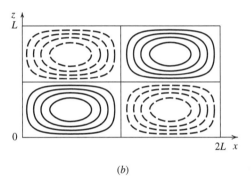

(b)

Figure 4.3.2   The displacement of a
rectangular membrane with $L_x/L_z = 2$
vibrating in a (2, 2) mode. (a) Isometric view.
(b) Contours of equal displacement. The
regions denoted by contours shown with
solid lines vibrate 180° out of phase from
those shown with dashed lines.

of $n = 1$ and $m = 1$ into (4.3.10). Overtones having $n = m$ will be harmonics of the
fundamental, while those for which $n \neq m$ may not be. Figure 4.3.1 illustrates a few
modes for a rectangular membrane. The normal modes are labelled by the ordered
pair (n, m). Figure 4.3.2 shows the displacement of a (2, 2) mode of a rectangular
membrane with fixed rim. Since the nodal lines have zero displacement, it is
possible to insert rigid supports along any of them without affecting the wave
pattern for the particular frequency involved.

4.4 FREE VIBRATIONS OF A CIRCULAR MEMBRANE WITH FIXED RIM

For a circular membrane fixed at $r = a$, the Helmholtz equation in cylindrical
coordinates

$$\frac{\partial^2 \Psi}{\partial r^2} + \frac{1}{r}\frac{\partial \Psi}{\partial r} + \frac{1}{r^2}\frac{\partial^2 \Psi}{\partial \theta^2} + k^2 \Psi = 0 \qquad (4.4.1)$$

can be solved by assuming that $\Psi(r, \theta)$ is the product of two terms, each a function of only one spatial variable,

$$\Psi = R(r)\Theta(\theta) \tag{4.4.2}$$

subject to the boundary condition

$$R(a) = 0 \tag{4.4.3}$$

In addition, Θ must be a smooth and continuous function of θ. Substitution into (4.2.9) gives

$$\Theta\frac{d^2R}{dr^2} + \frac{\Theta}{r}\frac{dR}{dr} + \frac{R}{r^2}\frac{d^2\Theta}{d\theta^2} + k^2R\Theta = 0 \tag{4.4.4}$$

where $k = \omega/c$. Multiplying this equation by $r^2/\Theta R$ and moving those terms containing r to one side of the equality sign and those containing θ to the other side results in

$$\frac{r^2}{R}\left(\frac{d^2R}{dr^2} + \frac{1}{r}\frac{dR}{dr}\right) + k^2r^2 = -\frac{1}{\Theta}\frac{d^2\Theta}{d\theta^2} \tag{4.4.5}$$

The left side of this equation, a function of r alone, cannot equal the right side, a function of θ alone, unless both functions equal the same constant. If we let this constant be m^2, then the right side becomes

$$\frac{d^2\Theta}{d\theta^2} = -m^2\Theta \tag{4.4.6}$$

which has harmonic solutions

$$\Theta(\theta) = \cos(m\theta + \gamma_m) \tag{4.4.7}$$

where the γ_m are determined by the (spatial factor in the) initial conditions. Since Θ must be smooth and single-valued, each m must be an integer. With m fixed in value, (4.4.5) is *Bessel's equation*,

$$\frac{d^2R}{dr^2} + \frac{1}{r}\frac{dR}{dr} + \left(k^2 - \frac{m^2}{r^2}\right)R = 0 \tag{4.4.8}$$

Solutions to this equation are the *Bessel functions* of *order m* of the *first kind* $J_m(kr)$ and *second kind* $Y_m(kr)$,

$$R(r) = AJ_m(kr) + BY_m(kr) \tag{4.4.9}$$

Some properties of Bessel functions are summarized in Appendixes A4 and A5. They are oscillatory functions of kr whose amplitudes diminish roughly as $1/\sqrt{kr}$. The $Y_m(kr)$ become unbounded in the limit $kr \to 0$.

While (4.4.9) is the general solution of (4.4.8), a membrane that extends across the origin must have finite displacement at $r = 0$. This requires $B = 0$ so that

$$R(r) = AJ_m(kr) \tag{4.4.10}$$

[If, however, the membrane were stretched between inner and outer rims, so that it did not span the origin, then both terms in (4.4.9) would have to be used to satisfy the two boundary conditions.]

Application of the boundary condition $\mathbf{R}(a) = 0$ requires $J_m(ka) = 0$. If the values of the argument of J_m that cause it to equal zero are denoted by j_{mn}, then k assumes the discrete values $k_{mn} = j_{mn}/a$. (See the Appendixes for values of, and formulas for, the arguments j_{mn}.)

The solutions are

$$\mathbf{y}_{mn}(r, \theta, t) = \mathbf{A}_{mn}J_m(k_{mn}r)\cos(m\theta + \gamma_{mn})e^{j\omega_{mn}t}$$

$$k_{mn}a = j_{mn}$$

(4.4.11)

and the natural frequencies are

$$f_{mn} = j_{mn}c/2\pi a \tag{4.4.12}$$

Recall that the physical motion of the (m, n)th solution is the real part of (4.4.11),

$$y_{mn}(r, \theta, t) = A_{mn}J_m(k_{mn}r)\cos(m\theta + \gamma_{mn})\cos(\omega_{mn}t + \phi_{mn}) \tag{4.4.13}$$

where $\mathbf{A}_{mn} = A_{mn}\exp(j\phi_{mn})$. The azimuthal phase angles γ_{mn} depend on the location of the initial excitation of the membrane.

Figure 4.4.1 illustrates some simpler modes of vibration for a circular membrane fixed at the rim. The integer m determines the number of *radial nodal lines* and the

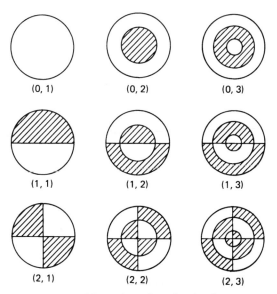

(0, 1)   (0, 2)   (0, 3)

(1, 1)   (1, 2)   (1, 3)

(2, 1)   (2, 2)   (2, 3)

Figure 4.4.1   Normal modes of a circular membrane with fixed rim. The modes are designated by the pair of integers (m, n). The hatched areas denote sections of the membrane that vibrate 180° out of phase with the unhatched areas. These areas are separated by nodal lines. The frequency of the modes increases down each column. (See Table 4.4.1.)

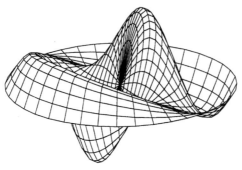

Figure 4.4.2   Isometric view of the displacement of a circular membrane with fixed rim vibrating in a $(1, 2)$ mode.

Table 4.4.1   Normal-mode frequencies of a circular membrane

$f_{01} = 1.0f_{01}$	$f_{11} = 1.593f_{01}$	$f_{21} = 2.135f_{01}$
$f_{02} = 2.295f_{01}$	$f_{12} = 2.917f_{01}$	$f_{22} = 3.500f_{01}$
$f_{03} = 3.598f_{01}$	$f_{13} = 4.230f_{01}$	$f_{23} = 4.832f_{01}$

second integer n determines the number of *nodal circles*. It should be noted that $n = 1$ is the minimum allowed value of n and corresponds to a mode of vibration in which the (only) nodal circle occurs at the fixed boundary of the membrane. Figure 4.4.2 shows the displacement of a circular membrane vibrating in a $(1, 2)$ mode.

For each m there exists a sequence of modes of increasing frequency. Table 4.4.1 lists a few of these frequencies f_{mn} expressed in terms of the fundamental frequency f_{01}. Note that none of the overtones are harmonics of the fundamental.

4.5 SYMMETRIC VIBRATIONS OF A CIRCULAR MEMBRANE WITH FIXED RIM

For many situations described by a circular membrane fixed at the rim, modes having circular symmetry are of greatest importance. Let us, therefore, confine our attention to those solutions that are independent of θ. Because $m = 0$ for these modes, we will suppress this subscript and retain only n,

$$\mathbf{y}_n(r, t) \equiv \mathbf{y}_{0n}(r, \theta, t) = \mathbf{A}_n J_0(k_n r)e^{j\omega_n t} \tag{4.5.1}$$

The natural frequencies are found from (4.4.12),

$$f_n/f_1 = j_{0n}/j_{01} \tag{4.5.2}$$

and the lowest three are given by the first column in Table 4.4.1. For all symmetric modes other than the fundamental, inner nodal circles will occur at radial distances for which $J_0(k_n r)$ vanishes.

The real part of $\mathbf{y}_n$ gives the displacement of the membrane in its nth symmetric mode, and the summation over all n gives the total displacement of the membrane when it is vibrating with circular symmetry,

$$y(r, t) = \sum_{n=1}^{\infty} A_n J_0(k_n r) \cos(\omega_n t + \phi_n) \tag{4.5.3}$$

where $A_n = |\mathbf{A}_n|$ is the displacement amplitude of the nth mode at $r = 0$.

Figure 4.4.1 shows that when the central part of the membrane is displaced up, the adjacent ring is displaced down, and vice versa. Consequently, a membrane vibrating at natural frequencies other than its fundamental produces little net displacement of the surrounding air. (For this reason, the vibrating head of a kettledrum has lower efficiencies of sound production for its overtone frequencies than for its fundamental.) One parameter for ranking the efficiency of each normal mode in producing sound is the average displacement amplitude of the mode. From (4.5.3), the average displacement amplitude $\langle \Psi_n \rangle_S$ of the nth symmetric normal mode is

$$\langle \Psi_n \rangle_S = \frac{1}{\pi a^2} \int_S A_n J_0(k_n r) \, dS = \frac{1}{\pi a^2} \int_0^a A_n J_0(k_n r) 2\pi r \, dr \tag{4.5.4}$$

$$= (2A_n / k_n a) J_1(k_n a)$$

where we have used the relationship $z J_0(z) = d[z J_1(z)]/dz$ from Appendix A4. [Note that for all modes other than the symmetric ones, the angular dependence $\cos(m\theta + \gamma_m)$ guarantees that the average displacement is zero.]

In many situations involving sources of sound with dimensions smaller than the radiated wavelength, the radiated pressure field depends primarily on the *amount* of air displaced, and not on the exact shape of the moving surface. A measure of the amount of air displaced is the *volume displacement amplitude*, defined as the surface area of the vibrating surface multiplied by the average displacement amplitude of that surface.

When vibrating in its lowest mode, the circular membrane (with fixed rim) has $k_1 a = 2.405$ and, from (4.5.4), the average displacement amplitude is

$$\langle \Psi_1 \rangle_S = (2A_1 / 2.405) J_1(2.405) = 0.432 A_1 \tag{4.5.5}$$

where A_1 is the displacement amplitude at the center. A simple piston of the same surface area and a displacement amplitude of $0.432 A_1$ will have the same volume displacement amplitude $0.432(\pi a^2) A_1$ as the membrane. If the membrane is vibrating in the mode of its first overtone, $\langle \Psi_2 \rangle_S = -0.123 A_2$. (The negative sign indicates that the average displacement is opposed to the displacement at the center.) If fundamental and first overtone have the same displacement amplitude at the center of the membrane, the fundamental would be about 3.5 times as effective as the first overtone in displacing air.

*4.6 THE DAMPED, FREELY VIBRATING MEMBRANE

Damping forces, such as those arising within the membrane from internal friction and external forces associated with the radiation of sound, cause the amplitude of each freely vibrating mode to decrease exponentially. As in Chapters 2 and 3, we will use a phenomenological approach. A generic loss term proportional to, and oppositely directed from, the velocity of the vibrating element is introduced into the wave equation. For convenience, let the proportionality constant be 2β so (4.2.4) becomes

$$\frac{\partial^2 y}{\partial t^2} + 2\beta \frac{\partial y}{\partial t} - c^2 \nabla^2 y = 0 \tag{4.6.1}$$

For calculational simplicity, assume oscillatory behavior and generalize y to be complex,

$$\mathbf{y} = \boldsymbol{\Psi} e^{j\omega t} \tag{4.6.2}$$

Since there are no applied driving forces, ω must be complex if damping is to occur. Substituting (4.6.2) into (4.6.1) and dividing out $\exp(j\omega t)$ results in the Helmholtz equation

$$\nabla^2 \boldsymbol{\Psi} + \mathbf{k}^2 \boldsymbol{\Psi} = 0 \tag{4.6.3}$$

with the complex separation constant $\mathbf{k}^2$ given by

$$\mathbf{k}^2 = (\omega/c)^2 - j2(\beta/c)(\omega/c) \tag{4.6.4}$$

In this case $\mathbf{k}$ must be real, since for membranes fixed at their edges the arguments of the normal modes must be real. Solution of (4.6.4) for ω is straightforward:

$$\begin{aligned}
\boldsymbol{\omega} &= \omega_d + j\beta \\
\omega_d &= (\omega^2 - \beta^2)^{1/2} \\
\omega &= kc
\end{aligned} \tag{4.6.5}$$

where ω is the natural angular frequency of the undamped case, ω_d the natural angular frequency of the damped case, and β the temporal absorption coefficient.

If the membrane is excited into motion and allowed to come naturally to rest, the resulting motion of the surface is a superposition of the excited normal modes, each with its own decay coefficient β and damped natural angular frequency ω_d:

$$\mathbf{y} = \sum_m \sum_n \boldsymbol{\Psi}_{mn} e^{-\beta_{mn} t} e^{j(\omega_d)_{mn} t} \tag{4.6.6}$$

Each normal mode $\boldsymbol{\Psi}_{mn}$ has a complex amplitude $\mathbf{A}_{mn}$ whose magnitude A_{mn} and phase angle ϕ_{mn} are determined by the initial conditions at $t = 0$. The decay coefficients are usually functions of frequency. Losses associated with the flexing of the membrane tend to increase with increasing frequency as the nodal pattern becomes more segmented. On the other hand, losses to the surrounding medium by the radiation of sound become smaller with more complicated modal patterns. (This reflects the observation that the volume displacement amplitudes are smaller for higher modes and zero for unsymmetric modes.) These two effects tend to offset each other, but as a general rule, higher modes damp out faster than do lower ones.

*4.7 THE KETTLEDRUM

Damping and inertial forces are two of the motion-induced forces that may act on the surface of a membrane. Another arises from the changes in pressure within a closed space behind the head of a drum or the diaphragm of a condenser microphone as the volume of the entrapped gas is altered by the motion of the membrane.

For example, the kettledrum has its head stretched tightly over the open end of a hemispherical cavity of volume V. As the head vibrates, the air in the cavity may be alternately compressed and expanded. If the phase speed of transverse waves on the membrane is considerably less than the speed of sound in air, the pressure resulting from any compression and expansion of the enclosed air is nearly uniform within the entire volume and thus depends only on the average instantaneous displacement $\langle y \rangle_S$. The incremental change in volume of the enclosed air is $dV = \pi a^2 \langle y \rangle_S$, where a is the radius of the drumhead. If the equilibrium volume inside the vessel is V_0 and the equilibrium pressure is $\mathcal{P}_0$, then for adiabatic changes in volume the new pressures $\mathcal{P}$ and volumes V are related by

$$\mathcal{P} V^\gamma = \mathcal{P}_0 V_0^\gamma \tag{4.7.1}$$

where γ is the ratio of the heat capacity of the entrapped air at constant pressure to that at constant volume (see Appendix A9). Differentiation shows that the excess pressure $d\mathcal{P}$ inside the kettle will be

$$d\mathcal{P} \approx -(\gamma \mathcal{P}_0 / V_0)\, dV = -\gamma (\mathcal{P}_0 / V_0) \pi a^2 \langle y \rangle_S \tag{4.7.2}$$

This generates an additional force $d\mathcal{P} r\, dr\, d\theta$ on each incremental area $r\, dr\, d\theta$ of the membrane. From the discussion of the previous section, the normal modes affected by this force must be just the symmetric ones. While these are relatively unimportant for the musical properties of the kettledrum, the effect of this induced force has interest in other applications, and so we shall pursue the analysis further. Including this force in the discussion of Section 4.2 and writing $\mathbf{y}$ as (4.6.2) with $\mathbf{\Psi}$ real leads to

$$\nabla^2 \Psi + k^2 \Psi = (\gamma \mathcal{P}_0 \pi a^2 / \rho_S c^2 V_0) \langle \Psi \rangle_S \tag{4.7.3}$$

for each symmetric normal mode Ψ. The subscripts 0 and n have been suppressed for economy of expression. Because it is proportional to displacement, the right side is a spring-like term; the allowed wave numbers k will, therefore, be increased. The homogeneous solutions to (4.7.3) will still be Bessel functions, but they may not have zeros at the rim. The boundary condition requires the presence of a particular solution, which in this case is a constant. Adding this to the homogeneous solution and satisfying the boundary condition gives

$$\Psi = A[J_0(kr) - J_0(ka)] \tag{4.7.4}$$

as a solution for each symmetric normal mode. The right side of (4.7.3) can now be evaluated with the help of

$$\pi a^2 \langle \Psi \rangle_S = \int_0^a \Psi 2\pi r\, dr = 2\pi A[(r/k)J_1(kr) - (r^2/2)J_0(ka)]\Big|_0^a \tag{4.7.5}$$

$$= \pi a^2 A \left[2J_1(ka)/ka - J_0(ka) \right] = \pi a^2 A\, J_2(ka)$$

Substitution shows that (4.7.4) is a solution of (4.7.3) if

$$J_0(ka) = -B J_2(ka)/(ka)^2$$
$$B = \pi a^4 \gamma \mathcal{P}_0 / \mathcal{T} V_0 \tag{4.7.6}$$

Solving (4.7.6) for ka determines the natural frequencies. The nondimensional parameter B measures the relative importance of the restoring force of the air in the vessel to the tension in the membrane.

Since the frequencies of only the modes Ψ_{0n} are affected by the pressure fluctuations within the vessel, the area πa^2 of the drumhead and the volume V_0 of the vessel are parameters that can be varied to alter the natural frequency distribution of the kettledrum. Variation of B affects the relative values of the f_{0n} frequencies. Altering a and V_0 such that a^4/V_0 remains constant will vary the nonsymmetric overtones f_{mn} ($m \neq 0$) with respect to the symmetric ones.

If damping is now considered, consistent with (4.6.5) each standing wave will have its angular frequency shifted from the value ω_{mn} for undamped motion to that with damping $(\omega_d)_{mn}$, and each standing wave will decay with its own decay constant β_{mn}. The form of each standing wave will be given by (4.6.6), with the symmetric Ψ_{0n} modes given by (4.7.4).

This development has not taken into consideration any inertance effects of the medium on the membrane. As the membrane vibrates, it radiates acoustic energy but also accelerates the surrounding medium locally, as if it were storing and recovering energy from the mass of the adjacent medium. This inertance is quite important in affecting the natural frequencies of the excited modes. In practice, the significant normal modes of the kettledrum are the lowest four or five of the asymmetric $(m, 1)$ family (beginning with $m = 1$). The inertance contributes an additional effective mass to the membrane, thereby lowering the frequency of the normal mode. The effect is greater for the lower modes, decreasing as the segmentations of the normal mode patterns increase. The natural frequencies are lowered with the lowest ones being most affected. The result brings the relative values close to 2:3:4:5 and this accounts for the distinctive timbre and clear pitch associated with the kettledrum. A quantitative treatment of inertance goes beyond our present purpose, but will be considered further starting in Chapter 7.

*4.8 FORCED VIBRATION OF A MEMBRANE

Introduction of a forcing function into the equation of motion is similarly straightforward. The units of each term in (4.6.1) are those of acceleration, so the forcing function must have the same. A suitable combination of terms is pressure divided by surface density. This gives the generalization of (4.6.1) that includes an external driving agent,

$$\frac{\partial^2 y}{\partial t^2} + 2\beta \frac{\partial y}{\partial t} - c^2 \nabla^2 y = \frac{P}{\rho_S} f(t) \tag{4.8.1}$$

where $f(t)$ is a dimensionless function of time. The pressure P can be a constant or any appropriate function of space, including a delta function. The function of time can be oscillatory, a delta function, or whatever is necessary to represent the temporal behavior of the applied force. For example, if both P and $f(t)$ were delta functions, this would approximate the stroke of a drumstick at a specific point on the membrane.

Here, we concentrate on applied oscillatory forces. Let $f(t) = \exp(j\omega t)$ and assume that the steady-state solution for $\mathbf{y}$ has the form

$$\mathbf{y} = \Psi e^{j\omega t} \tag{4.8.2}$$

with the angular frequency ω real. (In the case of forced motion, where there is a steady-state solution, ω cannot have an imaginary component.) Substitution into (4.8.1) and cancellation of the exponentials gives

$$(-\omega^2 + j2\beta\omega - c^2\nabla^2)\Psi = P/\rho_S \tag{4.8.3}$$

The solution of (4.8.3) consists of the sum of the solution to the homogeneous equation and a solution to the particular equation. The homogeneous equation can be written as

$$\nabla^2 \mathbf{\Psi} + \mathbf{k}^2 \mathbf{\Psi} = 0$$

$$\mathbf{k} = k - j\alpha$$

$$k = (\omega/c)[1 + (\beta/\omega)^2]^{1/2} \tag{4.8.4}$$

$$\alpha/k = (\beta/\omega)/[1 + (\beta/\omega)^2] \approx \beta/\omega$$

The top equation is the familiar Helmholtz equation, but with complex $\mathbf{k}$ rather than real k. This means that whatever functions solve the Helmholtz equation for lossless conditions are still solutions, but with k replaced with $\mathbf{k}$. For the cases we have studied (rectangular and circular membranes), the functions now have complex arguments and cannot satisfy the boundary condition of a fixed rim without the help of the particular solution.

For the case of uniform pressure P distributed over the circular membrane with fixed rim at $r = a$, the azimuthal symmetry of the problem restricts the homogeneous solution $\mathbf{\Psi}_h$ to the zeroth order Bessel function $J_0(\mathbf{k}r)$. The appropriate particular solution $\mathbf{\Psi}_p$ to (4.8.3) is a constant,

$$\mathbf{\Psi}_p = -(P/\rho_S)/(\mathbf{k}c)^2 \tag{4.8.5}$$

Adding this to the homogeneous solution $\mathbf{\Psi}_h$ and requiring that the sum vanish when evaluated at the rim results in the desired solution,

$$\mathbf{\Psi} = (P/\mathcal{T}\mathbf{k}^2)\big[J_0(\mathbf{k}r)/J_0(\mathbf{k}a) - 1\big] \tag{4.8.6}$$

The tension $\mathcal{T}$ has replaced $\rho_S c^2$. The values for $\mathbf{k} = k - j\alpha$ are determined from (4.8.4). Inspection of (4.8.6) shows that the amplitude of the displacement is directly proportional to that of the driving force and inversely proportional to the tension $\mathcal{T}$. The dependence on frequency of the amplitude of vibration at any location is given by the relatively complicated expression within the square bracket. When the driving frequency matches any natural frequency [found from $J_0(\mathbf{k}a) = 0$], then $J_0(\mathbf{k}a)$ has very small magnitude and $|\mathbf{\Psi}|$ may be very large, depending on the damping.

*4.9 THE DIAPHRAGM OF A CONDENSER MICROPHONE

An important case of a driven membrane is the circular diaphragm of a condenser microphone. The incident sound wave, acting on a tightly stretched metallic membrane placed above a metal plate, produces a nearly uniform driving force. As the membrane is displaced, the electrical capacitance between the membrane and the adjacent metal plate is changed. This generates an output voltage that is (for small motion) a linear function of the averaged displacement amplitude of the membrane,

$$\langle \mathbf{\Psi} \rangle_S = \frac{1}{\pi a^2}\frac{P}{\mathcal{T}}\frac{1}{\mathbf{k}^2}\int_0^a \left(\frac{J_0(\mathbf{k}r)}{J_0(\mathbf{k}a)} - 1\right)2\pi r\, dr = \frac{Pa^2}{\mathcal{T}}\frac{1}{(\mathbf{k}a)^2}\frac{J_2(\mathbf{k}a)}{J_0(\mathbf{k}a)} \tag{4.9.1}$$

If the frequency is below the region of the lowest resonance, $\mathbf{k}$ can be replaced with the wave number k and use of the small-argument approximations for the Bessel functions gives

$$\langle \mathbf{\Psi} \rangle_S \approx \tfrac{1}{8}(Pa^2/\mathcal{T})[1 + (ka)^2/6] \tag{4.9.2}$$

Thus, $\langle \mathbf{\Psi} \rangle_S$ is nearly constant for $ka < 1$, or for frequencies

$$f < c/2\pi a = (\mathcal{T}/\rho_S)^{1/2}/2\pi a \tag{4.9.3}$$

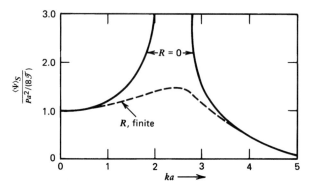

Figure 4.9.1   Average displacement $\langle \psi \rangle_S$ of a driven circular membrane with and without resistance.

Below this frequency limit, $\langle \Psi \rangle_S$ resembles the displacement amplitude of a stiffness-controlled harmonic oscillator. This upper frequency limit may be increased by increasing the tension or by decreasing either the radius or the surface density of the membrane. However, an increase in $\mathcal{T}$ or a decrease in a reduces $\langle \Psi \rangle_S$ and thus the voltage output of the microphone.

With sufficient damping, the response at the first resonance, $k_1 a = 2.405$, is reduced considerably and the region of fairly uniform response can be extended up to, and somewhat beyond, the first resonance. In the immediate vicinity of resonance, the term $J_0(\mathbf{k}a)$ in the denominator of (4.9.1) can be expanded in a Taylor's series about $k_1 a$,

$$J_0(\mathbf{k}a) = -J_1(k_1 a)(\mathbf{k}a - k_1 a) + \cdots \tag{4.9.4}$$

Writing $\mathbf{k} = k(1 - j\beta/\omega)$ from (4.8.4), substituting the quality factor $Q = \omega_1/2\beta$ from (1.10.7) for the resonance at ω_1, and restricting ω to values close to ω_1 casts (4.9.1) into the form

$$|\langle \Psi \rangle_S| = \frac{2Pa^2}{\mathcal{T}} \frac{1}{(ka)^3} \frac{J_2(j_{01})}{J_1(j_{01})} \frac{1}{[(\omega/\omega_1 - \omega_1/\omega)^2 + 1/Q^2]^{1/2}} \tag{4.9.5}$$

This displays the same behavior in the vicinity of its resonance as does a damped harmonic oscillator. The resonance peak and bandwidth are controlled by the quality factor in the same way.

Response curves showing the normalized average displacement amplitudes of a driven membrane with and without losses are given in Fig. 4.9.1. Note that (4.9.1) indicates minimal response at the frequencies for which $J_2(ka) = 0$. From (4.8.6), at frequencies for which $ka > 3.83$, a nodal circle appears within the rim of the membrane. With increasing frequency the radius of this circle decreases. The displacements within this circle are out of phase with those between it and the rim. As the nodal circle continues to shrink there is increasing cancellation leading to nearly zero response when $ka \approx 5.136$.

*4.10 NORMAL MODES OF MEMBRANES

Orthogonality was developed in Section 2.13 for a set of one-dimensional normal modes describing the vibration of a string for certain simple boundary conditions. It is appropriate here to extend that discussion to a treatment of normal modes on a two-dimensional surface. For the free vibration of each membrane we have studied, a set of normal modes Ψ that satisfy the Helmholtz equation and the boundary conditions has been obtained. For each normal mode Ψ_{mn} of the set, there is an associated separation constant k_{mn}^2 determined by the boundary conditions. If the Helmholtz equation for each of two normal modes Ψ_{mn} and $\Psi_{m'n'}$ is multiplied by the other of the pair and one equation subtracted from the other, the result is

$$\Psi_{m'n'}\nabla^2\Psi_{mn} - \Psi_{mn}\nabla^2\Psi_{m'n'} + (k_{mn}^2 - k_{m'n'}^2)\Psi_{mn}\Psi_{m'n'} = 0 \tag{4.10.1}$$

Integration over the surface S of the membrane gives

$$(k_{mn}^2 - k_{m'n'}^2)\int_S \Psi_{mn}\Psi_{m'n'}\,dS = \int_S \nabla\cdot(\Psi_{mn}\nabla\Psi_{m'n'} - \Psi_{m'n'}\nabla\Psi_{mn})\,dS \tag{4.10.2}$$

where use has been made of the identity in Appendix A8. Application of Gauss's theorem in two dimensions now gives the desired result,

$$(k_{mn}^2 - k_{m'n'}^2)\int_S \Psi_{mn}\Psi_{m'n'}\,dS = \int_{rim}\left[\Psi_{mn}(\hat{n}\cdot\nabla\Psi_{m'n'}) - \Psi_{m'n'}(\hat{n}\cdot\nabla\Psi_{mn})\right]dl \tag{4.10.3}$$

where the line integral is over the perimeter (rim) of the surface S and $\hat{n}$ is the unit normal in the plane of the membrane directed outward at each point on the rim. This equation is the generalization of the right side of (2.13.7) to a two-dimensional situation. There are two cases of interest to us here: (1) for a free rim, the gradient of Ψ at the rim is at right angles to the normal; and (2) for a fixed rim, Ψ is zero on the rim. In both cases, the right side vanishes and the normal modes form an orthogonal set.

One complication arises when there are two or more normal modes with the same natural frequency. When this happens, the separation constants are the same and the left side of (4.10.3) vanishes identically whether or not the degenerate modes are orthogonal. This means that if the membrane is excited into motion at this frequency and then left to vibrate freely, the shape of the resulting standing wave depends on the details of the excitation. This can result in standing waves of substantially different shapes depending on the relative phases and amplitudes of the degenerate normal modes. While these cases may need a little special attention, they present few problems. Mathematically, if the modes are not orthogonal, it is possible to choose two combinations of the pair that will be.

Solving for the motion of the membrane after excitation by some initial distribution of displacement and velocity proceeds just as for the string. The initial conditions at $t = 0$ are written as sums of the normal modes with unspecified amplitudes and phases. These equations are then multiplied by each of the normal modes and integrated over the surface, orthogonality applied, and the resulting reduced set of integrals evaluated to obtain the amplitudes and phases.

(a) The Rectangular Membrane with Fixed Rim

Since the rim is fixed, the normal modes for the rectangular membrane

$$\Psi_{nm}(x, z) = A_{nm}\sin k_{xn}x \sin k_{zm}z \tag{4.10.4}$$

form an orthogonal set. The separation constant in the Helmholtz equation (4.3.3) for each of these normal modes is

$$k_{nm}^2 = k_{xn}^2 + k_{zm}^2 \tag{4.10.5}$$

Thus,

$$\int_0^{L_z}\int_0^{L_x}\Psi_{nm}\Psi_{n'm'}\,dx\,dz = \frac{4A_{nm}^2}{L_xL_z}\delta_{n'n}\delta_{m'm} \tag{4.10.6}$$

If the membrane is struck at some position $\vec{r}_0$ with an impulse at $t = 0$, then the initial transverse speed of the membrane can be approximated by $v(\vec{r}, 0) = \mathcal{V}\delta(\vec{r} - \vec{r}_0)$. The two-dimensional delta function is given by

$$\int_{S_0} \delta(\vec{r} - \vec{r}_0)\, dS = \begin{cases} 1 & \vec{r}_0 \in S_0 \\ 0 & \vec{r}_0 \notin S_0 \end{cases} \tag{4.10.7}$$

where $\delta(\vec{r} - \vec{r}_0)$ has dimension m^{-2} and $\mathcal{V}$ has dimension $(\mathrm{m/s})\mathrm{m}^2$. If the point is at (x_0, z_0), then it can be seen (Problem 4.10.2) that $\delta(\vec{r} - \vec{r}_0)$ can be represented as a product of one-dimensional delta functions

$$\delta(\vec{r} - \vec{r}_0) = \delta(x - x_0)\,\delta(z - z_0) \tag{4.10.8}$$

The real standing wave associated with the (n, m) mode can be written as

$$y_{nm}(x, z, t) = A_{nm} \sin k_{xn}x \sin k_{zm}z \sin(\omega_{nm}t + \phi_{nm}) \tag{4.10.9}$$

At $t = 0$ the membrane is at rest, $y(x, z, 0) = 0$. This can be satisfied by choosing $\phi_{nm} = 0$ for all modes. With the ϕ's established, the particle speed when $t = 0$ provides that

$$\sum_{n,m} \omega_{nm} A_{nm} \sin k_{xn}x \sin k_{zm}z = \mathcal{V}\delta(x - x_0)\,\delta(z - z_0) \tag{4.10.10}$$

Since the normal modes are orthogonal, use of (4.10.6) gives the values of each A_{nm}, and we have

$$y(x, z, t) = \frac{4\mathcal{V}}{L_x L_z} \sum_{n,m} \frac{1}{\omega_{nm}} \sin k_{xn}x_0 \sin k_{zm}z_0 \sin k_{xn}x \sin k_{zm}z \sin \omega_{nm}t \tag{4.10.11}$$

Using the delta function introduces some convergence problems in this expression. Practically, time should be restricted to finite values, $t \geq t_0$, and the summations should be truncated at realistic values of the indices, $n < N$ and $m < M$, where t_0, N, and M are based on the true duration of the impact and the finite area of impact of the drumstick. Suitable decay can also be included as discussed earlier by introducing an exponential decay factor $\exp(-\beta_{nm}t)$ and shifting the natural angular frequency according to (4.6.5) for each normal mode.

(b) The Circular Membrane with Fixed Rim

Conceptually, analysis is exactly the same as for the rectangular membrane. The two-dimensional delta function must be expressed in terms of one-dimensional delta functions in the coordinates r and θ,

$$\delta(\vec{r} - \vec{r}_0) = \frac{1}{r}\delta(r - r_0)\,\delta(\theta - \theta_0) \tag{4.10.12}$$

(see Problem 4.10.6). We can orient the axes so that $\theta = 0$ corresponds to the azimuthal direction of the blow (which now requires $\theta_0 = 0$) and assert that the normal modes must be maximized in this direction. This requires that all $\gamma_{mn} = 0$ in (4.4.13). For a strike at $t = 0$ on a stationary membrane, we must have $\phi_{mn} = -\pi/2$. The individual standing waves can now be extracted,

$$y_{mn}(r, \theta, t) = A_{mn} J_m(k_{mn}r) \cos m\theta \sin \omega_{mn}t \tag{4.10.13}$$

Solution proceeds as before with the help of

$$\int_0^a \int_0^{2\pi} \left[J_m(k_{mn}r) \cos m\theta \right]^2 r\, dr\, d\theta = \begin{cases} \pi a^2 \left[J_m'(k_{mn}a) \right]^2 & m = 0 \\[2ex] \dfrac{\pi a^2}{2} \left[J_m'(k_{mn}a) \right]^2 & m > 0 \end{cases} \tag{4.10.14}$$

where $k_{mn}a = j_{mn}$. The resulting standing wave is

$$y(r, \theta, t) = \frac{\mathcal{V}}{\pi a^2} \sum_{m,n} \frac{\varepsilon_m}{\omega_{mn}} \frac{J_m(k_{mn}r_0)}{\left[J_m'(k_{mn}a) \right]^2} J_m(k_{mn}r) \cos m\theta \sin \omega_{mn}t \tag{4.10.15}$$

for a strike at the point $(r_0, 0)$ at time $t = 0$. The quantity ε_m is 1 for $m = 0$ and 2 for all other m. As before, times should be restricted to $t \geq t_0$ and the summation truncated at appropriate N and M. Notice that if the membrane is struck exactly at the center, then only terms with $m = 0$ contribute.

*4.11 VIBRATION OF THIN PLATES

There is an essential difference between the vibration of a membrane and of a thin plate. In a membrane, the restoring force arises entirely from the tension applied to the membrane, whereas in a thin plate the restoring force results from the stiffness of the diaphragm. This same difference exists between the transverse restoring forces in strings and bars. Analysis of the plate will be limited to the *symmetric vibrations of a uniform circular diaphragm*. A rigorous development of the equation of motion lies beyond our interests. The equation is

$$\frac{\partial^2 y}{\partial t^2} = -\frac{\kappa^2 Y}{\rho(1 - \sigma^2)} \nabla^2(\nabla^2 y) \tag{4.11.1}$$

where ρ is the *volume density* of the material, σ the *Poisson's ratio*, Y the *Young's modulus*, and κ the *radius of gyration* given by $\kappa = d/\sqrt{12}$, where d is the plate thickness.

In partial explanation, since the restoring force acting on a plate depends on its elastic response to bending, the coefficient of the right term in (4.11.1) should resemble that for the transverse vibration of a bar (3.9.4), $\kappa^2 Y/\rho$. However, like a bar, a sheet curls transversely when it is bent lengthwise, but the lateral extent of the sheet hampers the curling. Thus, there should be a slight decrease in the resultant strain of the sheet for the impressed bending stress and therefore a slight increase in the effective stiffness of the sheet. Analysis provides the factor $1/(1 - \sigma^2)$. Values of Poisson's ratio for various materials are given in Appendix A10. Note that $\sigma \sim 0.3$ for most materials.

Assume periodic vibration,

$$\mathbf{y} = \mathbf{\Psi} e^{j\omega t} \tag{4.11.2}$$

where, for circular symmetry, $\mathbf{\Psi}$ is a function only of r. Substitution into (4.11.1) yields

$$\nabla^2(\nabla^2\mathbf{\Psi}) - g^4\mathbf{\Psi} = 0$$
$$g^4 = \omega^2 \rho(1 - \sigma^2)/\kappa^2 Y \tag{4.11.3}$$

Now, direct substitution shows that (4.11.3) can be satisfied by

$$\nabla^2\mathbf{\Psi} \pm g^2\mathbf{\Psi} = 0 \tag{4.11.4}$$

In polar coordinates and with circular symmetry, (4.11.4) with the + sign is satisfied by $J_0(gr)$ and $Y_0(gr)$ and the solutions for the − sign are the Bessel functions of imaginary argument, $J_0(jgr) \equiv I_0(gr)$ and $Y_0(jgr)$. As before, the solutions involving Y_0 can be discarded since they have singularities at $r = 0$, so we have

$$\Psi = A J_0(gr) + B I_0(gr) \tag{4.11.5}$$

Some properties and tables of values of the *modified Bessel functions of the first kind* I_m are given in Appendixes A4 and A5.

To evaluate the constants **A** and **B** we must know how the diaphragm is supported. The most common type of support is rigid clamping of the diaphragm around its circumference at $r = a$. This is equivalent to

$$\Psi = 0 \quad \text{and} \quad \frac{\partial \Psi}{\partial r} = 0 \quad \text{at } r = a \tag{4.11.6}$$

These conditions give

$$\begin{aligned} A J_0(ga) &= -B I_0(ga) \\ A J_1(ga) &= B I_1(ga) \end{aligned} \tag{4.11.7}$$

and dividing one by the other gives the transcendental equation for the allowed values of g,

$$J_0(ga)/J_1(ga) = -I_0(ga)/I_1(ga) \tag{4.11.8}$$

Since both I_0 and I_1 are positive for all values of ga, solutions occur only when J_0 and J_1 are of opposite sign. The tables of Bessel functions show that this equation is satisfied by $g_n a = 3.20,\ 6.30,\ 9.44,\ 12.57, \ldots \approx n\pi$ with $n = 1,\ 2,\ 3, \ldots$. The approximation improves with increasing n.

Solving (4.11.3) for the lowest natural frequency f_1 gives

$$f_1 = \frac{g_1^2}{2\pi a^2} \frac{d}{\sqrt{12}} \left(\frac{Y}{\rho(1 - \sigma^2)} \right)^{1/2} = 0.47 \frac{d}{a^2} \left(\frac{Y}{\rho(1 - \sigma^2)} \right)^{1/2} \tag{4.11.9}$$

The frequencies of the other symmetric modes are not harmonics of the fundamental: $f_2/f_1 = (g_2/g_1)^2 = 3.88$, $f_3/f_1 = 8.70$, and so forth. The natural frequencies are spread much farther apart than those of the circular membrane.

The displacement of a thin circular plate vibrating in its fundamental mode is

$$y_1 = A_1 \left[J_0(3.2r/a) + 0.0555 I_0(3.2r/a) \right] \cos(\omega_1 t + \phi_1) \tag{4.11.10}$$

where the ratio of coefficients is obtained from (4.11.7). Note that the amplitude at the center $|y_1(0)|$ is not A_1 but $1.0555 A_1$. Comparing the displacement of the thin circular plate vibrating in its fundamental mode with that of a membrane vibrating at its fundamental shows that the relative displacement of the plate near its edge is much smaller than that of the membrane. Consequently, we should expect the ratio of its average amplitude to that at the center to be less than the same ratio for the membrane. The average displacement amplitude is $\langle \Psi_1 \rangle_S = 0.326 A_1$, or

$$\langle \Psi_1 \rangle_S = 0.309 |y_1(0)| \tag{4.11.11}$$

This is smaller by a factor of $(0.432/0.309) = 1.40$ than the averaged displacement for the circular fixed membrane (4.5.5) for the same amplitude at the center.

Treatments of loaded and driven plates are analogous to those for membranes, and the response curves for a uniform driving force are similar to those in Fig. 4.9.1, with large amplitudes at the fundamental resonance frequency unless there is considerable damping. Condenser microphones may be constructed with a thin circular plate instead of a stretched membrane for greater strength. However, the reduced sensitivity usually restricts such microphones to high-intensity applications where strength is necessary.

The most important utilization of the thin plate is in the diaphragms of ordinary telephone microphones and receivers. Although the responses of these devices are not uniform over a wide range of frequencies, they give adequate intelligibility and are simple and rugged. Another application is in sonar transducers used for producing sounds in water at frequencies below 1 kHz; sound is generated by the motion of relatively thin circular steel plates driven by alternations in the magnetic field of an adjacent electromagnet.

PROBLEMS

Except when otherwise noted, all membranes should be assumed fixed at their rims.

4.3.1. A square membrane of width a vibrates at its fundamental frequency with an amplitude A at its center. (*a*) Derive a general expression for its average displacement amplitude. (*b*) Derive a general expression for locating points on the membrane having an amplitude of $0.5A$. (*c*) Compute and plot a few points given by the equation derived in part (*b*). Do they form a circle?

4.3.2. A rectangular membrane has width a and length b. If $b = 2a$, compute the ratio of each of the first four overtone frequencies relative to the fundamental frequency.

4.3.3. A square membrane with sides of length L, uniform surface density ρ_S, and uniform tension $\mathcal{T}$ is fixed on three sides and free on the other. (*a*) Find the frequency of the fundamental mode. (*b*) Write a general expression for the natural frequencies and one for the normal modes. (*c*) Sketch the nodal patterns for the three normal modes with the lowest natural frequencies.

4.3.4. A square membrane of sides L and phase speed c is fixed on two of its opposed sides ($x = 0$ and $x = L$) and free on the other two ($z = 0$ and $z = L$). (*a*) Write the equation for the displacement of the membrane valid for all normal modes. (*b*) What are the frequencies for the five lowest modes? (*c*) Sketch the nodal patterns for these five modes.

4.4.1. Show that the total energy of a circular membrane when vibrating in its fundamental mode is given by $0.135(\pi a^2)\rho_S(\omega A_1)^2$, where a is its radius, ρ_S the area density, ω the angular frequency of vibration, and A_1 the amplitude at its center.

4.4.2. Although it may be hard to do physically, it is not hard to imagine a circular membrane with a free rim. (*a*) Write the general expression for the normal modes. (*b*) Sketch the nodal patterns for the three normal modes with the lowest natural frequencies. (*c*) Find the frequencies of these three normal modes in terms of the tension and surface density.

4.4.3. The maximum tensile stress that may be applied to aluminum is 2×10^8 Pa and to steel is 10^9 Pa. What is the maximum fundamental frequency (*a*) of a stretched aluminum membrane of 0.01 m radius and (*b*) of a steel membrane of equal radius? (For thin membranes these frequencies are independent of thickness.)

4.4.4. A circular membrane of 0.25 m radius has an area density of 1.0 kg/m^2 and is stretched to a tension of 25,000 N/m. (*a*) Compute the four lowest frequencies of free vibration. (*b*) For each of these frequencies locate any nodal circles.

4.5.1. A circular membrane of 1 cm radius and 0.2 kg/m^2 area density is stretched to a linear tension of 4000 N/m. When vibrating in its fundamental mode, the amplitude at the center is observed to be 0.01 cm. (*a*) What is its fundamental frequency? (*b*) What is the maximum volume of air displaced by the membrane?

4.5.2. At what fraction of the radius of a circular membrane does the nodal circle of the second symmetric mode occur?

4.5.3. A steel membrane of 0.02 m radius and 0.0001 m thickness is stretched to a tension of 20,000 N/m. (*a*) For circularly symmetric vibration, what is the frequency of the second overtone mode? (*b*) What are the radii of the two nodal circles when the membrane is vibrating at the above frequency? (*c*) When the membrane is vibrating at the above frequency, the displacement amplitude at the center is observed to be 0.0001 m. What is the average displacement amplitude?

4.5.4C. Plot the displacement as a function of radius and angle for the modes of a circular membrane shown in Fig. 4.4.1.

4.5.5C. Plot J_0 and J_1 for argument $0 < x < 10$ and compare to the small- and large-argument approximations. Comment on the range of x for which each approximation is good.

4.6.1. A circular membrane is acted on uniformly over its surface by a damping force per unit area of $-R(\partial y/\partial t)$. Introduce this term into (4.2.7) in a manner consistent with the dimensions, and solve the resulting equation to show that the amplitudes of the resulting free vibrations are damped exponentially as $\exp(-Rt/2\rho_S)$.

4.7.1. The circular membrane of a kettledrum has a radius of 0.25 m, an area density of 1.0 kg/m^2, and is stretched to a tension of 10,000 N/m. (*a*) What is its fundamental frequency without the kettle? (*b*) What is its fundamental frequency if the kettle is a hemispherical bowl of 0.25 m radius? Assume the kettle is filled with air at a pressure of 10^5 Pa, and the ratio of heat capacities is 1.4.

4.7.2. For the kettledrum, calculate the effect of B in (4.7.6) in changing the natural resonance frequencies associated with the lowest three symmetric normal modes. Calculate the values of ka for $B = 0, 1, 2, 5, 10$. Which frequency is the most changed?

4.7.3. (*a*) Find the values of ka for the lowest five members of the $(m, 1)$ family (beginning with $m = 1$) of the freely vibrating circular membrane. (Because these modes have no volume displacement amplitudes, they can represent those for a kettledrum.) (*b*) Assuming that f_{51} is not changed, calculate the fractional reduction in each of the lower frequencies to bring the series into the ratios 2:3:4:5:6. How uniform is the shifting of frequencies?

4.8.1. Find the resonance frequencies of a circular membrane with a free rim (but still under tension) driven by a uniform pressure $P \exp(j\omega t)$.

4.8.2. (*a*) Compute and plot the shape of the circular membrane when driven at one-half its fundamental frequency. (*b*) Similarly, compute and plot the shape of the membrane when driven at twice its fundamental frequency.

4.8.3. An undamped circular membrane of 0.02 m radius, 1.5 kg/m^2 area density, and 950 N/m tension is driven by a pressure of $6000 \cos(\omega t)$ Pa. (*a*) Compute and plot the amplitude of the displacement at the center as a function of frequency from 0 to 1 kHz. (*b*) Compute and plot the shape of the membrane when driven at 400 Hz. (*c*) Repeat part (*b*) for 1 kHz.

4.8.4C. For the forced vibration of a circular membrane, (*a*) plot the shape of the membrane for $ka = 1$ to $ka = 8$ in steps of 1.0. (*b*) Plot the displacement amplitude of the center for the same range of ka.

4.9.1. Perform the integration of (4.8.6) to obtain (4.9.1). *Hints:* Make a change of variable to $z = \mathbf{k}r$. Use the formula in the Appendix for the differentiation of $[zJ_1(z)]$ to determine the integral of $[zJ_0(z)]$. Relate J_1 to the appropriate combination of J_2 and J_0.

4.9.2. The diaphragm of a condenser microphone is a circular sheet of aluminum of 0.03 m diameter and 0.00002 m thickness. It may be stretched to a maximum tensile stress of 2×10^8 Pa. (*a*) What is the maximum tension (N/m)? (*b*) What will be its fundamental frequency when stretched to this tension? (*c*) What will be the displacement amplitude at its center when acted on by a sound wave of 500 Hz having a pressure amplitude of 2.0 Pa? (*d*) What will be the average displacement amplitude under these conditions?

4.9.3. If the volume of air trapped behind the diaphragm of the condenser microphone of Problem 4.9.2 is 3×10^{-7} m^3, by what percentage will its fundamental frequency be raised? Assume $\mathcal{P}_0 = 10^5$ Pa and $\gamma = 1.4$.

4.9.4. (*a*) Obtain the Taylor's expansion (4.9.4) with the help of the Appendix. (*b*) Show that (4.9.1) can be approximated by (4.9.5) for angular frequencies close to the lowest resonance. *Hint:* Show that $[(\omega/\omega_1) - (\omega_1/\omega)] \approx 2\Delta\omega/\omega_1$ for $\Delta\omega/\omega_1 \ll 1$ and use these relationships to simplify the angular frequency terms. (*c*) Compare the square root with that in Problem 1.10.1 to show that near resonance the average diaphragm displacement behaves like the displacement of a damped oscillator with the same resonance and damping.

4.9.5C. (*a*) Plot the average displacement of the driven circular membrane as a function of $\log(ka)$ for $0.01 < ka < 10$ using the exact solution and the low-frequency approximation. (*b*) Find the ratio of the frequency for which the low-frequency approximation is within 10% of the exact value to the frequency for which $J_0(ka) = 0$.

4.10.1. Show by direct integration of (4.10.4) over the surface area that the normal modes of the rectangular membrane form an orthogonal set. Find the values of $\mathbf{A}_{nm}$ that would make them an orthonormal set.

4.10.2. Show by direct application of (4.10.7) that $\delta(\vec{r} - \vec{r}_0) = \delta(x - x_0)\delta(z - z_0)$ is an appropriate representation, where $\vec{r}_0$ is directed from $(0,0)$ to (x_0, z_0).

4.10.3. A rectangular membrane has dimensions such that the (3, 1) and (1, 2) modes are degenerate. (*a*) What is the ratio of lengths L_z/L_x? (*b*) If the membrane is set into motion at the degenerate frequency f_{31} at the point $(L_x/2, L_z/2)$, which of the pair is excited? (*c*) Repeat (*b*) for locations of $(L_x/2, L_z/3)$, $(L_x/3, L_z/2)$, and $(L_x/3, L_z/3)$. (*d*) Find three other degenerate pairs of frequencies as multiples of f_{31}.

4.10.4. Verify (4.10.6) for $\mathbf{y}_{nm}(x, z, t) = \sin(n\pi x/L_x)\sin(m\pi z/L_z)\exp(j\omega_{nm}t)$.

4.10.5. Verify (4.10.11) by writing the displacement as a summation of standing waves and applying the initial condition that the membrane is at rest and given an initial impulse so that its transverse speed is described by $\partial y/\partial t = \mathcal{V}\delta(x - x_0)\delta(z - z_0)$ at $t = 0$, where $0 < x_0 < L_x$ and $0 < z_0 < L_z$.

4.10.6. Show by direct application of (4.10.7) that $\delta(\vec{r} - \vec{r}_0) = (1/r)\delta(r - r_0)\delta(\theta - \theta_0)$ is an appropriate representation of the two-dimensional delta function as a product of one-dimensional delta functions in polar coordinates, where the vector $\vec{r}_0$ has magnitude r_0 and polar angle θ_0. *Hint:* Let $\delta(\vec{r} - \vec{r}_0)$ be written as $f(r)g(\theta)$, integrate over an elemental area $dS = r\,dr\,d\theta$, separate the integrals into a product of one on r and the other on θ, and note the form of the integrands.

4.10.7. Fill in the mathematical steps to verify the steps from (4.10.13) to (4.10.15) for the fixed-rim circular membrane struck at a point a distance r_0 from its center.

4.10.8. Obtain the normal mode expansion for a circular membrane fixed at its rim and struck at its center.

4.10.9C. Design a program to show plots of the displacement of a rectangular membrane spanning $(0, L_x)$ and $(0, L_z)$ at various times after it has been struck at a point (x_0, z_0).

4.10.10C. Design a program to show plots at various times of the displacement of a circular membrane of radius a after it has been struck at a point $(r_0, 0)$, where $r_0 < a$.

4.11.1. The diaphragm of a telephone receiver consists of a circular sheet of steel 4 cm in diameter and 0.02 cm thick. (*a*) If it is rigidly clamped at its rim, what is its fundamental frequency of vibration? What will be the effect on this frequency (*b*) of doubling the thickness of the diaphragm and (*c*) of doubling the diameter?

4.11.2. To what tension would the diaphragm of Problem 4.11.1 need to be stretched if its fundamental frequency, considered as resulting from the restoring forces of tension alone, were to equal that resulting from stiffness forces alone?

4.11.3. (*a*) Determine the ratio of the constants $\mathbf{B}_2/\mathbf{A}_2$ for a thin circular plate clamped at its rim and vibrating in its first overtone mode. (*b*) Express the resulting motion by an equation analogous to (4.11.10). (*c*) Plot the shape function of the diaphragm. (*d*) What is the ratio of the radius of the nodal circle to the radius of the plate?

4.11.4. The vibrating circular steel plate of an electromagnetic sonar transducer of radius 0.1 m and thickness 0.005 m is clamped at its rim. What is its fundamental frequency of vibration?

4.11.5. For a circular plate of thickness d, (*a*) show that the surface radius of gyration is $\kappa = d/\sqrt{12}$. (*b*) If the thickness of the plate is doubled, what happens to the frequencies of the normal modes?

4.11.6. (*a*) By direct integration obtain (4.11.11). (*b*) Show that the average displacement amplitude is $0.309A$, where A is the displacement amplitude at the center.

4.11.7. Find the frequencies of the symmetric normal modes for a circular plate fixed at both center and rim.

4.11.8C. Plot the modified Bessel functions of the first three orders for arguments $0 < x < 6$.

4.11.9C. Plot the shape of a thin circular plate clamped at the rim when it is vibrating in each of its first three symmetric normal modes.

Chapter 5

THE ACOUSTIC WAVE EQUATION AND SIMPLE SOLUTIONS

5.1 INTRODUCTION

Acoustic waves constitute one kind of pressure fluctuation that can exist in a compressible fluid. In addition to the audible pressure fields of moderate intensity, the most familiar, there are also *ultrasonic* and *infrasonic* waves whose frequencies lie beyond the limits of hearing, *high-intensity* waves (such as those near jet engines and missiles) that may produce a sensation of pain rather than sound, *nonlinear* waves of still higher intensities, and *shock* waves generated by explosions and supersonic aircraft.

Inviscid fluids exhibit fewer constraints to deformations than do solids. The restoring forces responsible for propagating a wave are the pressure changes that occur when the fluid is compressed or expanded. Individual elements of the fluid move back and forth in the direction of the forces, producing adjacent regions of compression and rarefaction similar to those produced by longitudinal waves in a bar.

The following terminology and symbols will be used:

$\vec{r}$ = equilibrium position of a fluid element

$$\vec{r} = x\hat{x} + y\hat{y} + z\hat{z} \tag{5.1.1}$$

($\hat{x}$, $\hat{y}$, and $\hat{z}$ are the unit vectors in the x, y, and z directions, respectively)

$\vec{\xi}$ = *particle displacement* of a fluid element from its equilibrium position

$$\vec{\xi} = \xi_x\hat{x} + \xi_y\hat{y} + \xi_z\hat{z} \tag{5.1.2}$$

$\vec{u}$ = *particle velocity* of a fluid element

$$\vec{u} = \frac{\partial \vec{\xi}}{\partial t} = u_x\hat{x} + u_y\hat{y} + u_z\hat{z} \tag{5.1.3}$$

ρ = *instantaneous density* at (x, y, z)
ρ_0 = *equilibrium density* at (x, y, z)
s = *condensation* at (x, y, z)

113

$$s = (\rho - \rho_0)/\rho_0 \tag{5.1.4}$$

$\rho - \rho_0 = \rho_0 s = $ *acoustic density* at (x, y, z)

$\mathscr{P} = $ *instantaneous pressure* at (x, y, z)

$\mathscr{P}_0 = $ *equilibrium pressure* at (x, y, z)

$p = $ *acoustic pressure* at (x, y, z)

$$p = \mathscr{P} - \mathscr{P}_0 \tag{5.1.5}$$

$c = $ *thermodynamic speed of sound* of the fluid

$\Phi = $ *velocity potential* of the wave

$$\vec{u} = \nabla\Phi \tag{5.1.6}$$

$T_K = $ temperature in kelvins (K)

$T = $ temperature in degrees Celsius (or centigrade) (°C)

$$T + 273.15 = T_K \tag{5.1.7}$$

The terms *fluid element* and *particle* mean an infinitesimal volume of the fluid large enough to contain millions of molecules so that the fluid may be thought of as a continuous medium, yet small enough that all acoustic variables are uniform throughout.

The molecules of a fluid do not have fixed mean positions in the medium. Even without the presence of an acoustic wave, they are in constant random motion with average velocities far in excess of any particle velocity associated with the wave motion. However, a small volume may be treated as an unchanging unit since those molecules leaving its confines are replaced (on the average) by an equal number with identical properties. The macroscopic properties of the element remain unchanged. As a consequence, it is possible to speak of particle displacements and velocities when discussing acoustic waves in fluids, as was done for elastic waves in solids. The fluid is assumed to be lossless so there are no dissipative effects such as those arising from viscosity or heat conduction. The analysis will be limited to waves of relatively small amplitude, so changes in the density of the medium will be small compared with its equilibrium value. These assumptions are necessary to arrive at the simplest equations for sound in fluids. It is fortunate that experiments show these simplifications are successful and lead to an adequate description of most common acoustic phenomena. However, there are situations where these assumptions are violated and the theory must be modified.

5.2   THE EQUATION OF STATE

For fluid media, the equation of state must relate three physical quantities describing the thermodynamic behavior of the fluid. For example, the *equation of state for a perfect gas*

$$\mathscr{P} = \rho r T_K \tag{5.2.1}$$

gives the general relationship between the total pressure $\mathscr{P}$ in pascals (Pa), the density ρ in kilograms per cubic meter (kg/m³), and the absolute temperature T_K in kelvins (K) for a large number of gases under equilibrium conditions. The quantity r is the *specific gas constant* and depends on the *universal gas constant* $\mathscr{R}$ and the *molecular weight M* of the particular gas. See Appendix A9. For air, $r \approx 287$ J/(kg·K).

Greater simplification can be achieved if the thermodynamic process is restricted. For example, if the fluid is contained within a vessel whose walls are highly thermally conductive, then slow variations in the volume of the vessel will result in thermal energy being transferred between the walls and the fluid. If the walls have sufficient thermal capacity, they and the fluid will remain at a constant temperature. In this case, the perfect gas is described by the *isotherm*

$$\mathscr{P}/\mathscr{P}_0 = \rho/\rho_0 \qquad \text{(perfect gas isotherm)} \tag{5.2.2}$$

In contrast, acoustic processes are nearly *isentropic* (adiabatic and reversible). The thermal conductivity of the fluid and the temperature gradients of the disturbance are small enough that no appreciable thermal energy transfer occurs between adjacent fluid elements. Under these conditions, the *entropy* of the fluid remains nearly constant. The acoustic behavior of the perfect gas under these conditions is described by the *adiabat*

$$\mathscr{P}/\mathscr{P}_0 = (\rho/\rho_0)^\gamma \qquad \text{(perfect gas adiabat)} \tag{5.2.3}$$

where γ is the *ratio of specific heats* (or *ratio of heat capacities*). Finite thermal conductivity results in a conversion of acoustic energy into random thermal energy so that the acoustic disturbance attenuates slowly with time or distance. This and other dissipative effects will be considered in Chapter 8.

For fluids other than a perfect gas, the adiabat is more complicated. In these cases it is preferable to determine experimentally the isentropic relationship between pressure and density fluctuations. This relationship can be represented by a Taylor's expansion

$$\mathscr{P} = \mathscr{P}_0 + \left(\frac{\partial \mathscr{P}}{\partial \rho}\right)_{\rho_0} (\rho - \rho_0) + \frac{1}{2}\left(\frac{\partial^2 \mathscr{P}}{\partial \rho^2}\right)_{\rho_0} (\rho - \rho_0)^2 + \cdots \tag{5.2.4}$$

wherein the partial derivatives are determined for the isentropic compression and expansion of the fluid about its equilibrium density. If the fluctuations are small, only the lowest order term in $(\rho - \rho_0)$ need be retained. This gives a linear relationship between the pressure fluctuation and the change in density

$$\mathscr{P} - \mathscr{P}_0 \approx \mathscr{B}(\rho - \rho_0)/\rho_0 \tag{5.2.5}$$

with $\mathscr{B} = \rho_0(\partial\mathscr{P}/\partial\rho)_{\rho_0}$ the *adiabatic bulk modulus* discussed in Appendix A11. In terms of acoustic pressure p and condensation s, (5.2.5) can be rewritten as

$$\boxed{p \approx \mathscr{B}s} \tag{5.2.6}$$

The essential restriction is that the condensation is small.

Another approach in expressing the adiabat of any fluid is to model it on the adiabat of the perfect gas. This is done by generalizing $\mathcal{P}_0$ and γ to be empirically determined coefficients for the fluid in question. Expanding (5.2.3) in a Taylor's series in s and rearranging to isolate the acoustic pressure $p = \mathcal{P} - \mathcal{P}_0$ yields

$$p = \mathcal{P}_0\left[\gamma s + \tfrac{1}{2}\gamma(\gamma - 1)s^2 + \cdots\right] \tag{5.2.7}$$

Comparing this with (5.2.4) and equating the coefficients through second order in s reveals that $\mathcal{P}_0$ and γ can be expressed thermodynamically in general as

$$\gamma\mathcal{P}_0 = \mathcal{B} \tag{5.2.8}$$

$$\gamma - 1 \equiv \frac{B}{A} = \frac{\rho_0}{\mathcal{B}}\left(\frac{\partial\mathcal{B}}{\partial\rho}\right)_{\rho_0} \tag{5.2.9}$$

[Both $\mathcal{B}$ and $(\partial\mathcal{B}/\partial\rho)_{\rho_0}$ are evaluated under adiabatic conditions.] The quantity B/A is the *parameter of nonlinearity* of the fluid. Thus, knowing $\mathcal{B}$ and its derivative, we can determine $\mathcal{P}_0$ and γ. The equality of coefficients fails for terms of third order and above in s, but it has been demonstrated that these higher order terms are completely negligible for situations of practical importance.[1] Use of standard thermodynamic relationships allows the right sides of the above two equations to be expressed in terms of other thermodynamic properties of the fluid that are much more easily determined experimentally.

For liquids like water, simple alcohols, liquid metals, and many organic compounds, γ lies between about 4 and 12 and $\mathcal{P}_0$ between about 1×10^3 and 5×10^3 atm. The constant $\mathcal{P}_0$ suggests a fictitious *adiabatic internal pressure*, as if the liquid in its acoustic behavior were a gas under this hydrostatic pressure. The coefficient γ is an empirical constant whose difference from unity measures the nonlinear relationship between acoustic pressure and condensation. (Elsewhere, unless explicitly stated otherwise, it is the ratio of specific heats.)

5.3 THE EQUATION OF CONTINUITY

To connect the motion of the fluid with its compression or expansion, we need a functional relationship between the particle velocity $\vec{u}$ and the instantaneous density ρ. Consider a small rectangular parallelepiped volume element $dV = dx\,dy\,dz$, which is *fixed in space* and through which elements of the fluid travel. The net rate with which mass flows into the volume through its surface must equal the rate with which the mass within the volume increases. Referring to Fig. 5.3.1, we see that the net influx of mass into this spatially fixed volume resulting from flow in the x direction is

$$\left[\rho u_x - \left(\rho u_x + \frac{\partial(\rho u_x)}{\partial x}dx\right)\right]dy\,dz = -\frac{\partial(\rho u_x)}{\partial x}dV \tag{5.3.1}$$

---

[1] Beyer, *Nonlinear Acoustics*, Naval Ship Systems Command (1974).

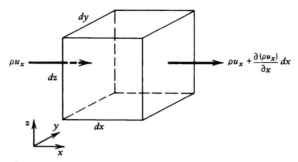

Figure 5.3.1   An elemental spatially fixed volume of fluid showing the rate of mass flow into and out of the volume resulting from fluid flowing in the x direction. A similar diagram can be drawn for fluid flowing in the y and z directions.

Similar expressions give the net influx for the y and z directions, so that the total influx must be

$$-\left(\frac{\partial(\rho u_x)}{\partial x} + \frac{\partial(\rho u_y)}{\partial y} + \frac{\partial(\rho u_z)}{\partial z}\right) dV = -\nabla \cdot (\rho \vec{u}) \, dV \tag{5.3.2}$$

The rate with which the mass increases in the volume is $(\partial \rho / \partial t) \, dV$. The net influx must equal the rate of increase,

$$\boxed{\frac{\partial \rho}{\partial t} + \nabla \cdot (\rho \vec{u}) = 0} \tag{5.3.3}$$

This is the *exact continuity equation*. The second term on the left involves the product of particle velocity and instantaneous density, both of which are acoustic variables. However, if we write $\rho = \rho_0(1 + s)$, require ρ_0 to be a sufficiently weak function of time, and assume that s is very small, (5.3.3) becomes

$$\rho_0 \frac{\partial s}{\partial t} + \nabla \cdot (\rho_0 \vec{u}) = 0 \tag{5.3.4}$$

the *linear continuity equation*. Furthermore, if ρ_0 is only a weak function of space

$$\boxed{\frac{\partial s}{\partial t} + \nabla \cdot \vec{u} = 0} \tag{5.3.5}$$

5.4 THE SIMPLE FORCE EQUATION: EULER'S EQUATION

In real fluids, the existence of viscosity and the failure of acoustic processes to be perfectly adiabatic introduce dissipative terms. As mentioned earlier, these effects will be investigated in Chapter 8.

Consider a fluid element $dV = dx\,dy\,dz$, which *moves with the fluid* and contains a mass dm of fluid. The net force $d\vec{f}$ on the element will accelerate it according to Newton's second law $d\vec{f} = \vec{a}\,dm$. In the absence of viscosity, the net force experienced by the element in the x direction is

$$df_x = \left[\mathscr{P} - \left(\mathscr{P} + \frac{\partial \mathscr{P}}{\partial x} dx \right) \right] dy\,dz = -\frac{\partial \mathscr{P}}{\partial x} dV \tag{5.4.1}$$

There are analogous expressions for df_y and df_z. The presence of the gravitational field introduces an additional force in the vertical direction of $\vec{g}\rho\,dV$, where $|\vec{g}| \approx 9.8 \text{ m/s}^2$ is the acceleration of gravity. Combination of these terms results in

$$d\vec{f} = -\nabla\mathscr{P}\,dV + \vec{g}\rho\,dV \tag{5.4.2}$$

The expression for the acceleration of the fluid element is a little more complicated. The particle velocity $\vec{u}$ is a function of both time and space. When the fluid element with velocity $\vec{u}(x, y, z, t)$ at position (x, y, z) and time t moves to a new location $(x + dx, y + dy, z + dz)$ at a later time $t + dt$, its new velocity is expressed by the leading terms of its Taylor expansion

$$\vec{u}(x + u_x\,dt, y + u_y\,dt, z + u_z\,dt, t + dt)$$

$$= \vec{u}(x, y, z, t) + \frac{\partial \vec{u}}{\partial x}u_x\,dt + \frac{\partial \vec{u}}{\partial y}u_y\,dt + \frac{\partial \vec{u}}{\partial z}u_z\,dt + \frac{\partial \vec{u}}{\partial t}\,dt \tag{5.4.3}$$

Thus the acceleration of the chosen element is

$$\vec{a} = \lim_{dt \to 0} \frac{\vec{u}(x + u_x\,dt, y + u_y\,dt, z + u_z\,dt, t + dt) - \vec{u}(x, y, z, t)}{dt} \tag{5.4.4}$$

or

$$\vec{a} = \frac{\partial \vec{u}}{\partial t} + u_x\frac{\partial \vec{u}}{\partial x} + u_y\frac{\partial \vec{u}}{\partial y} + u_z\frac{\partial \vec{u}}{\partial z} \tag{5.4.5}$$

If we define the vector operator $(\vec{u} \cdot \nabla)$ as

$$(\vec{u} \cdot \nabla) \equiv u_x\frac{\partial}{\partial x} + u_y\frac{\partial}{\partial y} + u_z\frac{\partial}{\partial z} \tag{5.4.6}$$

then $\vec{a}$ can be written more conveniently as

$$\vec{a} = \frac{\partial \vec{u}}{\partial t} + (\vec{u} \cdot \nabla)\vec{u} \tag{5.4.7}$$

Since the mass dm of the element is $\rho\,dV$, substitution into $d\vec{f} = \vec{a}\,dm$ gives

$$-\nabla\mathscr{P} + \vec{g}\rho = \rho\left(\frac{\partial \vec{u}}{\partial t} + (\vec{u} \cdot \nabla)\vec{u} \right) \tag{5.4.8}$$

This nonlinear, inviscid force equation is *Euler's equation* with gravity. In the case of no acoustic excitation, $\vec{g}\rho_0 = \nabla \mathcal{P}_0$, and thus $\nabla \mathcal{P} = \nabla p + \vec{g}\rho_0$ so that (5.4.8) becomes

$$-\frac{1}{\rho_0}\nabla p + \vec{g}s = (1+s)\left(\frac{\partial \vec{u}}{\partial t} + (\vec{u} \cdot \nabla)\vec{u}\right) \tag{5.4.9}$$

If we now make the assumptions that $|\vec{g}s| \ll |\nabla p|/\rho_0$, that $|s| \ll 1$, and that $|(\vec{u} \cdot \nabla)\vec{u}| \ll |\partial\vec{u}/\partial t|$, then

$$\boxed{\rho_0 \frac{\partial \vec{u}}{\partial t} = -\nabla p} \tag{5.4.10}$$

This is the *linear Euler's equation*, valid for acoustic processes of small amplitude.

5.5 THE LINEAR WAVE EQUATION

The linearized equations (5.2.6), (5.3.4), and (5.4.10) can be combined to yield a single differential equation with one dependent variable. First, take the divergence of (5.4.10),

$$\nabla \cdot \left(\rho_0 \frac{\partial \vec{u}}{\partial t}\right) = -\nabla^2 p \tag{5.5.1}$$

where $\nabla \cdot \nabla = \nabla^2$ is the three-dimensional Laplacian. Next, take the time derivative of (5.3.4) and use the facts that space and time are independent and ρ_0 is no more than a weak function of time,

$$\rho_0 \frac{\partial^2 s}{\partial t^2} + \nabla \cdot \left(\rho_0 \frac{\partial \vec{u}}{\partial t}\right) = 0 \tag{5.5.2}$$

Elimination of the divergence term between these two equations gives

$$\nabla^2 p = \rho_0 \frac{\partial^2 s}{\partial t^2} \tag{5.5.3}$$

Equation (5.2.6) allows the condensation to be expressed as $s = p/\mathcal{B}$, and with $\mathcal{B}$ no more than a weak function of time,

$$\boxed{\nabla^2 p = \frac{1}{c^2}\frac{\partial^2 p}{\partial t^2}} \tag{5.5.4}$$

where c is the *thermodynamic speed of sound* defined by

$$c^2 = \mathcal{B}/\rho_0 \tag{5.5.5}$$

Equation (5.5.4) is the *linear, lossless wave equation* for the propagation of sound in fluids with phase speed c. Since the above derivation never required a restriction on $\mathscr{B}$ or ρ_0 with respect to *space*, (5.5.4) is valid for propagation in media with sound speeds that are functions of space, such as found in the atmosphere or the ocean.

Use of (5.5.5) shows that the adiabat can be written as

$$p = \rho_0 c^2 s \qquad (5.5.6)$$

If ρ_0 and c are only weak functions of space, then p and s are essentially proportional and the condensation satisfies the wave equation.

Since the curl of the gradient of a function must vanish, $\nabla \times \nabla f = 0$, (5.4.10) shows that the particle velocity is irrotational, $\nabla \times \vec{u} = 0$. This means that it can be expressed as the gradient of a scalar function Φ,

$$\vec{u} = \nabla \Phi \qquad (5.5.7)$$

which was previously identified as the velocity potential. The physical meaning of this useful result is that the acoustic excitation of an *inviscid* fluid involves no rotational flow. A real fluid has finite viscosity and the particle velocity is not curl-free everywhere. For most acoustic processes, rotational effects are small and confined to the vicinity of boundaries. They exert little influence on the propagation of sound, so that (5.5.7) can be assumed true to very high accuracy in acoustic propagation.

Substitution of (5.5.7) into (5.4.10) and requiring ρ_0 to be no more than a gradual function of space gives

$$\nabla \left(\rho_0 \frac{\partial \Phi}{\partial t} + p \right) = 0 \qquad (5.5.8)$$

The quantity in parentheses can be chosen to vanish identically if there is no acoustic excitation so that

$$p = -\rho_0 \frac{\partial \Phi}{\partial t} \qquad (5.5.9)$$

Thus, Φ satisfies the wave equation within the same approximations.

5.6 SPEED OF SOUND IN FLUIDS

By combining (5.2.5) and (5.5.5), we get an expression for the thermodynamic speed of sound

$$c^2 = \left(\frac{\partial \mathscr{P}}{\partial \rho} \right)_{adiabat} \qquad (5.6.1)$$

This is a characteristic property of the fluid and depends on the equilibrium conditions.

When a sound wave propagates through a perfect gas, the adiabat may be utilized to derive an important special form of (5.6.1). Direct differentiation of (5.2.3) leads to

$$\left(\frac{\partial \mathscr{P}}{\partial \rho}\right)_{adiabat} = \gamma \frac{\mathscr{P}}{\rho} \tag{5.6.2}$$

Evaluating this expression at ρ_0 and substituting into (5.6.1), we obtain

$$c^2 = \gamma \mathscr{P}_0/\rho_0 \tag{5.6.3}$$

Substitution of the appropriate values for air from Appendix A10 gives

$$c_0 = (1.402 \times 1.01325 \times 10^5/1.293)^{1/2} = 331.5 \text{ m/s} \tag{5.6.4}$$

as the theoretical value for the speed of sound in air at 0°C and 1 atm pressure. This is in excellent agreement with measured values and supports the assumption that acoustic processes in a fluid are adiabatic. For most real gases at constant temperature, the ratio $\mathscr{P}_0/\rho_0$ is nearly independent of pressure so that the speed of sound is a function only of temperature. An alternate expression for the speed of sound in a perfect gas is found from (5.2.1) and (5.6.3) to be

$$c^2 = \gamma r T_K \tag{5.6.5}$$

The speed is proportional to the square root of the absolute temperature. In terms of the speed c_0 at 0°C, this becomes

$$c = c_0(T_K/273)^{1/2} = c_0(1 + T/273)^{1/2} \tag{5.6.6}$$

Theoretical prediction of the speed of sound for liquids is considerably more difficult than for gases. However, it is possible to show theoretically that $\mathscr{B} = \gamma \mathscr{B}_T$, where $\mathscr{B}_T$ is the isothermal bulk modulus. Since $\mathscr{B}_T$ is much easier to measure experimentally than $\mathscr{B}$, a convenient expression for the speed of sound in liquids is obtained from (5.5.5) and $\mathscr{B}_T$,

$$c^2 = \gamma \mathscr{B}_T/\rho_0 \tag{5.6.7}$$

where γ, $\mathscr{B}_T$, and ρ_0 all vary with the equilibrium temperature and pressure of the liquid. Since no simple theory is available for predicting these variations, they must be measured experimentally and the resulting speed of sound expressed as a numerical formula. For example, in distilled water a simplified formula for c in m/s is

$$c(\mathscr{P}, t) = 1402.7 + 488t - 482t^2 + 135t^3 + (15.9 + 2.8t + 2.4t^2)(\mathscr{P}_G/100) \tag{5.6.8}$$

where $\mathscr{P}_G$ is the gauge pressure in bar (1 bar = 10^5 Pa) and $t = T/100$, with T in degrees Celsius. A gauge pressure $\mathscr{P}_G$ of zero means an equilibrium pressure $\mathscr{P}_0$ of 1 atm (1.01325 bar). This equation is accurate to within 0.05% for $0 < T < 100$°C and $0 \le \mathscr{P}_G \le 200$ bar.

5.7 HARMONIC PLANE WAVES

In this and the next few sections, discussion will be restricted to homogeneous, isotropic fluids in which the speed of sound c is a constant throughout. Propagation in fluids having spatially dependent sound speeds will be deferred until Section 5.14.

The characteristic property of a *plane wave* is that each acoustic variable has constant amplitude and phase on any plane perpendicular to the direction of propagation. Since the surfaces of constant phase for any diverging wave become nearly planar far from their source, we may expect that the properties of diverging waves will, at large distances, become very similar to those of plane waves.

If the coordinate system is chosen so that the plane wave propagates along the *x* axis, the wave equation reduces to

$$\frac{\partial^2 p}{\partial x^2} = \frac{1}{c^2}\frac{\partial^2 p}{\partial t^2} \tag{5.7.1}$$

where $p = p(x,t)$. Direct comparison with (2.3.6) shows that the mathematical development of the solutions for transverse waves in Sections 2.4 and 2.5 can be applied here and need not be repeated. Let us therefore proceed directly to harmonic plane waves and the relationships among the acoustic variables.

The complex form of the harmonic solution for the acoustic pressure of a plane wave is

$$\mathbf{p} = \mathbf{A}e^{j(\omega t - kx)} + \mathbf{B}e^{j(\omega t + kx)} \tag{5.7.2}$$

and the associated particle velocity, from (5.4.10),

$$\vec{u} = u\hat{x} = [(\mathbf{A}/\rho_0 c)e^{j(\omega t - kx)} - (\mathbf{B}/\rho_0 c)e^{j(\omega t + kx)}]\hat{x} \tag{5.7.3}$$

is parallel to the direction of propagation.

If we use a subscript " + " to designate a wave traveling in the +*x* direction and a subscript "−" for a wave traveling in the −*x* direction, then

$$\mathbf{p}_+ = \mathbf{A}e^{j(\omega t - kx)} \quad \text{and} \quad \mathbf{p}_- = \mathbf{B}e^{j(\omega t + kx)} \tag{5.7.4}$$

$$\mathbf{u}_\pm = \pm\mathbf{p}_\pm/\rho_0 c \tag{5.7.5}$$

$$\mathbf{s}_\pm = \mathbf{p}_\pm/\rho_0 c^2 \tag{5.7.6}$$

$$\mathbf{\Phi}_\pm = -\mathbf{p}_\pm/j\omega\rho_0 \tag{5.7.7}$$

For a plane wave traveling in some *arbitrary* direction, it is plausible to try a solution of the form

$$\mathbf{p} = \mathbf{A}e^{j(\omega t - k_x x - k_y y - k_z z)} \tag{5.7.8}$$

Substitution into (5.5.4) shows that this is acceptable if

$$(\omega/c)^2 = k_x^2 + k_y^2 + k_z^2 \tag{5.7.9}$$

Definition of the *propagation vector* $\vec{k}$,

$$\vec{k} = k_x\hat{x} + k_y\hat{y} + k_z\hat{z} \tag{5.7.10}$$

which has magnitude ω/c, and a position vector $\vec{r}$,

$$\vec{r} = x\hat{x} + y\hat{y} + z\hat{z} \tag{5.7.11}$$

that gives the location of the point (x, y, z) with respect to the origin of the coordinate system, allows the trial solution (5.7.8) to be expressed as

$$\mathbf{p} = \mathbf{A}e^{j(\omega t - \vec{k}\cdot\vec{r})} \tag{5.7.12}$$

The surfaces of constant phase are given by $\vec{k}\cdot\vec{r} = constant$. Since, from the definition of the gradient, $\vec{k} = \nabla(\vec{k}\cdot\vec{r})$ is a vector perpendicular to the surfaces of constant phase, $\vec{k}$ points in the direction of propagation. The magnitude of $\vec{k}$ is the *wave number* (or *propagation constant*) k and $k_x/k, k_y/k$, and k_z/k are the direction cosines of $\vec{k}$ with respect to the x, y, and z axes.

As a special case, let us examine a plane wave whose surfaces of constant phase are parallel to the z axis. Equation (5.7.8) reduces to

$$\mathbf{p} = \mathbf{A}e^{j(\omega t - k_x x - k_y y)} \tag{5.7.13}$$

The surfaces of constant phase are given by

$$y = -(k_x/k_y)x + constant \tag{5.7.14}$$

which describes plane surfaces parallel to the z axis with a slope of $-(k_x/k_y)$ in the x-y plane. If we examine $\mathbf{p}$ as a function of x and t for $y = 0$, we have

$$\mathbf{p}(x, 0, t) = \mathbf{A}e^{j(\omega t - k_x x)} \tag{5.7.15}$$

This oblique "slice" of the wave has an apparent wavelength $\lambda_x = 2\pi/k_x$ measured in the x direction. From Fig. 5.7.1 we see that $\lambda/\lambda_x = \cos\phi$ so that $k_x = k\cos\phi$. The same argument applies in the y direction for fixed x and yields $k_y = k\sin\phi$. Thus,

$$\vec{k} = k\cos\phi\,\hat{x} + k\sin\phi\,\hat{y} \tag{5.7.16}$$

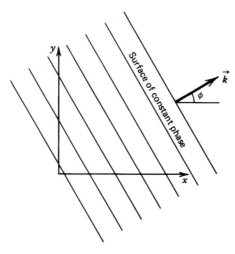

Figure 5.7.1 Surfaces of constant phase for a plane wave with wave number k traveling perpendicular to the z axis in a direction ϕ from the x axis.

and $\vec{k}$ is perpendicular to the z axis, pointing into the first quadrant of the x-y plane with an angle ϕ measured counterclockwise from the x axis. Substitution of k into (5.7.12) yields the convenient form

$$\mathbf{p} = \mathbf{A}e^{j(\omega t - kx\cos\phi - ky\sin\phi)} \tag{5.7.17}$$

5.8 ENERGY DENSITY

The energy transported by acoustic waves through a fluid medium is of two forms: (1) the *kinetic energy* of the moving elements and (2) the *potential energy* of the compressed fluid. Consider a small fluid element that moves with the fluid and occupies volume V_0 of the undisturbed fluid. The mass of the element is $\rho_0 V_0$ and its kinetic energy is

$$E_k = \tfrac{1}{2}\rho_0 V_0 u^2 \tag{5.8.1}$$

The change in potential energy associated with a volume change from V_0 to V is

$$E_p = -\int_{V_0}^{V} p\,dV \tag{5.8.2}$$

The negative sign indicates that the potential energy will increase (work is done *on* the element) when its volume is decreased by a positive acoustic pressure p. To carry this out, it is necessary to express all variables under the integral sign in terms of one variable—p, for example. From conservation of mass we have $\rho V = \rho_0 V_0$ so that

$$dV = -(V/\rho)\,d\rho \tag{5.8.3}$$

Now, with the use of $dp/d\rho = c^2$,

$$dV = (V/\rho c^2)\,dp \tag{5.8.4}$$

Substitution into (5.8.2) and integration of the acoustic pressure from 0 to p gives

$$E_p = \tfrac{1}{2}(p^2/\rho_0 c^2)V_0 \tag{5.8.5}$$

within the linear approximations. The total acoustic energy of the volume element is then

$$E = E_k + E_p = \tfrac{1}{2}\rho_0 V_0[u^2 + (p/\rho_0 c)^2] \tag{5.8.6}$$

and the *instantaneous energy density* $\mathscr{E}_i = E/V_0$ in joules per cubic meter (J/m^3) is

$$\mathscr{E}_i = \tfrac{1}{2}\rho_0[u^2 + (p/\rho_0 c)^2] \tag{5.8.7}$$

Both the pressure p and the particle speed u must be the *real* quantities obtained from the superposition of all acoustic waves present.

The instantaneous particle speed and acoustic pressure are functions of both position and time, and consequently the instantaneous energy density $\mathscr{E}_i$ is not necessarily constant throughout the fluid. The time average of $\mathscr{E}_i$ gives the *energy density* $\mathscr{E}$ at any point in the fluid:

$$\mathscr{E} = \langle \mathscr{E}_i \rangle_T = \frac{1}{T} \int_0^T \mathscr{E}_i \, dt \tag{5.8.8}$$

where the time interval T is one period of a harmonic wave.

The above expressions apply to any linear acoustic wave. To proceed further, it is necessary to know the relationship between p and u. For a plane harmonic wave traveling in the $\pm x$ direction, reference to (5.7.5) shows that $p = \pm \rho_0 c u$ so that (5.8.7) gives

$$\mathscr{E}_i = \rho_0 u^2 = p^2 / \rho_0 c^2 \tag{5.8.9}$$

and if P and U are the amplitudes of the acoustic pressure and particle speed,

$$\mathscr{E} = PU/2c = P^2/2\rho_0 c^2 = \rho_0 U^2/2 \tag{5.8.10}$$

In more complicated cases, there is no guarantee that $p = \pm \rho_0 c u$ nor that the energy density is given by $\mathscr{E} = PU/2c$. However, (5.8.10) is approximately correct for progressive waves when the radii of curvature of the surfaces of constant phase are much greater than a wavelength. This occurs, for example, for spherical or cylindrical waves at distances of many wavelengths from their sources.

5.9 ACOUSTIC INTENSITY

The *instantaneous intensity* $I(t)$ of a sound wave is the instantaneous rate per unit area at which work is done by one element of fluid on an adjacent element. It is given by $I(t) = pu$ in watts per square meter (W/m^2). The *intensity* I is the *time average* of $I(t)$, the time-averaged rate of energy transmission through a unit area normal to the direction of propagation,

$$I = \langle I(t) \rangle_T = \langle pu \rangle_T = \frac{1}{T} \int_0^T pu \, dt \tag{5.9.1}$$

where for a monofrequency wave T is the period.

For a plane harmonic wave traveling in the $\pm x$ direction, $p = \pm \rho_0 c u$, so that

$$I = \pm P^2/2\rho_0 c \tag{5.9.2}$$

There is a similarity between (5.9.2) and corresponding equations for electromagnetic waves and voltage waves on transmission lines. First, reexpress (5.9.2) in terms of effective (root-mean-square) amplitudes. If we define F_e as the *effective amplitude* of a periodic quantity $f(t)$, then

$$F_e = \left(\frac{1}{T} \int_0^T f^2(t) \, dt \right)^{1/2} \tag{5.9.3}$$

where T is the period of the motion. For harmonic waves this yields

$$P_e = P/\sqrt{2} \quad \text{and} \quad U_e = U/\sqrt{2} \tag{5.9.4}$$

so that

$$I_\pm = \pm P_e U_e = \pm P_e^2/\rho_0 c \tag{5.9.5}$$

for a plane wave traveling in either the $+x$ or $-x$ direction. It must be emphasized that, while (5.9.1) is completely general, $I_\pm = \pm P_e U_e$ is exact only for plane harmonic waves and is approximately true for diverging waves at great distances from their sources.

5.10 SPECIFIC ACOUSTIC IMPEDANCE

The ratio of acoustic pressure to the associated particle speed in a medium is the *specific acoustic impedance*

$$\mathbf{z} = \mathbf{p}/\mathbf{u} \tag{5.10.1}$$

For plane waves this ratio is

$$\mathbf{z} = \pm \rho_0 c \tag{5.10.2}$$

The choice of sign depends on whether propagation is in the plus or minus x direction. The MKS unit of specific acoustic impedance is the Pa·s/m, often called the *rayl* (1 MKS rayl = 1 Pa· s/m) in honor of John William Strutt, Baron Rayleigh (1842–1919). The product $\rho_0 c$ often has greater acoustical significance as a characteristic property of the medium than does either ρ_0 or c individually. For this reason $\rho_0 c$ is called the *characteristic impedance* of the medium.

Although the specific acoustic impedance of the medium is a real quantity for progressive plane waves, this is not true for standing plane waves or for diverging waves. In general, $\mathbf{z}$ will be complex

$$\mathbf{z} = r + jx \tag{5.10.3}$$

where r is the *specific acoustic resistance* and x the *specific acoustic reactance* of the medium for the particular wave being considered.

The characteristic impedance of a medium for acoustic waves is analogous to the wave impedance $\sqrt{\mu/\varepsilon}$ of a dielectric medium for electromagnetic waves and to the characteristic impedance Z_0 of an electric transmission line. Numerical values of $\rho_0 c$ for some fluids and solids are given in Appendix A10.

For air at a temperature of 20°C and atmospheric pressure, the density is 1.21 kg/m^3 and the speed of sound is 343 m/s, giving

$$\rho_0 c = 415 \text{ Pa·s/m} \quad \text{(air at 20°C)} \tag{5.10.4}$$

In distilled water at 20°C and 1 atm, the speed of sound is 1482.1 m/s and its density is 998.2 kg/m^3, resulting in a characteristic impedance of

$$\rho_0 c = 1.48 \times 10^6 \text{ Pa·s/m} \quad \text{(water at 20°C)} \tag{5.10.5}$$

5.11 SPHERICAL WAVES

Expressed in spherical coordinates, the Laplacian operator is

$$\nabla^2 = \frac{\partial^2}{\partial r^2} + \frac{2}{r}\frac{\partial}{\partial r} + \frac{1}{r^2 \sin\theta}\frac{\partial}{\partial\theta}\left(\sin\theta\frac{\partial}{\partial\theta}\right) + \frac{1}{r^2 \sin^2\theta}\frac{\partial^2}{\partial\phi^2} \qquad (5.11.1)$$

where $x = r\sin\theta\cos\phi, y = r\sin\theta\sin\phi$, and $z = r\cos\theta$ (see Appendix A7). If the waves have spherical symmetry, the acoustic pressure p is a function of radial distance and time but not of the angular coordinates, and this equation simplifies to

$$\nabla^2 = \frac{\partial^2}{\partial r^2} + \frac{2}{r}\frac{\partial}{\partial r} \qquad (5.11.2)$$

The wave equation for spherically symmetric pressure fields is then

$$\frac{\partial^2 p}{\partial r^2} + \frac{2}{r}\frac{\partial p}{\partial r} = \frac{1}{c^2}\frac{\partial^2 p}{\partial t^2} \qquad (5.11.3)$$

Conservation of energy and the relationship $I = P^2/2\rho_0 c$ lead us to expect that the pressure amplitude might fall off as $1/r$, so that the quantity rp would have amplitude independent of r. Rewriting (5.11.3) with rp treated as the dependent variable results in

$$\frac{\partial^2(rp)}{\partial r^2} = \frac{1}{c^2}\frac{\partial^2(rp)}{\partial t^2} \qquad (5.11.4)$$

If the product rp in this equation is considered as a single variable, the equation is the same as the plane wave equation with the general solution

$$p = \frac{1}{r}f_1(ct - r) + \frac{1}{r}f_2(ct + r) \qquad (5.11.5)$$

for all $r > 0$. The solution fails at $r = 0$. The first term represents a spherical wave diverging from the origin with speed c; the second term represents a wave converging on the origin. For the outgoing wave, the solution fails at the origin because some source of sound is required to supply the energy carried away, and our wave equation does not contain any term representing this energy source. (See Sections 5.15 and 5.16.) In practice, this means that the medium must be excluded from some volume of space including the origin, and this volume must be occupied by whatever vibrating body serves as the sound source. For the incoming waves, energy is being focused at the origin and the small-amplitude approximations will fail. This failure will manifest itself in a *nonlinear* wave equation and strong acoustic losses limiting the attainable amplitudes.

The most important diverging spherical waves are harmonic. Such waves are represented in complex form by

$$\mathbf{p} = \frac{\mathbf{A}}{r}e^{j(\omega t - kr)} \qquad (5.11.6)$$

Use of the relationships developed in Section 5.5 for a general wave allows the other acoustic variables to be expressed in terms of the pressure

$$\mathbf{\Phi} = -\mathbf{p}/j\omega\rho_0 \tag{5.11.7}$$

$$\tilde{\mathbf{u}} = \nabla\mathbf{\Phi} = \hat{r}(1 - j/kr)\mathbf{p}/\rho_0 c \tag{5.11.8}$$

The observed acoustic variables are obtained by taking the real parts of (5.11.6)–(5.11.8).

It is apparent from (5.11.8) that, in contrast with plane waves, the particle speed is *not* in phase with the pressure. The specific acoustic impedance is not $\rho_0 c$, but rather

$$\mathbf{z} = \rho_0 c \frac{kr}{\left[1 + (kr)^2\right]^{1/2}} e^{j\theta} \tag{5.11.9}$$

or

$$\mathbf{z} = \rho_0 c \cos\theta\, e^{j\theta} \tag{5.11.10}$$

$$\cot\theta = kr \tag{5.11.11}$$

A geometric representation of θ is given in Fig. 5.11.1. As is true with many other acoustic phenomena, the product kr is the determining factor, rather than k or r separately. Since $kr = 2\pi r/\lambda$, the angle θ is a function of the ratio of the source distance to the wavelength. When the distance from the source is only a small fraction of a wavelength, the phase difference between the complex pressure and particle speed is large. At distances corresponding to a considerable number of wavelengths, $\mathbf{p}$ and $\mathbf{u}$ are very nearly in phase and the spherical wave assumes the characteristics of a plane wave. This is to be expected, since the wave fronts become essentially planar at great distances from the source.

Separating (5.11.9) into real and imaginary parts, we have

$$\mathbf{z} = \rho_0 c \frac{(kr)^2}{1 + (kr)^2} + j\rho_0 c \frac{kr}{1 + (kr)^2} \tag{5.11.12}$$

The first term is the specific acoustic resistance and the second term the specific acoustic reactance. Both approach zero for very small values of kr, but for very large values of kr the resistive term approaches $\rho_0 c$ and the reactive term approaches zero.

The absolute magnitude z of the specific acoustic impedance is equal to the ratio of the pressure amplitude P of the wave to its speed amplitude U,

$$z = P/U = \rho_0 c \cos\theta \tag{5.11.13}$$

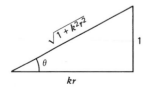

Figure 5.11.1   The relationship between θ and kr at a distance r from the source of a spherical wave of wave number k.

and the relationship between pressure and speed amplitude may be written as

$$P = \rho_0 c \, U \cos \theta \qquad (5.11.14)$$

For large values of kr, $\cos \theta$ approaches unity and the relationship between pressure and speed is that for a plane wave. As the distance from the source of a spherical acoustic wave to the point of observation is decreased, both kr and $\cos \theta$ decrease, so that larger and larger particle speeds are associated with a given pressure amplitude. For very small distances from a point source, the particle speed corresponding to even very low acoustic pressures becomes impossibly large: a source small compared to a wavelength is inherently incapable of generating waves of large intensity.

Let us rewrite (5.11.6) as

$$\mathbf{p} = \frac{\mathbf{A}}{r} e^{j(\omega t - kr)} \qquad (5.11.15)$$

where we have chosen a new origin of time so that $\mathbf{A}$ is a real constant A. Then A/r is the *pressure amplitude* of the wave. The pressure amplitude in a spherical wave is not constant, as it is for a plane wave, but decreases inversely with the distance from the source. The actual pressure is the real part of (5.11.15),

$$p = \frac{A}{r} \cos(\omega t - kr) \qquad (5.11.16)$$

Since $\mathbf{u} = \mathbf{p}/\mathbf{z}$, the corresponding complex expression for the particle speed is

$$\mathbf{u} = \frac{A}{r\mathbf{z}} e^{j(\omega t - kr)} \qquad (5.11.17)$$

Replacing $\mathbf{z}$ by (5.11.10) and then taking the real part of the resulting expression gives the actual particle speed,

$$u = \frac{1}{\rho_0 c} \frac{A}{r} \frac{1}{\cos \theta} \cos(\omega t - kr - \theta) \qquad (5.11.18)$$

It is apparent that, since θ is a function of kr, the speed amplitude

$$U = \frac{1}{\rho_0 c} \frac{A}{r} \frac{1}{\cos \theta} \qquad (5.11.19)$$

is not inversely proportional to the distance from the source.

For a harmonic spherical wave (5.9.1) yields

$$I = \frac{1}{T} \int_0^T P \cos(\omega t - kr) \, U \cos(\omega t - kr - \theta) \, dt = \frac{PU \cos \theta}{2} = \frac{P^2}{2\rho_0 c} \qquad (5.11.20)$$

where the factor $\cos \theta$ is analogous to the power factor of an alternating-current circuit. Note that the formula $I = P^2/2\rho_0 c$ is *exactly* true for both plane and spherical waves.

The average rate at which energy flows through a closed spherical surface of radius r surrounding a source of symmetric spherical waves is

$$\Pi = 4\pi r^2 I = 4\pi r^2 P^2 / 2\rho_0 c \qquad (5.11.21)$$

or since $p = A/r$

$$\Pi = 2\pi A^2 / \rho_0 c \qquad (5.11.22)$$

The average rate of energy flow through any spherical surface surrounding the origin is independent of the radius of the surface, a statement of energy conservation in a lossless medium.

5.12 DECIBEL SCALES

It is customary to describe sound pressures and intensities using logarithmic scales known as *sound levels*. One reason for this is the very wide range of sound pressures and intensities encountered in the acoustic environment; audible intensities range from approximately 10^{-12} to $10 \, \text{W}/\text{m}^2$. Using a logarithmic scale compresses the range of numbers required to describe this wide range of intensities and is also consistent with the fact that humans judge the relative loudness of two sounds by the ratio of their intensities.

The most generally used logarithmic scale for describing sound levels is the *decibel* (dB) scale. The *intensity level IL* of a sound of intensity I is defined by

$$IL = 10 \log(I/I_{ref}) \qquad (5.12.1)$$

where I_{ref} is a reference intensity, IL is expressed in *decibels referenced to I_{ref}* (dB re I_{ref}), and "log" represents the logarithm to base 10.

We have shown in Sections 5.9 and 5.11 that intensity and effective pressure of progressive plane and spherical waves are related by $I = P_e^2/\rho_0 c$. Consequently, the intensities in (5.12.1) may be replaced by expressions for pressure, leading to the sound pressure level

$$SPL = 20 \log(P_e/P_{ref}) \qquad (5.12.2)$$

where SPL is expressed in dB *re P_{ref}* with P_e the measured effective pressure amplitude of the sound wave and P_{ref} the reference effective pressure amplitude. If we choose $I_{ref} = P_{ref}^2/\rho_0 c$, then IL re $I_{ref} = SPL$ re P_{ref}.

Throughout the scientific disciplines a number of units are used to specify pressures, and many of these are found in acoustics. In addition, reference levels of various degrees of antiquity are encountered. Let us first catalog a few units:

CGS units

1 dyne/cm^2, also called the microbar (μbar). (The microbar was originally 10^{-6} atm but is now *defined* as 1 dyne/cm^2.)

MKS units

1 pascal (Pa), *defined* as 1 N/m^2 in the SI system of units

Others

$$1 \text{ atmosphere (atm)} \equiv 1.01325 \times 10^5 \text{ Pa} = 1.01325 \times 10^6 \mu\text{bar}$$
$$1 \text{ kilogram/cm}^2 \text{ (kgf/cm}^2) \equiv 0.980665 \times 10^5 \text{ Pa} = 0.967841 \text{ atm}$$

Equivalents

$$1 \mu\text{bar} \equiv 0.1 \text{ N/m}^2 \equiv 10^5 \text{ } \mu\text{Pa}$$

The reference standard for airborne sounds is 10^{-12} W/m^2, which is approximately the intensity of a 1 kHz pure tone that is just barely audible to a person with unimpaired hearing. Substitution of this intensity into (5.9.2) shows that it corresponds to a peak pressure amplitude of

$$P = (2\rho_0 c I)^{1/2} = 2.89 \times 10^{-5} \text{ Pa} \tag{5.12.3}$$

or a corresponding effective (root-mean-square) pressure of

$$P_e = P/\sqrt{2} = 20.4 \text{ } \mu\text{Pa} \tag{5.12.4}$$

This latter pressure, rounded to 20 μPa, is the reference for sound pressure levels in air. Essentially identical numerical results are obtained in air using either 10^{-12} W/m^2 in (5.12.1) or 20 μPa in (5.12.2) for plane or spherical progressive waves. However, in certain more complex sound fields, such as standing waves, intensity and pressure are no longer simply related by (5.9.5) and (5.11.20) and consequently (5.12.1) and (5.12.2) will not yield identical results. Since the voltage outputs of microphones and hydrophones commonly used in acoustic measurements are proportional to pressure, sound pressure levels are used more widely than intensity levels.

Three different pressures are encountered as reference pressures in underwater acoustics. One is an effective pressure of 20 μPa (the same as the reference pressure in air). The second reference pressure is 1 μbar and the third is 1 μPa. The last is now the standard.

This abundance of reference pressures can lead to confusion unless care is taken to always specify the reference pressure being used: *SPL re* 20 μPa, *re* 1 μPa, or *re* 1 μbar. Table 5.12.1 summarizes the various conventions.

From the above discussion, note that a given acoustic pressure in air corresponds to a much higher intensity than does the same acoustic pressure in water. Since (5.9.5) or (5.11.20) shows that, for a given pressure amplitude, intensity is inversely

Table 5.12.1   References and conversions for sound pressure levels

Medium	Reference	Nearly equivalent to
Air	10^{-12} W/m^2 20 μPa = 0.0002 μbar	20 μPa 10^{-12} W/m^2
Water	1 μbar = $10^5 \mu$Pa 0.0002 μbar = 20μPa 1 μPa	6.76×10^{-9} W/m^2 2.70×10^{-16} W/m^2 6.76×10^{-19} W/m^2

SPL re 1 μbar + 100 = *SPL re* 1 μPa
SPL re 0.0002 μbar − 74 = *SPL re* 1 μbar
SPL re 0.0002 μbar + 26 = *SPL re* 1 μPa

proportional to the characteristic impedance of the medium, the ratio of the intensity in air to that in water for the *same acoustic pressure* is $(1.48 \times 10^6)/415 = 3570$. On the other hand, if we compare two acoustic waves of the *same frequency and particle displacement*, the ratio of the intensity in air to that in water is $1/3570$.

Because of the conveniences afforded by decibel scales, electrical quantities are often specified in terms of levels. For example, the voltage level VL is defined as

$$VL(re\ V_{ref}) = 20\log(V/V_{ref}) \tag{5.12.5}$$

where V is the effective voltage and V_{ref} is some convenient reference effective voltage.

By convention, the subscript "e" and the adjective "effective" are omitted when specifying effective amplitudes of electrical quantities. Two common reference voltages are 1 V and 0.775 V. (This latter stems from an old reference, the voltage required to dissipate 1 mW of electrical power in a 600 ohm resistor.) Comparison of voltage levels referenced to the two common reference voltages reveals that

$$VL(re\ 0.775\ \text{V}) = VL(re\ 1\ \text{V}) + 2.21 \tag{5.12.6}$$

The abilities of electroacoustic sources and receivers to convert between electrical and acoustic quantities can be expressed in terms of *sensitivities*. For example, the *open circuit receiving sensitivity* $\mathcal{M}_o$ of a microphone is defined as

$$\mathcal{M}_o = (V/P_e)_{I=0} \tag{5.12.7}$$

where V is the output voltage produced (with negligible output current I) when the microphone is placed at a point where the effective pressure amplitude was P_e in the absence of the microphone. This is one of a number of sensitivities that can be defined for a microphone; more detail will be found in Chapter 14. A sensitivity $\mathcal{M}$ is usually expressed in terms of the associated *sensitivity level* $\mathcal{ML}$

$$\mathcal{ML}(re\ \mathcal{M}_{ref}) = 20\log(\mathcal{M}/\mathcal{M}_{ref}) \tag{5.12.8}$$

where $\mathcal{M}_{ref}$ is a reference sensitivity such as 1 V/μbar or 1 V/Pa.

Relationships among P, V, and $\mathcal{M}_o$ can be expressed in terms of either the fundamental quantities or the associated levels. For example, assume that a microphone of known sensitivity level $\mathcal{ML}$ dB *re* $\mathcal{M}_{ref}$ gives an output level VL dB *re* V_{ref}, and we wish to know the sound pressure level SPL dB *re* P_{ref} of the sound field. Algebraic manipulation reveals

$$SPL(re\ P_{ref}) = VL(re\ V_{ref}) - \mathcal{ML}(re\ \mathcal{M}_{ref}) + 20\log\left(\frac{V_{ref}/P_{ref}}{\mathcal{M}_{ref}}\right) \tag{5.12.9}$$

In complete analogy, an acoustic source is characterized by a source sensitivity $\mathcal{S} = P_e/V$ and a source sensitivity level $\mathcal{SL}$

$$\mathcal{SL}\left(re\ \mathcal{S}_{ref}\right) = 20\log\left(\frac{P_e/V}{\mathcal{S}_{ref}}\right) \tag{5.12.10}$$

where V is the voltage applied to the electrical input of the source, P_e is the effective pressure at some specified location (usually on the acoustic axis of the source extrapolated back from large distances to 1 m from the face of the source), and $\mathscr{S}_{ref}$ is a reference sensitivity such as 1 μPa/V or 1 μbar/V.

*5.13 CYLINDRICAL WAVES

Three-dimensional cylindrical waves have significant applications in atmospheric and underwater propagation. The wave equation for cylindrical propagation is (5.5.4) with the Laplacian expressed in cylindrical coordinates,

$$\left(\frac{\partial^2}{\partial r^2} + \frac{1}{r}\frac{\partial}{\partial r} + \frac{1}{r^2}\frac{\partial^2}{\partial \theta^2} + \frac{\partial^2}{\partial z^2}\right)p = \frac{1}{c^2}\frac{\partial^2 p}{\partial t^2} \tag{5.13.1}$$

Recall that the physical interpretation of r depends on the coordinates being used. In spherical coordinates, r denotes the *radial distance from the origin* to the field point in any direction. In cylindrical coordinates it refers to the *perpendicular distance from the z axis* to the field point.

Assuming harmonic solutions and separation of variables,

$$\mathbf{p}(r,\theta,z,t) = \mathbf{R}(r)\Theta(\theta)\mathbf{Z}(z)e^{j\omega t} \tag{5.13.2}$$

allows (5.13.1) to be decomposed into three differential equations and provides a relationship for the separation constants,

$$\frac{d^2\mathbf{R}}{dr^2} + \frac{1}{r}\frac{d\mathbf{R}}{dr} + \left(k_r^2 - \frac{m^2}{r^2}\right)\mathbf{R} = 0$$

$$\frac{d^2\mathbf{Z}}{dz^2} + k_z^2\mathbf{Z} = 0 \tag{5.13.3}$$

$$\frac{d^2\Theta}{d\theta^2} + m^2\Theta = 0$$

$$(\omega/c)^2 = k^2 = k_r^2 + k_z^2$$

The equation for Θ is the same as for the circular membrane. If we assume azimuthal symmetry, then $m = 0$. The equation for $\mathbf{Z}$ is solved by sinusoidal or complex exponential functions and corresponds to oblique waves whose propagation vectors have a projection on the z axis of k_z. The simplest case is $k_z = 0$, which describes waves whose surfaces of constant phase are cylinders concentric with the z axis. These two simplifications leave us with the z-independent, cylindrically symmetric solutions of the radial wave equation

$$\frac{d^2\mathbf{R}}{dr^2} + \frac{1}{r}\frac{d\mathbf{R}}{dr} + k^2\mathbf{R} = 0 \tag{5.13.4}$$

Reference to Section 4.4 and use of $m = 0$ gives the general solution

$$\mathbf{p}(r,t) = [\mathbf{A}J_0(kr) + \mathbf{B}Y_0(kr)]e^{j\omega t} \tag{5.13.5}$$

Since Y_0 diverges as $r \to 0$, (5.13.5) fails at $r = 0$ unless $\mathbf{B} = 0$. The reasons for this failure are identical with those discussed for spherical waves in Section 5.11, so when $\mathbf{B} \neq 0$ the z-axis must be excluded from the volume within which (5.13.5) can be applied.

Examination of (5.13.5) reveals that if $\mathbf{p}$ is to be a traveling wave, it must be a complex function of space. Furthermore, assuming that $I = P^2/2\rho_0 c$ is at least approximately true at large distances and using conservation of energy suggests that the pressure $\mathbf{p}(r,t)$ should be proportional to

$$\frac{1}{\sqrt{r}}\, e^{j(\omega t \pm kr)} \tag{5.13.6}$$

The $\pm$ sign in the exponent gives incoming or outgoing waves. The combinations of **A** and **B** that will produce (5.13.6) in the limit $r \to \infty$ can be found from the large-argument asymptotic forms of J_0 and Y_0,

$$J_0(kr) \to (2/\pi kr)^{1/2} \cos(kr - \pi/4)$$
$$Y_0(kr) \to (2/\pi kr)^{1/2} \sin(kr - \pi/4) \tag{5.13.7}$$

Equation (5.13.5) will take on the form (5.13.6) if $\mathbf{B} = \pm j\mathbf{A}$. These combinations are the Bessel functions of the third kind, or *Hankel functions*,

$$H_0^{(1)}(kr) = J_0(kr) + jY_0(kr)$$
$$H_0^{(2)}(kr) = J_0(kr) - jY_0(kr) \tag{5.13.8}$$

For an outgoing harmonic cylindrical wave with azimuthal symmetry and independent of z, the appropriate solution of (5.13.4) is

$$\mathbf{p}(r,t) = \mathbf{A}H_0^{(2)}(kr)e^{j\omega t} \tag{5.13.9}$$

While (5.13.9) was developed by imposing the asymptotic behavior (5.13.6) and using the asymptotic form of the Hankel function for large kr, it is an exact solution of (5.13.4) for all $r > 0$. (This is often referred to as imposing a *radiation boundary condition at infinity*.) For large kr this solution has asymptotic behavior

$$\mathbf{p}(r,t) \to \mathbf{A}(2/\pi kr)^{1/2}e^{j(\omega t - kr + \pi/4)} \tag{5.13.10}$$

Generating the velocity potential $\mathbf{\Phi}$ with (5.5.9), and then using (5.5.7) gives the particle speed

$$\mathbf{u}(r,t) = -j\left(\mathbf{A}/\rho_0 c\right)H_1^{(2)}(kr)e^{j\omega t} \tag{5.13.11}$$

with the help of Appendix A4. The specific acoustic impedance **z** follows at once:

$$\mathbf{z} = j\rho_0 c H_0^{(2)}(kr)/H_1^{(2)}(kr) \tag{5.13.12}$$

In the limit $kr \gg 1$, the asymptotic approximations of the Hankel functions show that $\mathbf{z} \to \rho_0 c$ at large distances. This is to be expected, since as kr increases beyond unity, the radii of curvature of the surfaces of constant phase become much larger than a wavelength and the waveform looks locally more and more like a plane wave.

Calculation of the acoustic intensity is a little more complicated. The instantaneous intensity is $I(r,t) = pu$. This yields

$$I(r,t) = (A^2/\rho_0 c)\left[J_0(kr)\cos\omega t + Y_0(kr)\sin\omega t\right]\left[J_1(kr)\sin\omega t - Y_1(kr)\cos\omega t\right] \tag{5.13.13}$$

where for ease we have chosen time so that $\mathbf{A} = A$. Taking the time average leaves us with the intensity

$$I(r) = (A^2/2\rho_0 c)\left[J_1(kr)Y_0(kr) - J_0(kr)Y_1(kr)\right] \tag{5.13.14}$$

The quantity in square brackets is the *Wronskian* of $J_0(kr)$ and $Y_0(kr)$ and has the known value $2/\pi kr$. Substitution gives us the result

$$I(r) = \frac{2A^2/\pi kr}{2\rho_0 c} = \frac{P_{as}^2}{2\rho_0 c} \tag{5.13.15}$$

where P_{as} is the *asymptotic* amplitude

$$P_{as} = A(2/\pi kr)^{1/2} \tag{5.13.16}$$

of $\mathbf{p}(r,t)$. The intensity falls off as $1/r$, as conservation of energy in a lossless fluid says it must for a cylindrically diverging wave, but the intensity is *not* simply $P^2/2\rho_0 c$ everywhere, as it was for plane and spherical waves.

*5.14 RAYS AND WAVES

Up to this point, we have considered the propagation of sound in a homogeneous medium having a constant speed of sound. The speed of sound is often a function of space and instead of plane, spherical, and cylindrical waves of infinite spatial extent we find waves whose directions of propagation change as they traverse the medium. One technique for studying this effect is based on the assumption that the energy is carried along reasonably well-defined paths through the medium, so that it is useful to think of *rays* rather than waves. In many cases, description in terms of rays is much easier than in terms of waves. However, rays are not exact replacements for waves, but only approximations that are valid under certain rather restrictive conditions.

(a) The Eikonal and Transport Equations

The wave equation with spatially dependent sound speed is

$$\left(\nabla^2 - \frac{1}{c^2(x,y,z)}\frac{\partial^2}{\partial t^2}\right)\mathbf{p}(x,y,z,t) = 0 \tag{5.14.1}$$

For sound traversing such a fluid, the amplitude varies with position and the surfaces of constant phase can be complicated. Assume a trial solution

$$\mathbf{p}(x,y,z,t) = A(x,y,z)e^{j\omega[t-\Gamma(x,y,z)/c_0]} \tag{5.14.2}$$

where Γ has units of length and c_0 is a reference speed to be defined later. The quantity Γ/c_0 is the *eikonal.* The values of (x,y,z) for which Γ is constant define the surfaces of constant phase. From the basic definition of the gradient, $\nabla\Gamma$ is everywhere perpendicular to these surfaces.

Substituting the trial solution into (5.14.1) and collecting real and imaginary parts gives

$$-\frac{\nabla^2 A}{A} + \left(\frac{\omega}{c_0}\right)^2 \nabla\Gamma \cdot \nabla\Gamma = \left(\frac{\omega}{c}\right)^2$$

$$2\frac{\nabla A}{A} \cdot \nabla\Gamma + \nabla^2\Gamma = 0 \tag{5.14.3}$$

These equations are difficult to solve because they are coupled and nonlinear. However, if we require

$$\left|\frac{\nabla^2 A}{A}\right| \ll \left(\frac{\omega}{c}\right)^2 \tag{5.14.4}$$

then the first of (5.14.3) assumes the simpler approximate form

$$\nabla\Gamma \cdot \nabla\Gamma = (c_0/c)^2 = n^2 \qquad (5.14.5)$$

where $n = c_0/c$ is the *index of refraction*. Equation (5.14.5) is the *eikonal equation*. It is immediately clear that

$$\nabla\Gamma = n\hat{s} \qquad (5.14.6)$$

where the unit vector $\hat{s}$ gives the local direction of propagation. Given $\hat{s}$ at a point in the sound field and then tracing how that specific $\hat{s}$ changes direction as it is advanced point to point within the fluid defines a *ray path*, the trajectory followed by the particular ray. Since according to (5.14.6) the local direction of propagation of the ray is perpendicular to the eikonal, in this approximation each ray is always perpendicular to the local surface of constant phase. Sufficient conditions for satisfying (5.14.4) are (1) the amplitude of the wave and (2) the speed of sound do not change significantly over distances comparable to a wavelength. If we consider a beam of sound with transverse dimensions much greater than a wavelength traveling through a fluid, (5.14.4) states that the eikonal equation may be applied over the central portion of the beam where A is not rapidly varying. At the edges of the beam, however, A may rapidly reduce to zero over distances on the order of a wavelength and the restriction (5.14.4) fails. The failure manifests itself in the *diffraction* of sound at the edges of the beam—analogous to the diffraction of light through a slit or pinhole. This means that (5.14.5) is accurate only in the limit of high frequencies—how high depends on the spatial variations of c and A. More rigorous *necessary* conditions can be stated, but their physical meanings are less direct. Indeed, there are propagating waves (Problem 5.14.10) that do not satisfy the sufficient conditions, but for which (5.14.5) is valid.

Analysis of the *transport equation*, the second of (5.14.3), will provide further justification for the concept of rays. Substitution of (5.14.6) into this equation and a little manipulation (Problem 5.14.4a) gives

$$\frac{d}{ds}\ln(nA^2) = -\nabla \cdot \hat{s} \qquad (5.14.7)$$

For distances more than a few wavelengths away from the source, the intensity is

$$I = P^2/2\rho_0 c = nA^2/2\rho_0 c_0 \qquad (5.14.8)$$

so that (5.14.7) becomes

$$\frac{1}{I}\frac{dI}{ds} = -\nabla \cdot \hat{s} \qquad (5.14.9)$$

The left side is the fractional change of intensity per unit distance *along a ray path* and $\nabla \cdot \hat{s}$ describes how the rays converge or diverge. Now apply Gauss's theorem to the volume defined by the bundle of rays shown in Fig. 5.14.1. The volume is chosen so that the rays pass only through the end caps. Integrate (5.14.9) over the volume $S\,\Delta h$. On the left side the volume integral becomes $(1/I)(dI/ds)S\,\Delta h = S[d(\ln I)/ds]\,\Delta h$. On the right side, use of Gauss's theorem converts the volume integral into a surface integral of $\hat{s} \cdot \hat{n}$. Since the rays enter and leave the volume only through the end caps, this integral yields the incremental change $-\Delta S$ in the cross-sectional area of the bundle of rays. Finally, recognize that ΔS is obtained along the ray path, so that $\Delta S = (dS/ds)\,\Delta h$. This gives us $d(\ln I)/ds = -d(\ln S)/ds$

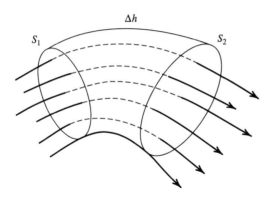

Figure 5.14.1   An elemental volume of a ray bundle with end caps of areas S_1 and S_2 separated by a distance Δh along the rays.

and the result

$$IS = constant \tag{5.14.10}$$

Thus, within the limitations of the eikonal equation, the energy within a ray bundle remains constant. This is the mathematical justification for the intuitive concept that energy in a sound wave travels along rays. Any mathematical or geometrical technique that allows a bundle of rays to be traced through space will allow calculation of the intensity throughout space.

(b) The Equations for the Ray Path

Solution of the eikonal equation (5.14.6) gives the direction $\hat{s}$ for each ray at every point along its path. The problem of obtaining the ray paths is equivalent to solving for the successive locations of $\hat{s}$. First, express $\hat{s}$ in terms of its direction cosines,

$$\hat{s} = \alpha\hat{x} + \beta\hat{y} + \gamma\hat{z}$$
$$\alpha^2 + \beta^2 + \gamma^2 = 1 \tag{5.14.11}$$

where the direction cosines are $\alpha = dx/ds$, $\beta = dy/ds$, and $\gamma = dz/ds$ with dx, dy, and dz the coordinate changes resulting from a step ds in the $\hat{s}$ direction along the ray path. If the change in any scalar along the ray

$$\frac{d}{ds} = \alpha\frac{\partial}{\partial x} + \beta\frac{\partial}{\partial y} + \gamma\frac{\partial}{\partial z} \tag{5.14.12}$$

is applied to both sides of the first of (5.14.11), the components become

$$\frac{d}{ds}(n\alpha) = \frac{\partial n}{\partial x}$$

$$\frac{d}{ds}(n\beta) = \frac{\partial n}{\partial y} \tag{5.14.13}$$

$$\frac{d}{ds}(n\gamma) = \frac{\partial n}{\partial z}$$

(See Problem 5.14.4b for details.) The eikonal equation relates the changes in the direction of propagation of a ray to the gradient of the local index of refraction. Given $n(x, y, z)$, it is possible to trace the trajectories of every element of a wave front through the medium. A simple example follows.

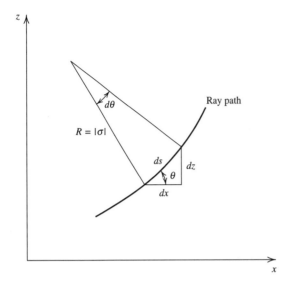

Figure 5.14.2   An element of a ray path in the x-z plane of length ds making an angle θ with the x axis will have a radius of curvature $R = |c/(g\cos\theta)|$, where c is the speed of sound and g is the sound speed gradient.

(c) The One-Dimensional Gradient

The speed of sound can often be considered a function of only one spatial dimension. In both the ocean and the atmosphere, for example, variations of the speed of sound with horizontal range are generally much weaker than the variations with depth or height.

Let the index of refraction be a function of z alone, where z is the vertical coordinate. Then (5.14.13) becomes

$$\frac{d}{ds}(n\alpha) = 0$$

$$\frac{d}{ds}(n\beta) = 0 \qquad\qquad (5.14.14)$$

$$\frac{d}{ds}(n\gamma) = \frac{dn}{dz}$$

If the coordinate axes are oriented so that a ray starts off in the x-z plane and makes an angle θ with the x axis (see Fig. 5.14.2), the initial value of β is zero and, according to the second of the above equations, β will remain zero and the ray path will stay in the x-z plane. We can then identify $\alpha = \cos\theta$ and $\gamma = \sin\theta$, and the remaining equations in (5.14.14) become

$$\frac{d}{ds}(n\cos\theta) = 0$$
$$\qquad\qquad (5.14.15)$$
$$\frac{d}{ds}(n\sin\theta) = \frac{dn}{dz}$$

The first of (5.14.15) reveals that $n\cos\theta$ must have the same value at every point along a particular ray path. If we specify the angle of elevation θ_0 where the ray path encounters the reference speed c_0, we then have a statement of *Snell's law*,

$$\boxed{\frac{\cos\theta}{c} = \frac{\cos\theta_0}{c_0}} \qquad\qquad (5.14.16)$$

From the definition of $n = c_0/c$ we see that dc/dz has the opposite sign from dn/dz. Then, the second of (5.14.15) shows that when the sound speed increases in the z direction, θ must decrease along the ray—the ray turns toward the lower sound speed. When the sound speed decreases in the z direction, θ increases along the ray—the ray still turns toward the lower sound speed. A ray *always bends toward the neighboring region of lower sound speed.* While this equation cannot be solved without knowing the dependence of c on z, it can be put into a geometrical form. With reference to Fig. 5.14.2, $dz = \sin\theta\,ds$ and $ds = \sigma\,d\theta$, where σ is a measure of the amount and orientation of the curvature of the ray path. For Fig. 5.14.2, $d\theta$ increases along the ray, so σ is *positive*. If the ray path were to curve the other way (with negative second derivative), σ would be *negative*. The magnitude of σ is the *radius of curvature* R. Use of these geometrical relationships along with (5.14.15) and (5.14.16) gives

$$\sigma = -\frac{1}{g}\frac{c_0}{\cos\theta_0}$$

$$g = \frac{dc}{dz}$$

(5.14.17)

where g is the *gradient of the sound speed.* The radius of curvature R of the ray is inversely proportional to $|g|$ at each point along the path. Each ray path must be computed separately since each has its own value of the Snell's law constant $(\cos\theta_0)/c_0$. See Chapter 15 for examples of ray tracing when g is piecewise constant.

(d) Phase and Intensity Considerations

Let I_1 be the acoustic intensity referred to a distance 1 meter from a source along a bundle of rays with initial angle of elevation θ_0. It is desired to know the intensity I of this bundle at some range x as shown in Fig. 5.14.3. For a lossless medium, the intensity multiplied by the cross-sectional area of the bundle must be constant. Let S_1 be the cross section of the bundle at 1 meter from the source and S the cross section at range x, where the intensity is I. Examination of the geometry of the figure reveals $S = x\,\Delta\phi\,\sin\theta\,dx$ and $S_1 = \Delta\phi\,d\theta_0\cos\theta_0$

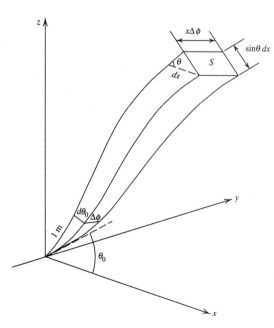

Figure 5.14.3   A ray bundle in the x-z plane that is used to determine intensity from conservation of energy. At $x = 1$ m the cross-sectional area of the bundle is $\Delta\phi\,d\theta_0\cos\theta_0$, where $\Delta\phi$ is the horizontal angular width of the bundle, $d\theta_0$ its initial vertical angular width, and θ_0 the initial angle of elevation. The area at range x where the ray makes an angle θ with the horizontal is $x\,\Delta\phi\sin\theta\,dx$, where $dx = (\partial x/\partial\theta_0)_z\,d\theta_0$.

so that conservation of energy provides $Ix \sin \theta \, dx = I_1 \cos \theta_0 \, d\theta_0$. The element dx can be expressed by $dx = (\partial x / \partial \theta_0)_z \, d\theta_0$, where the range x must be written as a function of θ_0 and z. Combination of the above equations results in

$$\frac{I}{I_1} = \frac{1}{x} \frac{\cos \theta_0}{\sin \theta} \frac{1}{(\partial x / \partial \theta_0)_z} \tag{5.14.18}$$

When neighboring rays from the source intersect at some field point, the partial derivative vanishes and the intensity becomes infinite. The *locus* of such neighboring points may form a surface of infinite intensity, called a *caustic*. The intensity does not really become infinite on a caustic, of course, because the conditions necessary for the validity of the eikonal equation fail. Caustics do, however, identify regions of high intensity where there is strong focusing of acoustic energy.

A different situation can occur when *nonadjacent* ray paths intersect at some point away from the source. An example would be reflection from a boundary, for which the direct and reflected ray paths intersect. For a continuous monofrequency signal generated by the source, there are two different approaches to this combination:

1. **Incoherent Summation.** If spatial irregularities and fluctuations in the boundary or the speed of sound profile are sufficient to randomize the relative phases of the signals propagating over intersecting ray paths, then we can make a *random phase approximation*. Under this approximation, a reasonable estimate of the average acoustic intensity where the different paths intersect is the sum of the intensities for the individual rays. The acoustic pressure amplitude is then the square root of the sum of the squares of the pressure amplitudes of the signals where they intersect.

2. **Coherent Summation.** If, however, the irregularities in propagation do not appreciably affect the phases of the signals, *phase coherence* is retained and it is necessary to calculate the travel time Δt of each signal along its path so that the relative phases can be obtained. The total pressure and phase of the combination is then obtained by adding the phasors with proper regard for the phases.

A typical case of continuous wave propagation may lie somewhere between these two idealizations. Coherence is favored by short-range, low-frequency, smooth boundaries, few boundary reflections, and a stable and smooth speed of sound profile. Random phasing is favored by the converse conditions. The travel time can be calculated in a number of ways, each simple to derive:

$$\Delta t = \int_0^s \frac{1}{c} \, ds = \int_{x_0}^x \frac{1}{c \cos \theta} \, dx = \int_{z_0}^z \frac{1}{c \sin \theta} \, dz = \int_{\theta_0}^\theta \frac{1}{g \cos \theta} \, d\theta \tag{5.14.19}$$

where each integrand must be expressed as a function of the variable of integration.

For very short transient acoustic signals, the travel times along the various ray paths may be so different that the individual arrivals do not overlap each other. This would then yield a combined signal in which each of the arrivals along a different ray path would be separate and distinct. As the transients become longer, however, partial overlapping would generate a complicated combination.

*5.15 THE INHOMOGENEOUS WAVE EQUATION

In previous sections we developed a wave equation that applied to regions of space devoid of any sources of acoustic energy. However, a source must be present to generate an acoustic field. Certain sources internal to the region of interest can be taken into account by introducing time-dependent boundary conditions, as described for strings, bars, and membranes. In Chapter 7, this is the procedure that will be used to relate the motion of the surface of a source to the sound field created by the source. However, there are times when it is more convenient to adopt an approach that builds the sources into the wave equation by modifying the fundamental equations to include source terms.

1. If mass is injected (or appears to be) into the space at a rate per unit volume $G(\vec{r}, t)$, the linearized equation of continuity becomes

$$\rho_0 \frac{\partial s}{\partial t} + \nabla \cdot (\rho_0 \vec{u}) = G(\vec{r}, t) \tag{5.15.1}$$

This $G(\vec{r}, t)$ is generated by a closed surface that changes volume, such as the outer surface of an explosion, an imploding evacuated glass sphere, or a loudspeaker in an enclosed cabinet.

2. If there are *body forces* present in the fluid, a body force per unit volume $\vec{F}(\vec{r}, t)$ must be included in Euler's equation. The linearized equation of motion becomes

$$\rho_0 \frac{\partial \vec{u}}{\partial t} + \nabla p = \vec{F}(\vec{r}, t) \tag{5.15.2}$$

Examples of this kind of force are those produced by a source that moves through the fluid without any change in volume, such as the cone of an unbaffled loudspeaker or a vibrating sphere of constant volume.

   If these two modifications are combined with the linearized equation of state, an inhomogeneous wave equation is obtained,

$$\nabla^2 p - \frac{1}{c^2} \frac{\partial^2 p}{\partial t^2} = -\frac{\partial G}{\partial t} + \nabla \cdot \vec{F} \tag{5.15.3}$$

3. A third type of sound source was first described by Lighthill[2] in 1952. Lighthill's result includes the effects of shear and bulk viscosity and its derivation is beyond the scope of this text. However, in virtually all cases of practical interest, the contributions from viscous forces are completely negligible and a simplified derivation can be made. The source of acoustic excitation lies in the convective term $(\vec{u} \cdot \nabla)\vec{u}$ of the acceleration. Retaining this term and discarding the terms involving viscosity and gravity in (5.4.8) gives

$$-\nabla p = \rho \left(\frac{\partial \vec{u}}{\partial t} + (\vec{u} \cdot \nabla)\vec{u} \right) = \frac{\partial (\rho \vec{u})}{\partial t} - \vec{u} \frac{\partial \rho}{\partial t} + \rho(\vec{u} \cdot \nabla)\vec{u} \tag{5.15.4}$$

Use the nonlinear continuity equation (5.3.3) to replace $\vec{u}(\partial \rho / \partial t)$ with $-\vec{u}\nabla \cdot (\rho \vec{u})$, take the time derivative of (5.3.3) and the divergence of (5.15.4), eliminate the common term, and use (5.5.6) to express ρ in terms of p in the linear term. The result is an inhomogeneous wave equation

$$\nabla^2 p - \frac{1}{c^2} \frac{\partial^2 p}{\partial t^2} = -\nabla \cdot [\vec{u}\nabla \cdot (\rho \vec{u}) + \rho(\vec{u} \cdot \nabla)\vec{u}] \tag{5.15.5}$$

The source term can be given direct physical meaning if it is rewritten

$$\nabla^2 p - \frac{1}{c^2} \frac{\partial^2 p}{\partial t^2} = -\frac{\partial^2 (\rho u_i u_j)}{\partial x_i \partial x_j} \tag{5.15.6}$$

Tensor notation has been used for economy of notation. The subscripts i and j take on the values 1, 2, and 3 and represent the x, y, and z directions. A *summation convention* is

[2]Lighthill, *Proc. R. Soc. (London) A*, **211**, 564 (1952).

used, wherein if any subscript appears more than once, it is assumed that subscript is summed over all its values. For example, $\partial u_i/\partial x_i$ is equivalent to $\nabla \cdot \vec{u}$ and $u_j(\partial u_i/\partial x_j)$ is equivalent to $(\vec{u} \cdot \nabla)\vec{u}$. Thus, there are nine quantities in the source term. This source term describes the spatial rates of change of momentum flux *within* the fluid, and Lighthill showed that it is responsible for the sounds produced by regions of *turbulence*, as in the exhaust of a jet engine. [See Problem 5.15.3 to show that the source term in (5.15.5) is equivalent to that in (5.15.6).]

It can be seen that each of these three source terms [described separately in (1), (2), and (3) above] arises independently, so that the complete inhomogeneous lossless wave equation accounting for mass injection, body forces, and turbulence is

$$\nabla^2 p - \frac{1}{c^2}\frac{\partial^2 p}{\partial t^2} = -\frac{\partial G}{\partial t} + \nabla \cdot \vec{F} - \frac{\partial^2 \left(\rho u_i u_j\right)}{\partial x_i \, \partial x_j} \tag{5.15.7}$$

The effects of gravity could be included by adding a term $\nabla \cdot (\rho_0 \vec{g} s)$ to the left side of (5.15.7) and speed of sound profiles by considering c a function of position. The sources on the right side of (5.15.7) will be related to *monopole*, *dipole*, and *quadrupole* radiation in Section 7.10.

*5.16 THE POINT SOURCE

The monofrequency spherical wave given by (5.11.15) is a solution to the homogeneous wave equation (5.5.4) everywhere except at $r = 0$. (This is consistent with the fact that there must be a source at $r = 0$ to generate the wave.) However, (5.11.15) does satisfy the inhomogeneous wave equation

$$\nabla^2 \mathbf{p} - \frac{1}{c^2}\frac{\partial^2 \mathbf{p}}{\partial t^2} = -4\pi A\delta(\vec{r})e^{j\omega t} \tag{5.16.1}$$

for all $\vec{r}$. The three-dimensional delta function $\delta(\vec{r})$ is defined by

$$\int_V \delta(\vec{r})\, dV = \begin{cases} 1 & \vec{r} = 0 \in V \\ 0 & \vec{r} = 0 \notin V \end{cases} \tag{5.16.2}$$

To prove this, multiply both sides of (5.16.1) by dV, integrate over a volume V that includes $\vec{r} = 0$, and use (5.16.2) to evaluate the delta function integral and Gauss's theorem to reduce the volume integral to a surface integral. This gives

$$\int_S \nabla \mathbf{p} \cdot \hat{n}\, dS - \frac{1}{c^2} \int_V \frac{\partial^2 \mathbf{p}}{\partial t^2}\, dV = -4\pi A e^{j\omega t} \tag{5.16.3}$$

where $\hat{n}$ is the unit outward normal to the surface S of V. Now, substitute (5.11.15) for p and carry out the surface integration over a sphere centered on $\vec{r} = 0$. See Problem 5.16.1.

To generalize to a point source located at $\vec{r} = \vec{r}_0$, make the appropriate change of variable in (5.11.15):

$$\mathbf{p} = \frac{A}{|\vec{r} - \vec{r}_0|} \exp\left[j\left(\omega t - k|\vec{r} - \vec{r}_0|\right)\right] \tag{5.16.4}$$

This is a solution of

$$\nabla^2 \mathbf{p} - \frac{1}{c^2}\frac{\partial^2 \mathbf{p}}{\partial t^2} = -4\pi A\delta(\vec{r} - \vec{r}_0)\, e^{j\omega t} \tag{5.16.5}$$

In the proper circumstances, incorporation of a point source directly into the wave equation provides considerable mathematical simplification. (See Sections 7.10 and 9.7–9.9 as examples.) We will, however, use this formalism only when necessary, utilizing in most cases methods more closely related to elementary physical intuition.

PROBLEMS

5.2.1. (a) Linearize (5.2.3) by assuming $s \ll 1$. Then, by comparing this result with (5.2.5), obtain the adiabatic bulk modulus of a perfect gas in terms of $\mathcal{P}_0$ and γ. (b) With the help of (5.2.1) applied to equilibrium conditions, obtain the temperature dependence of $\mathcal{B}$ at constant volume.

5.2.2. Another form of the perfect gas law is $\mathcal{P}V = n\mathcal{R}T_K$, where n is the number of moles and $\mathcal{R} = 8.3143 \text{ J}/(\text{mol·K})$ is the universal gas constant (the mole is the molecular weight M in grams). Obtain a relationship between r and $\mathcal{R}$. Evaluate $\mathcal{R}$ in J/(kmol·K) (the kilomole is the molecular weight in kilograms).

5.2.3. If the adiabat for a fluid is presented in the form $\mathcal{P} = \mathcal{P}_0 + A[(\rho - \rho_0)/\rho_0] + \frac{1}{2}B[(\rho - \rho_0)/\rho_0]^2$ and is to be written as $(\mathcal{P}/\mathcal{P}_0) = (\rho/\rho_0)^a$, find an approximate expression for the exponent a. *Hint:* Expand $(\rho/\rho_0)^a$ about ρ_0 through second order and equate coefficients. Relate the results to (5.2.9) and (5.2.7).

5.2.4. The major constituents in *standard air* and the percentage and molecular weight in grams of each are: nitrogen (N_2), 78.084, 28.0134; oxygen (O_2), 20.948, 31.9988; argon (Ar), 0.934, 39.948; carbon dioxide (CO_2), 0.031, 44.010. (a) Calculate the effective molecular weight of air. (b) Obtain the specific gas constant $r = \mathcal{R}/M$ for air and compare with the value listed in Appendix A1.

5.3.1. From the linear continuity equation (5.3.4), show that the condensation and particle displacement are related by $s = -\nabla \cdot \vec{\xi}$. *Hint:* Assume ρ_0 is independent of time. The integral of (5.3.4) over time must yield a constant independent of the forms of s and $\vec{u}$. Evaluate the constant when there is no sound.

5.4.1. Show that the change in the density of a particular fluid element moving with velocity $\vec{u}$ is given by $(\partial \rho/\partial t) + \vec{u} \cdot \nabla \rho$.

5.4.2. A flow is incompressible if a fluid element does not change its density as the element moves. From Problem 5.4.1, this means $(\partial \rho/\partial t) + \vec{u} \cdot \nabla \rho = 0$. (a) Show that for an incompressible fluid the equation of continuity reduces to $\nabla \cdot \vec{u} = 0$. (b) Write Euler's equation for the flow of an incompressible fluid. (c) What is c for an incompressible fluid?

5.5.1. Use the adiabat and the linearized equations of continuity and motion to show that all the scalar acoustic variables obey the wave equation $\nabla^2 - (1/c^2) \partial^2/\partial t^2 = 0$ within the accuracy of the linearizing approximations and the near constancy of ρ_0 and c.

5.5.2. (a) Use the adiabat and the linearized equations of continuity and motion (and the near constancy of ρ_0 and c) to show that $\nabla(\nabla \cdot \vec{u}) = (1/c^2) \partial^2 \vec{u}/\partial t^2$. (b) Show that, since $\vec{u}$ is irrotational, this is equivalent to $\nabla^2 \vec{u} = (1/c^2) \partial^2 \vec{u}/\partial t^2$. (c) Write the latter equation in spherical coordinates with spherical symmetry and compare it with the wave equation for the pressure in the same coordinates. (See Appendix A7 for $\nabla^2 \vec{u}$.)

5.6.1. (a) Find the speed of sound in hydrogen at 1 atm and 0°C from its values of $\mathcal{P}_0$, ρ_0, and γ. (b) Compare with the result given in Appendix A10. Is your agreement within the round-off of the tabulated values? (c) What error in temperature would give the same disagreement?

5.6.2. (a) By means of (5.6.8), determine the speed of sound in distilled water at atmospheric pressure and a temperature of 30°C. (b) What is the rate of change of the speed of sound in water with respect to temperature at this temperature?

5.6.3. (a) For a perfect gas, does c vary with the equilibrium pressure? With the instantaneous pressure in an acoustic process? (b) Find c for a perfect gas that obeys the isotherm (5.2.2). (c) Compare the value of c from (b) to that for air at 20°C.

5.7.1. If $\vec{u} = \hat{x} U \exp[j(\omega t - kx)]$, show that $|(\vec{u} \cdot \nabla)\vec{u}| / |\partial\vec{u}/\partial t| = U/c$, the acoustic Mach number. Relate this to the relevant assumption made to obtain the linear Euler's equation (5.4.10).

5.7.2. For an acoustic wave with propagation constant k, show that the mathematical assumption made to obtain (5.5.8) is equivalent to requiring $|(1/\rho_0)\nabla\rho_0| \ll k$. Physically, what does this mean?

5.7.3. For a plane wave $\mathbf{u} = U \exp[j(\omega t - kx)]$, find expressions for the acoustic Mach number U/c (a) in terms of P, ρ_0, and c and (b) in terms of s.

5.7.4. Using (5.7.8) for an oblique wave, obtain the velocity potential and then the acoustic particle velocity, and show that the velocity is parallel to the propagation vector.

5.7.5. (a) Show that if the density is not approximated by ρ_0 in the gravity term in Euler's equation, the wave equation for acoustic pressure contains a term $\nabla \cdot (\vec{g}\rho_0 s)$. (b) Show for a plane wave that this term is negligible as long as $\omega \gg |\vec{g}|/c$. Evaluate $|\vec{g}|/c$ for water and air.

5.7.6. For an acoustic wave of angular frequency ω, find a condition justifying ignoring any time dependence in ρ_0 in the linear equation of continuity.

5.8.1. Two parallel traveling plane waves have different angular frequencies ω_1 and ω_2 and pressure amplitudes P_1 and P_2. (a) Show that the instantaneous energy density $\mathscr{E}_i$ at a point in space varies between $(P_1 + P_2)^2/\rho_0 c^2$ and $(P_1 - P_2)^2/\rho_0 c^2$. (b) Show that the total energy density $\mathscr{E}$ at the point averages to the sum of the individual energy densities of each wave alone. Hint: Let the averaging time be much greater than $2\pi/|\omega_1 - \omega_2|$.

5.9.1. If $\mathbf{p} = P \exp[j(\omega t - kx)]$, find (a) the acoustic density, (b) the particle speed, (c) the velocity potential, (d) the energy density, and (e) the intensity.

5.9.2. (a) Derive an equation expressing the adiabatic temperature rise ΔT produced in a gas by an acoustic pressure p. (b) What is the amplitude of the temperature fluctuation produced by a sound of intensity 10 W/m² in air at 20°C and standard atmospheric pressure?

5.9.3. Repeat Problem 5.9.1 for the standing wave $p = P \cos(\omega t) \cos(kx)$.

5.10.1. For a wave consisting of two waves traveling in the $+x$ direction but with different frequencies, show that the specific acoustic impedance is $\rho_0 c$.

5.10.2. Show that for any plane wave traveling in the $+x$ direction, the specific acoustic impedance is $\rho_0 c$. Hint: Let $\Phi = f(ct - x)$ and generate $\mathbf{p}$ and $\mathbf{u}$ from Φ.

5.10.3. Find the specific acoustic impedance for a standing wave $\mathbf{p} = P \sin kx \exp(j\omega t)$.

5.11.1C. For values of kr between 0.1 and 10, plot the specific acoustic resistance and reactance. In what range of kr do these quantities make transitions between low- and high-frequency behaviors? What are their maximum values?

5.11.2. Given a small source of spherical waves in air, at a radial distance of 10 cm compute the difference in phase angle between pressure and particle velocity for 10 Hz,

100 Hz, 1 kHz, and 10 kHz. Compute the magnitude of the specific acoustic impedance for each frequency at this location.

5.11.3. For a spherical wave $\mathbf{p} = (A/r)\cos(kr)\exp(j\omega t)$, find (a) the particle speed, (b) the specific acoustic impedance, (c) the instantaneous intensity, and (d) the intensity.

5.11.4. Show that the specific acoustic reactance of a spherical wave is a maximum for $kr = 1$.

5.11.5. At some location, the pressure amplitude and particle speed of a 100 Hz sound wave in air are measured to be 2 Pa and 0.0100 m/s. Assuming that this is a spherical wave, find the distance from the source. What additional measurements could be made at this same place to determine the direction of the source?

5.11.6C. Plot the magnitude and the phase of the specific acoustic impedance (normalized by dividing by $\rho_0 c$) of a spherical wave as a function of kr. Above what value of kr does the spherical wave approximate the behavior of a plane wave within about 10%?

5.12.1C. A spherical wave in air has a sound pressure amplitude of 100 dB re 20 μPa at 1 m from the origin. (a) Plot the ratio of amplitudes of the pressure P and the particle speed U as a function of r for various frequencies. (b) Is the distance at which the ratio P/U comes to within 10% of $\rho_0 c$ independent of frequency? (c) If not, plot this distance as a function of frequency.

5.12.2. For a 171 Hz plane traveling wave in air with a sound pressure level of 40 dB re 20 μPa, find (a) the acoustic pressure amplitude, (b) the intensity, (c) the acoustic particle speed amplitude, (d) the acoustic density amplitude, (e) the particle displacement amplitude, and (f) the condensation amplitude.

5.12.3. A plane sound wave in air of 100 Hz has a peak acoustic pressure amplitude of 2 Pa. (a) What is its intensity and its intensity level? (b) What is its peak particle displacement amplitude? (c) What is its peak particle speed amplitude? (d) What is its effective or rms pressure? (e) What is its sound pressure level re 20 μPa?

5.12.4. An acoustic wave has a sound pressure level of 80 dB re 1 μbar. Find (a) the sound pressure level re 1 μPa and (b) the sound pressure level re 20 μPa.

5.12.5. (a) Show that a plane wave having an effective acoustic pressure of 1 μbar in air has an intensity level of 74 dB re 10^{-12} W/m^2. (b) Find the intensity (W/m^2) produced by an acoustic plane wave in water of $SPL(1 \ \mu$bar$) = 120$ dB. (c) What is the ratio of the acoustic pressure in water for a plane wave to that of a similar wave in air of equal intensity?

5.12.6. (a) Determine the energy density and effective pressure amplitude of a plane wave in air of intensity level 70 dB re 10^{-12} W/m^2. (b) Determine the energy density and effective pressure amplitude of a plane wave in water if its sound pressure level is 70 dB re 1 μbar.

5.12.7. (a) Show that at constant $\mathcal{P}_0$ the characteristic impedance of a gas is inversely proportional to the square root of its absolute temperature T_K. (b) What is the characteristic impedance of air at 0°C? At 80°C? (c) If the pressure amplitude of a sound wave remains constant, what is its percent change in intensity as the temperature increases from 0°C to 80°C? (d) What would be the corresponding change in intensity level? In pressure level?

5.12.8. Cavitation may take place at the face of a sonar transducer when the sound pressure amplitude being produced exceeds the hydrostatic pressure in the water. (a) For a hydrostatic pressure of 200,000 Pa, what is the highest intensity that may be radiated without producing cavitation? (b) What is the sound pressure level of this sound *re* 1 μbar? (*c*) What is the condensation amplitude? (*d*) At what depth in the ocean would this hydrostatic pressure be found?

5.12.9. A transmitter generates a sound pressure level at 1 m of 100 dB *re* 1 μbar for a driving voltage of 100 V (rms). Find the sensitivity level in dB *re* 1 μbar/V.

5.12.10. A transmitter has a sensitivity level of 60 dB *re* 1 μbar/V. Find its sensitivity level *re* 1 μPa/V and *re* 20 μPa/V.

5.12.11. The receiving sensitivity level of a hydrophone is -80 dB *re* 1 V/μbar. (*a*) Express this level *re* 1 V/μPa. (*b*) What will be the (rms) output voltage if the pressure field is 80 dB *re* 1 μbar?

5.12.12. A microphone reads 1 mV for an incident effective pressure level of 120 dB *re* 20 μPa. Find the sensitivity level of the microphone *re* 1 V/μbar and 1 V/20 μPa.

5.13.1C. Compare the magnitude of $H_0^{(2)}(kr)$ with its asymptotic expression and find the value of *kr* beyond which the disagreement is within 10%.

5.13.2C. Find the value of *kr* beyond which $|\mathbf{z}|$ in (5.13.12) is within 10% of $\rho_0 c$.

5.13.3. For various z, test the assertion that the Wronskian of $J_0(z)$ and $Y_0(z)$ is $2/\pi z$.

5.13.4. Find the fractional change in pressure amplitude for each doubling of the propagation distance for (*a*) spherical waves, (*b*) cylindrical waves for $kr \gg 1$, (*c*) plane waves.

5.13.5. Assume that $k_z \neq 0$ in (5.13.3). (*a*) Show that

$$\mathbf{p} = H_0^{(2)}(k_r r) \sin k_z z \, e^{j\omega t}$$

is a solution of (5.13.3). (*b*) Write $\sin k_z z$ in terms of complex exponentials and show that $\mathbf{p}$ consists of two outward-traveling waves, each having conical surfaces of constant phase. (*c*) Find the angles of elevation and depression with respect to the $z = 0$ plane of the propagation vectors.

5.14.1. (*a*) If c is a function only of z, show that $d\theta/ds = -[(\cos\theta_0)/c_0]\, dc/dz$, with θ_0 the angle of elevation of the ray where $c = c_0$. (*b*) If the gradient $g = dc/dz$ is a constant, find the radius of curvature R of the ray in terms of g, c, and θ. Is R a constant? (*c*) If the temperature of air decreases linearly with height z, verify that $c(z) = c_0 - gz$, where $g > 0$. If the temperature decreases 5 C°/km, find the radius of curvature of a ray that is horizontal at $z = 0$ (assume $c_0 = 340$ m/s). At what horizontal range will this ray have risen to a height of 10 m?

5.14.2. Assume the speed of sound is given by the quasi-parabolic profile $c(z) = c_0[1 - (\varepsilon z)^2]^{-1/2}$. Let the depth $z = 0$, which defines the axis of the sound channel, lie well below the surface of the ocean. (*a*) Find the equation $z(x)$ for rays emitted by a source at $(x, z) = (0, 0)$ with angles of elevation or depression $\pm\theta_0$. *Hint:* use Snell's law, $dz/dx = \tan\theta$, and $\int (a^2 - u^2)^{-1/2}\, du = \sin^{-1}(u/a)$. (*b*) For a given ray, find an expression for the average speed with which energy propagates out to a distance x lying on the channel axis. Explain why there is no dependence on the parameter ε. For small angles, approximate your expression through the first nonzero term in θ_0. (*c*) For $|\theta_0| \leq \pi/8$, show that $c(z)$ is a good approximation of the parabolic profile $c_0[1 + \frac{1}{2}(\varepsilon z)^2]$. What is the percentage discrepancy at 22°? (*d*) A certain ocean channel with axis more than about 1 km below the ocean surface can be approximated by $c(z)$ with $c_0 = 1475$ m/s and $\varepsilon = 1.5 \times 10^{-4}$ m^{-1}. Calculate the travel speeds of (*c*) for $\theta_0 = 0, 1, 2, 5$, and 10°. (*e*) For each of the angles in (*d*), determine the greatest height above the channel axis reached by the ray and the distance between successive axis crossings. (*f*) Explain why the results of (*d*) and (*e*) are not inconsistent.

5.14.3. Assume the speed of sound is given by the quasi-linear profile $c(z) = c_0[1 - \varepsilon|z|]^{-1/2}$. Let the depth $z = 0$, which defines the axis of the sound channel, lie well below the surface of the ocean. (*a*) Find the equation $z(x)$ for rays emitted by a source at $(x, z) = (0, 0)$ with angles of elevation or depression $\pm\theta_0$. *Hint:* use Snell's law and $dz/dx = \tan\theta$. (*b*) For a ray with initial angle θ_0, find the distance Δx between x-axis intercepts and the maximum distance Δz it attains above or below $z = 0$. (*c*) For a given ray, find an expression for the average speed with which energy propagates out to a distance x lying on the channel axis. Explain why there is no dependence on the parameter ε. For small angles, approximate your expression through the first nonzero term in θ_0. (*d*) For $|\theta_0| \le \pi/8$, show that $c(z)$ is a good approximation of the linear profile $c_0(1 + \frac{1}{2}\varepsilon|z|)$. What is the percentage discrepancy at 22°? (*d*) A certain ocean channel with axis more than about 1 km below the ocean surface can be approximated by a quasi-bilinear profile with $c_0 = 1467$ m/s and $\varepsilon_1 = 4.0 \times 10^{-5}$ m^{-1} above the axis and $\varepsilon_2 = 2.0 \times 10^{-5}$ m^{-1} below. Calculate the travel speeds of (*c*) for $\theta_0 = 0$, 1, 2, 5, and 10°. (*e*) For each of the angles in (*d*), determine the greatest distances above and below the channel axis reached by the ray and the two distances between successive axis crossings. (*f*) Explain why the results of (*d*) and (*e*) are not inconsistent.

5.14.4. (*a*) Verify (5.14.7). *Hint:* Substitute (5.14.6) into (5.14.3) and note that $\nabla A \cdot (n\hat{s}) = n(dA/ds)$ and $\nabla \cdot (n\hat{s}) = dn/ds + n\nabla \cdot \hat{s}$. (*b*) Obtain (5.14.13) from (5.14.12). Deal with this component by component. Show that the x component of $d(\nabla\Gamma)/ds$ can be written as $d(n\alpha)/ds$ from (5.14.6) and as $(\alpha\, \partial/\partial x + \beta\, \partial/\partial y + \gamma\, \partial/\partial z)(\partial\Gamma/\partial x)$ from (5.14.12). In the latter expression, exchange orders of differentiation, use (5.14.6), expand the derivatives, and regroup using $\alpha^2 + \beta^2 + \gamma^2 = 1$.

5.14.5. If the speed of sound in water is 1500 m/s at the surface and increases linearly with depth at a rate of 0.017/s, find the range at which a ray emitted horizontally from a source at 100 m depth will reach the surface.

5.14.6. For the conditions of Problem 5.14.5, calculate the ratio of the intensity when the ray reaches the surface to that at 1 meter from the source. Compare this to the intensity if the spreading were spherical.

5.14.7. Plot, as a function of time, the phase coherent sum of two sinusoidal signals of equal frequency and amplitude for phase differences from 0° to 360° in steps of 45°. For each case, calculate the ratio of the intensity of the summed signal to the intensity of the individual signals and compare to results obtained from the plots.

5.14.8. Sound from a single source arrives at some point over two separate paths. Let the two signals have pressures $p_1 = P_1 \cos(\omega t)$ and $p_2 = P_2 \cos(\omega t + \phi)$ at the point. (*a*) Under the assumption that the waves are plane and essentially parallel, show that the intensity at that point is $I = [(P_1 + P_2 \cos\phi)^2 + (P_2 \sin\phi)^2]/2\rho_0 c$. (*b*) If incoherence effects now cause ϕ to be a slowly varying function of time compared to the period of the waves, show that the total intensity at the point is the sum of the individual intensities of the two waves. *Hint:* Let the accumulated values of ϕ be distributed with equal probability over the interval $0 \le \phi < 2\pi$.

5.14.9C. Plot, as a function of time, the sum of two quasi-random signals of about equal intensity. Verify that the intensity of the summed signals is the sum of the intensities of the individual signals. *Hint:* Construct the signals from sinusoids, each with its phase independently randomized at each time step.

5.14.10. Show that a spherically symmetric outward-traveling wave in an isospeed medium satisfies (5.14.5) identically for all r.

5.14.11C. The sound speed in deep water can be approximated by two layers: an upper layer in which the sound speed decreases linearly from 1500 m/s at the surface to 1475 m/s at 1000 m, and an infinitely deep layer in which the sound speed increases at a constant rate of 0.017 s^{-1}. For a source at the surface, (*a*) plot the distance at which a ray returns to the surface for depression angles between 0°

and 10°. (*b*) Find the depression angle and range of the ray that reaches the surface closest to the source. (*c*) The region where rays with different source angles reach the surface at the same range is the *resweep zone*. Find the width of this region. (*d*) Find the greatest depth attained by a ray that contributes to this region.

5.14.12. The sound speed in air varies linearly from 343 m/s at the ground to 353 m/s at 100 m altitude and then decreases above this. For a source at ground level, find (*a*) the maximum elevation angle for a ray that returns to ground level, and (*b*) the range at which this ray returns to ground level.

5.14.13C. At the range found in Problem 5.14.12, find the difference in the times of arrival of the ray that leaves the source horizontally and the ray that leaves at the maximum elevation angle.

5.14.14. (*a*) Show that, within the approximations yielding the eikonal equation, $\nabla p = pk\nabla\Gamma$. (*b*) The intensity, written explicitly as a vector, is $\vec{I} = \langle p\vec{u}\rangle_T$. Using the relationship between p and $\vec{u}$, show that $\vec{I}$ is parallel to the ray path.

5.15.1. Express in vector notation: (*a*) $u_i v_i$, (*b*) $\partial u_j/\partial x_j$, (*c*) $u_i\,\partial f/\partial x_i$, (*d*) $f\,\partial u_j/\partial x_j + u_i\,\partial f/\partial x_i$.

5.15.2. Express in subscript (tensor) notation: (*a*) $(\vec{u}\cdot\nabla)f$, (*b*) $\nabla\cdot[(\vec{u}\cdot\nabla)\vec{u}]$, (*c*) $\nabla\cdot[\vec{u}(\nabla\cdot\vec{u})]$.

5.15.3. Show that the right sides of (5.15.5) and (5.15.6) are equivalent. *Hint:* Take $(\partial/\partial x_j)$ of $(\rho u_j)u_i$ and convert to vector notation.

5.16.1. Prove the equality given by (5.16.3). *Hint:* Use the indefinite integral relation $\int x\exp(-jx)\,dx = \exp(-jx) + jx\exp(-jx)$.

5.16.2. Show that in spherical coordinates with spherical symmetry the three-dimensional delta function can be written as $\delta(\vec{r}) = (4\pi r^2)^{-1}\delta(r)$, where $\delta(r)$ is the one-dimensional delta function.

5.16.3. Show that in cylindrical coordinates with radial symmetry for a source lying on the z axis at $z = z_0$, the three-dimensional delta function $\delta(\vec{r} - \vec{r}_0)$ can be written as $\delta(\vec{r} - \vec{r}_0) = (2\pi r)^{-1}\delta(r)\delta(z - z_0)$.

5.16.4. (*a*) Show that $p = (A/r)f(t - r/c)$ is a solution of the inhomogeneous wave equation $\nabla^2 p - (1/c^2)\partial^2 p/\partial t^2 = -4\pi A\delta(\vec{r})f(t)$. (*b*) Show that $p = (1/r)\delta(t - r/c)$ is a solution of this equation when $f(t) = \delta(t)$.

Chapter 6

REFLECTION AND TRANSMISSION

6.1 CHANGES IN MEDIA

When an acoustic wave traveling in one medium encounters the boundary of a second medium, reflected and transmitted waves are generated. Discussion of this phenomenon is greatly simplified if it is assumed that both the incident wave and the boundary between the media are planar and that all media are fluids. The complications that arise when one of the media is a solid will be left to Section 6.6. However, it is worthwhile to note that for normal incidence many solids obey the same equations developed for fluids. The only modification needed is that the speed of sound in the solid must be the *bulk speed of sound*, based on both the *bulk* and *shear* moduli since, unlike the bar of Chapter 3, an extended solid is not free to change its transverse dimensions. See Appendix A11. Values of the bulk speeds of sound in some solids are listed in Appendix A10.

The ratios of the pressure amplitudes and intensities of the reflected and transmitted waves to those of the incident wave depend on the characteristic acoustic impedances and speeds of sound in the two media and on the angle the incident wave makes with the interface. Let the incident and reflected waves travel in a fluid of characteristic acoustic impedance $r_1 = \rho_1 c_1$, where ρ_1 is the equilibrium density of the fluid (the subscript "0" has been suppressed for economy of notation) and c_1 the speed of sound in the fluid. Let the transmitted wave travel in a fluid of characteristic acoustic impedance $r_2 = \rho_2 c_2$. If the complex pressure amplitude of the incident wave is $\mathbf{P}_i$, that of the reflected wave $\mathbf{P}_r$, and that of the transmitted wave $\mathbf{P}_t$, then we can define the *pressure transmission and reflection coefficients*:

$$\mathbf{T} = \mathbf{P}_t/\mathbf{P}_i \tag{6.1.1}$$

$$\mathbf{R} = \mathbf{P}_r/\mathbf{P}_i \tag{6.1.2}$$

Since the intensity of a harmonic plane progressive wave is $P^2/2r$, the *intensity transmission and reflection coefficients* are real and are defined by

$$T_I = I_t/I_i = (r_1/r_2)|\mathbf{T}|^2 \tag{6.1.3}$$

$$R_I = I_r/I_i = |\mathbf{R}|^2 \qquad (6.1.4)$$

Most real situations have beams of sound with *finite* cross-sectional area. As we have seen, a beam can be described locally by nearly parallel rays and thus can be approximated by a plane wave of finite extent. While there can be some anomalies resulting from diffraction at the edges of the beam, if the cross-sectional area is sufficiently great compared to a wavelength, they can be ignored and the equations developed in this chapter applied.

The power carried by a beam of sound is the acoustic intensity multiplied by the cross-sectional area of the beam. If an incident beam of cross-sectional area A_i is obliquely incident on a boundary, the cross-sectional area A_t of the transmitted beam generally is not the same as that of the incident beam. It will be shown later that the cross-sectional areas of the incident and reflected beams are equal under all circumstances. The *power transmission and reflection coefficients* are defined by

$$T_\Pi = (A_t/A_i)T_I = (A_t/A_i)(r_1/r_2)|\mathbf{T}|^2 \qquad (6.1.5)$$

$$R_\Pi = R_I = |\mathbf{R}|^2 \qquad (6.1.6)$$

From conservation of energy, the power in the incident beam must be shared between reflected and transmitted beams so that

$$R_\Pi + T_\Pi = 1 \qquad (6.1.7)$$

Cases more complicated than those included in this chapter are available in specialized textbooks.[1]

6.2 TRANSMISSION FROM ONE FLUID TO ANOTHER: NORMAL INCIDENCE

As indicated in Fig. 6.2.1, let the plane $x = 0$ be the boundary between fluid 1 of characteristic acoustic impedance r_1 and fluid 2 of characteristic acoustic impedance r_2. Let there be an incident wave traveling in the $+x$ direction,

$$\mathbf{p}_i = \mathbf{P}_i e^{j(\omega t - k_1 x)} \qquad (6.2.1)$$

which, when striking the boundary, generates a reflected wave

$$\mathbf{p}_r = \mathbf{P}_r e^{j(\omega t + k_1 x)} \qquad (6.2.2)$$

and a transmitted wave

$$\mathbf{p}_t = \mathbf{P}_t e^{j(\omega t - k_2 x)} \qquad (6.2.3)$$

All the waves must have the same frequency, but, since the speeds c_1 and c_2 are different, the wave numbers $k_1 = \omega/c_1$ in fluid 1 and $k_2 = \omega/c_2$ in fluid 2 are different.

---

[1]Officer, *Introduction to the Theory of Sound Transmission*, McGraw-Hill (1958). Ewing, Jardetzky, and Press, *Elastic Waves in Layered Media*, McGraw-Hill (1957). Brekhovskikh, *Waves in Layered Media*, Academic Press (1960).

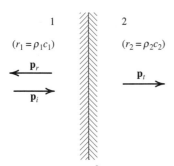

Figure 6.2.1 Reflection and transmission of a plane wave normally incident on the planar boundary between fluids with different characteristic impedances.

There are two boundary conditions to be satisfied for all times at all points on the boundary: (1) the acoustic pressures on both sides of the boundary must be equal and (2) the normal components of the particle velocities on both sides of the boundary must be equal. The first condition, *continuity of pressure*, means that there can be no net force on the (massless) plane separating the fluids. The second condition, *continuity of the normal component of velocity*, requires that the fluids remain in contact. The pressure and normal particle velocity in fluid 1 are $\mathbf{p}_i + \mathbf{p}_r$ and $(\mathbf{u}_i + \mathbf{u}_r)\hat{x}$ so that the boundary conditions are

$$\mathbf{p}_i + \mathbf{p}_r = \mathbf{p}_t \qquad \text{at } x = 0 \tag{6.2.4}$$

$$\mathbf{u}_i + \mathbf{u}_r = \mathbf{u}_t \qquad \text{at } x = 0 \tag{6.2.5}$$

Division of (6.2.4) by (6.2.5) yields

$$\frac{\mathbf{p}_i + \mathbf{p}_r}{\mathbf{u}_i + \mathbf{u}_r} = \frac{\mathbf{p}_t}{\mathbf{u}_t} \qquad \text{at } x = 0 \tag{6.2.6}$$

which is a statement of the *continuity of normal specific acoustic impedance* across the boundary.

Since a plane wave has $\mathbf{p}/\mathbf{u} = \pm r$, the sign depending on the direction of propagation, (6.2.6) becomes

$$r_1 \frac{\mathbf{p}_i + \mathbf{p}_r}{\mathbf{p}_i - \mathbf{p}_r} = r_2 \tag{6.2.7}$$

which leads directly to the reflection coefficient

$$\mathbf{R} = \frac{r_2 - r_1}{r_2 + r_1} = \frac{r_2/r_1 - 1}{r_2/r_1 + 1} \tag{6.2.8}$$

Then, since (6.2.4) is equivalent to $1 + \mathbf{R} = \mathbf{T}$, we have

$$\mathbf{T} = \frac{2r_2}{r_2 + r_1} = \frac{2r_2/r_1}{r_2/r_1 + 1} \tag{6.2.9}$$

The intensity reflection and transmission coefficients follow directly from (6.1.3) and (6.1.4),

$$R_I = \left(\frac{r_2 - r_1}{r_2 + r_1}\right)^2 = \left(\frac{r_2/r_1 - 1}{r_2/r_1 + 1}\right)^2 \tag{6.2.10}$$

and

$$T_I = \frac{4r_2 r_1}{(r_2 + r_1)^2} = \frac{4r_2/r_1}{(r_2/r_1 + 1)^2} \tag{6.2.11}$$

Since the cross-sectional areas of all the beams are equal, the power coefficients in (6.1.5) and (6.1.6) are equal to the intensity coefficients.

From (6.2.8), $\mathbf{R}$ is always real. It is positive when $r_1 < r_2$ and negative when $r_1 > r_2$. Consequently, at the boundary the acoustic pressure of the reflected wave is either in phase or 180° out of phase with that of the incident wave. When the characteristic acoustic impedance of fluid 2 is greater than that of fluid 1 (a wave in air incident on the air–water interface), a positive pressure in the incident wave is reflected as a positive pressure. On the other hand, if $r_1 > r_2$ (a wave in water incident on the water–air interface), a positive pressure is reflected as a negative pressure. Note that when $r_1 = r_2$ then $\mathbf{R} = 0$, and there is complete transmission.

From (6.2.9), it is seen that $\mathbf{T}$ is real and positive regardless of the relative magnitudes of r_1 and r_2. Consequently, at the boundary the acoustic pressure of the transmitted wave is *always* in phase with that of the incident wave. Study of (6.2.11) reveals that whenever r_1 and r_2 have strongly dissimilar values, the intensity transmission coefficient is small. In addition, from the symmetries of (6.2.10) and (6.2.11), it is apparent that the intensity reflection and transmission coefficients are *independent* of the direction of the wave. For example, they are the same from water into air as from air into water. This is a special case of *acoustic reciprocity*.

In the limit $r_1/r_2 \to 0$, the wave is reflected with no reduction in amplitude and no change in phase. The transmitted wave has a pressure amplitude twice that of the incident wave, and the normal particle velocity at the boundary is zero. Because of this latter fact, the boundary is termed *rigid*.

For $r_1/r_2 \to \infty$, the amplitude of the reflected wave is again equal to that of the incident wave, and the transmitted wave has zero pressure amplitude. Since the acoustic pressure at the boundary is zero, the boundary is termed *pressure release*.

6.3 TRANSMISSION THROUGH A FLUID LAYER: NORMAL INCIDENCE

Assume that a plane fluid layer of uniform thickness L lies between two dissimilar fluids and that a plane wave is normally incident on its boundary, as indicated in Fig. 6.3.1. Let the characteristic impedances of the fluids be r_1, r_2, and r_3, respectively.

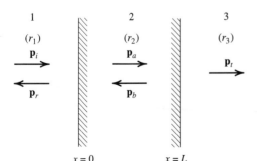

Figure 6.3.1   Reflection and transmission of a plane wave normally incident on a layer of uniform thickness.

When an incident signal in fluid 1 arrives at the boundary between fluids 1 and 2, some of the energy is reflected and some transmitted into the second fluid. The portion of the wave transmitted will proceed through fluid 2 to interact with the boundary between fluids 2 and 3, where again some of the energy is reflected and some transmitted. The reflected wave travels back to the boundary between fluids 1 and 2, and the whole process is repeated. If the duration of the incident signal is less than $2L/c_2$, an observer in either fluid 1 or 3 will see a series of echoes separated in time by $2L/c_2$ whose amplitudes can be calculated by applying the results of the previous section the appropriate number of times. Otherwise, if the incident wave train has a monofrequency carrier and has duration much greater than $2L/c_2$, it can be assumed to be

$$\mathbf{p}_i = \mathbf{P}_i e^{j(\omega t - k_1 x)} \tag{6.3.1}$$

The various transmitted and reflected waves now combine so that in the steady state the wave reflected back into fluid 1 is

$$\mathbf{p}_r = \mathbf{P}_r e^{j(\omega t + k_1 x)} \tag{6.3.2}$$

the transmitted and reflected waves in fluid 2 are

$$\mathbf{p}_a = \mathbf{A} e^{j(\omega t - k_2 x)} \tag{6.3.3}$$

$$\mathbf{p}_b = \mathbf{B} e^{j(\omega t + k_2 x)} \tag{6.3.4}$$

and the wave transmitted into fluid 3 is

$$\mathbf{p}_t = \mathbf{P}_t e^{j(\omega t - k_3 x)} \tag{6.3.5}$$

Continuity of the normal specific acoustic impedance at $x = 0$ and at $x = L$ gives

$$\frac{\mathbf{P}_i + \mathbf{P}_r}{\mathbf{P}_i - \mathbf{P}_r} = \frac{r_2}{r_1}\frac{\mathbf{A} + \mathbf{B}}{\mathbf{A} - \mathbf{B}} \qquad \frac{\mathbf{A}e^{-jk_2 L} + \mathbf{B}e^{jk_2 L}}{\mathbf{A}e^{-jk_2 L} - \mathbf{B}e^{jk_2 L}} = \frac{r_3}{r_2} \tag{6.3.6}$$

and algebraic manipulation yields the pressure reflection coefficient

$$\mathbf{R} = \frac{(1 - r_1/r_3)\cos k_2 L + j(r_2/r_3 - r_1/r_2)\sin k_2 L}{(1 + r_1/r_3)\cos k_2 L + j(r_2/r_3 + r_1/r_2)\sin k_2 L} \tag{6.3.7}$$

The intensity transmission coefficient is found by using (6.1.3)–(6.1.7) and noting that $A_t = A_i$:

$$T_I = \frac{4}{2 + (r_3/r_1 + r_1/r_3)\cos^2 k_2 L + (r_2^2/r_1 r_3 + r_1 r_3/r_2^2)\sin^2 k_2 L} \tag{6.3.8}$$

A few special forms of (6.3.8) are of particular interest.

1. If the final fluid is the same as the initial fluid, $r_1 = r_3$,

$$T_I = \frac{1}{1 + \frac{1}{4}(r_2/r_1 - r_1/r_2)^2 \sin^2 k_2 L} \tag{6.3.9}$$

If, in addition, $r_2 \gg r_1$, (6.3.9) further simplifies to

$$T_I = \frac{1}{1 + \frac{1}{4}(r_2/r_1)^2 \sin^2 k_2 L} \tag{6.3.10}$$

This latter situation applies, for example, to the transmission of sound from air in one room through a solid wall into air in an adjacent room. The solid materials forming the walls of rooms have such large characteristic impedances relative to air that $(r_2/r_1) \sin k_2 L \gg 2$ for all reasonable frequencies and thicknesses of walls. Therefore, when the fluid medium is air, (6.3.10) reduces to

$$T_I \approx \left(\frac{2 r_1}{r_2 \sin k_2 L} \right)^2 \tag{6.3.11}$$

Finally, for all situations except those of high frequencies and very thick walls, $k_2 L \ll 1$ and $\sin k_2 L \approx k_2 L$ so that (6.3.11) becomes

$$T_I = \left(\frac{2}{k_2 L} \frac{r_1}{r_2} \right)^2 \tag{6.3.12}$$

(At 1 kHz the value of $k_2 L$ for a 0.1 m thick concrete wall is $2\pi \times 1000 \times 0.1/3100 = 0.20$.) Note that the transmitted pressure is inversely proportional to the thickness L and, therefore, also inversely proportional to the mass per unit area of the wall. This behavior is observed to be approximately true for certain kinds of commonly encountered walls. In the case of solid panels in water, both terms occurring in the denominator of (6.3.10) usually are significant, so that the complete equation must be used. However, for either thin panels or low frequencies such that $(r_2/r_1) \sin k_2 L \ll 1$, (6.3.10) simplifies to

$$T_I \approx 1 \tag{6.3.13}$$

This behavior is used in the design of free-flooding streamlined domes for sonar transducers.

2. Another special form of (6.3.8) is obtained by assuming that the intermediate fluid has a larger characteristic impedance than either fluid 1 or fluid 3 but such small thickness that $r_2 \sin k_2 L \ll 1$ and $\cos k_2 L \approx 1$. Then, (6.3.8) reduces to

$$T_I = \frac{4 r_3 r_1}{(r_3 + r_1)^2} \tag{6.3.14}$$

This is equivalent to (6.2.11), which gives the intensity transmission coefficient for a wave moving directly from fluid 1 into fluid 3. Thus, a thin membrane of solid material of appropriate characteristic impedance may be used in preventing two gases or two liquids from mixing and yet not interfere with sound transmission between them. In particular, note that if $r_1 = r_3$ then there is total transmission from fluid 1 to fluid 3 as if fluid 2 did not exist.

3. Returning to the general form of T_I in (6.3.8), we see that if $k_2 L = n\pi$, (6.3.8) reduces to (6.3.14) for frequencies

$$f \approx n c_2 / 2L \tag{6.3.15}$$

For these frequencies, $L \approx n\lambda_2/2$ and the intermediate layer is an integral number of half-wavelengths thick. Again, it is as if fluid 2 did not exist.

4. Finally, if $k_2 L \approx (n - \frac{1}{2})\pi$, where n is any integer, then we have $L \approx (2n - 1)\lambda_2/4$ so that $\cos k_2 L \approx 0$ and $\sin k_2 L \approx 1$, and (6.3.8) becomes

$$T_I = \frac{4r_1 r_3}{(r_2 + r_1 r_3/r_2)^2} \tag{6.3.16}$$

for frequencies very close to $f = (n - \frac{1}{2})c_2/2L$. As an interesting special case, note that (6.3.16) yields $T_I \approx 1$ when $r_2 = \sqrt{r_1 r_3}$. It is therefore possible to obtain total transmission of acoustic power from one medium to another through the use of an intermediate medium whose characteristic impedance is the geometric mean of the other two. However, this action is selective, since it occurs only for narrow bands of frequencies centered about these particular values. This technique of obtaining complete transmission of acoustic power through the use of a quarter-wavelength intermediate layer is similar to the method of making nonreflective glass lenses by coating them with a quarter-wavelength layer of some suitable material. Another example is the use of quarter-wavelength sections to match an antenna to an electrical transmission line.

The impedance $\mathbf{z}_2$ presented to fluid 1 by any number of sequential layers can be expressed in terms of the pressure reflection coefficient. The boundary between fluid 1 and fluid 2 corresponds to an impedance given by

$$\mathbf{z}_2 = \left. \frac{\mathbf{p}_i + \mathbf{p}_r}{\mathbf{u}_i + \mathbf{u}_r} \right|_{x=0} \tag{6.3.17}$$

Division of numerator and denominator by $\mathbf{p}_i$ and use of the relation $\mathbf{p}_\pm = \pm r\mathbf{u}_\pm$ results in

$$\mathbf{z}_2 = r_1 \frac{1 + \mathbf{R}}{1 - \mathbf{R}} \tag{6.3.18}$$

In this way a multilayered fluid boundary to the right of fluid 1 can be replaced by a single boundary at $x = 0$ whose impedance may have real and imaginary components.

6.4 TRANSMISSION FROM ONE FLUID TO ANOTHER: OBLIQUE INCIDENCE

Assume that the boundary separating two fluids is the plane $x = 0$ and that the incident, reflected, and transmitted waves make the respective angles θ_i, θ_r, and θ_t with the x axis, as shown in Fig. 6.4.1. For propagation vectors lying in the x-y plane these waves can be written as

$$\mathbf{p}_i = \mathbf{P}_i e^{j(\omega t - k_1 x \cos\theta_i - k_1 y \sin\theta_i)} \tag{6.4.1}$$

$$\mathbf{p}_r = \mathbf{P}_r e^{j(\omega t + k_1 x \cos\theta_r - k_1 y \sin\theta_r)} \tag{6.4.2}$$

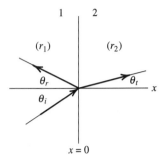

Figure 6.4.1   Reflection and transmission of a plane wave obliquely incident on the planar boundary between fluids with different characteristic impedances.

and

$$\mathbf{p}_t = \mathbf{P}_t e^{j(\omega t - k_2 x \cos \theta_t - k_2 y \sin \theta_t)} \tag{6.4.3}$$

The reason for writing θ_t as a complex quantity will emerge shortly.
   Applying continuity of pressure at the boundary $x = 0$ yields

$$\mathbf{P}_i e^{-jk_1 y \sin \theta_i} + \mathbf{P}_r e^{-jk_1 y \sin \theta_r} = \mathbf{P}_t e^{-jk_2 y \sin \theta_t} \tag{6.4.4}$$

Since this must be true for all y, the exponents must all be equal. This means that

$$\boxed{\sin \theta_i = \sin \theta_r} \tag{6.4.5}$$

so that the angle of incidence is equal to the angle of reflection, and

$$\boxed{\frac{\sin \theta_i}{c_1} = \frac{\sin \theta_t}{c_2}} \tag{6.4.6}$$

a statement of Snell's law. The presence of sines rather than cosines results from the convention of measuring the angle with respect to the normal to the boundary when dealing with reflection and transmission in air. For ray tracing in underwater and atmospheric acoustics, the convention in Section 5.14 is usually adopted. Since the exponents in (6.4.4) are all equal, this equation reduces to

$$1 + \mathbf{R} = \mathbf{T} \tag{6.4.7}$$

Continuity of the normal component of particle velocity at the boundary gives

$$\mathbf{u}_i \cos \theta_i + \mathbf{u}_r \cos \theta_r = \mathbf{u}_t \cos \theta_t \tag{6.4.8}$$

Replacing each $\mathbf{u}$ with the appropriate value of $\pm \mathbf{p}/r$, and recalling that $\theta_i = \theta_r$ we have

$$1 - \mathbf{R} = \frac{r_1 \cos \theta_t}{r_2 \cos \theta_i} \mathbf{T} \tag{6.4.9}$$

Equations (6.4.7) and (6.4.9) can be combined to eliminate **T**, giving

$$R = \frac{r_2/r_1 - \cos\boldsymbol{\theta}_t/\cos\theta_i}{r_2/r_1 + \cos\boldsymbol{\theta}_t/\cos\theta_i} = \frac{r_2/\cos\boldsymbol{\theta}_t - r_1/\cos\theta_i}{r_2/\cos\boldsymbol{\theta}_t + r_1/\cos\theta_i} \qquad (6.4.10)$$

where Snell's law reveals

$$\cos\boldsymbol{\theta}_t = (1 - \sin^2\boldsymbol{\theta}_t)^{1/2} = [1 - (c_2/c_1)^2 \sin^2\theta_i]^{1/2} \qquad (6.4.11)$$

Equation (6.4.10) is known as the *Rayleigh reflection coefficient*. It is important to note three consequences of this equation.

1. If $c_1 > c_2$, the angle of transmission $\boldsymbol{\theta}_t$ is *real* and *less* than the angle of incidence. The transmitted beam is bent *toward* the normal for all angles of incidence.
2. If $c_1 < c_2$, and $\theta_i < \theta_c$, where the *critical angle* θ_c is defined by

$$\boxed{\sin\theta_c = c_1/c_2} \qquad (6.4.12)$$

the angle of transmission is again *real* but *greater* than the angle of incidence; the transmitted beam is bent *away* from the normal for all angles of incidence less than the critical angle.

3. If $c_1 < c_2$, and $\theta_i > \theta_c$, the transmitted wave assumes a peculiar form. From (6.4.11), we see that $\sin\boldsymbol{\theta}_t$ is real and greater than unity, so that $\cos\boldsymbol{\theta}_t$ is now pure imaginary,

$$\cos\boldsymbol{\theta}_t = -j[(c_2/c_1)^2 \sin^2\theta_i - 1]^{1/2} \qquad (6.4.13)$$

Examination of (6.4.3) then reveals that the transmitted pressure is

$$\mathbf{p}_t = \mathbf{P}_t e^{-\gamma x} e^{j(\omega t - k_1 y \sin\theta_i)} \qquad (6.4.14)$$
$$\gamma = k_2[(c_2/c_1)^2 \sin^2\theta_i - 1]^{1/2}$$

The transmitted wave propagates in the y direction, *parallel* to the boundary, and has an amplitude that decays *perpendicular* to the boundary. [Had we chosen the *positive* imaginary root in (6.4.13), γ would have been negative, and the amplitude would have *increased* exponentially with increasing x, a physical impossibility.] Because $\boldsymbol{\theta}_t$ is pure imaginary, the numerator of **R** in (6.4.10) is the complex conjugate of the denominator. Both have the same magnitude, but they have opposite phase. Solving for the phase angle of the ratio and using (6.4.13) to express $\cos\boldsymbol{\theta}_t$ in terms of the angle of incidence and the critical angle gives

$$\mathbf{R} = e^{j\phi}$$
$$\phi = 2\tan^{-1}[(\rho_1/\rho_2)\sqrt{(\cos\theta_c/\cos\theta_i)^2 - 1}] \qquad (6.4.15)$$

under the restriction $\theta_i > \theta_c$. For all angles greater than the critical angle, the reflected wave has the same amplitude as the incident wave. The incident wave is totally reflected and in the steady state no energy propagates away from the boundary into the second medium. The transmitted wave possesses energy, but its propagation vector is parallel to the boundary so that the wave "clings" to the interface. For angles of incidence just exceeding critical, ϕ is close to zero, the reflection coefficient is $+1$, and the interface resembles a rigid boundary. As θ_i increases toward extreme grazing, ϕ approaches π, the reflection coefficient approaches -1, and the interface resembles a pressure release boundary.

If we return to the general case, Snell's law shows that the reflected and incident beams have the same cross-sectional area, as asserted earlier. The power transmission coefficient (6.1.5) can be most readily computed from (6.1.7), giving

$$T_\Pi = \left(4\frac{r_2}{r_1}\frac{\cos\theta_t}{\cos\theta_i}\right) \bigg/ \left(\frac{r_2}{r_1} + \frac{\cos\theta_t}{\cos\theta_i}\right)^2 \qquad \theta_t \text{ real} \qquad (6.4.16)$$

$$T_\Pi = 0 \qquad\qquad\qquad\qquad\qquad \theta_t \text{ imaginary} \qquad (6.4.17)$$

The first equality (6.4.16) applies when either $c_1 > c_2$ or $\theta_i < \theta_c$. When $r_2/r_1 = \cos\theta_t/\cos\theta_i$, the power reflection coefficient is zero and all the incident power is transmitted. If this condition is combined with (6.4.6) to eliminate θ_t, then

$$\sin\theta_I = \left(\frac{(r_2/r_1)^2 - 1}{(r_2/r_1)^2 - (c_2/c_1)^2}\right)^{1/2} = \left(\frac{1 - (r_1/r_2)^2}{1 - (\rho_1/\rho_2)^2}\right)^{1/2} \qquad (6.4.18)$$

defines the *angle of intromission* θ_I, the angle of incidence for which there is no reflection and, therefore, complete transmission. This angle can exist under only two circumstances: (1) $r_1 < r_2$ and $c_2 < c_1$ or (2) $r_1 > r_2$ and $c_2 > c_1$. In this second circumstance there is a critical angle, and it is greater than the angle of intromission.

At *grazing incidence*, $\theta_i \to 90°$, $\cos\theta_i \to 0$ and (6.4.10) is reduced to $R \approx -1$. Consequently, at grazing incidence there is complete reflection of the incident acoustic energy irrespective of the relative characteristic acoustic impedances of the two fluids.

Figures 6.4.2–6.4.5 show typical behaviors for the reflection coefficients for all possible conditions.

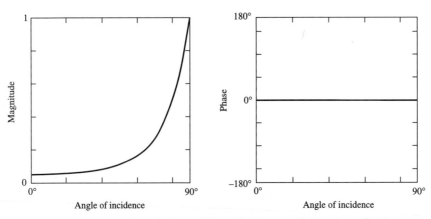

Figure 6.4.2   Magnitude and phase of the reflection coefficient for reflection from a slow bottom with $c_2/c_1 = 0.9$ and $r_2/r_1 = 0.9$.

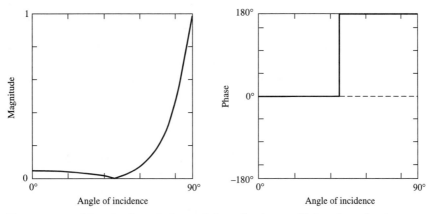

Figure 6.4.3 Magnitude and phase of the reflection coefficient for reflection from a slow bottom with $c_2/c_1 = 0.9$ and $r_2/r_1 = 1.1$. Note angle of intromission at $46.4°$.

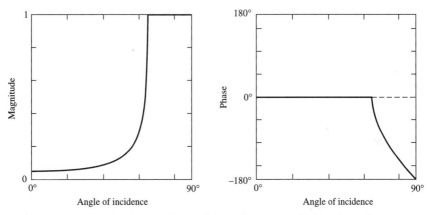

Figure 6.4.4 Magnitude and phase of the reflection coefficient for reflection from a fast bottom with $c_2/c_1 = 1.1$ and $r_2/r_1 = 1.1$. Note critical angle at $65.6°$.

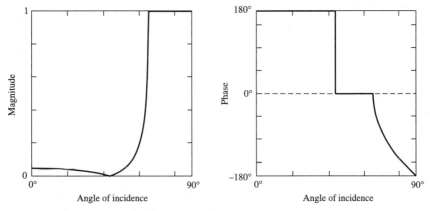

Figure 6.4.5 Magnitude and phase of the reflection coefficient for reflection from a fast bottom with $c_2/c_1 = 1.1$ and $r_2/r_1 = 0.9$. Note angle of intromission at $43.2°$ and critical angle at $65.6°$.

The reflection that takes place in seawater from a sand or silt bottom is a good example of reflection associated with two fluids in contact. Such behavior is to be expected since the saturated sand or silt is more like a fluid than a solid in its inability to transmit shear waves. As a first approximation, (6.4.10) may be used for computing a reflection coefficient. Measured values of ρ_2 and c_2 for sand and silt yield $\rho_2/\rho_1 = 1.5$ to 2.0 and $c_2/c_1 = 0.9$ to 1.1, where ρ_1 and c_1 are the values for seawater.

*6.5 NORMAL SPECIFIC ACOUSTIC IMPEDANCE

Satisfying the boundary conditions at the interface of two fluids amounts to requiring continuity of pressure and continuity of the normal component of particle velocity across the boundary. This is equivalent to requiring continuity of the *normal specific acoustic impedance* $\mathbf{z}_n$:

$$\mathbf{z}_n = \frac{\mathbf{p}}{\vec{\mathbf{u}} \cdot \hat{n}} = \frac{\mathbf{p}}{\mathbf{u} \cos \theta_i} \qquad (6.5.1)$$

where $\hat{n}$ is the unit vector perpendicular to the interface and θ_i is the appropriate angle. The normal specific acoustic impedance at the boundary can be expressed in terms of the properties of the incident and reflected waves at the boundary,

$$\mathbf{z}_n = \frac{r_1}{\cos \theta_i} \frac{1 + \mathbf{R}}{1 - \mathbf{R}} \qquad (6.5.2)$$

and this solved for the pressure reflection coefficient gives

$$\mathbf{R} = \frac{\mathbf{z}_n - r_1 / \cos \theta_i}{\mathbf{z}_n + r_1 / \cos \theta_i} \qquad (6.5.3)$$

Note that for normal incidence $\mathbf{z}_n = r_2$ and $\cos \theta_i = 1$, so this equation reduces to (6.2.8). For oblique incidence $\mathbf{z}_n = r_2 / \cos \boldsymbol{\theta}_t$ and (6.5.3) is identical with (6.4.10). Because the incident and reflected pressures are not always exactly in or out of phase, the normal specific acoustic impedance may be a complex quantity:

$$\mathbf{z}_n = r_n + j x_n \qquad (6.5.4)$$

where r_n and x_n are the normal specific acoustic resistance and reactance, respectively. The reflected wave at the boundary may either lead or lag the incident wave by angles ranging from 0° to 180°.

*6.6 REFLECTION FROM THE SURFACE OF A SOLID

Solids can support two types of elastic waves—longitudinal and shear. In an isotropic solid (amorphous materials like glass, hardened clays, concrete, and polycrystalline substances) of transverse dimensions much larger than the wavelength of the acoustic wave, the appropriate phase speed for the longitudinal waves is not the *bar speed* $\sqrt{Y/\rho_0}$, but rather the *bulk speed*

$$c^2 = (\mathcal{B} + \tfrac{4}{3}\mathcal{G})/\rho_0 \qquad (6.6.1)$$

where $\mathscr{B}$ and $\mathscr{G}$ are the *bulk* and *shear moduli* of the solid and ρ_0 its density. (See Appendix A11. Values of the bulk speed for various solids are given in Appendix A10.) The bulk speed for each material is always higher than that for longitudinal waves in thin bars.

(a) Normal Incidence

For this case, we have $\cos\theta_i = 1$ and (6.5.3) becomes

$$\mathbf{R} = \frac{(r_n - r_1) + jx_n}{(r_n + r_1) + jx_n} \tag{6.6.2}$$

The intensity reflection coefficient is

$$R_I = \frac{(r_n - r_1)^2 + x_n^2}{(r_n + r_1)^2 + x_n^2} \tag{6.6.3}$$

and the intensity transmission coefficient is

$$T_I = \frac{4r_n r_1}{(r_n + r_1)^2 + x_n^2} \tag{6.6.4}$$

(b) Oblique Incidence

No single simple method is available for analyzing the reflection of plane waves obliquely incident on the surface of a solid. Because of the differences in the porosity and internal elastic structure of various solids, the nature of the process varies. For instance, the wave transmitted into the solid may be refracted (1) so that it is propagated effectively only perpendicular to the surface, (2) in a manner similar to plane waves entering a second fluid, or (3) into two waves, a longitudinal wave traveling in one direction and a transverse (shear) wave traveling at a lower speed in a different direction.

1. The first type of refraction occurs for *normally-reacting* or *locally acting* surfaces. One example of this occurs in *anisotropic* solids, where waves propagated parallel to the surface travel with a much lower speed than those propagated perpendicular to the surface. This is typical of solids having a honeycomb structure in which the speed of compressional waves through the fluid contained in capillary pores perpendicular to the surface is much higher than that from pore to pore through the solid material of the structure. This type of refraction also will occur in an *isotropic* solid when the speed of longitudinal wave propagation in the solid is small compared with that in the adjacent fluid. Many sound-absorbing materials used in buildings (acoustic tile, perforated panels, etc.) behave as normally-reacting surfaces. When $c_2 \ll c_1$, Snell's law requires that $\theta_t \ll \theta_i$, and a reasonable approximation is to set $\cos\theta_t = 1$. This yields (6.5.3), which can be rewritten as

$$\mathbf{R} = \frac{(r_n - r_1/\cos\theta_i) + jx_n}{(r_n + r_1/\cos\theta_i) + jx_n} \tag{6.6.5}$$

This has the same form as (6.6.2) but with r_1 replaced by $(r_1/\cos\theta_i)$. The intensity reflection and transmission coefficients consequently are given by (6.6.3) and (6.6.4), respectively, with the same replacement. See Problem 6.6.4.

For most solid materials $r_n > r_1$, so that as θ_i increases an angle will be reached where $r_n = (r_1/\cos\theta_i)$. When this occurs the power reflection coefficient $R_\Pi = R_I$ will be near its minimum value. In particular, if x_n were zero, R_Π would be zero and T_Π would be unity. For $\theta_i \to 90°$, R_Π approaches unity. Plotted in Fig. 6.6.1 are curves for the reflection coefficient R_Π as a function of the angle of incidence θ_i for a few assumed values of the nondimensional parameters r_n/r_1 and x_n/r_1.

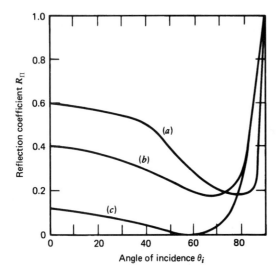

Figure 6.6.1   The reflection coefficient for a typical normally-reacting solid. (a) $r_n/r_1 = x_n/r_1 = 4$. (b) $r_n/r_1 = x_n/r_1 = 2$. (c) $r_n/r_1 = 2$ and $x_n/r_1 = 0$.

2. The second type of refraction is similar to the reflection and refraction occurring between two fluids as discussed in Section 6.4.

3. The third type of refraction occurs for rigid elastic solids. A detailed discussion requires consideration of the coupling of acoustic energy from the incident wave into both shear and longitudinal waves in the solid. Interested readers are referred to the sources listed at the beginning of this chapter.

*6.7 TRANSMISSION THROUGH A THIN PARTITION: THE MASS LAW

A case of practical importance in architectural acoustics is the transmission of sound through a thin partition between two enclosures, as found in many office or temporary working spaces. The partition is often a material whose motion is normal to the interface regardless of the angle of incidence of the sound, and whose thickness L is much smaller than a wavelength ($k_2 L \ll 1$) for the frequency range of interest. Since both media 1 and 3 are the same, by Snell's law any wave transmitted into 3 must have the same direction of propagation as the incident wave in 1. The angles are the same, so that continuity of the normal component of particle velocity is equivalent to

$$\mathbf{u}_i + \mathbf{u}_r = \mathbf{u}_t \tag{6.7.1}$$

If the intervening layer 2 is thin and completely flexible with surface density ρ_S, the layer can be treated as an interface possessing mass so that the difference in pressures across the interface equals the product of the surface density ρ_S of the interface and its acceleration,

$$\mathbf{p}_i + \mathbf{p}_r - \mathbf{p}_t = j\omega \rho_S \mathbf{u}_t \cos\theta \tag{6.7.2}$$

After (6.7.1) is multiplied by r_1 and converted into pressures, and both (6.7.1) and (6.7.2) are divided by the incident pressure amplitude, these equations yield

$$1 - \mathbf{R} = \mathbf{T}$$
$$1 + \mathbf{R} = \mathbf{T} + j\frac{\omega \rho_S}{r_1}\mathbf{T}\cos\theta \tag{6.7.3}$$

Solution for the power transmission coefficient results in

$$T_{\Pi}(\theta) = |\mathbf{T}(\theta)|^2 = \frac{1}{1 + \left[(\omega \rho_s/2r_1)\cos\theta\right]^2} \tag{6.7.4}$$

For an incident wave falling normally on the surface, (6.7.4) reduces to the equivalent case given by (6.3.10) with fluids 1 and 3 having the same characteristic impedances and fluid 2 a thin layer with $r_2 \gg r_1$. (See Problem 6.7.1.)

In most practical situations in air for moderate frequencies the quantity $\omega \rho_s/r_1$ is relatively large. For example, a light partition between two work spaces made out of a sheet or two of gypsum or thick plywood will have a nominal surface density $\rho_S \sim 10 \text{ kg/m}^2$. For frequencies above about 60 Hz, $\omega \rho_s/r_1 > 9$. For this case, (6.7.4) can be approximated by

$$T_{\Pi}(\theta) \sim (2r_1/\omega \rho_S \cos\theta)^2 \tag{6.7.5}$$

as long as θ does not exceed about 70°. This approximation, which fails for near-grazing incidence, expresses a form of the *mass law*: the power transmission coefficient is reduced fourfold for each doubling of the surface density. Further properties of the transmission of sound from a space containing a *diffuse sound field* through a partition to another space, and the effects of *coincidence* on the transmission of acoustic power through an elastic partition, will be deferred until Chapter 13.

6.8 METHOD OF IMAGES

Up to now, we have discussed the reflection and transmission of *plane* waves at plane interfaces. In this section we will investigate the reflection of *spherical* waves at plane boundaries, beginning with boundaries that are perfectly reflecting. (An approximation of such a boundary is the air–water interface.) This problem is amenable to the *method of images*. This approach is also used in electrostatics and in optics. A familiar example of application of the method of images in optics is the analysis of the interference resulting from the reflection of a light source from a single mirror (*Lloyd's mirror*).

In the case of a single plane boundary, fluid 2 is replaced by fluid 1 and an *image* is introduced whose strength and location are selected to satisfy the boundary conditions on the plane of the former interface. Because a solution of the wave equation is unique if the boundary conditions are satisfied, the acoustic field in the real fluid 1 is the same as that for the original situation. The acoustic field in the space containing fluid 2 will *not* be correctly represented.

(a) Rigid Boundary

Let a source of spherical waves be placed in a fluid of medium 1, which extends throughout all space. If this source is located on the z axis a distance $+d$ from the origin, as shown in Fig. 6.8.1, a spherical wave exists in all space given by

$$\mathbf{p}_i = \frac{A}{r_-} e^{j(\omega t - kr_-)} \tag{6.8.1}$$

$$r_- = [(z-d)^2 + y^2 + x^2]^{1/2}$$

where r_- is the distance from the point $(0, 0, d)$. If a second source, the *image*, of equal strength, frequency, and initial phase angle is placed at $(0, 0, -d)$,

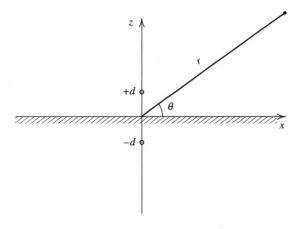

Figure 6.8.1   The use of image theory for calculating the acoustic field of a source of spherical waves near a rigid planar boundary. The source is located at $(0, 0, +d)$ and an image of equal strength and the same phase is located at $(0, 0, -d)$. The field point is located at (r, θ).

$$\mathbf{p}_r = \frac{A}{r_+} e^{j(\omega t - k r_+)}$$

$$r_+ = [(z + d)^2 + y^2 + x^2]^{1/2}$$

(6.8.2)

it is easy to show that the normal component of the particle velocity vanishes on the x-y plane. Therefore, the fluid on the negative side of the x-y plane can be replaced by a rigid boundary at $z = 0$. (The components of the particle velocity parallel to the x-y plane do not cancel, so there is a velocity parallel to the boundary. Including the effects of viscosity would introduce some very small acoustic losses at the boundary, but these are negligible for our purposes here.)

The pressure in the region $z > 0$ is given by the sum of (6.8.1) and (6.8.2):

$$\mathbf{p} = \mathbf{p}_i + \mathbf{p}_r = A \left(\frac{1}{r_-} e^{-jkr_-} + \frac{1}{r_+} e^{-jkr_+} \right) e^{j\omega t}$$

(6.8.3)

While it is instructive to plot the pressure amplitude in the region of fluid 1 without making any approximations (see Problem 6.8.1C), greater insight can be gained by looking for an analytical solution for distances $r \gg d \cos \theta$, where θ is the *grazing* angle with respect to the boundary. In this approximation, a little geometry shows

$$\Delta r \approx d \sin \theta$$

$$r_- \approx r - \Delta r$$

$$r_+ \approx r + \Delta r$$

(6.8.4)

so that (6.8.3) becomes

$$\mathbf{p}(r, \theta, t) \approx \frac{A}{r} e^{j(\omega t - kr)} \left(\frac{e^{jk\Delta r}}{1 - \Delta r / r} + \frac{e^{-jk\Delta r}}{1 + \Delta r / r} \right)$$

(6.8.5)

The term $\Delta r / r$ gives the differences in amplitudes resulting from the slightly different distances to the field point from the source and its image. They yield minor contributions as long as $r \gg \Delta r$. The $k \Delta r$ in the exponent is another matter. Unless $k \Delta r \ll 1$, there will be significant phase interference between the pressures received from the source and its image. Discarding the terms $\Delta r / r$ in

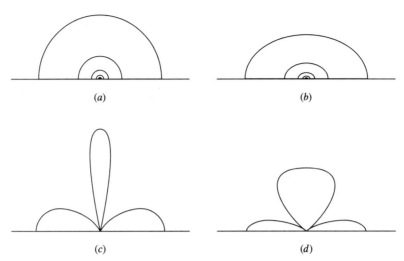

Figure 6.8.2 Contours of equal pressure amplitude for a source of spherical waves with wave number k at a distance d from a planar rigid surface. (a) $kd = 0.40$. (b) $kd = 0.80$. (c) $kd = 1.60$. (d) $kd = 3.20$.

the denominators in (6.8.5) and using standard exponential and trigonometric relationships gives

$$\mathbf{p}(r, \theta, t) \approx \frac{2A}{r} \cos(kd \sin \theta)e^{j(\omega t - kr)} \quad \text{(rigid boundary)} \quad (6.8.6)$$

The pressure field is an outgoing spherical wave with the amplitude depending on the angle θ. The pressure field for reflection of a spherical wave from a rigid boundary is sketched in Fig. 6.8.2 for several values of kd.

(b) Pressure Release Boundary

The pressure field for reflection of a spherical wave from a pressure release surface is found by using an image of the same amplitude as the source, but of *opposite* phase. The proof of this and the derivation of the result,

$$\mathbf{p}(r, \theta, t) \approx j\frac{2A}{r} \sin(kd \sin \theta)e^{j(\omega t - kr)} \quad \text{(pressure release boundary)} \quad (6.8.7)$$

are left as exercises.

(c) Extensions

There are a number of extensions to these simple examples that can be made easily.

1. The method does not depend on the source vibrating at a single frequency. If the source emits a spherical wave

$$\mathbf{p}_i = \frac{1}{r_-}f(ct - r_-) \quad (6.8.8)$$

the pressure field of the image would be the same function with r_+ replacing r_- and multiplied by ± 1 depending on whether the boundary is rigid or pressure release. The resultant total acoustic pressure is

$$\mathbf{p} = \frac{1}{r_-}f(ct - r_-) \pm \frac{1}{r_+}f(ct - r_+) \tag{6.8.9}$$

2. There may be several elements comprising the source (as an array of point sources or a loudspeaker). Elementary application of superposition shows that the image will be the mirror reflection of the source with amplitude multiplied by ± 1 depending on the boundary condition.

3. If the boundary is not rigid or pressure release, then the method of images can be used as a reasonably good approximation if the source is many wavelengths away from the boundary, so that at the boundary the radii of curvature of the wave fronts are much greater than a wavelength. Under this condition, the waves incident on the boundary look locally like plane waves and the local reflection coefficient will be very similar to that for an incident plane wave. Then, for example, the spherical monofrequency source radiating the pressure field (6.8.1) will generate a total field

$$\mathbf{p} = \mathbf{p}_i + \mathbf{p}_r = A\left(\frac{1}{r_-}e^{-jkr_-} + \frac{\mathbf{R}(\theta)}{r_+}e^{-jkr_+}\right)e^{j\omega t} \tag{6.8.10}$$

where the reflection coefficient is evaluated at the angle θ defined for the *specular reflection* $(\theta_r = \theta_i)$ between source and field point. The resulting field will be missing certain features that a more exact, and considerably more complicated, analysis would provide, but under the geometrical restriction stated above these effects will be relatively small. This restriction must also be applied to the location of the field point, as explained below.

4. More than one reflecting surface can be present (a hall of mirrors). Under the same geometrical limitation as in extension 3, each boundary behaves like a mirror with reflectivity determined along each possible path from the source to the field point.

All applications of the method of images for rigid or pressure release plane boundaries exhibit a very important feature. If we have a point source at $(0, 0, d)$ and a point receiver at (x, y, z), examination of the expressions for r_- and r_+ shows that exchanging the positions of source and receiver does not change the value of the acoustic pressure at the receiver. The pressure fields in the medium for the two different geometries may have different interference patterns, but the signal observed by the receiver will be unaffected by the exchange. This means, for example, that the geometrical restrictions made in extension 3 concerning d/λ to obtain a simple approximation must also be applied to the distance between the field point (receiver) and the boundary.

For directional sources and receivers, and for other than rigid and pressure release boundaries, the conditions of the exchange and the relative orientations of the source and receiver must be handled more carefully, but similar results can be obtained. This will be dealt with further in Chapter 7 when we develop a more general expression of *acoustic reciprocity*.

PROBLEMS

6.2.1. A 1 kHz plane wave in water of 50 Pa effective (rms) pressure is incident normally on the water–air boundary. (*a*) What is the effective pressure of the plane wave transmitted into the air? (*b*) What is the intensity of the incident wave in the water and of the wave transmitted into the air? (*c*) Express, as a decibel reduction, the ratio of the intensity of the transmitted wave in air to that of the incident wave in water. (*d*) Answer the same three questions for the above sound wave incident on a thick layer of ice. (*e*) What is the power reflection coefficient from the layer of ice?

6.2.2. If a plane wave is reflected from the ocean floor at normal incidence with a level 20 dB below that of the incident wave, what are the possible values of the specific acoustic impedance of the fluid bottom material?

6.2.3. (*a*) A plane wave in seawater is normally incident on the water–air interface. Find the pressure and intensity transmission coefficients. (*b*) Repeat (*a*) for a wave in air normally incident on the air-water interface. (*c*) For (*a*) and (*b*) find the change in pressure and intensity levels if P_{ref} and I_{ref} are the same in both media.

6.2.4. Assume a reflection coefficient **R** = 0.5 for a normally incident wave in air at 500 Hz and pressure amplitude 2 Pa. (*a*) What are the intensities of the incident, reflected, and transmitted waves? (*b*) Calculate the intensity of the total pressure field in fluid 1. (*c*) Write the total field in fluid 1 as the sum of a traveling and a standing wave. (*d*) Calculate the intensities for each of the two waves in part (*c*). (*e*) Are your results consistent with conservation of energy?

6.2.5C. For Problem 6.2.4, plot the total pressure amplitude in fluid 1 as a function of the distance from the boundary. Derive an equation that relates the ratio of the pressure amplitude at the antinodes to that at the nodes in terms of the pressure amplitudes of the incident and reflected waves and compare to your graph.

6.2.6C. Plot the pressure reflection and transmission coefficients and the intensity reflection and transmission coefficients for normal incidence of a plane wave on a fluid–fluid boundary for $0 < r_1/r_2 < 10$. Comment on the results for $r_1/r_2 = 0$, $r_1/r_2 = 1$, and $r_1/r_2 \to \infty$.

6.3.1. Show that when $r_2 = r_3$ the pressure reflection, intensity reflection, and intensity transmission coefficients all reduce to those for Section 6.2.

6.3.2. (*a*) What must be the thickness of, and the speed of sound in, a plastic layer having a density of 1500 kg/m^3 if it is to transmit plane waves at 20 kHz from water into steel with no reflection? (*b*) What would be the intensity reflection coefficient back into water for normally incident waves impinging on an infinitely thick layer of this plastic?

6.3.3. For a 2 kHz plane wave in water impinging normally on a steel plate of 1.5 cm thickness, (*a*) what is the transmission loss, expressed in dB, through the steel plate into water on the opposite side? (*b*) What is the power reflection coefficient of this plate? (*c*) Repeat (*a*) and (*b*) for a 1.5 cm thick slab of sponge rubber having a density of 500 kg/m^3 and a longitudinal wave speed of 1000 m/s.

6.3.4. Given the task of maximizing the transmission of sound waves from water into steel, (*a*) what is the optimum characteristic impedance of the material to be placed between the water and the steel? (*b*) What must be the density of, and sound speed in, a layer of 1 cm thickness that will produce 100% transmission at 20 kHz?

6.3.5. For normal incidence on a layer between fluids 1 and 3 and with $r_2 = r_1$, (*a*) show that the magnitude of the pressure reflection coefficient reduces to that in Section 6.2 for the appropriate fluids. (*b*) Interpret the phase angle of the reflection coefficient in terms of the time of flight of the signal in the layer.

6.3.6. Assume a layer of fluid 2 separates fluid 1 from fluid 3. (*a*) Compare the pressure reflection coefficient for a plane wave traveling in fluid 1 that reflects normally from the layer with that for a plane wave traveling in fluid 3 that reflects normally from the layer. (*b*) Are the pressure reflection coefficients the same for the two cases? (*c*) Repeat (*a*) and (*b*) for the power reflection and transmission coefficients. (*d*) What do these results suggest in terms of energy transmission?

6.3.7C. Plot the intensity transmission coefficient as a function of the scaled layer thickness $k_2 L$ for $r_1/r_2 = 2$ and $1 < r_1/r_3 < 9$. Comment on the conditions required to obtain minimum and maximum transmissions.

6.3.8. A plane wave pulse consisting of 10 cycles of 10 kHz carrier is normally incident from water onto a 50 m thick layer of red clay overlaying a thick sedimentary bottom with $c_3 = 2300$ m/s and $\rho_3 = 2210$ kg/m^3. For the first two reflections received back in the water, calculate (*a*) the time interval between arrivals, (*b*) the amplitudes relative to the incident pulse, and (*c*) the relative phase between the arrivals assuming they were continuous waves rather than pulses.

6.4.1. (*a*) As a function of θ_i, plot the amplitude and phase of the pressure reflection coefficient for the case $c_2 = c_1$ and $\rho_2 > \rho_1$ and identify any significant features. (*b*) Plot the amplitude and phase of the pressure reflection coefficient for the case $\rho_2 = \rho_1$ and $c_2 > c_1$ and identify any significant features.

6.4.2. For plane wave reflection from a fluid–fluid interface it is observed that at normal incidence the pressure amplitude of the reflected wave is one-half that of the incident wave (no phase information is recorded). As the angle of incidence is increased, the amplitude of the reflected wave first decreases to zero and then increases until at $30°$ the reflected wave is as strong as the incident wave. Find the density and sound speed in the second medium if the first medium is water.

6.4.3. A plane wave traveling from air into hydrogen gas through a thin separating membrane is refracted by $40°$ from its original direction. (*a*) What is the angle of incidence in the air? (*b*) What is the sound power transmission coefficient?

6.4.4. A plane wave in water of 100 Pa peak pressure amplitude is incident at $45°$ on a mud bottom having $\rho_2 = 2000$ kg/m^3 and $c_2 = 1000$ m/s. Compute (*a*) the angle of the ray transmitted into the mud, (*b*) the peak pressure amplitude of the transmitted ray, (*c*) the peak pressure amplitude of the reflected ray, and (*d*) the sound power reflection coefficient.

6.4.5. Plane waves in water of 100 Pa effective (rms) pressure are incident normally on a sand bottom. The sand has density 2000 kg/m^3 and sound speed 2000 m/s. (*a*) What is the effective pressure of the wave reflected back into the water? (*b*) What is the effective pressure of the wave transmitted into the sand? (*c*) What is the power reflection coefficient? (*d*) What is the smallest angle of incidence at which all of the incident energy will be reflected?

6.4.6. A sand bottom in seawater is characterized by $\rho_2 = 1700$ kg/m^3 and $c_2 = 1600$ m/s. (*a*) What is the critical angle of incidence corresponding to total reflection? (*b*) For what angle of incidence is the power reflection coefficient equal to 0.25? (*c*) What is the power reflection coefficient for normal incidence?

6.4.7C. For the oblique reflection of a plane wave at a fluid–fluid interface, plot the magnitude and phase of the pressure reflection coefficient as a function of θ_i for (*a*) $r_2/r_1 = 0.5$ and $c_2/c_1 = 0.5, 1, 1.5$; (*b*) $r_2/r_1 = 1$ and $c_2/c_1 = 0.5, 1, 1.5$; and (*c*) $r_2/r_1 = 1.5$ and $c_2/c_1 = 0.5, 1, 1.5$.

6.4.8. Derive the approximate behavior (lowest order in small angle) of **R** (amplitude and phase) (*a*) in terms of the *grazing* angle $\beta = (\pi/2 - \theta_i)$ for near-grazing incidence, (*b*) in terms of $(\theta_c - \theta_i)$ for θ_i slightly less than θ_c, and (*c*) in terms of $(\theta_i - \theta_c)$ for θ_i

slightly more than θ_c. (d) With the help of the approximation $\exp(\delta) \approx 1 + \delta$ for small δ, approximate the results through lowest order for these cases in exponential form.

6.6.1. An acoustic tile panel is characterized by a normal specific acoustic impedance of $900 - j1200$ Pa·s/m. (a) For what angle of incidence in air will the power reflection coefficient be a minimum? (b) What is the power reflection coefficient for an angle of incidence of $80°$? (c) What is the power reflection coefficient for normal incidence?

6.6.2. A wall reflects plane waves like a normally-reacting surface of normal specific acoustic impedance $\mathbf{z}_n = r_1 + j\omega\rho_S$, where r_1 is the characteristic impedance of the air and ρ_S is the area density of the wall (kg/m²). Derive a general equation for the power reflection coefficient as a function of the incident angle θ_i. For $\rho_S = 2$ kg/m², compute and plot the power reflection coefficient at 100 Hz as a function of θ_i.

6.6.3C. Plot the magnitude and phase of the pressure reflection coefficient as a function of incident angle for a normally-reacting solid for (a) $r_n/r_1 = 2$ and $x_n/r_1 = 0$, (b) $r_n/r_1 = x_n/r_1 = 2$, and (c) $r_n/r_1 = x_n/r_1 = 4$. Comment on the conditions for minimum reflection coefficient.

6.6.4. Starting with (6.6.5), show that the power reflection and transmission coefficients for oblique reflection from a normally-reacting solid are

$$R_\Pi = \frac{(r_n \cos\theta_i - r_1)^2 + (x_n \cos\theta_i)^2}{(r_n \cos\theta_i + r_1)^2 + (x_n \cos\theta_i)^2}$$

$$T_\Pi = \frac{4r_1 r_n \cos\theta_i}{(r_n \cos\theta_i + r_1)^2 + (x_n \cos\theta_i)^2}$$

6.7.1. (a) With the help of the speed of sound c_2 for the material of the partition, show that $\omega\rho_S/r_1 = (r_2/r_1)k_2 L$, where L is the thickness of the partition. (b) For normal incidence, find the inequality necessary for (6.3.10) to reduce to (6.7.4). (c) Is the inequality of (b) consistent with the approximations made in obtaining (6.3.12)?

6.8.1C. A source of spherical waves of wave number k is a distance d above an infinite, plane, rigid boundary. (a) For $kd = 0.1$, plot contours of equal pressure amplitude for the exact solution (6.8.3) and the approximation (6.8.6). Comment on the differences, if any. (b) Repeat for $kd = 10$.

6.8.2. (a) Show that in the limit $kd \ll 1$ the pressure field (6.8.6) reduces to

$$\mathbf{p}(r, \theta, t) \approx \frac{2A}{r} e^{j(\omega t - kr)}$$

(b) What effect does the presence of the boundary have on the sound pressure level observed (many wavelengths away) for the same source in the absence of the boundary?

6.8.3. (a) Show from (6.8.3) that the use of an image source of the same amplitude but opposite phase satisfies the pressure release boundary condition on the plane halfway between them. (b) Show that for $r \gg d\cos\theta$ the pressure above a pressure release surface is given by (6.8.7) and (c) obtain an expression for the locations of the pressure nodal surfaces.

6.8.4. In the limit $kd \ll 1$ show that (6.8.7) reduces to

$$\mathbf{p}(r, \theta, t) \approx -j\frac{2Akd}{r} \sin\theta \, e^{j(\omega t - kr)}$$

6.8.5. (a) A source of spherical waves of frequency f and pressure amplitude A at 1 meter is a distance d above a plane, rigid boundary. Calculate the amplitude of the pressure on the boundary as a function of d and r. (*b*) For the same source at the same distance above a plane, pressure release boundary, calculate the normal component of the particle velocity on the boundary. (*c*) Make plots of (*a*) as a function of R, the distance along the boundary measured from the point closest to the source, for various values of d. (*d*) Repeat (*c*) for (*b*).

6.8.6C. (*a*) A source of spherical waves of frequency f and pressure amplitude A at 1 meter is located in water a distance d above a flat, quartz sand bottom. For $kd = 20\pi$, plot the amplitude of the pressure at the same distance d above the boundary as a function of kr, where r is the distance between source and receiver. (*b*) Repeat for a red clay bottom.

6.8.7C. A source of monofrequency spherical waves is located in air midway between parallel rigid surfaces a distance H apart. (*a*) Assuming incoherent summation, design a program to calculate the pressure at a receiver also located at the midpoint for distances greater than $10d$ away from the source. (*b*) Determine if the pressure approaches an asymptotic functional dependence proportional to $1/\sqrt{r}$ as the number of images is increased. (*c*) How well do your results for larger numbers of images satisfy the relation $SPL(1) - SPL(r) = 10 \log r + 10 \log(H/\pi)$?

6.8.8. Assume that the near-shore ocean can be modeled by two nonparallel planes: a horizontal pressure release top and a sloping rigid bottom. (*a*) Sketch the position and indicate the phases of the images representing the paths from a source in the layer that reflect once off the top, once off the bottom, first off the top then off the bottom, and first off the bottom and then off the top. (*b*) Show that all images lie on a circle passing through the source and centered at the shore.

6.8.9C. A horizontal plane, pressure release surface makes an angle of 20° with a plane, rigid bottom. A source of spherical waves of frequency f and pressure amplitude A at 1 meter is located at mid-depth in this wedge at a distance R from the vertex. (*a*) Assuming incoherent summation, calculate the pressure amplitude as a function of distance along the mid-depth of the wedge. Cover the distance from the vertex out to at least twice R (to avoid overflow, omit distances in the immediate vicinity of the source). (*b*) Repeat assuming coherent summation. Note that there are a total of 17 images in addition to the source.

6.8.10C. Repeat Problem 6.8.9C for a line passing through the source and parallel to the vertex. Cover distances out to at least $2R$ from the source.

Chapter **7**

RADIATION
AND RECEPTION
OF ACOUSTIC WAVES

7.1 RADIATION FROM A PULSATING SPHERE

The acoustic source simplest to analyze is a pulsating sphere—a sphere whose radius varies sinusoidally with time. While pulsating spheres are of little practical importance, their analysis is useful for they serve as the prototype for an important class of sources referred to as *simple sources*.

In a medium that is infinite, homogeneous, and isotropic, a pulsating sphere will produce an outgoing spherical wave

$$\mathbf{p}(r,t) = (\mathbf{A}/r)e^{j(\omega t - kr)} \tag{7.1.1}$$

where $\mathbf{A}$ is determined by an appropriate boundary condition.

Consider a sphere of average radius a, vibrating radially with complex speed $U_0 \exp(j\omega t)$, where the displacement of the surface is much less than the radius, $U_0/\omega \ll a$. The acoustic pressure of the fluid in contact with the sphere is given by (7.1.1) evaluated at $r = a$. (This is consistent with the small-amplitude approximation of linear acoustics.) The radial component of the velocity of the fluid in contact with the sphere is found using the specific acoustic impedance for the spherical wave (5.11.10) also evaluated at $r = a$,

$$\mathbf{z}(a) = \rho_0 c \cos \theta_a \, e^{j\theta_a} \tag{7.1.2}$$

where $\cot \theta_a = ka$. The pressure at the surface of the source is then

$$\mathbf{p}(a,t) = \rho_0 c U_0 \cos \theta_a \, e^{j(\omega t - ka + \theta_a)} \tag{7.1.3}$$

Comparing (7.1.3) with (7.1.1) gives

$$\mathbf{A} = \rho_0 c U_0 a \cos \theta_a \, e^{j(ka + \theta_a)} \tag{7.1.4}$$

so the pressure at any distance $r > a$ is

$$\mathbf{p}(r,t) = \rho_0 c U_0 (a/r) \cos \theta_a \, e^{j[\omega t - k(r-a) + \theta_a]} \tag{7.1.5}$$

171

The acoustic intensity, found from (5.11.20), is

$$I = \tfrac{1}{2}\rho_0 c U_0^2 (a/r)^2 \cos^2\theta_a \qquad\qquad (7.1.6)$$

If the radius of the source is small compared to a wavelength, $\theta_a \to \pi/2$ and the specific acoustic impedance near the surface of the sphere is strongly reactive. (This reactance is a symptom of the strong radial divergence of the acoustic wave near a small source and represents the storage and release of energy because successive layers of the fluid must stretch and shrink circumferentially, altering the outward displacement. This inertial effect manifests itself in the mass-like reactance of the specific acoustic impedance.) In this long wavelength limit the pressure

$$\mathbf{p}(r, t) = j\rho_0 c U_0 (a/r) ka\, e^{j(\omega t - kr)} \qquad ka \ll 1 \qquad\qquad (7.1.7)$$

is nearly $\pi/2$ out of phase with the particle speed (pressure and particle speed are not *exactly* $\pi/2$ out of phase, since that would lead to a vanishing intensity), and the acoustic intensity is

$$I = \tfrac{1}{2}\rho_0 c U_0^2 (a/r)^2 (ka)^2 \qquad ka \ll 1 \qquad\qquad (7.1.8)$$

For constant U_0 this intensity is proportional to the square of the frequency and depends on the fourth power of the radius of the source. Thus, we see that sources small with respect to a wavelength are inherently poor radiators of acoustic energy.

In the next section, it will be shown that all *simple sources*, no matter what their shapes, will produce the same acoustic field as a pulsating sphere provided the wavelength is greater than the dimensions of the source and the sources have the same *volume velocity*.

7.2 ACOUSTIC RECIPROCITY AND THE SIMPLE SOURCE

Acoustic reciprocity is a powerful concept that can be used to obtain some very general results. Let us begin by deriving one of the more commonly encountered statements of acoustic reciprocity.

Consider a space occupied by two sources, as suggested by Fig. 7.2.1. By changing which source is active and which passive, it is possible to set up different sound fields. Choose two situations having the same frequency and denote them as 1 and 2. Establish a volume V of space that does not itself contain any sources but bounds them. Let the surface of this volume be S. The volume V and the surface S remain the same for both situations. Let the velocity potential be Φ_1 for situation 1 and Φ_2 for situation 2. Green's theorem (see Appendix A8) gives the general relation

$$\int_S (\Phi_1 \nabla \Phi_2 - \Phi_2 \nabla \Phi_1) \cdot \hat{n}\, dS = \int_V (\Phi_1 \nabla^2 \Phi_2 - \Phi_2 \nabla^2 \Phi_1)\, dV \qquad\qquad (7.2.1)$$

where $\hat{n}$ is the unit outward normal to S. Since the volume does not include any sources, and since both velocity potentials are for excitations of the same

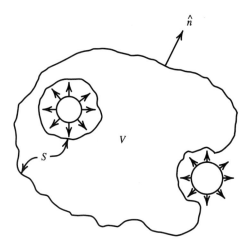

Figure 7.2.1   Geometry used in deriving the theorem of acoustical reciprocity.

frequency, the wave equation yields

$$\nabla^2 \Phi_1 = -k^2 \Phi_1$$
$$\nabla^2 \Phi_2 = -k^2 \Phi_2$$

(7.2.2)

so that the right side of (7.2.1) vanishes identically throughout V. Furthermore, recall that the pressure is $\mathbf{p} = -j\omega \rho_0 \Phi$ and the particle velocity for irrotational motion is $\vec{\mathbf{u}} = \nabla \Phi$. Substitution of these expressions into the left side of (7.2.1) gives

$$\int_S (\mathbf{p}_1 \vec{\mathbf{u}}_2 \cdot \hat{n} - \mathbf{p}_2 \vec{\mathbf{u}}_1 \cdot \hat{n})\, dS = 0$$

(7.2.3)

This is one form of the *principle of acoustic reciprocity*. This principle states that, for example, if the locations of a small source and a small receiver are interchanged in an unchanging environment, the received signal will remain the same.

To obtain information about simple sources, let us develop a more restrictive but simpler form of (7.2.3). Assume that some portion of S is removed a great distance from the enclosed source. In any real case there is always some absorption of sound by the medium so the intensity at this surface will decrease faster than $1/r^2$. Since the area of the surface increases as r^2, the product of intensity and area vanishes in the limit $r \to \infty$. In addition, if each of the remaining portions of S is either (1) perfectly rigid so that $\vec{\mathbf{u}} \cdot \hat{n} = 0$, (2) pressure release so that $\mathbf{p} = 0$, or (3) normally reacting so that $\mathbf{p}/(\vec{\mathbf{u}} \cdot \hat{n}) = z_n$, then the surface integrals over these surfaces must vanish. Under these conditions, (7.2.3) reduces to an integral over only those portions of S that correspond to sources active in situations 1 or 2:

$$\int_{sources} (\mathbf{p}_1 \vec{\mathbf{u}}_2 \cdot \hat{n} - \mathbf{p}_2 \vec{\mathbf{u}}_1 \cdot \hat{n})\, dS = 0$$

(7.2.4)

This simple result will now be applied to develop some important general properties of sources that are small compared to a wavelength.

Consider a region of space in which there are two irregularly shaped sources, as shown in Fig. 7.2.2. Let source A be active and source B be perfectly rigid in

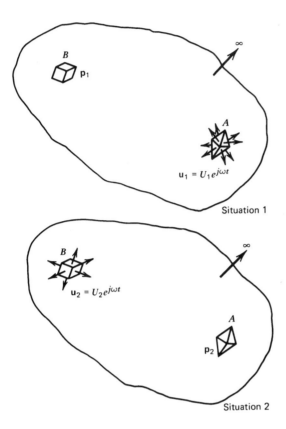

Figure 7.2.2   Reciprocity theorem applied to simple sources.

situation 1, and vice versa in situation 2. If we define $\mathbf{p}_1$ as the pressure at B when source A is active with $\vec{\mathbf{u}}_1$ the velocity of its radiating element, and $\mathbf{p}_2$ as the pressure at A when source B is active with $\vec{\mathbf{u}}_2$ the velocity of its radiating element, application of (7.2.4) yields

$$\int_{S_A} \mathbf{p}_2 \vec{\mathbf{u}}_1 \cdot \hat{n}\, dS = \int_{S_B} \mathbf{p}_1 \vec{\mathbf{u}}_2 \cdot \hat{n}\, dS \tag{7.2.5}$$

If the sources are small with respect to a wavelength and several wavelengths apart, then the pressure is uniform over each source so that

$$\frac{1}{\mathbf{p}_1} \int_{S_A} \vec{\mathbf{u}}_1 \cdot \hat{n}\, dS = \frac{1}{\mathbf{p}_2} \int_{S_B} \vec{\mathbf{u}}_2 \cdot \hat{n}\, dS \tag{7.2.6}$$

Assume that the moving elements of a source have complex vector displacements

$$\vec{\xi} = \vec{\Xi} e^{j(\omega t + \phi)} \tag{7.2.7}$$

where $\vec{\Xi}$ gives the magnitude and direction of the displacement and ϕ the temporal phase of each element. If $\hat{n}$ is the unit outward normal to each element dS of the surface, the source will displace a volume of the surrounding medium

$$\mathbf{V} = \int_S \vec{\Xi} e^{j(\omega t + \phi)} \cdot \hat{n}\, dS = V e^{j(\omega t + \theta)} \tag{7.2.8}$$

where $\mathbf{V}$ is the *complex volume displacement*, V the generalization of the volume displacement amplitude discussed in Section 4.5, and θ the accumulated phase over the surface of the element. The time derivative $\partial\mathbf{V}/\partial t$, the *complex volume velocity*, defines the *complex source strength* $\mathbf{Q}$

$$\mathbf{Q}e^{j\omega t} = \frac{\partial\mathbf{V}}{\partial t} = \int_S \vec{\mathbf{u}} \cdot \hat{n}\, dS \tag{7.2.9}$$

where $\vec{\mathbf{u}} = \partial\vec{\xi}/\partial t$ is the complex velocity distribution of the source surface. The complex source strength of the pulsating sphere has only a real part,

$$\mathbf{Q} = Q = 4\pi a^2 U_0 \tag{7.2.10}$$

Substitution of (7.2.9) and $\mathbf{p} = P(r)\exp[j(\omega t - kr)]$ into (7.2.6) gives

$$\mathbf{Q}_1, P_1(r) = \mathbf{Q}_2/P_2(r) \tag{7.2.11}$$

which shows that the ratio of the source strength to the pressure amplitude at distance r from the source is the same for all simple sources (at the same frequency) in the same surroundings. This allows us to calculate the pressure field of any irregular simple source since it must be identical with the pressure field produced by a small pulsating sphere of the same source strength. If the simple sources are in free space, (7.1.7) and (7.2.10) show that the ratio of (7.2.11) is

$$\mathbf{Q}/P(r) = -j\,2\lambda r/\rho_0 c \tag{7.2.12}$$

This is the *free field reciprocity factor*.

Rewriting (7.1.7) with the help of (7.2.10) results in

$$\mathbf{p}(r,t) = \tfrac{1}{2}j\rho_0 c(Q/\lambda r)e^{j(\omega t - kr)} \tag{7.2.13}$$

which, from the above, must be true for all simple sources. The pressure amplitude is

$$P = \tfrac{1}{2}\rho_0 cQ/\lambda r \quad \text{(simple source)} \tag{7.2.14}$$

and the intensity is

$$I = \tfrac{1}{8}\rho_0 c(Q/\lambda r)^2 \tag{7.2.15}$$

Integration of the intensity over a sphere centered at the source gives the power radiated,

$$\Pi = \tfrac{1}{2}\pi\rho_0 c(Q/\lambda)^2 \tag{7.2.16}$$

Another case of practical interest is that of a simple source mounted on or very close to a rigid plane boundary. If the dimensions of the boundary are much greater than a wavelength of sound, the boundary can be considered a plane of infinite extent. This kind of boundary is termed a *baffle*. As shown in Section

6.8, the pressure field in the half-space occupied by the source will be twice that generated by the source (with the same source strength) in free space,

$$P = \rho_0 cQ/\lambda r \qquad \text{(baffled simple source)} \qquad (7.2.17)$$

The intensity is increased by a factor of four,

$$I = \tfrac{1}{2}\rho_0 c(Q/\lambda r)^2 \qquad (7.2.18)$$

and integration of the intensity over a hemisphere (there is no acoustic penetration of the space behind the baffle) gives twice the radiated power,

$$\Pi = \pi\rho_0 c(Q/\lambda)^2 \qquad (7.2.19)$$

A doubling of the power output of the source may seem surprising but results from the fact that the source has the same source strength in both cases: the source face is moving with the same velocity in both cases, but in the baffled case it is working into twice the force and therefore must expend twice the power to maintain its own motion in the presence of the doubled pressure.

7.3 THE CONTINUOUS LINE SOURCE

As an example of a distribution of point sources used to describe an extended source, consider a long, thin cylindrical source of length L and radius a. This configuration, suggested in Fig. 7.3.1, is termed a *continuous line source*. Let the surface vibrate radially with speed $U_0 \exp(j\omega t)$. Consider the source to be made up of a large number of cylinders of length dx. Each of these elements can be considered an unbaffled simple source of strength $dQ = U_0 2\pi a\, dx$. Each generates the increment of pressure given by (7.2.13) with r replaced by the distance r' from

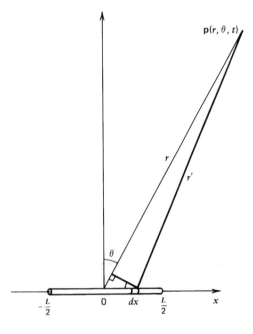

Figure 7.3.1   The far field acoustic field at (r, θ) of a continuous line source of length L and radius a is found by summing the contributions of simple sources of length dx and radius a.

the element to the field point at (r, θ). The total pressure is found by integrating $d\mathbf{p}$ over the length of the source,

$$\mathbf{p}(r, \theta, t) = \frac{j}{2}\rho_0 c U_0 ka \int_{-L/2}^{L/2} \frac{1}{r'}e^{j(\omega t - kr')}dx \qquad (7.3.1)$$

The acoustic field close to the source is complicated, but a simple expression can be obtained in the *far field approximation*. Under the assumption $r \gg L$, the denominator of the integrand can be replaced by its approximate value r, which amounts to making very small errors in the amplitudes of the acoustic fields at (r, θ) generated by each of the simple sources. In the exponent, however, this simplification cannot always be made because the relative phases of the elements will be very strong functions of angle when kL approaches or exceeds unity. Then the more accurate approximation $r' \approx r - x \sin \theta$ must be used, and the integral takes the form

$$\mathbf{p}(r, \theta, t) = \frac{j}{2}\rho_0 c U_0 \frac{ka}{r}e^{j(\omega t - kr)} \int_{-L/2}^{L/2} e^{jkx \sin \theta}dx \qquad (7.3.2)$$

Evaluation is immediate,

$$\mathbf{p}(r, \theta, t) = \frac{j}{2}\rho_0 c U_0 \frac{a}{r}kL \left(\frac{\sin\left(\frac{1}{2}kL \sin \theta\right)}{\frac{1}{2}kL \sin \theta}\right)e^{j(\omega t - kr)} \qquad (7.3.3)$$

The acoustic pressure amplitude in the far field can be written

$$\boxed{P(r, \theta) = P_{ax}(r)H(\theta)} \qquad (7.3.4)$$

where

$$H(\theta) = \left|\frac{\sin v}{v}\right| \qquad v = \frac{1}{2}kL \sin \theta \qquad (7.3.5)$$

is the *directional factor* and

$$P_{ax}(r) = \frac{1}{2}\rho_0 c U_0 (a/r)kL \qquad (7.3.6)$$

is the amplitude of the *far field axial pressure*.

Separating the far field pressure amplitude into one factor that depends only on angle and has maximum value of unity on the *acoustic axis* and another that depends only on the distance from the source is common practice in describing the sound fields of complicated sources. Note that in the far field the axial pressure is proportional to $1/r$, as for a simple source. This is a feature common to all acoustic sources.

The behavior of $(\sin v)/v$ is shown in Fig. 7.3.2. This function is known as the *sinc function* or the *zeroth order spherical Bessel function of the first kind*. The corresponding *beam pattern* $b(\theta) = 20 \log H(\theta)$ is plotted in Fig. 7.3.3 for the case $kL = 24$. There are *nodal surfaces* (cones in the present case) at angles where $H(\theta) = 0$, for which $\frac{1}{2}kL \sin \theta_n = \pm n\pi$, with $n = 1, 2, 3, \ldots$. These nodal surfaces are separated by *lobes*

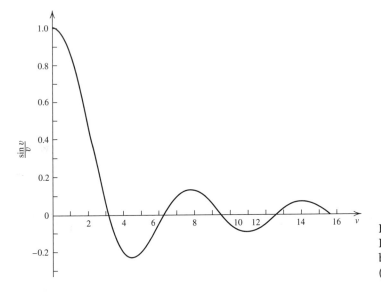

Figure 7.3.2 Functional behavior of $(\sin v)/v$.

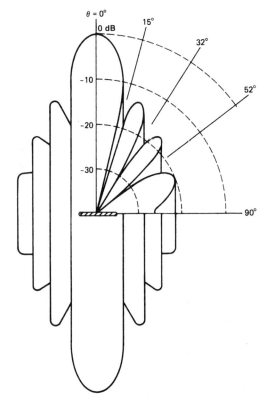

Figure 7.3.3 Beam pattern $b(\theta)$ for a continuous line source of length L radiating sound of wave number k with $kL = 24$.

where the acoustic energy is nonzero. Most of the acoustic energy is projected in the *major lobe*, contained within the angles given by $n = 1$ and centered on a plane perpendicular to the line source. The amplitudes of the *minor lobes* are less than unity and tend to decrease away from this plane. Clearly, the larger the value of kL the more narrowly directed will be the major lobe and the greater the number of minor lobes.

Note that the pressure, when expressed in terms of the source strength $Q = U_0\, 2\pi aL$, becomes

$$\mathbf{p}(r,\theta,t) = \frac{j}{2}\rho_0 c \frac{Q}{\lambda r}\frac{\sin v}{v}e^{j(\omega t - kr)} \tag{7.3.7}$$

Comparison with (7.2.13) shows that the pressure field is the product of that generated by a simple source of source strength Q and a directional factor $\sin v/v$.

7.4 RADIATION FROM A PLANE CIRCULAR PISTON

An acoustic source of practical interest is the plane circular piston, which is the model for a number of sources, including loudspeakers, open-ended organ pipes, and ventilation ducts. Consider a piston of radius a mounted on a flat rigid baffle of infinite extent. Let the radiating surface of the piston move uniformly with speed $U_0 \exp(j\omega t)$ normal to the baffle. The geometry and coordinates are sketched in Fig. 7.4.1.

The pressure at any field point can be obtained by dividing the surface of the piston into infinitesimal elements, each of which acts as a baffled simple source of strength $dQ = U_0\, dS$. Since the pressure generated by one of these sources is given by (7.2.17), the total pressure is

$$\mathbf{p}(r,\theta,t) = j\rho_0 c \frac{U_0}{\lambda}\int_S \frac{1}{r'}e^{j(\omega t - kr')}\, dS \tag{7.4.1}$$

where the surface integral is taken over the region $\sigma \le a$. While this integral is difficult to solve for a general field point, closed-form solutions are possible for two regions: (a) along a line perpendicular to the face of the piston and passing through its center (the acoustic axis), and (b) at sufficiently large distances, in the *far field*.

(a) Axial Response

The field along the acoustic axis (the z axis) is relatively simple to calculate. With reference to Fig. 7.4.1, we have

$$\mathbf{p}(r,0,t) = j\rho_0 c \frac{U_0}{\lambda}e^{j\omega t}\int_0^a \frac{\exp\!\left(-jk\sqrt{r^2 + \sigma^2}\right)}{\sqrt{r^2 + \sigma^2}}2\pi\sigma\, d\sigma \tag{7.4.2}$$

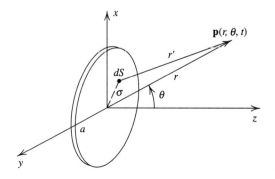

Figure 7.4.1 Geometry used in deriving the acoustic field of a baffled circular plane piston of radius a radiating sound of wave number k.

The integrand is a perfect differential,

$$\frac{\sigma \exp\left(-jk\sqrt{r^2 + \sigma^2}\right)}{\sqrt{r^2 + \sigma^2}} = -\frac{d}{d\sigma}\left(\frac{\exp\left(-jk\sqrt{r^2 + \sigma^2}\right)}{jk}\right) \qquad (7.4.3)$$

so the complex acoustic pressure is

$$\mathbf{p}(r, 0, t) = \rho_0 c\, U_0 \left\{1 - \exp\left[-jk\left(\sqrt{r^2 + a^2} - r\right)\right]\right\} e^{j(\omega t - kr)} \qquad (7.4.4)$$

The pressure amplitude on the axis of the piston is the magnitude of the above expression,

$$P(r, 0) = 2\rho_0 c U_0 \left|\sin\left\{\tfrac{1}{2}kr\left[\sqrt{1 + (a/r)^2} - 1\right]\right\}\right| \qquad (7.4.5)$$

For $r/a \gg 1$, the square root can be simplified to

$$\sqrt{1 + (a/r)^2} \approx 1 + \tfrac{1}{2}(a/r)^2 \qquad (7.4.6)$$

If also $r/a > ka/2$, the pressure amplitude on the axis has asymptotic form

$$P_{ax}(r) = \tfrac{1}{2}\rho_0 c U_0 (a/r)ka \qquad (7.4.7)$$

which reveals the expected spherical divergence at sufficiently large distances. (The inequality $r/a > ka/2$ can be rewritten as $r > \pi a^2/\lambda$. In general, the quantity S/λ, where S is the moving area of the source, is called the *Rayleigh length*.)

Study of (7.4.5) reveals that the axial pressure exhibits strong interference effects, fluctuating between 0 and $2\rho_0 c U_0$ as r ranges between 0 and ∞. These extremes of pressure occur for values of r satisfying

$$\tfrac{1}{2}kr\left[\sqrt{1 + (a/r)^2} - 1\right] = m\pi/2 \qquad m = 0, 1, 2, \ldots \qquad (7.4.8)$$

Solution of the above for the values of r at the extrema yields

$$r_m/a = a/m\lambda - m\lambda/4a \qquad (7.4.9)$$

Moving in toward the source from large r, one encounters the first local maximum in axial pressure at a distance r_1 given by

$$r_1/a = a/\lambda - \lambda/4a \qquad (7.4.10)$$

For still smaller r, the pressure amplitude falls to a local minimum at r_2 given by

$$r_2/a = a/2\lambda - \lambda/2a \qquad (7.4.11)$$

and then continues to fluctuate until the face of the piston is reached. A sketch of this behavior is shown in Fig. 7.4.2.

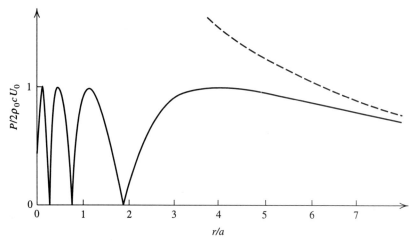

Figure 7.4.2 Axial pressure amplitude for a baffled circular plane piston of radius a radiating sound of wave number k with $ka = 8\pi$. Solid line is calculated from the exact theory. Dashed line is the far field approximation extrapolated into the near field. For this case, the far field approximation is accurate only for distances beyond about seven piston radii.

For $r > r_1$ the axial pressure decreases monotonically, approaching an asymptotic $1/r$ dependence. For $r < r_1$ the axial pressure displays strong interference effects, suggesting that the acoustic field close to the piston is complicated. The distance r_1 serves as a convenient demarcation between the complicated *near field* found close to the source and the simpler *far field* found at large distances from the source. The quantity r_1 has physical meaning only if the ratio a/λ is large enough that $r_1 > 0$. Indeed, if $a = \lambda/2$, then $r_1 = 0$ and there is no near field. At still lower frequencies the radiation from the piston approaches that of a simple source.

(b) Far Field

To aid in the evaluation of the far field, additional coordinates are introduced as indicated in Fig. 7.4.3. Let the x and y axes be oriented so the field point (r, θ) lies in the x-z plane. This allows the piston surface to be divided into an array of continuous line sources of differing lengths, each parallel to the y axis, so the field point is on the acoustic axis of each line source. The far field radiation pattern is found by imposing the restriction $r \gg a$ so the contribution to the field point from each of the line sources is simply its far field axial pressure. Since each line

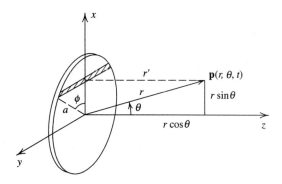

Figure 7.4.3 Geometry used in deriving the far field at (r, θ) of a baffled circular plane piston of radius a.

is of length $2a \sin \phi$ and width dx, the source strength from one such source is $dQ = 2U_0 a \sin \phi \, dx$ and the incremental pressure $d\mathbf{p}$ for this *baffled* source is, from (7.3.7),

$$d\mathbf{p} = j\rho_0 c \frac{U_0}{\pi r'} ka \sin \phi \, e^{j(\omega t - kr')} \, dx \tag{7.4.12}$$

For $r \gg a$, the value of r' is well approximated by

$$r' \approx r + \Delta r = r - a \sin \theta \cos \phi \tag{7.4.13}$$

and the acoustic pressure is

$$\mathbf{p}(r, \theta, t) = j\rho_0 c \frac{U_0}{\pi r'} ka \, e^{j(\omega t - kr')} \int_{-a}^{a} e^{jka \sin \theta \cos \phi} \sin \phi \, dx \tag{7.4.14}$$

where $r' \to r$ in the denominator, but $r' = r + \Delta r$ in the phase in accordance with the far field approximation. Using $x = a \cos \phi$, we can convert the integration from dx to $d\phi$:

$$\mathbf{p}(r, \theta, t) = j\rho_0 c \frac{U_0}{\pi} \frac{a}{r} ka \, e^{j(\omega t - kr)} \int_{0}^{\pi} e^{jka \sin \theta \cos \phi} \sin^2 \phi \, d\phi \tag{7.4.15}$$

By symmetry, the imaginary part of the integral vanishes. The real part is tabulated in terms of a Bessel function,

$$\int_{0}^{\pi} \cos(z \cos \phi) \sin^2 \phi \, d\phi = \pi \frac{J_1(z)}{z} \tag{7.4.16}$$

so that

$$\mathbf{p}(r, \theta, t) = \frac{j}{2} \rho_0 c U_0 \frac{a}{r} ka \left[\frac{2J_1(ka \sin \theta)}{ka \sin \theta} \right] e^{j(\omega t - kr)} \tag{7.4.17}$$

All the angular dependence is in the bracketed term. Since this term goes to unity as θ goes to 0, we can make the identifications

$$|\mathbf{p}(r, \theta)| = P_{ax}(r) H(\theta)$$

$$H(\theta) = \left| \frac{2J_1(v)}{v} \right| \qquad v = ka \sin \theta \tag{7.4.18}$$

Note that the axial pressure amplitude is identical with the asymptotic expression (7.4.7). A plot of $2J_1(v)/v$ is given in Fig. 7.4.4, and numerical values are in Appendix A6. It is well worth comparing and contrasting Figs. 7.3.2 and 7.4.4.

The angular dependence of $H(\theta)$ reveals that there are pressure nodes at angles θ_m given by

$$ka \sin \theta_m = j_{1m} \qquad m = 1, 2, 3, \ldots \tag{7.4.19}$$

where j_{1m} designates the values of the argument of J_1 that reduce this Bessel function to zero, $J_1(j_{1m}) = 0$. (See Appendix A5.) Note that the form of $H(\theta)$ yields

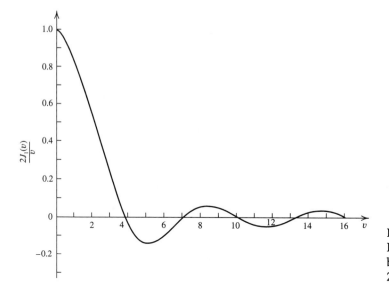

Figure 7.4.4 Functional behavior of $2J_1(v)/v$.

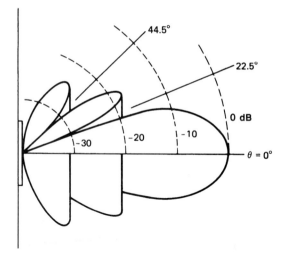

Figure 7.4.5 Beam pattern $b(\theta)$ for a circular plane piston of radius a radiating sound with $ka = 10$.

a maximum along $\theta = 0$. The angles θ_m define conical nodal surfaces with vertices at $r = 0$. Between these surfaces lie pressure lobes, as suggested in Fig. 7.4.5. The relative strengths and angular locations of the acoustic pressure maxima in the lobes are given by the relative maxima of $H(\theta)$. Thus, for constant r, if the intensity level on the axis is set at 0 dB, then the level of the maximum of the first side lobe is about -17.5 dB.

For wavelengths much smaller than the radius of the piston ($ka \gg 1$) the radiation pattern has many side lobes and the angular width of the major lobe is small. If the wavelength is sufficiently large ($ka < 3.83$) only the major lobe will be present. For $ka \ll 1$, the directional factor is nearly unity for all angles, so that the piston becomes a baffled simple source with source strength $Q = \pi a^2 U_0$.

The radiation patterns produced by a piston-type loudspeaker differ to some extent from these idealized patterns for reasons including the following: (1) The area of the baffle in which the speaker is mounted is finite. At low frequencies the wavelength of the sound may be the same as, or greater than, the linear dimensions of the baffle and the assumption that each element of the piston radiates with hemispherical divergence will be in error. (2) If the loudspeaker cabinet is not closed, the radiation from the back of the speaker may propagate into the region in front of the speaker, resulting in a radiation pattern approximating an acoustic doublet rather than a piston in an infinite baffle. (3) The material of a loudspeaker cone is not perfectly rigid. Driving the speaker at its center establishes velocity amplitudes higher near the center of the cone than near its rim at low frequencies, and at high frequencies the cone may vibrate in standing waves. Under these circumstances, U_0 may become a complex function $\mathbf{U}_0$ of the radial distance σ and angle ϕ. By a suitable choice of the relation between U_0 and σ, a wide variety of radiation patterns can be obtained. Altering radiation patterns with the help of a flexible radiating surface is an important consideration in loudspeaker design. Even in small rooms, loudspeakers that project higher frequencies into narrow major lobes often sound "sharp" or "edgy" to listeners on the acoustic axis and "dull" to listeners off the axis. Broadening the major lobes for higher frequencies helps to counteract this beaming of sound. In small rooms, avoiding high-frequency absorption at the walls is another aid in scattering high-frequency energy. When public address systems are used outdoors or in large auditoriums, scattering is negligible and uniform distribution of higher frequencies must be obtained by employing multidirectional clusters of speakers or groups of speakers aimed in different directions.

7.5 RADIATION IMPEDANCE

In Chapter 2 it was found useful to define the input mechanical impedance of a string as the force applied to the string divided by the resulting speed of the string at the point where the force is applied. If the force is not applied *directly* to the string, but to some device attached to the string, then it was shown in Problem 2.9.2 that the force applied to the device divided by the speed of the device was equal to the mechanical impedance of the device plus the input mechanical impedance of the string as seen by the device. Similarly, in the discussion of acoustic sources it will be useful to express the input mechanical impedance of the source in terms of the *mechanical impedance* of the source radiating into a vacuum and the *radiation impedance* of the acoustic wave propagated into the fluid.

Consider a transmitter whose active face (*diaphragm*) of area S moves with a normal velocity component $\mathbf{u}$ whose magnitude and phase may be a function of position. If $d\mathbf{f}_S$ is the normal component of force on an element dS of the active face, the radiation impedance is

$$\mathbf{Z}_r = \int \frac{d\mathbf{f}_S}{\mathbf{u}} \tag{7.5.1}$$

If the diaphragm has mass m, mechanical resistance R_m, and stiffness s and moves *uniformly* with a normal component of velocity $\mathbf{u}_0 = U_0 \exp(j\omega t) = j\omega\boldsymbol{\xi}_0$ under the externally applied force $\mathbf{f} = F\exp(j\omega t)$, Newton's law of motion yields

$$\mathbf{f} - \mathbf{f}_S - R_m\frac{d\boldsymbol{\xi}_0}{dt} - s\boldsymbol{\xi}_0 = m\frac{d^2\boldsymbol{\xi}_0}{dt^2} \tag{7.5.2}$$

where the force of the diaphragm on the fluid is $\mathbf{f}_S = \mathbf{Z}_r\mathbf{u}_0$. Recalling that $\mathbf{Z}_m = \mathbf{R}_m + j(\omega m - s/\omega)$ and solving for $\mathbf{u}_0$ gives

$$\mathbf{u}_0 = \mathbf{f}/(\mathbf{Z}_m + \mathbf{Z}_r) \tag{7.5.3}$$

Thus, in the presence of fluid loading, the applied force encounters the sum of the mechanical impedance of the source and the radiation impedance. The radiation impedance can be expressed as

$$\mathbf{Z}_r = Z_r e^{j\theta} = R_r + jX_r \tag{7.5.4}$$

where R_r is the *radiation resistance* and X_r is the *radiation reactance*.

A positive R_r will increase the total resistance, increasing the power dissipated by the source by an amount equal to the power radiated into the fluid,

$$\Pi = \frac{1}{T}\int_0^T \mathrm{Re}\{\mathbf{f}_S\}\mathrm{Re}\{\mathbf{u}_0\}\,dt \tag{7.5.5}$$

or

$$\Pi = \tfrac{1}{2}U_0^2 Z_r \cos\theta = \tfrac{1}{2}U_0^2 R_r \tag{7.5.6}$$

The radiation resistance can be found directly from the power radiated into the fluid. For example, use of (7.2.16) and (7.2.19) shows that for a simple source

$$R_r = \rho_0 c(kS)^2/4\pi \quad \text{(simple source)} \tag{7.5.7}$$

$$R_r = \rho_0 c(kS)^2/2\pi \quad \text{(baffled simple source)} \tag{7.5.8}$$

where in each case S is the surface area of the relevant source.

A positive X_r will manifest itself as a mass loading that decreases the resonance frequency ω_0 of the oscillator from $\sqrt{s/m}$ to $\sqrt{s/(m + m_r)}$, where $m_r = X_r/\omega$ is the *radiation mass*. The effect of the *radiation mass* can be slight for sources operating in light media such as air, but for a dense fluid like water the decrease in resonance frequency resulting from the presence of the medium may be quite marked.

(a) The Circular Piston

To calculate the radiation impedance of a baffled circular piston of radius a and normal complex velocity $\mathbf{u}_0 = U_0\exp(j\omega t)$, consider an infinitesimal area dS of the surface of the piston (Fig. 7.5.1) and let $d\mathbf{p}$ be the incremental pressure that the motion of dS produces at some other element of area dS' of the piston. The total pressure $\mathbf{p}$ at dS' can be obtained by integrating (7.4.1) over the surface of the piston,

$$\mathbf{p} = j\rho_0 c\,\frac{U_0}{\lambda}\int_S \frac{1}{r}e^{j(\omega t - kr)}\,dS \tag{7.5.9}$$

where r is the distance between dS and dS'. The total force $\mathbf{f}_S$ on the piston from the pressure is the integral of $\mathbf{p}$ over dS', so that $\mathbf{f}_S = \int \mathbf{p}\,dS'$. The integrations over dS to get $\mathbf{p}$ and then over dS' to get $\mathbf{f}_S$ include both the force on dS' resulting from the motion of dS and vice versa. But from acoustic reciprocity, these two forces must be the same. Consequently, the result of the double integration is twice what

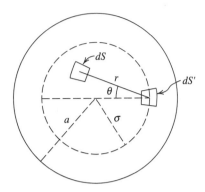

Figure 7.5.1 Surface elements dS and dS' used in obtaining the reaction force on a radiating plane circular piston.

would be obtained if the limits of integration were chosen to include the force between each pair of elements only once. This latter choice of limits leads to a considerable simplification of the problem. Refer to Fig. 7.5.1. With σ the radial distance from the center of the piston to dS', each pair of elements is used only once by integrating over the area of the piston within this circle of radius σ. The maximum distance from dS' to any point within the circle is $2\sigma \cos \theta$, so the entire area within the circle will be covered if we integrate r from 0 to $2\sigma \cos \theta$ and then integrate θ from $-\pi/2$ to $\pi/2$. The integration of dS' is now extended over the entire surface of the piston by setting $dS' = \sigma \, d\sigma \, d\psi$ and integrating ψ from 0 to 2π and then σ from 0 to a. After multiplying this by two, we have our desired expression,

$$\mathbf{f}_S = 2j\rho_0 c \frac{U_0}{\lambda} e^{j\omega t} \int_0^a \int_0^{2\pi} \int_{-\pi/2}^{\pi/2} \int_0^{2\sigma \cos \theta} \sigma e^{-jkr} dr \, d\theta \, d\psi \, d\sigma \tag{7.5.10}$$

The details of the integration are left to Problem 7.5.2. The result for the radiation impedance $\mathbf{Z}_r = \mathbf{f}_S/\mathbf{u}_0$ is

$$\mathbf{Z}_r = \rho_0 c S[R_1(2ka) + jX_1(2ka)] \tag{7.5.11}$$

where $S = \pi a^2$ is the area of the piston face. The *piston resistance function* R_1 and *piston reactance function* X_1 are given by

$$R_1(x) = 1 - \frac{2J_1(x)}{x} = \frac{x^2}{2 \cdot 4} - \frac{x^4}{2 \cdot 4^2 \cdot 6} + \frac{x^6}{2 \cdot 4^2 \cdot 6^2 \cdot 8} - \cdots$$

$$X_1(x) = \frac{2\mathbf{H}_1(x)}{x} = \frac{4}{\pi}\left(\frac{x}{3} - \frac{x^3}{3^2 \cdot 5} + \frac{x^5}{3^2 \cdot 5^2 \cdot 7} - \cdots\right) \tag{7.5.12}$$

with $\mathbf{H}_1(x)$ the *first order Struve function*, described in Appendix A4. Sketches of R_1 and X_1 are shown in Fig. 7.5.2 and numerically tabulated in Appendix A6.

In the low-frequency limit ($ka \ll 1$) the radiation impedance can be approximated by the first terms of the power expansions. The radiation resistance becomes

$$R_r \approx \tfrac{1}{2}\rho_0 c S(ka)^2 \tag{7.5.13}$$

and the radiation reactance becomes

$$X_r \approx (8/3\pi)\rho_0 c S k a \tag{7.5.14}$$

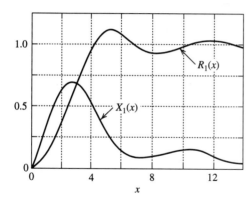

Figure 7.5.2 Radiation resistance and reactance for a plane circular piston of radius a radiating sound of wave number k ($x = 2ka$).

Note that, in the low-frequency limit, the radiation resistance for the piston is identical with that for a baffled simple source of the same surface area S. The low-frequency reactance is that of a mass

$$m_r = X_r/\omega = \rho_0 S(8a/3\pi) \tag{7.5.15}$$

Thus, the piston appears to be loaded with a cylindrical volume of fluid whose cross-sectional area S is that of the piston and whose effective height is $8a/3\pi \approx 0.85a$.

In the high-frequency limit $ka \gg 1$, we have $X_1(2ka) \rightarrow (2/\pi)/(ka)$ and $R_1(2ka) \rightarrow 1$, so that $Z_r \rightarrow R_r \approx S\rho_0 c$. This yields

$$\Pi \approx \tfrac{1}{2}\rho_0 c S U_0^2 \tag{7.5.16}$$

which is the same as the power that would be carried by a plane wave of particle speed amplitude U_0 in a fluid of characteristic impedance $\rho_0 c$ through a cross-sectional area S.

(b) The Pulsating Sphere

The radiation impedance of the pulsating sphere is easily found from (7.1.2) to be

$$\mathbf{Z}_r = \rho_0 c S \cos \theta_a \, e^{j\theta_a} \tag{7.5.17}$$

where $S = 4\pi a^2$ is the surface area of the sphere. For high frequencies ($ka \gg 1$), this reduces to a pure radiation resistance $\mathbf{Z}_r = R_r$, where

$$R_r = \rho_0 c S \tag{7.5.18}$$

For low frequencies ($ka \ll 1$), $\mathbf{Z}_r$ becomes

$$\mathbf{Z}_r \approx \rho_0 c S(ka)^2 + j\rho_0 c S ka \tag{7.5.19}$$

The radiation resistance is much less than the radiation reactance, and the radiation reactance is again like a mass,

$$m_r = X_r/\omega = 3\rho_0 V \tag{7.5.20}$$

where $V = 4\pi a^3/3$ is the volume of the sphere. In the low-frequency limit the radiation mass is three times the mass of the fluid displaced by the sphere.

7.6 FUNDAMENTAL PROPERTIES OF TRANSDUCERS

Several definitions are used to describe the more important aspects of the field without the necessity of displaying the entire radiation pattern.

(a) Directional Factor and Beam Pattern

We have shown that the far field radiation for each of two uncomplicated sources (continuous line and piston) can be expressed as a product of an axial pressure $P_{ax}(r)$ and a directional factor $H(\theta)$. For sources of lower symmetry, this same separation is possible, although the directional factor may depend on two angles, $H(\theta, \phi)$. The directional factor is always normalized so its maximum value is unity, as illustrated by (7.3.5) and (7.4.18). The directions for which $H = 1$ determine the acoustic axes. An acoustic "axis" may be a line, a plane, or a conical surface. The normalized far field pressure along any radial line designated by angles θ and ϕ is simply $H(\theta, \phi)/r$.

The variation of intensity level (or sound pressure level) with angle is the *beam pattern*

$$b(\theta, \phi) = 10 \, \log[I(r, \theta, \phi)/I_{ax}(r)] = 20 \, \log[P(r, \theta, \phi)/P_{ax}(r)]$$
$$= 20 \, \log H(\theta, \phi) \tag{7.6.1}$$

(b) Beam Width

No single definition has been agreed upon for determining the angles that mark the effective extremities of the major lobe. Hence, the criterion must be clearly stated when beam widths are specified. The values of $I(r, \theta, \phi)/I_{ax}(r)$ used to delineate the effective width of a major lobe range from a maximum of 0.5 (down 3 dB or "half-power"), through 0.25 (down 6 dB or "quarter-power"), to a minimum of 0.1 (down 10 dB). As an illustration of the ambiguity that arises if the ratio of intensities is not specified, consider a piston that is radiating sound of wavelength $\lambda = a/4$. The calculated beam widths corresponding to the three ratios given above are 7.4° (down 3 dB), 10.1° (down 6 dB), and 12.9° (down 10 dB), whereas the beam width corresponding to the first null is 17.3°. Even when the outer limit of the major lobe is defined as being down 10 dB relative to the axial level, it is still some 7.5 dB higher than the maximum level of the first minor lobe.

(c) Source Level

A measure of the axial output of a source is the *source level SL*. Assume that the acoustic axis of the source has been determined and the pressure amplitude along this line is measured in the far field (where the pressure varies as $1/r$). The curve of $P_{ax}(r)$ versus $1/r$ can be extrapolated from large r to a position $r = 1$ m from the source to give

$$P_{ax}(1) = \lim_{r \downarrow 1} P_{ax}(r) \tag{7.6.2}$$

[Note that $P_{ax}(1)$ is not necessarily the actual axial pressure at 1 m. It is simply a convenient extrapolation from the far field behavior.] Since $P_{ax}(1)$ is a peak pressure amplitude, it must be reduced to an effective (or rms) value $P_e(1)$ by dividing by $\sqrt{2}$. The source level is then

$$SL(re\ P_{ref}) = 20\ \log[P_e(1)/P_{ref}] \qquad P_e(1) = P_{ax}(1)/\sqrt{2} \qquad (7.6.3)$$

where the reference effective pressure P_{ref} is either 1 μPa, 20 μPa, or 1 μbar as discussed in Section 5.12.

(d) Directivity

Given the amplitude $P(r, \theta, \phi)$ of the pressure in the far field, the total radiated power is obtained by integrating the intensity over a sphere enclosing the source,

$$\Pi = \frac{1}{2\rho_0 c} \int_{4\pi} P^2(r, \theta, \phi) r^2\, d\Omega \qquad (7.6.4)$$

Recalling that $P(r, \theta, \phi) = P_{ax}(r)H(\theta, \phi)$ and noting that r is constant for the integration, we can write

$$\Pi = \frac{1}{2\rho_0 c} r^2 P_{ax}^2(r) \int_{4\pi} H^2(\theta, \phi)\, d\Omega \qquad (7.6.5)$$

For a *simple source* that generates the *same acoustic power*, the pressure amplitude $P_s(r)$ to be found at the distance r is given by

$$\Pi = 4\pi r^2 P_S^2(r)/2\rho_0 c \qquad (7.6.6)$$

Clearly, for the same acoustic power the directional source will have greater intensity at a distance r on the acoustic axis than will the simple source. The ratio of these intensities reveals how much more efficiently a directional source concentrates the available acoustic power into a preferred direction. This ratio defines the *directivity D*,

$$D = I_{ax}(r)/I_S(r) = P_{ax}^2(r)/P_S^2(r) \qquad (7.6.7)$$

Substitution of (7.6.5) and (7.6.6) into (7.6.7) results in

$$D = 4\pi \left/ \int_{4\pi} H^2(\theta, \phi)\, d\Omega \right. \qquad (7.6.8)$$

Thus, the directivity D of a source is the reciprocal of the average of $H^2(\theta, \phi)$ over solid angle. Now, (7.6.5) becomes

$$\Pi = 4\pi P_e^2(1)/D\rho_0 c \qquad (7.6.9)$$

and substitution for $P_e(1)$ into (7.6.3) gives

$$SL(re\ P_{ref}) = 10 \log(D\rho_0 c\Pi/4\pi P_{ref}^2) \qquad (7.6.10)$$

(1) CONTINUOUS LINE SOURCE. The directional factor for a continuous line source is (7.3.5). A study of the cylindrical geometry reveals

$$D = 4\pi \Big/ 2 \int_0^{\pi/2} H^2(\theta) 2\pi \cos\theta \, d\theta \qquad (7.6.11)$$

and the change of variable $v = \frac{1}{2}kL\sin\theta$ gives

$$D = \frac{kL}{2} \Big/ \int_0^{kL/2} \left(\frac{\sin v}{v}\right)^2 dv \qquad (7.6.12)$$

If the line is long ($kL \gg 1$), the upper limit can be taken arbitrarily large with little loss in accuracy. The resulting definite integral is known,

$$\int_0^\infty \left(\frac{\sin v}{v}\right)^2 dv = \frac{\pi}{2} \qquad (7.6.13)$$

so the directivity of a long continuous line source is approximately

$$D \approx kL/\pi = 2L/\lambda \qquad (7.6.14)$$

(2) PISTON SOURCE. The directivity of a piston is determined from the directional factor (7.4.18) by

$$D = 4\pi \Big/ \int_0^{\pi/2} \left[\frac{2J_1(ka\sin\theta)}{ka\sin\theta}\right]^2 2\pi\sin\theta \, d\theta \qquad (7.6.15)$$

where $2\pi\sin\theta \, d\theta$ is the incremental solid angle $d\Omega$ for this axisymmetric case. While this integral can be evaluated, another approach is used in Problem 7.6.1. The result is

$$D = \frac{(ka)^2}{1 - J_1(2ka)/ka} \qquad (7.6.16)$$

For low frequencies ($ka \to 0$), the Bessel function can be replaced by the first two terms of its series expansion, and in this limit $D \to 2$, which is the same as a hemispherical source on an infinite baffle. For high frequencies the Bessel function becomes small and

$$D \approx (ka)^2 \qquad ka \gg 1 \qquad (7.6.17)$$

which shows that the piston is highly directive at higher frequencies.

(e) Directivity Index

The *directivity index DI* is given by

$$\boxed{DI = 10\log D} \qquad (7.6.18)$$

For water and with a reference pressure of 1 μPa,

$$SL(re\ 1\ \mu Pa) = 10\ \log\ \Pi + DI + 171 \qquad \text{(water)} \qquad (7.6.19)$$

where the acoustic power must be in watts. In air, the conventional reference pressure is 20 μPa, and the source level becomes

$$SL(re\ 20\ \mu Pa) = 10\log\Pi + DI + 109 \qquad \text{(air)} \qquad (7.6.20)$$

The ability of a directional receiver to ignore isotropic noise is determined by the directivity index DI. However, if the noise field has directionality, such as the noise from a busy freeway or noise from distant shipping in the ocean (which tends to arrive from directions close to horizontal), a more general measure, the *array gain AG*, must be introduced. If $N(\theta, \phi)$ is the effective pressure amplitude of the noise arriving from the (θ, ϕ) direction, then the array gain for a directional receiver is

$$AG = 10\log\left(\frac{\int_{4\pi}|N(\theta,\phi)|^2\,d\Omega}{\int_{4\pi}|N(\theta,\phi)|^2 H^2(\theta,\phi)\,d\Omega}\right) \qquad (7.6.21)$$

The numerator measures the noise power received by an omnidirectional receiver and the denominator measures that received by the directional receiver. In a nonisotropic noise field, the array gain depends on the properties of the field and the orientation of the receiver. If the noise field is isotropic then $N(\theta, \phi)$ is constant for all angles, the argument of the log reduces to (7.6.8), and the array gain becomes identical to the directivity index.

(f) Estimates of Radiation Patterns

For reasonably directive sources of simple geometry, the properties of the radiation fields can be estimated from the size and geometry of the source and the wavelength of the excitation. The source may be one of those previously discussed or may be a mosaic, or array, of such sources. The requirement that the source be reasonably directive is $\lambda \ll L$, where L is the greatest dimension of the source.

(1) EXTENT OF THE NEAR FIELD. Let r_{max} be the distance from the furthest element of the source and r_{min} the distance from the nearest element to a field point in the far field on the acoustic axis. As the field point approaches the source on the axis, the difference $\Delta r = r_{max} - r_{min}$ will gradually increase above the asymptotic value Δr_∞ at large distances. When the increase approaches a half-wavelength, $r_{max} - r_{min} = \Delta r_\infty - \lambda/2$, then the phases of the signals from the individual points on the source combining at the field point will have shifted sufficiently from those observed in the far field to alter the axial pressure amplitude from P_{ax}. See Fig. 7.6.1 for a flat source. If the greatest extent of the source transverse to the acoustic axis is L, a little geometry shows that the value of r_{min} demarking the beginning of the far field is given roughly by

$$r_{min}/L \sim L/4\lambda \qquad (7.6.22)$$

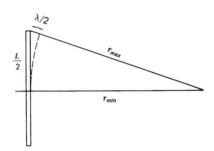

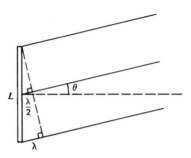

Figure 7.6.1   Geometry used in estimating the extent of the near field of a source of maximum extent L radiating sound of wavelength λ.

Figure 7.6.2   Geometry used for estimating the beam width for a source of maximum extent L radiating sound of wavelength λ.

(2) MAJOR LOBE ANGULAR WIDTH. The major lobe corresponds to that portion of the far field radiation pattern in which the source elements are phased for maximum constructive interference. As the angle off the acoustic axis increases, destructive interference is increased and the edge of the major lobe is approached. Very approximately, when θ has increased until about half of the elements are shifted in phase by $\pi/2$ with respect to the other half, a nodal surface will be encountered. From the simple one-dimensional example shown in Fig. 7.6.2, it can be seen that this occurs at an angle of about λ/L. Thus, the half-angle subtended by the major lobe can be estimated by

$$\sin \theta_1 \sim \lambda/L \qquad (7.6.23)$$

The reader should verify that (7.6.22) and (7.6.23) are in agreement with the quantitative predictions for the circular piston and that (7.6.23) agrees with the major-lobe width calculated for the continuous line source.

For more complicated sources with major dimensions L_1 and L_2 transverse to the acoustic axis, the major lobe will have angular widths $2\theta_{11} \sim 2\lambda/L_1$ in the one direction and $2\theta_{12} \sim 2\lambda/L_2$ in the other.

(3) ESTIMATION OF DIRECTIVITY. Since an exact evaluation of the integral expression (7.6.8) may be too difficult or more accurate than required by the problem at hand, it is useful to be able to estimate the directivity D. If the source is reasonably directive and is designed so the side lobes are considerably weaker than the major lobe, D can be estimated by setting the integrand to unity over the strong central portion of the major lobe and to zero otherwise. The expression for D is

$$\boxed{D = 4\pi/\Omega_{eff}} \qquad (7.6.24)$$

Evaluation of D is thus reduced to the geometrical problem of obtaining a good approximation of the effective solid angle Ω_{eff} subtended by the central portion of the major lobe. This reduces to calculating the *effective* angle θ' describing the half-angular beam width of the major lobe. For highly directive sources θ_1 tends to overestimate θ'. A better approximation is to take that portion of the major lobe over which the directional factor H falls from its maximum value of 1 to a value

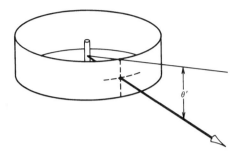

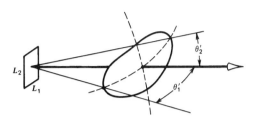

Figure 7.6.3   Area of the unit sphere ensonified by a line-like source at the origin.

Figure 7.6.4   Area of the unit sphere ensonified by a plane piston source of arbitrary shape.

of 0.5 (quarter-power point) and assume H is unity within that region and zero outside. For the cases studied so far, this means obtaining the value θ' solving $(\sin v)/v \approx \frac{1}{4}$ with $v = \frac{1}{2}kL \sin\theta$. A little numerical estimation gives

$$\theta' \sim 2\theta_1/\pi \tag{7.6.25}$$

For a line-like source, the central portion of the major lobe is distributed over the surface of the unit sphere as shown in Fig. 7.6.3. The height of this belt is approximated by $2\theta'$ and the circumference is 2π, so that $\Omega_{eff} \approx 4\pi\theta'$ and $D \sim 1/\theta'$.

For a piston-like source, the central portion of the major lobe is the roughly elliptical patch shown in Fig. 7.6.4. On the unit sphere, the area of this patch is approximated by $\Omega_{eff} \approx \pi\theta_1'\theta_2'$, where $2\theta_1'$ is the effective angular beam width pertinent to the length L_1 and $2\theta_2'$ is that pertinent to L_2. The resultant directivity is $D \sim 4/\theta_1'\theta_2'$.

Comparison of these estimates with (7.6.14) and (7.6.17) shows that they are reasonably good approximations.

*7.7 DIRECTIONAL FACTORS OF REVERSIBLE TRANSDUCERS

While the details of operation of a few of the more common acoustic sources and receivers will be discussed in Chapter 14, it is appropriate here to develop an important relationship between the transmitting and receiving directional properties of a *reversible* transducer. A reversible transducer is one that can be used either as a source or as a receiver of acoustic energy. The common office intercom incorporates such devices. The acoustic element, usually a small loudspeaker, can be switched from acting as an acoustic source (to generate a message) to acting as a receiver (to detect the response to the message).

If a reversible transducer exhibits directionality as a source, it will also be directional as a receiver. For example, a plane wave falling obliquely on the surface of a large plane piston will cause the piston to move with a normal component of velocity proportional to the spatially averaged pressure on the piston. Thus, if the wavelength of the sound is comparable to or smaller than the dimensions of the piston, the response of the piston to the incident plane wave will depend on the angle of arrival of the wave. The measure of this response is the receiving directional factor H_r. We will show that the transmitting and receiving directional factors for a reversible transducer are identical.

Consider plane waves incident on a receiver from a direction specified by θ and ϕ. Let $\langle \mathbf{p}_B \rangle_S$ be the average of the incident sound pressure over the diaphragm of the receiver, measured with the diaphragm held perfectly still (*blocked*). The receiving directional factor

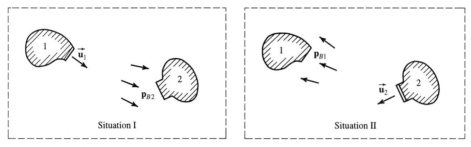

Figure 7.7.1   The reciprocity theorem applied to reversible transducers.

is then defined as

$$H_r(\theta, \phi) = \left| \frac{\langle \mathbf{p}_B(\theta, \phi) \rangle_S}{\langle \mathbf{p}_{Bax} \rangle_S} \right| \tag{7.7.1}$$

This measures the phase cancellation of the incident wave over the blocked diaphragm of the receiver as a function of θ and ϕ and thus gives the directional sensitivity of the receiver. (The receiving directional factor is defined with the diaphragm blocked to eliminate any field radiated by the motion of the diaphragm; more on this in Chapter 14.)

A relationship can be established between H_r and H for reversible transducers with the help of the reciprocity theorem. Consider the situation represented in Fig. 7.7.1. There are two reversible transducers (with all surfaces other than their diaphragms perfectly rigid) a large distance r apart in otherwise free space. (The requirement of large r ensures that near field effects are avoided.) Situation I requires one of the transducers to be active and the other passive with its diaphragm blocked. Situation II reverses the roles of the two transducers. Application of (7.2.4) yields

$$\int_{S_2} \mathbf{p}_{B2} \vec{\mathbf{u}}_2 \cdot \hat{n} \, dS = \int_{S_1} \mathbf{p}_{B1} \vec{\mathbf{u}}_1 \cdot \hat{n} \, dS \tag{7.7.2}$$

where $\mathbf{p}_B$ is the pressure distribution over each blocked diaphragm and S is the area of the diaphragm of each of transducers 1 and 2. If each diaphragm moves as a unit, so that $\vec{\mathbf{u}}_1$ and $\vec{\mathbf{u}}_2$ are constant over S_1 and S_2, then this simplifies to

$$\mathbf{u}_2 \langle \mathbf{p}_{B2} \rangle S_2 = \mathbf{u}_1 \langle \mathbf{p}_{B1} \rangle S_1 \tag{7.7.3}$$

where $\mathbf{u}_1$ and $\mathbf{u}_2$ are the components of the particle velocities perpendicular to the diaphragms.

Now, if transducer 2 is sufficiently small, it does not appreciably disturb the pressure field $\mathbf{p}_1$, which is radiated by transducer 1, so that $\mathbf{p}_{B2} = \mathbf{p}_1(r, \theta, \phi, t)$. Furthermore, the pressure $\mathbf{p}_{B1}$ is uniform over the active surface of transducer 2, so that (7.7.3) becomes

$$\mathbf{u}_2 \mathbf{p}_1(r, \theta, \phi, t) S_2 = \mathbf{u}_1 \langle \mathbf{p}_{B1}(\theta, \phi, t) \rangle_{S_1} S_1 \tag{7.7.4}$$

Now, if transducers 1 and 2 are rotated so that they are on each other's acoustic axis, (7.7.4) gives the additional equality

$$\mathbf{u}_2 \mathbf{p}_{1ax}(r, t) S_2 = \mathbf{u}_1 \langle \mathbf{p}_{B1ax}(t) \rangle_{S_1} S_1 \tag{7.7.5}$$

The magnitude of the ratio of the above pair of equations yields

$$\left| \frac{\mathbf{p}_1(r, \theta, \phi, t)}{\mathbf{p}_{1ax}(r, t)} \right| = \left| \frac{\langle \mathbf{p}_{B1}(\theta, \phi, t) \rangle_{S_1}}{\langle \mathbf{p}_{B1ax}(t) \rangle_{S_1}} \right| \tag{7.7.6}$$

The left side of (7.7.6) is H and the right side is H_r. Thus,

$$\boxed{H(\theta, \phi) = H_r(\theta, \phi)} \qquad (7.7.7)$$

and a reversible acoustic transducer has the same directional properties whether it is transmitting or receiving.

*7.8 THE LINE ARRAY

Consider a line of N simple sources with adjacent elements spaced distance d apart, as shown in Fig. 7.8.1. If all sources have the same source strength and radiate waves with the same phase, then the ith source generates a pressure wave of the form $(A/r_i')\exp[j(\omega t - kr_i')]$, where r_i' is the distance from this source to (r, θ). The resultant pressure at the field point is the summation

$$\mathbf{p}(r, \theta, t) = \sum_{i=1}^{N} \frac{A}{r_i'} e^{j(\omega t - kr')} \qquad (7.8.1)$$

If we restrict attention to the far field [specified by $r \gg L$, where $L = (N-1)d$ is the length of the array], all r_i' are approximately parallel. Then $r_i = r_1 - (i-1)\Delta r$, where $\Delta r = d \sin\theta$. The distance to the center of the array can be expressed as $r = r_1 - \frac{1}{2}(L/d)\,\Delta r$. In the far field, r_i' in the denominator of (7.8.1) can be replaced with r and (7.8.1) takes the form

$$\mathbf{p}(r, \theta, t) = \frac{A}{r} e^{-j(L/2d)k\Delta r} e^{j(\omega t - kr)} \sum_{i=1}^{N} e^{j(i-1)k\Delta r} \qquad (7.8.2)$$

Use of the trigonometric identities in Appendix A3 results in

$$\mathbf{p}(r, \theta, t) = \frac{A}{r} e^{j(\omega t - kr)} \left(\frac{\sin[(N/2)k\Delta r]}{\sin[(1/2)k\Delta r]} \right) \qquad (7.8.3)$$

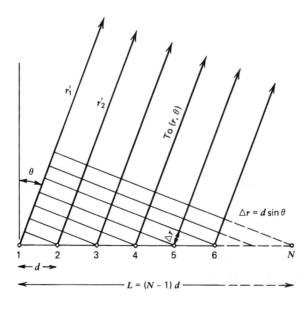

Figure 7.8.1 Geometry used in deriving the far field acoustic field at (r, θ) of a line array of N in-phase elements, spaced distance d apart.

The pressure on the axis ($\theta = 0$) is

$$\mathbf{p}(r, 0, t) = N(A/r)e^{j(\omega t - kr)} \tag{7.8.4}$$

and has the maximum possible pressure amplitude

$$P_{ax}(r) = NA/r \tag{7.8.5}$$

Identification of the directional factor

$$H(\theta) = \left| \frac{1}{N} \frac{\sin[(N/2)kd\sin\theta]}{\sin[(1/2)kd\sin\theta]} \right| \tag{7.8.6}$$

allows us to write the amplitude of the pressure in the familiar form

$$P(r, \theta) = P_{ax}(r)H(\theta) \tag{7.8.7}$$

The denominator of H may vanish if $\frac{1}{2}kd|\sin\theta| = m\pi$, but the numerator vanishes also, and the pressure amplitude becomes $P_{ax}(r)$. Thus, we can have more than one major lobe. The angles of these occur for

$$|\sin\theta| = m\lambda/d \qquad m = 0, 1, 2, \ldots, [d/\lambda] \tag{7.8.8}$$

(This result can be restated as $|\Delta r| = m\lambda$, which reveals that the radiated pressure is maximized at those angles for which the distances from the field point to the adjacent array elements differ by integral numbers of wavelengths.)

There are additional zeros in the numerator at angles given by

$$|\sin\theta| = (n/N)\lambda/d \qquad n \neq mN \qquad n = 0, 1, 2, \ldots, [Nd/\lambda] \tag{7.8.9}$$

where the integer n is neither zero nor a multiple of N. Since the denominator is not zero, the pressure vanishes and these values of θ determine the nodal surfaces in the far field. There are also secondary maxima of H that designate the directions and magnitudes of the minor lobes. The directions of these side lobes are given approximately by

$$|\sin\theta| = \left[(n + \tfrac{1}{2})/N\right]\lambda/d \qquad n \neq mN \quad \text{and} \quad n \neq mN - 1 \tag{7.8.10}$$

and the amplitudes by

$$P_n(r) = \frac{P_{ax}(r)}{N\sin\left[(n + \tfrac{1}{2})\pi/N\right]} \tag{7.8.11}$$

A sketch of a representative beam pattern for a linear array is given in Fig. 7.8.2.

Certain loudspeaker systems contain such line arrays, mounted vertically so that vertical directivity is large and horizontal directivity small.

In some applications it is desired to have a single narrow major lobe. A simple requirement, which results in one major lobe *almost* as narrow as possible, is to have $\theta = \pi/2$ when $n = N - 1$. This gives

$$\lambda/d = N/(N-1) \tag{7.8.12}$$

or $kd = 2\pi(N-1)/N$, and the beam pattern terminates at the null adjacent to the second major lobe. While not quite exact, we shall refer to this as a single *narrowest* major lobe. This major lobe is contained within angles $\pm\theta_1$ found from

$$\sin\theta_1 = 1/(N-1) \tag{7.8.13}$$

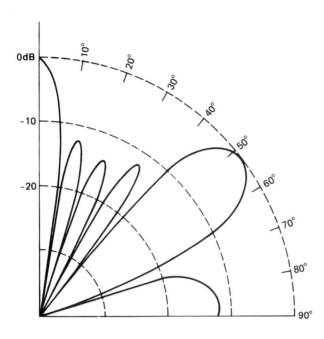

Figure 7.8.2 Beam pattern $b(\theta)$ for a line array of in-phase elements radiating sound of wave number k with $kd = 8$ and $N = 5$.

For an array of many elements, this equation reveals that, if only one narrowest major lobe is to occur, the approximate angular width of the major lobe and the directivity are

$$2\theta_1 \approx 2/N \qquad D \approx (\pi/2)N \tag{7.8.14}$$

For very large arrays it is often desirable to transmit or receive in various directions without physically rotating the array. This can be accomplished by electronic steering. If a time delay $i\tau$ is inserted into the electronic signal for the ith element of the array, (7.8.1) becomes

$$\mathbf{p}(r, \theta, t) = \sum_{i=1}^{N} \frac{A}{r'_i} e^{j[\omega(t+i\tau)-kr'_i]} \tag{7.8.15}$$

and the directional factor becomes

$$H(\theta) = \left| \frac{1}{N} \frac{\sin[(N/2)kd(\sin\theta - \sin\theta_0)]}{\sin[(1/2)kd(\sin\theta - \sin\theta_0)]} \right| \tag{7.8.16}$$

where the major lobe now points in the direction θ_0 given by

$$\sin\theta_0 = c\tau/d \tag{7.8.17}$$

Thus, the introduction of a progressive time delay across the array steers the major lobe off the $\theta = 0$ plane into a cone determined by θ_0. Note that (7.8.17) is independent of frequency. In practice, this steering can be accomplished by inserting time delays in the electrical signals driving the sources or generated by the receivers either through hardwired circuits or by the use of computer software.

Figure 7.8.3 shows the beam pattern for a steered line array designed to have a single narrowest major lobe when steered to $\theta_0 = 0$ (*broadside*). Note that as the beam is steered toward $\theta_0 = \pi/2$ (*endfired*), a second major lobe develops. The only way to avoid this second major lobe is to design the array to give one narrowest major lobe when the beam is steered to $\pi/2$. This requires placement of the last null before encountering a second major lobe at $\theta = -\pi/2$, which results in

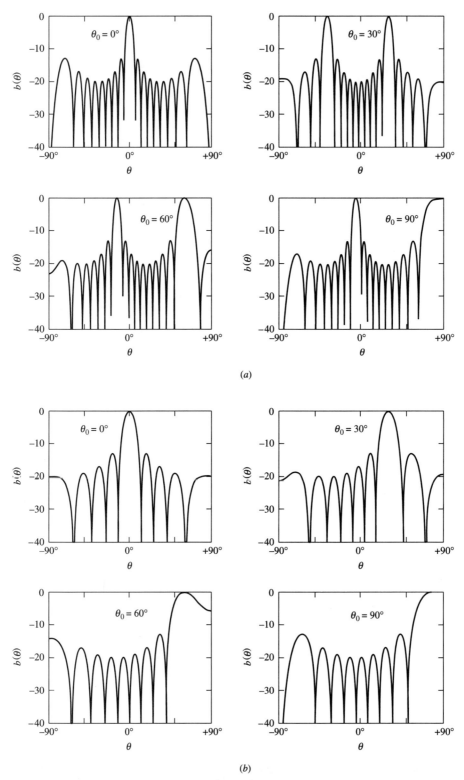

Figure 7.8.3 The beam pattern $b(\theta)$ for a linear array of 10 elements spaced distance d apart. (a) The value of kd is $2\pi(N-1)/N$ so that there is a single narrowest major lobe when $\theta_0 = 0$. As the beam is steered toward $\theta_0 = 90°$, a second major lobe comes in from $\theta = -90°$. (b) The same array but with $kd = \pi(N-1)/N$ so that there is a single narrowest major lobe in the endfire condition ($\theta_0 = 90°$). Although the beam width of the major lobe is larger than before, there is now only one major lobe for all steering angles.

$$\lambda/d = 2N/(N-1) \tag{7.8.18}$$

or $kd = \pi(N-1)/N$. If the maximum angle into which the major lobe is to be steered is $\pm\theta_0$, then the condition required to get only one narrowest major lobe is the placement of the last null before encountering a second major lobe at $\theta = \mp\pi/2$. This results in

$$\lambda/d = [N/(N-1)](1 + |\sin\theta_0|) \tag{7.8.19}$$

The directivity of a steered line array with one narrowest major lobe for all angles of steering (from broadside to endfired) can be determined by estimating the area ensonified on a unit sphere by the major lobe. A glance at Fig. 7.8.3 shows that when the beam is in the endfire position, the ensonified area is a spherical cap, and a simple calculation shows that (for large N)

$$D \sim (\pi/2)^2 N \quad \text{(endfire)} \tag{7.8.20}$$

However, as the beam is steered away from endfire, the ensonified area resembles a belt, and calculation for large N gives

$$D \sim (\pi/4)N \quad \text{(steered)} \tag{7.8.21}$$

During the transition from steered to endfired beam, the main lobe takes on a complicated shape and further analysis is required to determine the way D changes from $(\pi/2)^2 N$ to $(\pi/4)N$.

Amplitude shading of an array is accomplished by applying different gains to the individual elements of the array. This replaces the amplitude A in each term of the summation (7.8.15) with A_i. Amplitude shading can be used to reduce, or even eliminate, side lobes, but at the expense of a wider major lobe. (Examples of amplitude shaded arrays are given in the problems for this section.)

*7.9 THE PRODUCT THEOREM

In the preceding discussion of an array, it was assumed that each element is a simple source so the individual pressure waveforms were spherically symmetric. It is straightforward to generalize the results to an array of identical directive sources all oriented in the same direction. If attention is restricted to the far field, the pressure generated by each element must contain the factor H_e, the directionality of each element of the array. Since all rays are parallel, this factor must be the same in each term of the summation over the elements. Given this, the pressure amplitude can be modified and generalized to

$$P(r,\theta,\phi) = P_{ax}(r)H_e(\theta,\phi)H(\theta,\phi) \tag{7.9.1}$$

where H is the directional factor for the array with simple sources at the position of each element and H_e is the directional factor for a single element. This is the *product theorem*. The directional factor of an array of identical directional sources is the product of the directional factor of an array with identical geometry but with simple sources and the directional factor of a single element of the array.

*7.10 THE FAR FIELD MULTIPOLE EXPANSION

Another approach to obtaining the far field radiation pattern of an acoustic source begins with the inhomogeneous wave equation for a point source. Comparison of (5.16.4) and (7.2.13) shows that, at large distances from the source, $\mathbf{A}$ and Q are related by $\mathbf{A} = j\omega\rho_0 Q/4\pi$. Substitution into (5.16.5) and expressing the acoustic pressure in terms of the velocity potential $\mathbf{p} = -j\omega\rho_0\boldsymbol{\Phi}$ (for notational simplicity) results in

$$(\nabla^2 + k^2)\boldsymbol{\Phi} = Q\delta(\vec{r} - \vec{r}_0)e^{j\omega t} \tag{7.10.1}$$

with the particular solution

$$\boldsymbol{\Phi} = -\frac{Q}{4\pi|\vec{r} - \vec{r}_0|}e^{j(\omega t - k|\vec{r} - \vec{r}_0|)} \tag{7.10.2}$$

for a point source of source strength Q located at $\vec{r}_0$. If we have a collection of sources all within a volume V_0, then this distribution can be described by a *source strength density* $\mathbf{q}(\vec{r}_0)$ and the inhomogeneous wave equation and its particular solution become

$$
\begin{aligned}
(\nabla^2 + k^2)\boldsymbol{\Phi} &= \mathbf{q}(\vec{r}_0)e^{j\omega t} \\
\boldsymbol{\Phi} &= -\frac{1}{4\pi}\int_{V_0} \frac{\mathbf{q}(\vec{r}_0)}{|\vec{r} - \vec{r}_0|}e^{j(\omega t - k|\vec{r} - \vec{r}_0|)}\, dV_0
\end{aligned}
\tag{7.10.3}
$$

The volume integral is over the variable $\vec{r}_0$ (the distance vector $\vec{r}$ from the origin of the coordinate system to the field point is a constant with respect to the integration).

If we assume that the field point is far away from the volume V_0, then the denominator in the integral can be approximated by r and the distance in the phase approximated by $|\vec{r} - \vec{r}_0| \approx r - \vec{r}_0 \cdot \hat{r}$ where $\hat{r} = \vec{r}/r$ is the unit vector in the $\vec{r}$ direction. A Taylor's expansion of the exponential gives

$$e^{jk\vec{r}_0 \cdot \hat{r}} = \sum_{n=0}^{\infty} \frac{1}{n!}(jk\vec{r}_0 \cdot \hat{r})^n = 1 + jk\vec{r}_0 \cdot \hat{r} - \frac{1}{2!}(k\vec{r}_0 \cdot \hat{r})^2 - \frac{j}{3!}(k\vec{r}_0 \cdot \hat{r})^3 + \cdots \tag{7.10.4}$$

and integrating term by term yields

$$\boldsymbol{\Phi} = -\frac{1}{4\pi r}e^{j(\omega t - kr)}\left(\int_{V_0} \mathbf{q}(\vec{r}_0)\, dV_0 + jk\int_{V_0} \mathbf{q}(\vec{r}_0)(\vec{r}_0 \cdot \hat{r})\, dV_0 - \frac{1}{2!}k^2\int_{V_0} \mathbf{q}(\vec{r}_0)(\vec{r}_0 \cdot \hat{r})^2\, dV_0 - \cdots\right) \tag{7.10.5}$$

Let the successive terms on the right in (7.10.5) be labeled $\boldsymbol{\Phi}_1, \boldsymbol{\Phi}_2, \boldsymbol{\Phi}_3, \ldots$. In what follows, we will use spherical coordinates. See Appendix A7.

1. The first term in (7.10.5) can be written as

$$\boldsymbol{\Phi}_1 = -(Q/4\pi r)e^{j(\omega t - kr)} \tag{7.10.6}$$

$$Q = \int_{V_0} \mathbf{q}(\vec{r}_0)\, dV_0$$

where $\mathbf{Q}$ is the *monopole strength* and $\boldsymbol{\Phi}_1$ the *monopole field* of a point source of source strength $\mathbf{Q}$ located at the origin and radiating a spherically symmetric field that falls off as $1/r$ (exterior to the volume V_0).

2. The second term $\boldsymbol{\Phi}_2$ is also easily interpreted,

$$\boldsymbol{\Phi}_2 = -jk(\vec{\mathbf{D}} \cdot \hat{r}/4\pi r)e^{j(\omega t - kr)} \tag{7.10.7}$$

$$\vec{\mathbf{D}} = \int_{V_0} \mathbf{q}(\vec{r}_0)\vec{r}_0\, dV_0$$

The vector $\vec{\mathbf{D}}$ is the first moment of the charge distribution and is called the *vector dipole strength*. The associated field Φ_2 falls off as $1/r$ in all directions with the amplitude in the $\hat{r}$ direction proportional to the scalar product $\vec{\mathbf{D}} \cdot \hat{r}$. This is a *dipole field* with two lobes of opposite phase separated by a nodal plane perpendicular to the direction of the dipole. For example, let the source strength density be

$$\mathbf{q}(\vec{r}_0) = Q\left[\delta(\vec{r}_0 - \hat{z}d) - \delta(\vec{r}_0 + \hat{z}d)\right] \tag{7.10.8}$$

This describes two monopoles, one at $(0,0,d)$ and the other at $(0,0,-d)$, each of source strength Q and with opposite phases. Direct substitution into (7.10.5) shows that Φ_1 and Φ_3 both vanish so that

$$\Phi_2 = -jk\left[Q(2d)/4\pi r\right]\cos\theta\, e^{j(\omega t - kr)} \tag{7.10.9}$$

This is a *dipole field* with vector dipole strength $\vec{\mathbf{D}} = Q(2d)\hat{z}$. See Problem 7.10.4 for a further analysis of the properties of this field. [Succeeding nonzero terms (Φ_4 et seq.) in the series expansion give higher order terms in kd. These provide the corrections giving the radiation pattern of a *doublet*, for which kd has finite value.]

3. A quantitative discussion of the third term in the series for Φ would go beyond the purposes of this book. However, just as a dipole could be constructed by placing two monopoles of opposite phases very close together, the juxtaposition of two dipoles whose strength vectors are equal in magnitude and opposite in direction will generate a *quadrupole*. There are two different geometries: (*a*) dipoles side-by-side (the *lateral* or *tesseral* quadrupole) and (*b*) dipoles head-to-head (the *axial* or *longitudinal* quadrupole). In either geometry, it can be seen easily that Φ_1 and Φ_2 vanish so that the first nonzero contribution is Φ_3.

(*a*) For the lateral geometry, place two sources of strengths Q at coordinates $(d,d,0)$ and $(-d,-d,0)$ and two of strengths $-Q$ at $(-d,d,0)$ and $(d,-d,0)$. Substitute the appropriate density function $\mathbf{q}(\vec{r}_0)$ into the integrand and evaluate the scalar product at each of the coordinates of the delta functions. This gives

$$\Phi_3 = -\frac{1}{4\pi r}\frac{1}{2!}k^2 Q\left[2\left(d\frac{x}{r} + d\frac{y}{r}\right)^2 - 2\left(d\frac{x}{r} - d\frac{y}{r}\right)^2\right]e^{j(\omega t - kr)}$$

$$= \frac{1}{4\pi r}k^2\left[Q(2d)^2\right]\frac{xy}{r^2}e^{j(\omega t - kr)} \tag{7.10.10}$$

This lateral quadrupole field has the form

$$\Phi_3 = k^2\left[Q(2d)^2/4\pi r\right]\sin\phi\cos\phi\sin^2\theta\, e^{j(\omega t - kr)} \tag{7.10.11}$$

where $Q(2d)^2$ is the *quadrupole strength*. The nodal surfaces are the two planes defined by $x = 0$ and $y = 0$ and the line corresponding to the z axis. A cross section in the plane of the sources ($z = 0$) shows that the directional factor $H(\pi/2, \phi)$ is in the shape of a four-leaf clover.

(*b*) For the axial quadrupole, position sources of strength Q on the z axis at $\pm d$ and a source of strength $-2Q$ at the origin. Straightforward analysis provides the axial quadrupole field

$$\Phi_3 = k^2\left[Q(2d)^2/4\pi r\right]\cos^2\theta\, e^{j(\omega t - kr)} \tag{7.10.12}$$

In this case, the directional factor has cylindrical symmetry around the z axis, there is a nodal surface perpendicular to the z axis through the origin, and there are two lobes of the *same* phase pointing in opposite directions along the z axis.

We can now relate the above discussion to the inhomogeneous wave equation (5.15.7) and identify the three source terms with appropriate far field multipole radiation. For monofrequency motion, rewriting (5.15.7) in terms of the velocity potential gives

$$(\nabla^2 + k^2)\Phi = -\frac{G}{\rho_0} + \frac{1}{j\omega\rho_0}\nabla\cdot\vec{F} - \frac{1}{j\omega\rho_0}\frac{\partial^2(\rho u_i u_j)}{\partial x_i \partial x_j} \tag{7.10.13}$$

where each nonzero term on the right has time dependence $\exp(j\omega t)$.

Let these source terms be functions of $\vec{r}_0$ contained within a small volume V_0 that is a large distance away from the field point. If only the first term on the right in (7.10.13) is nonzero, then

$$\mathbf{q}(\vec{r}_0)e^{j\omega t} = -\mathbf{G}/\rho_0 \tag{7.10.14}$$

and any of the integrals in (7.10.5) can be nonzero. Source terms corresponding to mass injection can generate any combination of multipole radiation terms. In particular, the monopole term can be excited, producing a sound field equivalent to a point source at the origin of strength Q given by (7.10.6). An example is the pulsating sphere discussed earlier in this chapter. (See Problem 7.10.7 for another case.)

If just the second source term in (7.10.13) is nonzero, then

$$\mathbf{q}(\vec{r}_0)e^{j\omega t} = (\nabla\cdot\vec{F})/j\omega\rho_0 \tag{7.10.15}$$

and the sound field is that of a dipole at the origin with vector dipole strength given by (7.10.7). (See Problem 7.10.8.) Note this is the first moment of the source strength. An example of this type of source is a sphere of constant radius a vibrating in the x direction with speed $\hat{x}U\exp(j\omega t)$. For $ka \ll 1$ and at large distances ($kr \gg 1$), the sound field is[1]

$$\Phi_2 = -\left[\rho_0 c U(ka)^2(a/2r)\cos\theta\right]e^{j(\omega t - kr)} \tag{7.10.16}$$

This is a dipole field with dipole strength magnitude $\rho_0 c U(ka)(2\pi a^2)$.

Finally, if just the third source term is nonzero, then

$$\mathbf{q}(\vec{r}_0)e^{j\omega t} = -\frac{1}{j\omega\rho_0}\frac{\partial^2(\rho u_i u_j)}{\partial x_i \partial x_j} \tag{7.10.17}$$

where the x_i are the components of $\vec{r}_0$. It can be shown that this source contributes no monopole and no dipole contributions to the acoustic field. The lowest nonzero contribution is quadrupolar with a quadrupole strength given by the second moment of the source distribution, $\frac{1}{2}\int (q\, x_i x_j)\, dV_0$.

An important property of multipole radiation is the *radiation efficiency* η_{rad} defined as

$$\eta_{rad} = R_r/\sqrt{R_r^2 + X_r^2} \tag{7.10.18}$$

For monopole radiation, the radiation efficiency of a pulsating sphere is found from the radiation impedance (7.5.19) to be $\eta_{rad} = ka$. For dipole radiation, the radiation impedance of a vibrating sphere is equal to its input mechanical impedance, so that[1]

$$Z_r = \frac{2}{3}\rho_0 c\pi a^2 \frac{jka(1 + jka)}{1 + jka - (ka)^2/2} \tag{7.10.19}$$

---

[1]Dowling, *Encyclopedia of Acoustics*, Chap. 9, Wiley (1997).

If $ka \ll 1$, then analysis of (7.10.19) shows that $\eta_{rad} = (ka)^3/2$. In general, it can be shown[2] that if the size of the source is much less than the wavelength of the radiated sound, then

$$\eta_{rad} = (ka)^{2m+1} / \{(m+1)[1 \cdot 3 \cdot 5 \cdots (2m-1)]\} \tag{7.10.20}$$

where a is the characteristic dimension of the source and m is the order of the multipole, $m = 0$ (monopole), 1 (dipole), 2 (quadrupole). Thus, we see that at low frequencies monopole radiation is the most efficient and will dominate. If monopole radiation is absent, then dipole radiation can become important. Quadrupole radiation is important only if there are no strong monopoles or dipoles.

*7.11 BEAM PATTERNS AND THE SPATIAL FOURIER TRANSFORM

Beam patterns can also be obtained by *spatial Fourier transforms* or *spatial filtering*. It is simple to show that the equation used for performing a Fourier transform is identical to that for calculating a far field beam pattern. To investigate this approach, let us revisit the continuous line source. If the individual incremental elements have source strengths $d\mathbf{Q} = \mathbf{g}(x)U_0 2\pi x\, dx$, then (7.3.2) becomes

$$\mathbf{p}(r, \theta, t) = \frac{j}{2}\rho_0 c U_0 \frac{ka}{r} e^{j(\omega t - kr)} \int_{-L/2}^{L/2} \mathbf{g}(x)e^{jkx\sin\theta}\, dx \tag{7.11.1}$$

If $\mathbf{g}(x)$ is zero for values of x exceeding the extent of the array, then (7.11.1) can be written as

$$\mathbf{p}(r, \theta, t) = \frac{j}{2}\rho_0 c U_0 \frac{ka}{r}\mathbf{f}(u)e^{j(\omega t - kr)} \tag{7.11.2}$$

$$\mathbf{f}(u) = \int_{-\infty}^{\infty} \mathbf{g}(x)e^{jux}\, dx \qquad u = k\sin\theta \tag{7.11.3}$$

Direct comparison of (7.11.3) with (1.15.1) and application of the Fourier transform with the pair (w, t) replaced by (x, u) shows that

$$\mathbf{g}(x) = \frac{1}{2\pi}\int_{-\infty}^{\infty} \mathbf{f}(u)e^{-jux}\, du \tag{7.11.4}$$

The quantity $\mathbf{g}(x)$ is the *aperture function*. The absolute magnitude of $\mathbf{f}(u)$, when normalized to have a maximum amplitude of unity and with u replaced by $k\sin\theta$, is the directional factor $H(\theta)$. Thus, given the amplitude and phase distribution along the line source, we can predict the directional factor in the far field from (7.11.3), and vice versa—given a desired far field directional factor, we can use (7.11.4) to determine the required distribution of amplitude and phase that the individual elements of the source must have to obtain a desired directional factor.

If just the amplitudes of the individual elements are modified, but all elements remain in phase, the line source is *amplitude shaded*. If the amplitudes are all kept at the same value but the phases of the individual elements are adjusted, the source is *phase shaded*. One example of phase shading is the steered line array discussed previously.

As an example of amplitude shading, consider a continuous line source of length L with triangular shading—the center element at $x = 0$ having amplitude $L/2$ and the successively

[2]Morse and Ingard, *Theoretical Acoustics*, Princeton (1986). Morse, *Vibration and Sound*, Acoustical Society of America (1976). Ross, *Mechanics of Underwater Noise*, p. 51, Pergamon (1976).

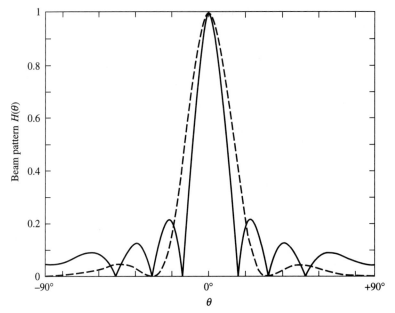

Figure 7.11.1   The directional factor for a continuous line source with $kL = 24$. The solid line is for the unshaded source and the dashed line is for the same source with symmetric triangular shading. This shading reduces the number and level of the side lobes while increasing the width of the major lobe.

more distant pairs of elements having amplitudes that decrease linearly with distance away from the center, falling to zero at $x = \pm L/2$. By symmetry, (7.11.3) becomes

$$\mathbf{f}(u) = 2\int_0^{L/2} (L/2 - x)\cos ux\,dx \qquad (7.11.5)$$

Evaluation is relatively easy, and after normalizing and taking the magnitude to obtain the directional factor, we have

$$H(\theta) = \left|\frac{\sin(v/2)}{v/2}\right|^2 \qquad (7.11.6)$$

where $v = \frac{1}{2}kL\sin\theta$. Comparison of (7.11.6) with (7.3.5) reveals that the first side lobe of the shaded line source is 26 dB below the peak compared to 13 dB for the unshaded line of the same length, but the main lobe is twice as wide. (See Fig. 7.11.1.) This trade-off of lower side lobes for a wider major lobe, and vice versa, is typical of most shading techniques. Although less useful, increasing the amplitude near the ends of the source will narrow the major lobe and increase the strength of the side lobes.

In the case of an array of point sources, the aperture function will be a summation of delta functions $\delta(x - x_i)$ representing the locations x_i of the individual elements. Amplitude shading is accomplished by multiplying each of the elements by its amplitude a_i or source strength Q_i, and phase shading introduces a factor $\exp(j\phi_i)$ for each of the elements.

Problems

7.1.1. A pulsating sphere of radius $a = 0.1$ m radiates spherical waves into air at 100 Hz and with an intensity of 50 mW/m^2 at a distance 1.0 m from the center of the sphere.

(a) What is the radiated acoustic power? (b) At the surface of the sphere, $r = a$, compute the intensity, the amplitudes of the acoustic pressure, particle speed, particle displacement Ξ, ratio Ξ/r, condensation, and acoustic Mach number U_0/c. (c) Repeat part (b) at a distance of 0.5 m from the center of the sphere.

7.1.2. A pulsating sphere of radius a vibrates with a surface velocity amplitude U_0 and at such a high frequency that $ka \gg 1$. Derive expressions for the pressure amplitude, the particle velocity amplitude, the intensity, and the total acoustic power radiated in the resulting acoustic wave.

7.1.3. (a) A spherical source of radius a is operated in water at a frequency for which $ka = 1$. Evaluate the specific acoustic impedance at the source radius for this frequency. Find the error in calculating the acoustic intensity by the formula valid for $ka \ll 1$. (b) If the source strength of a small ($ka \ll 1$) spherical source is kept constant, find the frequency dependence of the radiated power. If this small source is operated with constant acceleration amplitude, find the frequency dependence of the radiated power.

7.1.4. A simple source of sound in air radiates an acoustic power of 10 mW at 400 Hz. At 0.5 m from the source, compute (a) the intensity, (b) the pressure amplitude, (c) the particle speed amplitude, (d) the particle displacement amplitude, and (e) the condensation amplitude.

7.1.5C. (a) Show that in the limit $ka \ll 1$ the specific acoustic impedance at the surface of a pulsating sphere can be approximated by $\mathbf{z}(a) \approx \rho_0 cka(j + ka)$. (b) As a function of kr, plot the resistance, reactance, and magnitude of the impedance to compare the approximation in (a) with (7.1.2) and find the values of ka within which the errors in the approximation are less than 10%.

7.2.1. A hemisphere of radius a and a piston of radius a are each mounted so that they radiate on one side of an infinite baffle. They are both vibrating with the same maximum speed amplitude U_0 and at the same frequency so that $ka \ll 1$. (a) For a distance such that $r \gg a$, what is the ratio of the axial intensity of the piston to that of the hemisphere? (b) What is the ratio of the total power radiated by the hemisphere to that radiated by the piston?

7.2.2. (a) For the pulsating sphere of radius a show that the pressure amplitude at distance r is $P(r) = \frac{1}{2}(\rho_0 cQ/\lambda r)\sin\theta_a$. (b) Does the result of (a) reduce to that for a simple source as $ka \to 0$?

7.2.3. Find an approximation for the pressure field given by (6.8.6) in the limit $kd \ll 1$. Is this result consistent with (7.2.14) and (7.2.17) for the simple source and baffled simple source?

7.3.1. (a) Show for the continuous line source that the number N of nodal surfaces is given by $N = [L/\lambda]$. (b) Find the number of major and minor lobes for $L/\lambda = 4.8, 5, 5.2$. (c) For which of the cases of (b) are the last minor lobes fully developed? Only partially developed?

7.3.2. A simple line source is designed so that $kL = 50$. (a) How many major lobes are there? (b) Find the total number of nodal surfaces. (c) Find the angular width in degrees of the major lobe that is centered at $\theta = 0$. (d) Estimate the relative strength in dB of the first side lobe.

7.3.3C. Assume a continuous line source with $kL = 24$. (a) Compute and plot contours of equal acoustic pressure amplitude in the near field by direct numerical integration of (7.3.1) before approximating r'. (b) Compare these contours with those obtained in the far field from (7.3.4). Describe the transition from near to far field behaviors.

7.4.1. For a baffled piston of radius a driven at angular frequency ω, (a) find the smallest angle θ_1 for which the pressure is zero in the far field, (b) find the greatest finite distance for which the pressure is zero on the acoustic axis, and (c) discuss the possibility of obtaining $\theta_1 \ll 1$ and $r_1/a \ll 1$ simultaneously.

7.4.2. A piston of radius a is mounted so as to radiate on one side of an infinite baffle into air. The piston is driven at a frequency such that $\lambda = \pi a$. (a) Compute and plot the relative axial intensities produced by the piston from its surface to a distance of $r = 3a$. (b) Over what range of distances is the divergence approximately spherical?

7.4.3. A circular piston sonar transducer of 0.5 m radius radiates 5000 W of acoustic power into water at 10 kHz. What is its beam width at the -10 dB direction?

7.4.4. Show that the nodal angles of the piston can be approximated by $\sin \theta_m \approx (m + \frac{1}{4})\pi/ka$. Estimate the error in θ_m for the first nodal surface given by $m = 1$.

7.4.5. By expanding $\exp(jka \sin \theta \cos \phi)$ in (7.4.15) as a power series show that (7.4.16) is correct and

$$\int_0^\pi e^{jka \sin \theta \cos \phi} \sin^2 \phi \, d\phi = \pi \frac{J_1(ka \sin \theta)}{ka \sin \theta}$$

7.4.6C. (a) For a circular piston, plot the on-axis pressure amplitude as a function of scaled distance r/a for several values of ka between 3 and 12. (b) Plot the range beyond which the pressure amplitude is within 10% of the asymptotic form (7.4.7). (c) For a piston of 20 cm radius operating at 4 kHz in water, find the distance corresponding to (b).

7.4.7C. For $ka = 3$, use numerical integration to plot the pressure amplitude of a circular piston (a) on axis and compare to (7.4.5), (b) on the face of the piston, and (c) in the near field off axis.

7.5.1. (a) Find the resonance frequency of a piston transducer with the mechanical properties m, s, and R_m radiating into a fluid with specific acoustic impedance $\rho_0 c$. Assume $ka \gg 2$. (b) Sketch the frequency dependence of the radiated power if the transducer is driven with a force of constant amplitude. Assume that the resonance frequency occurs well above the lower limit of the approximations implicit in $ka \gg 2$. Indicate where the transducer is mass controlled and where it is stiffness controlled.

7.5.2. Evaluate the integrals in (7.5.10) and obtain the radiation impedance for the piston. *Hints:* (a) Perform the integration over r directly. (b) Use the integral forms of the Bessel and Struve functions

$$\frac{2}{\pi} \int_0^{\pi/2} \left\{ \begin{array}{c} \cos(x \cos \theta) \\ \sin(x \cos \theta) \end{array} \right\} d\theta = \left\{ \begin{array}{c} J_0(x) \\ \mathbf{H}_0(x) \end{array} \right\}$$

to integrate over θ. (c) Use the integral relations

$$\int_0^b \left\{ \begin{array}{c} J_0(x) \\ \mathbf{H}_0(x) \end{array} \right\} x \, dx = b \left\{ \begin{array}{c} J_1(b) \\ \mathbf{H}_1(b) \end{array} \right\}$$

to evaluate the integral over σ.

7.5.3. (a) Find the radiation impedance of a pulsating hemisphere. (b) Find the radiation resistance for high frequencies, and from this find the radiated power and compare to the results in Section 7.1. (c) Find the radiation reactance for low frequencies, and from this find the ratio of the radiation mass to the mass of the fluid displaced by the hemisphere.

7.6.1. Obtain (7.6.16) for the baffled piston directivity as follows: (*a*) Use (7.5.6) and (7.6.9) to relate the directivity D, extrapolated axial pressure amplitude $P_{ax}(1)$, and radiation resistance R_r of the piston. (*b*) Use (7.4.7) to eliminate $P_{ax}(1)$ and then (7.5.11) and (7.5.12) to express R_r in terms of $J_1(2ka)/ka$. (*c*) Solve the result for D.

7.6.2. A plane circular piston in an infinite baffle operates into water. The radius of the piston is 1 m. At $6/\pi$ kHz the sound pressure level on axis at 1 km is 100 dB *re* 1 μbar. (*a*) Find all angles at which the pressure amplitude in the far field is zero. (*b*) Find the rms speed of the piston. (*c*) If the frequency were doubled while keeping the speed amplitude of the piston constant, what would be the dB change in the sound pressure level on the axis in the far field and what would be the dB change in the directivity index?

7.6.3. A flat piston of 0.2 m radius radiates 100 W of acoustic power at 20 kHz in water. (*a*) Assuming the radiation to be equivalent to that of a piston mounted in an infinite baffle and radiating on only one side, what is the velocity amplitude of the piston? (*b*) What is the radiation mass loading of the piston? (*c*) What is the beam width at the down 10 dB direction? (*d*) What is the directivity index of the beam?

7.6.4. A piston is mounted so as to radiate on one side of an infinite baffle into air. The radius of the piston is a, and it is driven at a frequency such that $\lambda = \pi a$. (*a*) If $a = 0.1$ m and the maximum displacement amplitude of the piston is 0.0002 m, how much acoustic power is radiated? (*b*) What is the axial intensity at a distance of 2.0 m? (*c*) What is the directivity index of the radiated beam? (*d*) What is the radiation mass?

7.6.5. It is desired to design a highly directive piston transducer that will produce a given acoustic pressure amplitude P on axis at a specified range r. The operating frequency must be f, and the total acoustic power output is fixed. Find the radius and speed amplitude of this transducer.

7.6.6. A flat piston of 0.15 m radius is mounted to radiate on one side of an infinite baffle into air at 330 Hz. (*a*) What must be the speed amplitude of the piston if it is to radiate 0.5 W of acoustic power? (*b*) If the piston has a mass of 0.015 kg, a stiffness constant of 2000 N/m, and negligible internal mechanical resistance, what force amplitude is required to produce this velocity amplitude?

7.6.7. A baffled piston transducer with radius 10 cm is normally operated at 15 kHz. If it is desired to operate this same transducer at 3.5 kHz while maintaining the same acoustic pressure on the axis in the far field, calculate the ratio of the total acoustic power output at 3.5 kHz to that at 15 kHz. Assume operation in water.

7.6.8. A continuous line source is designed so that $kL = 50$. (*a*) If the length of the array is 100 m, estimate the distance to the far field. (*b*) Estimate the directivity index from (7.6.14). (*c*) Repeat (*b*) but using (7.6.25). (*d*) Compare the discrepancies in D and DI between (*b*) and (*c*) and comment on their significance.

7.6.9C. (*a*) For a circular piston, as a function of ka, plot the exact directivity and its high-frequency approximation, and their percent difference. (*b*) If the radius of a piston operating in water is 20 cm, below what frequency will the approximate directivity differ by more than 10% from the exact value?

7.8.1. Show for the line array that $P_{ax}(r) = (N/2)\rho_0 c Q / \lambda r$.

7.8.2. It is desired to design an underwater linear array of 30 equally spaced elements. (The array is not steered or shaded.) (*a*) If there is to be a single narrowest major lobe at 300 Hz, find the spacing between elements. (*b*) What is the angular width in degrees of the major lobe? (*c*) What is the geometrical shape of the axis of the major lobe? (*d*) Estimate the directivity index of the array.

7.8.3. (*a*) For an endfired line array, derive the condition (7.8.18) that guarantees there will be only one major lobe. (*b*) Derive the similar condition (7.8.19) for a line array steered into a maximum angle θ_0 away from broadside.

7.8.4. An array with a large number N of elements is designed to be steered through all angles and to have a single narrowest major lobe when endfired. (*a*) With the help of the estimation techniques of Section 7.6, derive the directivity (7.8.20) of the array when endfired. (*b*) Derive the directivity (7.8.21) when this array is steered into the angle θ_0 not close to endfired.

7.8.5. It is desired to design an endfired linear array of 30 equally spaced elements to be operated in water at 300 Hz. (*a*) Find the spacing between elements if there is to be only one major lobe. (*b*) What is the geometric shape of the major lobe axis? (*c*) Find the angular width in degrees of the major lobe. (*d*) Estimate the directivity index of this endfired array.

7.8.6. Find the minimum number of array elements for which $\theta_1 = 1/N$ is within about 20% of the value given by (7.8.13), and estimate the error in DI from that discrepancy alone.

7.8.7. (*a*) Verify (7.8.20) and (7.8.21). *Hint:* Obtain the major lobe angular width with the help of (7.8.16)–(7.8.19), and the approximation for small δ that $\sin(\theta_0+\delta) \approx \sin\theta_0 + \delta\cos\theta_0$. Then use the techniques of Section 7.6 to estimate the directivity. (*b*) Find a rough criterion in terms of the major lobe half-angle for the maximum steering angle around which this estimate becomes poor.

7.8.8. Assume a three-element array with no steering has amplitude weighting $A_1 = A_3 = 1$ and $A_2 = 2$. (This is called *binomial shading* because the amplitudes are proportional to the binomial coefficients.) Find (*a*) the condition on kd for there to be just a single major lobe, and (*b*) the equation for $H(\theta)$ under the condition of (*a*).

7.8.9C. Estimate the directivity of a steered line array with one narrowest major lobe when endfired for steering angles bridging the gap from endfire with $D = (\pi/2)^2 N$ to strongly steered where $D = (\pi/4)N$.

7.8.10C. An unsteered four-element array is designed to have a single narrowest major lobe. If $A_2 = A_3 = 1$ and $A_1 = A_4$ can vary from 2 to -2, make plots of the directional factors and describe the effects on the lobes and nulls.

7.8.11. A "shotgun" microphone is constructed with N parallel tubes whose respective lengths measured from the diaphragm are L, $L - d$, $L - 2d$, $L - 3d, \ldots, L - (N - 1)d$. Show that the directional factor for such a microphone is

$$H(\theta) = \left| \frac{\sin[N\,kd\,\sin^2(\theta/2)]}{N\sin[kd\,\sin^2(\theta/2)]} \right|$$

where θ is the off-axis angle measured from the axes of the tubes.

7.9.1. Write the directional factor $H(\theta,\phi)$ for a rectangular piston transducer in terms of the angles, k, and the dimensions L_x and L_y of the piston.

7.9.2C. A twin-line array consists of two identical linear arrays of N elements and spacing d. The linear arrays are parallel and separated by a distance D. The arrays operate in the endfired mode with one narrowest major lobe. If $N = 15$ and $kD = \pi$, plot the beam patterns of an individual array and of the twin-line array.

7.9.3C. Three identical piston sources of radius a are set in a line on an infinite baffle with their centers a distance d apart and their axes perpendicular to the baffle. For $kd = 2\pi$, plot the far-field beam patterns for $d/a = 2$, 3, and 4. Comment on the effect of the directionality of the piston sources on the beam pattern of the array.

7.10.1. Show that if $\mathbf{q}(\vec{r}_0) = Q\delta(0)$ then (7.10.3) reduces immediately to (7.10.2) with $\vec{r}_0 = 0$.

7.10.2. Show by direct integration of the infinite series of (7.10.5) that a monopole located at $\vec{r} = \vec{a}$ gives the same velocity potential as (7.10.2) under the far field approximation $|\vec{r} - \vec{a}| \approx r - \vec{a} \cdot \hat{r}$.

7.10.3. Assume three point sources of equal amplitudes and phases are located along the x axis at $x = d, 0, -d$. (a) Obtain the multipole expansion for the array from (7.10.5). (b) Show that it can be put into the form

$$\Phi = \frac{Q}{4\pi r}\left[3 - 4\sin^2\left(\tfrac{1}{2}kd\sin\theta\right)\right]e^{j(\omega t - kr)}$$

(c) Show that (b) is identical with the result for the three-element line array obtained from (7.8.3). *Hint:* Verify that $4\sin^3\beta = 3\sin\beta - \sin 3\beta$ and use this to simplify.

7.10.4. Convert (7.10.9) into an expression for the acoustic pressure and compare the result with (6.8.7) in the limit $kd \ll 1$. Note that the angle used in Section 6.8 is the complement of θ defined for spherical coordinates.

7.10.5. Derive (7.10.12). *Hint:* Verify that the source strength density is given by $\mathbf{q}(\vec{r}_0) = Q[\delta(\vec{r}_0 - \hat{z}d) + \delta(\vec{r}_0 + \hat{z}d) - 2\delta(0)]$.

7.10.6. Show that three identical axial quadrupoles each oriented along a different coordinate axis and all with their centers at the origin generate a spherically symmetric field of order $O(kd)^2$.

7.10.7. Find the monopole, dipole, and quadrupole fields for the source strength distribution $q = A(a - r)$ for $0 < r \le a$ and $q = 0$ for $r > a$, where A is a constant and r is the distance from the origin.

7.10.8. Show that for the source strength given by (7.10.15) the monopole contribution is zero. *Hint:* Apply Gauss's theorem to a volume integral that includes all the sources.

7.10.9. Show that for the source strength given by (7.10.17) the monopole contribution is zero. *Hint:* Write (7.10.17) in vector form with the help of (5.15.5).

7.10.10C. (a) Make a sketch (representing a three-dimensional view) and plot the directional factor of a dipole in representative planes. (b) Repeat (a) for a lateral quadrupole. (c) Repeat (a) for an axial quadrupole.

7.10.11. (a) Show that the radiation efficiency of a pulsating sphere is $\eta_{rad} = ka$. (b) Show that the radiation efficiency of a vibrating sphere is $\eta_{rad} = (ka)^3/2$.

7.11.1. Assume three point sources of equal amplitudes and phases are located along the x axis at $x = d, 0, -d$. (a) Write down an appropriate form for the aperture function. (b) Obtain the directional factor using (7.11.3). (c) Show that this result is identical with that for the three-element line array obtained from (7.8.3).

7.11.2C. (a) Plot the far-field directional factor of an unsteered, unshaded, 9-element array with spacing that gives one narrowest major lobe. (b) Repeat for the same spacing, but with amplitude shading (1,2,3,4,5,4,3,2,1). (c) Repeat for shading (5,4,3,2,1,2,3,4,5). (d) Comment on the effects of these shadings.

Chapter **8**

ABSORPTION
AND ATTENUATION
OF SOUND

8.1 INTRODUCTION

The wave equation was derived in Chapter 5 under the assumption that all losses of acoustic energy could be neglected. While in many situations dissipation is so slight that it can be ignored for the distances or times of interest, ultimately all acoustic energy is converted into random thermal energy. The sources of this dissipation may be divided into two general categories: (1) those intrinsic to the medium and (2) those associated with the boundaries of the medium. Losses in the medium may be further subdivided into three basic types: viscous losses, heat conduction losses, and losses associated with internal molecular processes. Viscous losses occur whenever there is relative motion between adjacent portions of the medium, such as during shear deformation or the compressions and expansions that accompany the transmission of a sound wave. Heat conduction losses result from the conduction of thermal energy from higher temperature condensations to lower temperature rarefactions. Molecular processes leading to absorption include the conversion of the kinetic energy of the molecules into (1) stored potential energy (as in a structural rearrangement of adjacent molecules in some cluster), (2) rotational and vibrational energies (for polyatomic molecules), and (3) energies of association and dissociation between different ionic species and complexes in ionized solutions (magnesium sulfate and boric acid in seawater).

So far in this book, we have assumed that the fluid is a continuum having directly observable properties, such as pressure, density, compressibility, specific heat, and temperature, and have not been concerned with its molecular structure. In this same spirit, by use of viscosity, Stokes developed the first successful theory of sound absorption. Subsequently, Kirchhoff utilized the property of thermal conductivity to develop an additional contribution to generate what is now called the classical absorption coefficient. In more recent times, as more accurate sound absorption measurements were made, it became evident that explanations of sound absorption from this viewpoint were inadequate in some fluids. Consequently, it became necessary in developing additional absorption mechanisms to adopt a microscopic view and consider such phenomena as the binding energies within and between molecules. These mechanisms are commonly referred to as *molecular* or *relaxational* sound absorption. (In point of fact, all loss mechanisms are relaxational in nature, but often certain effects of the relaxations are not observed in the range of frequencies and temperatures usually encountered.) For a more complete discussion, the reader is referred to the literature.[1]

8.2 ABSORPTION FROM VISCOSITY

If the effects of viscosity are retained in developing the force equation, it is necessary to perform some fairly elaborate tensor analysis. This lies beyond our interests.[2] The result of this more general derivation is the nonlinear *Navier–Stokes equation*, which, in the absence of external body forces, is

$$\rho\left(\frac{\partial \vec{u}}{\partial t} + (\vec{u} \cdot \nabla)\vec{u}\right) = -\nabla p + (\tfrac{4}{3}\eta + \eta_B)\nabla(\nabla \cdot \vec{u}) - \eta\nabla \times \nabla \times \vec{u} \qquad (8.2.1)$$

The viscosity coefficients η and η_B have units of pascal-seconds (Pa·s).

The *coefficient of shear viscosity* η can be measured directly. While η is manifested clearly in shear flow, it is actually a measure of the diffusion of momentum by molecular collisions between regions of the fluid possessing different net velocities; it is therefore active in producing absorption even in pure longitudinal motion. Experimentally, it is observed to be independent of frequency and to depend only on temperature for almost all fluids over the range of physical parameters of practical interest. (This has been supported by the kinetic theory for a perfect gas.) Because the temperature fluctuations in acoustic propagation are very small, η can be assumed to be a function only of the equilibrium temperature.

The *coefficient of bulk viscosity* η_B is zero in monatomic gases but can be finite in other fluids. It appears to be a measure of some conversions of energy between molecular motion, internal molecular states, and structural potential energy states. Bulk viscosity is often called *expansive* or *volume* viscosity.

These viscous processes require time for the system to approach equilibrium when the density and temperature of the fluid are changed by an expansion or compression. These time delays generate conversion of acoustic energy to random thermal energy.

The term $\eta\nabla \times \nabla \times \vec{u}$ represents the dissipation of acoustic energy involving turbulence, laminar flow, vorticity, and so forth. While these effects can be dominant in nonacoustic situations, in linear acoustics they are usually confined to small regions near boundaries and are of lesser importance.

When the left side of (8.2.1) is linearized, then use of the linearized equation of continuity

$$\nabla \cdot \vec{u} = -\frac{\partial s}{\partial t} \qquad (8.2.2)$$

---

[1]Markham, Beyer, and Lindsay, *Rev. Mod. Phys.*, **23**, 533 (1951). Herzfeld and Litovitz, *Absorption and Dispersion of Ultrasonic Waves*, Academic Press (1959). *Physical Acoustics*, Vol. IIA, ed. Mason, Academic Press (1965).

[2]Development is available in many books, including Temkin, *Elements of Acoustics*, Wiley (1981), and Morse and Ingard, *Theoretical Acoustics*, Princeton University Press (1986).

and the adiabat

$$p = \rho_0 c^2 s \tag{8.2.3}$$

gives a lossy wave equation

$$\left(1 + \tau_S \frac{\partial}{\partial t}\right)\nabla^2 p = \frac{1}{c^2}\frac{\partial^2 p}{\partial t^2}$$

$$\tag{8.2.4}$$

$$\tau_S = \left(\tfrac{4}{3}\eta + \eta_B\right)/\rho_0 c^2$$

where τ_S is a *relaxation time* and c is the thermodynamic speed of sound, determined from $c^2 = (\partial\mathcal{P}/\partial\rho)_{ad}$: it is not necessarily the phase speed c_p because of the term containing τ_S.

If we assume monofrequency motion $\exp(j\omega t)$, this wave equation is reduced to a (lossy) Helmholtz equation

$$\nabla^2\mathbf{p} + \mathbf{k}^2\mathbf{p} = 0$$

$$\tag{8.2.5}$$

$$\mathbf{k} = k - j\alpha_S = (\omega/c)/(1 + j\omega\tau_S)^{1/2}$$

Solution for α_S and c_p gives, after some manipulation,

$$\alpha_S = \frac{\omega}{c}\frac{1}{\sqrt{2}}\left[\frac{\sqrt{1 + (\omega\tau_S)^2} - 1}{1 + (\omega\tau_S)^2}\right]^{1/2}$$

$$\tag{8.2.6}$$

$$c_p = \frac{\omega}{k} = c\sqrt{2}\left[\frac{1 + (\omega\tau_S)^2}{\sqrt{1 + (\omega\tau_S)^2} + 1}\right]^{1/2}$$

For a plane wave traveling in the $+x$ direction, the solution to (8.2.5) is

$$\mathbf{p} = P_0 e^{j(\omega t - \mathbf{k}x)}$$

$$= P_0 e^{-\alpha_S x} e^{j(\omega t - kx)} \tag{8.2.7}$$

Since the amplitude decays as $\exp(-\alpha_S x)$, α_S is the *spatial absorption coefficient*, and the *phase speed* is c_p. Since c_p is a function of frequency, the propagation is *dispersive*.

Calculation of τ_S for representative fluids shows that typical values are about 10^{-10} s in gases and about 10^{-12} s for all liquids except highly viscous ones like glycerin. Thus, frequencies for which the assumption $\omega\tau_S \ll 1$ fails lie in the very high ultrasonic range.

From kinetic theory, the average speed of the molecules in a perfect gas is

$$v = \sqrt{8rT_K/\pi} \tag{8.2.8}$$

where r is the specific gas constant and the coefficient of shear viscosity is

$$\eta = \tfrac{1}{3}\rho_0 l v \tag{8.2.9}$$

where l is the mean free path between successive collisions of the gas atoms. Combination shows that in a perfect gas the shear viscosity is proportional to

$\sqrt{T_K}$. Straightforward manipulation gives

$$v/c = \sqrt{8/\pi\gamma} \sim 1 \qquad (8.2.10)$$

so the average speed of the molecules is very similar to the speed of sound. Since the mean time between collisions is $\tau_C = l/v$,

$$\tau_C/\tau_S = \tfrac{9}{32}\pi\gamma \sim 1 \qquad (8.2.11)$$

The relaxation time for viscous absorption is similar to the mean time between collisions. Thus, for frequencies approaching the relaxation frequency, the wavelength is about the same size as the mean free path. This violates the assumption of a fluid continuum, which underlies the Navier–Stokes equation, so that the model for thermoviscous attenuation cannot be trusted, at least in perfect gases, for frequencies approaching the relaxation frequency.

Continuum mechanics predicts the existence of the absorption mechanisms. However, it does not provide a means of predicting either the value of the viscosity coefficient or its temperature dependence. These quantities can be calculated for simple fluids from statistical mechanics.

Since the theory is appropriate only for $\omega\tau_S \ll 1$, (8.2.6) yields the more useful approximations

$$\alpha_S \approx \tfrac{1}{2}(\omega/c)\omega\tau_S = (\omega^2/2\rho_0 c^3)(\tfrac{4}{3}\eta + \eta_B)$$
$$c_p \approx c\left[1 + \tfrac{3}{8}(\omega\tau_S)^2\right] \qquad (8.2.12)$$

The absorption coefficient is proportional to the square of the frequency. As a consequence, in experimental measurements of absorption, data are usually plotted as α_S/f^2 against f so that any departure from a horizontal line signals a deviation from the prediction of (8.2.12). The dispersion is of $O(\omega\tau_S)^2$, so it is only slight and the phase speed is virtually identical with c. Combination of α and c_p reveals the dimensionless forms

$$\alpha_S/k \approx \tfrac{1}{2}\omega\tau_S$$
$$c_p/c \approx 1 + \tfrac{3}{2}(\alpha_S/k)^2 \qquad (8.2.13)$$

8.3 COMPLEX SOUND SPEED AND ABSORPTION

Before proceeding to a discussion of other specific absorption mechanisms, it is appropriate to develop the *complex speed of sound* for a monofrequency acoustic wave. It is easy to show (see Problem 8.3.6) that if we postulate a *dynamic* (or monofrequency) equation relating **p** and **s**

$$\mathbf{p} = \rho_0 \mathbf{c}^2 \mathbf{s} \qquad (8.3.1)$$

where **c**, the complex sound speed, is a function of frequency to be determined from the thermodynamics of the particular loss mechanism, then combination of (8.3.1) with the linearized Euler's equation (5.4.10) and equation of continuity (5.3.5) gives a lossy Helmholtz equation with a damped traveling wave solution,

$$\nabla^2 \mathbf{p} + \mathbf{k}^2 \mathbf{p} = 0$$

$$\mathbf{k} = k - j\alpha = \omega/\mathbf{c} \qquad (8.3.2)$$

$$\mathbf{p} = P_0 e^{-\alpha x} e^{j(\omega t - kx)}$$

The condensation is obtained from substituting (8.2.7) into (8.3.1). Use of the continuity equation (8.2.2) yields the associated particle speed and thence the specific acoustic impedance for this wave,

$$\mathbf{z} = \mathbf{p}/\mathbf{u} = \rho_0 c_p / (1 - j\alpha/k) \qquad (8.3.3)$$

or, for $\alpha/k \ll 1$,

$$\mathbf{z} \approx e^{j\alpha/k} \rho_0 c \qquad (8.3.4)$$

Calculation of the acoustic intensity from its definition (5.9.1) gives, through the same order of accuracy,

$$I(x) = (P_0 e^{-\alpha x})^2 / 2\rho_0 c = I(0) e^{-2\alpha x} \qquad (8.3.5)$$

The absorption coefficient α is expressed in nepers per meter (Np/m), where the neper (Np) is a dimensionless unit. For the plane wave, when $x = 1/\alpha$ the pressure amplitude has dropped to $1/e = 0.368$ of its initial value P_0 and the intensity has fallen by $1/e^2 = 0.135$.

The loss in intensity level with distance of the attenuated plane wave, expressed in dB, is given by

$$IL(0) - IL(x) = 10 \log[I(0)/I(x)] = 10 \log e^{2\alpha x}$$
$$= 8.7 \, \alpha x \equiv ax \qquad (8.3.6)$$

where $a = 8.7\alpha$ is the absorptive loss in dB/m. It is left as an exercise to show that for a spherically symmetric wave an analogous expression for the loss of intensity level with distance is

$$IL(1) - IL(r) \approx 20 \log r + ar \qquad (8.3.7)$$

when $\alpha \ll 0.1$ Np/m. (While incorrect, it is conventional to leave unwritten the division of r by 1 m in the argument of the logarithm.)

This section and the previous one have shown that acoustic losses can be introduced into the wave equation either by including dissipative forces in Euler's equation or by generating a phase angle between the pressure and density. The latter approach, although somewhat more restricted since it assumes monofrequency motion, is nevertheless very useful in considering molecular effects, since it is often more convenient to modify the adiabat relating acoustic pressure and condensation than to introduce extra forces in the force equation. Furthermore, once the monofrequency behavior of the Helmholtz equation is known, more complicated waveforms can be developed using Fourier synthesis.

8.4 ABSORPTION FROM THERMAL CONDUCTION

Another mechanism producing absorption is thermal conduction. A general derivation of the absorption coefficient for thermal conduction losses requires rather extensive use of thermodynamics. Instead, we will develop thermal absorption more heuristically from physical arguments. For calculational simplicity, we will restrict the development to perfect gases.

When a fluid is subjected to an acoustic process, the compressed regions will have higher temperatures than will the rarefied regions. If we assume a plane wave p of angular frequency ω and propagation constant k traveling in the $+x$ direction, then in a *lossless* perfect gas with equilibrium absolute temperature T_{eq} the temperature can be found from the equation of state and the adiabat to be

$$T = T_{eq} + T_{eq}(\gamma - 1)s \tag{8.4.1}$$

where $s = p/\rho_0 c^2$. The subscript K on the absolute temperature is suppressed throughout this section for notational convenience. For a wave with pressure amplitude P, the magnitude of the temperature fluctuation is

$$|T - T_{eq}| = T_{eq}(\gamma - 1)P/\rho_0 c^2 \tag{8.4.2}$$

In a *lossy* gas, the amplitude of the temperature fluctuation will decay as does the pressure amplitude. However, with a little judicious manipulation we can base our calculations of losses on the lossless expression (8.4.1). From kinetic theory, the kinetic energy of translation for a perfect gas is proportional to the temperature. The molecules in hotter regions have greater kinetic energies that diffuse into the surrounding cooler regions through intermolecular collisions. As energy leaves the region, it is lost from the acoustic process, converted to random thermal energy of molecular motion. The change in thermal energy is related to the change in temperature by

$$\frac{\Delta q}{\Delta t} = c_\mathcal{P} \rho_0 \frac{\partial T}{\partial t} \tag{8.4.3}$$

where $c_\mathcal{P}$ is the *specific heat at constant pressure* [in J/(kg·K)] and Δq is the gain in thermal energy of a unit volume of the gas. [See Appendix A9 and note that $c_\mathcal{P} = C_\mathcal{P}/M$, where M is the molecular weight (kg). Use of (A9.4) is justified because the acoustic pressure fluctuations are much less than the equilibrium pressure.] The diffusion process is described by a *diffusion equation*, which can be written for temperature as

$$\frac{\partial T}{\partial t} = \frac{\kappa}{c_\mathcal{P} \rho_0} \nabla^2 T \tag{8.4.4}$$

where κ is the *thermal conductivity* and has units W/(m·K). Combining these, we have

$$\frac{\Delta q}{\Delta t} = \kappa \nabla^2 T \tag{8.4.5}$$

Integrating (8.4.5) over a volume of the gas gives the *instantaneous* rate of loss of acoustic energy in the volume. The time average of this over a period of the motion will give the *average* rate of acoustic energy loss.

Lossless acoustic approximations cannot be used in (8.4.5) as it stands, since the result will have a time average of zero. Instead, to use the lossless (8.4.1), we must reexpress the right side of (8.4.5) in a form that will allow us to isolate the oscillatory part from the accumulating part for a fluid with losses. This is accomplished by the identity

$$\nabla^2 T = \frac{1}{T}\nabla T \cdot \nabla T + T\nabla \cdot \left(\frac{1}{T}\nabla T\right) \tag{8.4.6}$$

Consistent with the linearizing acoustic approximations, the T's in the denominators can be approximated by the equilibrium temperature T_{eq}. The second term on the right is an oscillatory term whose time average is zero. The first term on the right is never negative and represents the accumulation over time of energy lost from the acoustic wave. The (time-averaged) rate of change in the acoustic energy density $\mathcal{E}$ is then

$$\frac{d\mathcal{E}}{dt} = -\frac{1}{V}\left\langle \int_V \frac{\Delta q}{\Delta t}dV \right\rangle_t = -\frac{\kappa}{T_{eq}}\frac{1}{V}\left\langle \int_V \nabla T \cdot \nabla T\, dV \right\rangle_t \tag{8.4.7}$$

Now, consider a cylindrical volume V of cross-sectional area S and length $\lambda = 2\pi/k$ with its axis parallel to the propagation vector $\vec{k}$ of the acoustic wave. To obtain the rate at which thermal energy is being lost from the acoustic wave, integrate (8.4.7) over this volume. The temperature gradient for the traveling acoustic wave is found from (8.4.1). Integration over x and averaging over one period of the motion are straightforward,

$$\frac{d\mathcal{E}}{dt} = -\frac{\kappa}{T_{eq}}\frac{S}{V}\left\langle \int_0^\lambda \nabla T \cdot \nabla T\, dx \right\rangle_t = -\tfrac{1}{2}\kappa T_{eq}\left(\gamma - 1\right)^2\left(\frac{kP}{\rho_0 c^2}\right)^2 \tag{8.4.8}$$

The absorption coefficient is found from $(d\mathcal{E}/dt)/\mathcal{E} = -2\alpha_\kappa c$. For this wave, $\mathcal{E} = \tfrac{1}{2}P^2/\rho_0 c^2$. Furthermore, for a perfect gas $T_{eq}(\gamma - 1) = c^2/c_{\mathcal{P}}$ so that the absorption coefficient for thermal conduction is

$$\alpha_\kappa = \frac{\omega^2}{2\rho_0 c^3}\frac{(\gamma - 1)\kappa}{c_{\mathcal{P}}} \tag{8.4.9}$$

We will state without proof that (8.4.9) applies to any fluid. This absorption coefficient has the same frequency dependence as that for viscous absorption. (Note that α_κ vanishes for $\gamma = 1$. This is expected since that value makes the adiabat the same as the isotherm, and in such a fluid there would be no thermal fluctuation accompanying the acoustic wave and thus no thermal conduction.) Expression (8.4.9) is valid for frequencies much less than the relaxation frequency. More advanced theoretical analysis shows that the relaxation time for thermal conduction in a fluid is

$$\tau_\kappa = \frac{1}{\rho_0 c^2}\frac{\kappa}{c_{\mathcal{P}}} \tag{8.4.10}$$

If we had tried to extract the relaxation time by assuming it is related to α_κ just as τ_s is related to α_S by (8.2.12), we would have the above result multiplied by $(\gamma - 1)$. The absence of this factor is plausible, since it is simply a measure of how much adiabatic and isothermal conditions differ. The *mechanism* for thermal conduction lies in collisions on the molecular level and so the relaxation time should not depend on γ. The *amount* of absorption depends on the deviation between adiabatic and isothermal conditions, and this should depend on $(\gamma - 1)$, as seen in (8.4.9). Finally, this development does not yield an evaluation of the phase speed. That is of little consequence, since the deviations are very small for frequencies far below the relaxation frequency.

8.5 THE CLASSICAL ABSORPTION COEFFICIENT

For gases, the absorption associated with heat conduction is somewhat less than that for viscous absorption but of the same magnitude. For most *nonmetallic* liquids the absorption produced by thermal conductivity is negligible compared with that from viscosity.

When losses are small, it is plausible and can be shown that for independent sources of acoustic losses the total absorption coefficient is the sum of the absorption coefficients of the individual loss mechanisms calculated as if each were operating alone,

$$\boxed{\alpha = \sum_i \alpha_i} \tag{8.5.1}$$

Quality factors can also be related fairly simply. Let Q_i be the quality factor that a resonator would have if there were only one absorptive mechanism characterized by α_i. Note from (1.10.7) and the relationship $\beta_i = \alpha_i c$ that $Q_i = \frac{1}{2}k/\alpha_i$. Then, if there are a number of mechanisms creating acoustic losses, the total quality factor Q of the system is given by

$$\frac{1}{Q} = \sum_i \frac{1}{Q_i} \tag{8.5.2}$$

The historical development of the study of absorptive processes in fluids led to the definition of the *classical absorption coefficient* α_c as the sum of the viscous and thermal absorption coefficients under the *Stokes assumption* $\eta_B = 0$,

$$\boxed{\alpha_c = \frac{\omega^2}{2\rho_0 c^3}\left(\frac{4}{3}\eta + \frac{(\gamma - 1)\kappa}{c_{\mathscr{P}}}\right)} \tag{8.5.3}$$

If use is made of the *Prandtl number,*

$$\mathrm{Pr} = \eta c_{\mathscr{P}}/\kappa \tag{8.5.4}$$

Table 8.5.1   Acoustic absorption in fluids

All Data for $T = 20°C$ and $\mathcal{P}_0 = 1$ atm	α/f^2 (Np · s²/m)			
	Shear Viscosity	Thermal Conductivity	Classical	Observed
Gases	Multiply all values by 10^{-11}			
Argon	1.08	0.77	1.85	1.87
Helium	0.31	0.22	0.53	0.54
Oxygen	1.14	0.47	1.61	1.92
Nitrogen	0.96	0.39	1.35	1.64
Air (dry)	0.99	0.38	1.37	α/f peaks at 40 Hz
Carbon dioxide	1.09	0.31	1.40	α/f peaks at 30 kHz
Liquids	Multiply all values by 10^{-15}			
Glycerin	3000.0	—	3000.0	3000.0
Mercury	—	6.0	6.0	5.0
Acetone	6.5	0.5	7.0	30.0
Water	8.1	—	8.1	25.0
Seawater	8.1	—	8.1	α/f peaks at 1.2 kHz and 136 kHz

which measures the importance of viscosity with respect to thermal conductivity, the classical absorption coefficient assumes the form

$$\alpha_c = \frac{\omega^2 \eta}{2\rho_0 c^3}\left(\frac{4}{3} + \frac{(\gamma - 1)}{\text{Pr}}\right) \tag{8.5.5}$$

For air at 20°C and 1 atm the Prandtl number is about 0.75. Comparison of the relaxation times under the Stokes assumption shows that

$$\tau_s/\tau_\kappa = \tfrac{4}{3}\text{Pr} \tag{8.5.6}$$

Table 8.5.1 contains comparative data on calculated and observed values of the absorption coefficient for representative gases and liquids. As expected, the absorption observed for the monatomic gases such as argon and helium is in good agreement with the classical absorption coefficient (shear viscosity and thermal conductivity). The classical absorption is also in good agreement for highly viscous liquids such as glycerin and highly conducting liquid metals such as mercury. However, the classical absorption falls short of the observed results in polyatomic gases and in most common liquids.

8.6 MOLECULAR THERMAL RELAXATION

Further mechanisms for acoustic absorption can be predicted by taking into account the internal structure of the molecules and the interactions between them that lead to internal vibrations, rotations, ionizations, and short-range ordering.

The oldest and most successful of the many theoretical approaches to these problems is that treating molecular thermal relaxation in gases composed of polyatomic molecules. In this theory, it is acknowledged that, in addition to the three degrees of translational freedom each molecule possesses, there are also internal degrees of freedom associated with the rotation and vibration. The time necessary for energy to be transferred from translational motion of the molecule into internal states compared to the period of the acoustic process determines how much acoustic energy will be converted to thermal energy during the transitions. If the period of the acoustic excitation is long compared to the relaxation time τ of the internal energy state ($\omega\tau \ll 1$), then the state can be fully populated; the phase lag is finite but small, so the fraction of energy lost is very small over each period of the motion. On the other hand, if the acoustic period is much shorter than the relaxation time ($\omega\tau \gg 1$), the internal energy state cannot be heavily populated before conditions are reversed, and the energy loss over each period will also be small. However, at periods close to the relaxation time ($\omega\tau \sim 1$), the energy loss per period should be maximized.

If the thermodynamic system is in equilibrium at some temperature T_K, the heat capacity C_V depends on the number of ways the molecules can store appreciable energy. These are called *degrees of freedom*. All molecules possess three degrees of translational freedom. Polyatomic molecules can also store energy in rotational and vibrational degrees of freedom. A *population function* $H_i(T_K)$ accounts for the fact that an energy state (other than the translational ones) cannot be significantly populated unless the temperature is above the *Debye temperature* T_D for that state. Each $H_i(T_K)$ asymptotically approaches zero at low temperatures and unity at sufficiently high temperatures. It changes rapidly only for temperatures close to the specific Debye temperature for that state. At room temperatures the population functions $H_i(T_K)$ will be nearly unity for most rotational states. They will be near unity or appreciably less for the lower vibrational states, and extremely small for higher vibrational levels. The heat capacity of a gas is

$$C_V = \mathfrak{R}\left(\frac{N}{2} + \frac{1}{2}\sum_{i=1}^{n} H_i(T_K)\right) \qquad (8.6.1)$$

where N enumerates the contributions of the fully excited degrees of freedom, the summation over n covers the partially excited degrees of freedom, and $H_i(T_K)$ is the fraction of the ith degree of freedom populated at T_K. Unless the ambient temperature is extraordinarily high, the energy of a monatomic gas can reside only in the three degrees of translation, so $N = 3$. For diatomic gases or linear polyatomic gases like carbon dioxide there are only two rotational states, both fully excited, so $N = 5$. (Molecular hydrogen is an exception in that its rotational states are not fully excited until temperatures in excess of room temperatures are encountered.) For nonlinear (kinked) polyatomic gases, $N = 6$. If there are no other excited states, then (with the help of Appendix A9) $\gamma = C_{\mathscr{P}}/C_V$ and $C_{\mathscr{P}} = C_V + \mathfrak{R}$ yield $\gamma = 1 + 2/N$ so that the values of γ are 1.67, 1.40, and 1.33, respectively. For vibrational states, possessing both kinetic and potential energies, there are two degrees of freedom available for each state. For most gases at normal temperatures the relaxation times for the rotational states are very short. Usually only a few collisions are necessary to bring the rotational states into equilibrium with the translational ones. Relaxation times for most vibrational states are much longer, requiring many collisions. However, these states generally require higher temperatures to be significantly excited and, therefore, with a few exceptions are not important at normal temperatures.

Acoustic absorption due to a given internal state will have greatest importance when the period of the sound is close to the relaxation time for the state. Then, a significant phase angle exists between acoustic pressure and condensation. This leads to a pressure–condensation relation of the form (8.3.1) for monofrequency acoustic motion. Since these internal energy states affect the heat capacity C_V, we need a relationship between the heat capacity and the complex speed of sound. The speed of sound for a perfect gas is $c = \sqrt{\gamma r T_K}$. The ratio of complex heat capacities is $\gamma = \mathbf{C}_{\mathscr{P}}/\mathbf{C}_V$ and $\mathbf{C}_{\mathscr{P}} = \mathbf{C}_V + \mathscr{R}$. Thus, for a complex heat capacity $\mathbf{C}_V$ the speed of sound is also complex and can be written as

$$\frac{\mathbf{c}}{c} = \left(\frac{\boldsymbol{\gamma}}{\gamma}\right)^{1/2} = \left(\frac{1 + \mathscr{R}/\mathbf{C}_V}{\gamma}\right)^{1/2} \tag{8.6.2}$$

Obtaining $\mathbf{C}_V$ provides the form of $\mathbf{c}$ in (8.3.1), from which α and k follow as discussed in Section 8.3.

To proceed, it is necessary to develop a little nonequilibrium thermodynamics. Restrict attention to the ith degree of freedom. The rate of change of the energy stored in this degree of freedom is proportional to the difference between the energy that would be stored under equilibrium conditions $E_i(eq)$ and the amount E_i that is stored at some instant of time. This is expressed mathematically as

$$\frac{dE_i}{dt} = \frac{1}{\tau}\left[E_i(eq) - E_i\right] \tag{8.6.3}$$

where τ is the proportionality factor. If the system is changed instantaneously from a thermodynamic configuration for which the equilibrium energy stored is E_0 to one in which the amount stored will be $E_0 + \Delta E_i$, then solution for E_i gives

$$\begin{aligned} E_i &= E_0 + (1 - e^{-t/\tau})\,\Delta E_i \qquad t > 0 \\ &= E_0 \qquad\qquad\qquad\qquad\quad t < 0 \end{aligned} \tag{8.6.4}$$

and τ is now identified as the relaxation time. The *equilibrium* thermodynamic heat capacity associated with this degree of freedom is given by

$$\Delta E_i = C_i\,\Delta T_K \tag{8.6.5}$$

where $C_i = \frac{1}{2}\mathscr{R}H_i(T_K)$ and ΔT_K and ΔE_i are the equilibrium values of the change in temperature and internal energy, reached in the limit $t/\tau \to \infty$. Assume the process is not allowed to attain equilibrium so that $E_i(eq)$ is a fluctuating quantity to which E_i is always trying to adjust. For a monofrequency acoustic process the temperature fluctuation will be $\Delta T_K = T_0 \exp(j\omega t)$, so that

$$\mathbf{E}_i(eq) = E_0 + C_i T_0 e^{j\omega t} \tag{8.6.6}$$

The oscillatory (particular) solution of (8.6.3) with $\mathbf{E}_i(eq)$ given by (8.6.6) is

$$\mathbf{E}_i = E_0 + \frac{C_i}{1 + j\omega\tau} T_0 e^{j\omega t} \tag{8.6.7}$$

This can be rewritten as

$$\mathbf{E}_i = E_0 + \mathbf{C}_i T_0 e^{j\omega t}$$
$$\mathbf{C}_i = C_i/(1 + j\omega\tau) \tag{8.6.8}$$

with $\mathbf{C}_i$ defined as the *complex heat capacity* for this degree of freedom.

For N degrees of freedom that are fully excited and n degrees of freedom that are partially excited, the combined dynamic heat capacity becomes

$$\mathbf{C}_V = \mathcal{R}\left(\frac{N}{2} + \frac{1}{2}\sum_{i=1}^{n} \frac{H_i(T_K)}{1 + j\omega\tau_i}\right) \tag{8.6.9}$$

which reduces to (8.6.1) if $\omega\tau_i \ll 1$.

A classic example of a gas having only one molecular thermal relaxation is dry carbon dioxide gas at normal temperatures. Carbon dioxide is a linear molecule. It has three translational and two rotational degrees of freedom that are fully excited, and one vibrational mode that is partially excited. The vibrational motion consists of two degenerate modes (because of the linearity of the molecule), and each degenerate vibrational mode has two degrees of freedom. Thus, in (8.6.9) we have $N = 5, n = 4$, and all four terms in the summation are the same,

$$\mathbf{C}_V = \mathcal{R}\left(\frac{5}{2} + 2\frac{H(T_K)}{1 + j\omega\tau_M}\right) \tag{8.6.10}$$

where τ_M is the relaxation time. For notational simplicity, rewrite this as

$$\mathbf{C}_V = C_e + C_i/(1 + j\omega\tau_M)$$
$$C_e = \tfrac{5}{2}\mathcal{R} \tag{8.6.11}$$
$$C_i = 2\mathcal{R}H(T_K)$$

The equilibrium thermodynamic heat capacity C_V is the limit of $\mathbf{C}_V$ for $\omega\tau_M \to 0$,

$$C_V = C_e + C_i \tag{8.6.12}$$

With $\mathbf{C}_V$ determined, substitution into (8.6.2) yields

$$\mathbf{c} = \frac{c}{\sqrt{\gamma}}\left(1 + \frac{\mathcal{R}}{C_e + C_i/(1 + j\omega\tau_M)}\right)^{1/2} \tag{8.6.13}$$

Considerable algebraic manipulation then reveals that the absorption coefficient for this vibrational excitation is found from

$$\frac{\alpha_M}{\omega/c} = \frac{1}{2}\frac{\mathcal{R}C_i}{C_e(C_e + \mathcal{R})}\frac{\omega\tau_M}{1 + (\omega\tau_M)^2} \tag{8.6.14}$$

and the phase speed $c_p = \omega/k$ from

$$\frac{c_p}{c} = \frac{1}{\sqrt{\gamma}}\left(1 + \mathcal{R}\frac{C_V + C_e(\omega\tau_M)^2}{C_V^2 + C_e^2(\omega\tau_M)^2}\right)^{1/2} \tag{8.6.15}$$

For frequencies far below the relaxation frequency, the absorption is proportional to f^2. As the frequency increases, the absorption coefficient levels off, and above the relaxation frequency the absorption approaches a constant value. Note also that α_M is proportional to τ_M when $\omega\tau_M \ll 1$. This means that for α_M to be important compared to other absorption coefficients, τ_M must be sufficiently large. For low frequencies, the phase speed is greater than but close to the lossless limit,

$$\frac{c_p(0)}{c} = \frac{1}{\sqrt{\gamma}}\left(1 + \frac{\mathcal{R}}{C_V}\right)^{1/2} = 1 \tag{8.6.16}$$

For frequencies above the relaxation value,

$$\frac{c_p(\infty)}{c} = \left(\frac{1 + \mathcal{R}/C_e}{1 + \mathcal{R}/C_V}\right)^{1/2} \tag{8.6.17}$$

Thus, the phase speed c_p is always greater than c unless $C_i = 0$.

In graphing the measured absorption caused by a molecular thermal relaxation, it is customary to plot the absorption per wavelength $\alpha_M\lambda = 2\pi\alpha_M/k$ against frequency. When this is done, curves similar to Fig. 8.6.1 are obtained. Since the peak value of $\alpha_M\lambda$ occurs at $\omega = \omega_M = 1/\tau_M$, the relaxation frequency $f_M = \omega_M/2\pi$ can be found directly from such a plot. The maximum value of $\alpha_M\lambda$ is

$$\mu_{max} = (\alpha_M\lambda)_{f_M} = \frac{\pi}{2}\frac{\mathcal{R}C_i}{C_e(C_e + \mathcal{R})} \tag{8.6.18}$$

Experimental determination of μ_{max} provides the relationship between C_e and C_i.

Combination of (8.6.18) and (8.6.14) gives

$$\alpha_M\lambda = 2\mu_{max}\frac{f/f_M}{1 + (f/f_M)^2} \tag{8.6.19}$$

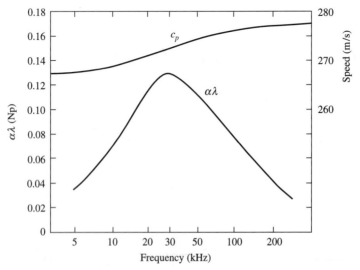

Figure 8.6.1   Absorption per wavelength and phase speed for CO_2 at 20°C.

$$\frac{\alpha_M}{f^2} = \frac{2\mu_{max}}{c} \frac{f_M}{f^2 + f_M{}^2} \qquad (8.6.20)$$

(Remember that both f_M and μ_{max} are functions of temperature.)

Measurements in dry carbon dioxide have shown good agreement with theory, with $\alpha_M\lambda$ peaking at about 30 kHz and α_M about 1200 times greater than the classical absorption coefficient at this frequency.

Equations (8.6.14)–(8.6.20) give the contribution to the absorption coefficient of a single molecular thermal relaxation. If more than one relaxation can be excited, the total absorption coefficient is essentially the sum (8.5.1) of the absorption coefficients calculated separately for each of the relaxations.

In certain cases, small concentrations of molecules of another species have considerable influence on the absorption coefficient of a gas. For example, the addition of water vapor to carbon dioxide gas has a profound effect. While the water vapor does not contribute any significant additional absorption mechanisms, the water molecules act as catalysts, lowering the average number of collisions for the transfer of kinetic energy into and out of the carbon dioxide vibrational states. This decreases τ_M and increases f_M. The presence of 1% water vapor in carbon dioxide shifts f_M from about 30 kHz to around 2 MHz. The absorption per wavelength at the relaxation frequency remains the same, but the decreased wavelength at the higher frequency greatly increases the absorption coefficient. At the relaxation frequency of the *moist* gas, the ratio of the absorption coefficients for moist and dry carbon dioxide gas is about 33. On the other hand, at frequencies far below the relaxation frequency of dry carbon dioxide, the absorption coefficient of the moist carbon dioxide is only 0.015 times that of the dry gas.

Another polyatomic gas that has been extensively studied is air.[3] Air consists of molecular oxygen and nitrogen with traces of other gases, including water vapor and carbon dioxide. The diatomic molecules and carbon dioxide each have two degrees of rotational freedom, fully excited at room temperatures. Water vapor has three rotational degrees, but because of its low concentration (even at high relative humidities) contributes only slightly to the rotational component of C_V. Figure 8.6.2 shows the absorption coefficients for different relative humidities in air, calculated from measured and assumed relaxation times and interaction rates. The appreciable increase in the absorption coefficient above that of the classical prediction for all frequencies below about 100 kHz is a consequence of molecular thermal relaxations. This excess absorption increases rapidly with temperature. Except in the driest air, water vapor appears to act as a catalyst, increasing the relaxation frequencies associated with the vibrational states of N_2 and O_2. Oxygen and water vapor collisions exciting the O_2 vibrational state assume greatest importance for absorption at frequencies lying between about 1 and 10 kHz and for relative humidities above a few percent. For smaller humidities this contribution to the total absorption becomes less important than that from the classical thermoviscous mechanisms. See Fig. 8.6.2. The absorption arising from excitation of the N_2 vibrational state by collisions between N_2 and water vapor dominates the total absorption at frequencies below about 1 kHz. In very dry air the collisions with water vapor again become unimportant, and collisions of N_2 with CO_2 become important. Quantitative evaluation is quite involved and left to the reference.

[3]Bass, Bauer, and Evans, *J. Acoust. Soc. Am.*, **52**, 821 (1972). Bass, Sutherland, Piercy, and Evans, *Absorption of Sound by the Atmosphere*, Physical Acoustics XVII, ed. Mason, Academic Press (1984). Bass, Sutherland, Zuckerwar, Blackstock, and Hester, *J. Acoust. Soc. Am.*, **97**, 680 (1995).

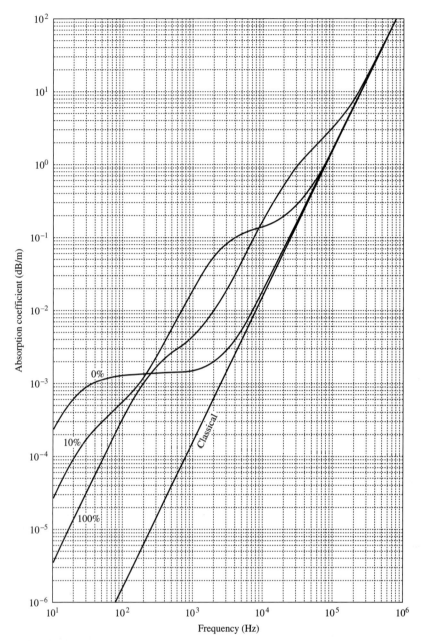

Figure 8.6.2   Absorption of sound in air at 20°C and 1 atm for various relative humidities. (After Bass et al., *op. cit.*)

8.7 ABSORPTION IN LIQUIDS

One type of excess absorption occurring in liquids is that associated with thermal conductivity. Thermal relaxation theory has been applied successfully to explain the excess absorption observed in many nonassociated nonpolar liquids such as carbon disulfide, benzene, and acetone. For instance, it explains the behavior in acetone where the measured absorption is some 4.3 times the value predicted by classical theory. (See Table 8.5.1.)

Thermal relaxation, however, has not been successful in accounting for the observed excess absorption in associated polar liquids, such as the alcohols and water. It appears that in these liquids the intermolecular forces are so strong that they cause any existing thermal relaxation time to be very short. Since the magnitude of the absorption coefficient is proportional to the relaxation time, the resulting absorption associated with the process is small. That the excess absorption in water does not result from thermal relaxation has been demonstrated through measurements made in the vicinity of 4°C.[4] If the measured excess absorption in water were caused by thermal relaxation, then this excess should vanish at 4°C, where the coefficient of thermal expansion is zero. At this temperature, compression or rarefaction will not change the temperature and thus thermal relaxation cannot take place. Measurements in water in the vicinity of 4°C give no evidence of any decrease in absorption at this temperature. Some other relaxation mechanism must be found to explain why the measured absorption is some three times the classical value. One such explanation is offered by a theory of *structural relaxation*, applied to water by Hall.[5] This theory attributes the excess absorption in water to a structural change directly related to a volume change (and not to a temperature change). Water is assumed to be a two-state liquid. The state of lower energy is the normal state and the state of higher energy is one in which the molecules have a more closely packed structure. Under ordinary static conditions of equilibrium, most of the molecules are in the first energy state. However, the passage of a compressional wave is assumed to promote the transfer of molecules from the more open first state to the more closely packed second state. The time delays in this process and in its reversal lead to a relaxational dissipation of acoustic energy. A detailed analysis indicates that structural relaxation may be taken into consideration by assuming the existence of a nonvanishing coefficient of bulk viscosity. The resulting expression for total absorption in water becomes

$$\alpha = (\omega^2/2\rho_0 c^3)(\tfrac{4}{3}\eta + \eta_B) \tag{8.7.1}$$

Direct measurement[6] of η_B for water indicates that it is approximately three times η. The resultant calculation of α is in satisfactory agreement with the measured value (Table 8.5.1).

Another liquid that has generated considerable investigation is seawater. Figure 8.7.1 displays the absorption of acoustic waves in freshwater and seawater at 5°C. The pronounced difference between the two curves at frequencies below 500 kHz is evidence of additional absorptive mechanisms in seawater. It is natural to attribute these to the dissolved salts. Laboratory measurements[7] have shown that the excess acoustic absorption in the midfrequency range is caused by dissolved magnesium sulfate ($MgSO_4$). This is a *chemical* relaxation. The acoustic process changes the concentrations of the associated and dissociated $MgSO_4$ ions. There is a relaxation time for the process, and hence an absorption.

Measurements of the absorption coefficient in seawater at lower frequencies, while difficult because of the small values encountered (0.001 dB/km at 100 Hz), reveal a second relaxation mechanism active below about 1 kHz. This has been shown to be a chemical relaxation involving boric acid. Although the boric

[4]Fox and Rock, *Phys. Rev.*, **70**, 68 (1946).
[5]Hall, *Phys. Rev.*, **73**, 775 (1948).
[6]Liebermann, *Phys. Rev.*, **75**, 1415 (1949).
[7]Leonard, Combs, and Skidmore, *J. Acoust. Soc. Am.*, **21**, 63 (1949).

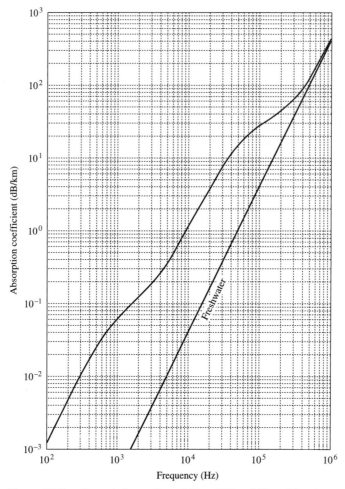

Figure 8.7.1   Sound absorption at $T = 5°C$ and $Z = 0$ km in freshwater and seawater (pH = 8, $S = 35$ ppt).

acid concentration in the ocean is only about 4 ppm (parts per million), the associated absorption at low frequencies is nearly 300 times greater than that of freshwater and 20 times greater than that of seawater without the boric acid. These two chemical relaxation mechanisms and the absorption mechanisms active in freshwater yield an absorption coefficient for seawater

$$a = \left(\frac{A}{f_1^2 + f^2} + \frac{B}{f_2^2 + f^2} + C \right) f^2 \qquad \text{dB/km} \qquad (8.7.2)$$

$$= a(\text{boric acid}) + a(\text{MgSO}_4) + a(\text{H}_2\text{O})$$

where f_1 and f_2 are the temperature-dependent relaxation frequencies associated with boric acid and $MgSO_4$, respectively and all frequencies are in Hz. The values of A, B, and C depend on temperature and hydrostatic pressure. Figure 8.7.2 shows the contribution to the total absorption of the two relaxation processes. Note that as the frequency exceeds that of a relaxation, the contribution of the process becomes increasingly less important.

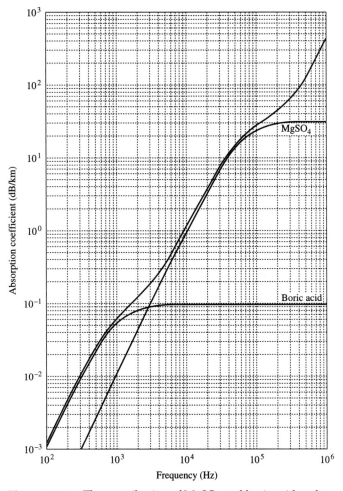

Figure 8.7.2 The contribution of $MgSO_4$ and boric acid to the total sound absorption in seawater (pH = 8, S = 35 ppt, T = 5°C, and Z = 0 km).

Since A and B must go to zero for freshwater, it is not unreasonable to assume they depend linearly on the salinity. Extended analyses of large amounts of experimental data for seawater have been made.[8] A simple approximation is

$$f_1 = 780 \exp(T/29)$$

$$f_2 = 42000 \exp(T/18)$$

$$A = 0.083(S/35) \exp\left[T/31 - Z/91 + 1.8(\mathrm{pH} - 8)\right] \qquad (8.7.3)$$

$$B = 22(S/35) \exp(T/14 - Z/6)$$

$$C = 4.9 \times 10^{-10} \exp(-T/26 - Z/25)$$

[8]Fisher and Simmons, *J. Acoust. Soc. Am.*, **62**, 558 (1977). Mellen, Scheifele, and Browning, *Global Model for Sound Absorption in Sea Water*, NUSC Scientific and Engineering Studies, New London, CT (1987). Fisher and Worcester, *Encyclopedia of Acoustics*, Chapt. 35, Wiley (1997).

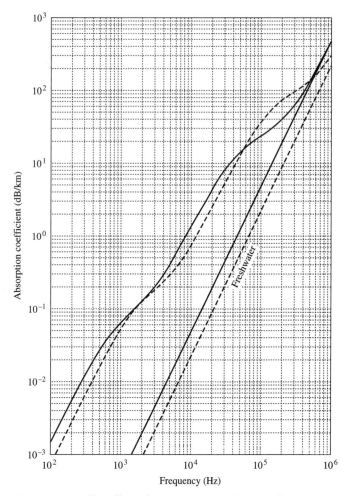

Figure 8.7.3   The effect of temperature on the sound absorption in seawater. The solid line is for $T = 0°C$ and the dashed line for $T = 20°C$ (pH = 8, $S = 35$ ppt, and $Z = 0$ km).

where T is in °C, the salinity S is in parts per thousand, and the effect of hydrostatic pressure is expressed in terms of the depth Z in km below the surface of the ocean. These estimates are accurate within a few percent for those combinations of parameters occurring naturally in the oceans and for depths less than 6 km. Figure 8.7.3 shows the effects of temperature on the absorption in seawater. It is left as an exercise to show that the effects of pressure and salinity are much less pronounced than that of temperature.

Other relaxations appear to be present, and experimental and theoretical investigations continue. For frequencies below about 1 kHz, scattering appears to be more important than absorption in attenuating underwater sound.

*8.8 VISCOUS LOSSES AT A RIGID WALL

In this section we develop a simple model for the viscous losses arising from a sound wave grazing a wall. Applications include the propagation losses experienced by sound in pipes and ducts of constant cross section. The approach is based on assuming a simple lossless plane wave, then inserting a boundary perpendicular to the surfaces of constant phase and solving for the additional acoustic field necessary to satisfy the boundary conditions.

As discussed after (8.2.1), viscous losses resulting from boundaries involve the curl term in the Navier–Stokes equation. Losses in the bulk of the fluid have already been developed, and we have observed that small independent dissipative mechanisms can be treated separately and, when taken together, yield losses that are additive. Consequently, in studying the absorption arising from shear at a boundary, we can discard $(4\eta/3 + \eta_B)\nabla(\nabla \cdot \vec{u})$ but retain $\eta\nabla \times \nabla \times \vec{u}$. The Navier–Stokes equation then becomes

$$\rho_0 \frac{\partial \vec{u}}{\partial t} + \nabla p = -\eta\nabla \times \nabla \times \vec{u} \tag{8.8.1}$$

Let an acoustic plane wave exist in the positive z space with propagation vector $\vec{k}$ parallel to the x axis. This *primary* wave has a particle velocity with only an x component u_x, which is a function of just x and t, and an acoustic pressure also a function of just x and t. This wave satisfies Euler's equation [(8.8.1) with $\eta = 0$] exactly. Now, introduce a rigid wall in the region $z \le 0$ with its boundary at $z = 0$. The presence of the wall introduces an additional particle velocity $u'\hat{x}$ parallel to its surface. This *secondary* wave is a function of x, z, and t. The sum of primary and secondary waves must satisfy (8.8.1) and the appropriate boundary conditions. In the presence of viscosity, the boundary condition at the stationary wall is "no slip," meaning that the velocity of the fluid must vanish on it. Furthermore, u' must vanish at large z so that only the primary wave remains. Thus, if we write the total particle velocity as

$$\vec{u} = u\hat{x} = (u_x + u')\hat{x} \tag{8.8.2}$$

then $u' = -u_x$ at the wall and $u' \to 0$ at large z. Substitution of (8.8.2) into (8.8.1) and recognition that the terms involving the original acoustic wave have $\nabla \times (u_x\hat{x}) = 0$ and must sum to zero on the left side of (8.8.1) gives

$$\rho_0 \frac{\partial u'}{\partial t} + \frac{\partial p'}{\partial x} = \eta \frac{\partial^2 u'}{\partial z^2}$$

$$\frac{\partial p'}{\partial y} = 0 \tag{8.8.3}$$

$$\frac{\partial p'}{\partial z} = \eta \frac{\partial^2 u'}{\partial x \, \partial z}$$

where p' is the pressure associated with the particle velocity $u'\hat{x}$. If $\partial p'/\partial x$ can be neglected (discussed below), the first equation can be approximated as

$$\frac{\partial u'}{\partial t} = \frac{\eta}{\rho_0} \frac{\partial^2 u'}{\partial z^2} \tag{8.8.4}$$

This is a one-dimensional diffusion equation for u', the counterpart in viscosity to (8.4.4) for thermal conductivity. Assume a frequency dependence $\exp(j\omega t)$. The complex solution for $\mathbf{u}'$ that satisfies the boundary conditions can be seen by direct substitution to be

$$\mathbf{u}' = -\mathbf{u}_x e^{-(1+j)z/\delta}$$

$$\delta = \sqrt{2\eta/\rho_0\omega} \tag{8.8.5}$$

The quantity δ is the *viscous penetration depth*, the *acoustic boundary layer thickness*, or more simply the *skin depth*. Thus, for $\mathbf{u}_x = U_0 \exp[j(\omega t - kx)]$,

$$\mathbf{u}' = -U_0 e^{-z/\delta} e^{j(\omega t - kx - z/\delta)} \tag{8.8.6}$$

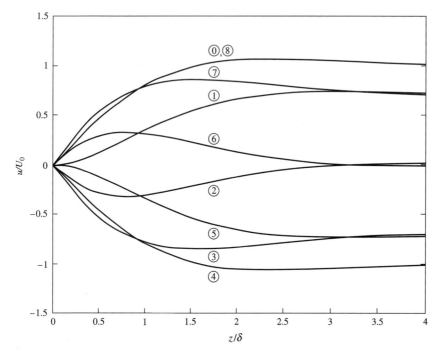

Figure 8.8.1   Profiles of the normalized particle speed $\text{Re}\{\mathbf{u}(z)\}/U_0$ in a viscous fluid in the vicinity of a wall for one cycle of the primary sound field at $x = 0$. The speed far from the wall is $U_0 \cos \omega t$ and that at the wall vanishes. Curve ⓪ $\omega t = 0$, ① $\pi/4$, ② $\pi/2$, ③ $3\pi/4$, ④ π, ⑤ $5\pi/4$, ⑥ $3\pi/2$, ⑦ $7\pi/4$, ⑧ 2π.

The contribution $\mathbf{u}'$ is a wave that attenuates in the z direction as it propagates from the boundary into the fluid with propagation vector components k in the x direction and $1/\delta$ in the z direction. Note that the attenuation coefficient and the propagation vector component in the $+z$ direction are both given by $1/\delta$. Some representative profiles for a full cycle of the primary field are shown in Fig. 8.8.1. For z exceeding a couple of skin depths, the total field reduces to the primary field.

Neglecting $\partial p'/\partial x$ in (8.8.3) is justified (see Problem 8.8.2) when we have $k\delta \ll 1$ or

$$\delta/\lambda \ll 1 \tag{8.8.7}$$

The acoustic wavelength must be much greater than the skin depth. For air with $\rho_0 \approx 1.3$ kg/m^3, $c \approx 340$ m/s, and $\eta \approx 1.7 \times 10^{-5}$ Pa·s, the inequality holds for all frequencies below several hundred MHz.

*8.9 LOSSES IN WIDE PIPES

Assume a monofrequency plane wave in a pipe of radius a such that $a \gg \delta$. Since the radius of curvature of the pipe wall is much greater than the skin depth, the results of Section 8.8 can be applied with little error.

(a) Viscosity

In the absence of a boundary layer, the force accelerating a disk of the fluid with thickness Δx and cross-sectional area $S = \pi a^2$ in the x direction is $\mathbf{f} = -S(\partial \mathbf{p}/\partial x)\,\Delta x$. The mechanical impedance this element presents to the force is $\mathbf{Z}_m = \mathbf{f}/\langle\mathbf{u}\rangle_S$, where

$$\langle\mathbf{u}\rangle_S = \frac{\mathbf{u}_x}{\pi a^2}\int_S \left(1 - e^{-(1+j)z/\delta}\right)dS \tag{8.9.1}$$

The first term integrates to πa^2 and the integration of the second term can be approximated,

$$\langle\mathbf{u}\rangle_S \approx \mathbf{u}_x\left(1 - \frac{2\pi a}{\pi a^2}\int_0^\infty e^{-(1+j)z/\delta}\,dz\right)$$
$$\approx \mathbf{u}_x\left(1 - 2\frac{\delta}{a}\frac{1}{1+j}\right) \tag{8.9.2}$$

The mechanical impedance of the fluid disk with the viscous boundary layer is

$$\mathbf{Z}_m = \frac{\mathbf{f}}{\langle\mathbf{u}\rangle_S} = \frac{j\omega m}{1 - 2(\delta/a)/(1+j)}$$
$$\approx \omega m\delta/a + j\omega m(1 + \delta/a) \tag{8.9.3}$$

with the help of $\delta/a \ll 1$. The fluid element has a mechanical resistance $R_m = \omega m(\delta/a)$, which accounts for the friction-like losses introduced by the viscous forces in the boundary layer. Also, the apparent mass of the fluid disk is increased slightly by the boundary layer.

(1) STANDING WAVES IN A PIPE. For a plane standing wave in a pipe, the particle velocity in the absence of any boundary is the real part of

$$\mathbf{u}_x\hat{x} = \hat{x}U\sin kx\,e^{j\omega t} \tag{8.9.4}$$

The instantaneous power dissipated by the motion of the fluid element in the presence of the boundary is $R_m\langle u\rangle_S^2$. Converting to unit volume, approximating $\langle u\rangle_S^2$ with $\langle u_x\rangle_S^2$, and integrating over the period T gives the energy density loss at each position x. Then, averaging over a wavelength gives the average energy density lost during one cycle of motion,

$$\mathscr{E}_w \approx \frac{1}{\lambda}\int_0^\lambda\int_0^T \rho_0\omega\frac{\delta}{a}\langle u_x\rangle_S^2\,dt\,dx = \frac{\pi}{2}\rho_0 U^2\frac{\delta}{a} \tag{8.9.5}$$

The total energy density is found from (5.8.7) and (5.8.8) using the pressure and particle velocity $u_x\hat{x}$ of the standing wave,

$$\mathscr{E} = \tfrac{1}{4}\rho_0 U^2 \tag{8.9.6}$$

The ratio $\mathscr{E}_w/\mathscr{E} = 2\pi/Q_{w\eta}$ gives the quality factor resulting from viscous losses at the wall,

$$Q_{w\eta} = a/\delta = a\sqrt{\rho_0\omega/2\eta} \tag{8.9.7}$$

and then use of (1.10.7) and $\alpha/k \approx \beta/\omega$ yields the absorption coefficient describing the boundary layer losses,

$$\alpha_{w\eta} = \frac{1}{ac}\left(\frac{\eta\omega}{2\rho_0}\right)^{1/2} \tag{8.9.8}$$

Note that the absorption coefficient for wall losses increases as $\sqrt{\omega}$. Since the thermoviscous absorption outside the boundary layer increases as ω^2, as the frequency is increased wall losses will eventually become less important than mainstream losses.

(2) Traveling Waves in a Pipe. A plane wave traveling in a pipe has particle velocity

$$u_x \hat{x} = \hat{x} U \cos(\omega t - kx) \tag{8.9.9}$$

Calculation of the energy density lost each cycle gives twice the result as before, but the energy density of a traveling wave is twice that of a standing wave with the same amplitude. The absorption coefficient is unchanged, given by (8.9.8).

(b) Thermal Conduction

Calculation of the absorption coefficient for thermal conductivity losses resulting from the isothermal wall of a pipe of radius $a \gg \delta$ is straightforward in concept but rather laborious in execution. The temperature field accompanying a traveling plane wave of pressure amplitude $P = \rho_0 c U$ is given by (8.4.1):

$$\mathbf{T} = T_{eq} + T_{eq}(\gamma - 1)\mathbf{s}$$
$$\mathbf{s} = (U/c)e^{j(\omega t - kx)} \tag{8.9.10}$$

(The subscript K on temperature has again been dropped for simplicity.) The presence of the pipe wall, assumed to be isothermal, requires an additional temperature field $\mathbf{T'}$ that must maintain the equilibrium temperature T_{eq} at the pipe wall and go to zero for large distance z away from the wall. The behavior of the temperature in this boundary layer region is described by the diffusion equation (8.4.4). When this is applied to the combined field $\mathbf{T} + \mathbf{T'}$ and the acoustic wavelength is assumed to be very large with respect to the skin depth of the thermal boundary layer, we are left with

$$\frac{\partial T'}{\partial t} = \frac{\kappa}{c_{\mathcal{P}} \rho_0} \frac{\partial^2 T'}{\partial z^2} \tag{8.9.11}$$

This is analogous with (8.8.4) and has analogous boundary conditions. Following the same mathematical steps, we find

$$\mathbf{T} - T_{eq} = (1 - e^{-(1+j)z/\delta_\kappa})T_{eq}(\gamma - 1)\mathbf{s}$$
$$\delta_\kappa = \sqrt{2\kappa/c_{\mathcal{P}}\rho_0\omega} \tag{8.9.12}$$

where δ_κ is the skin depth for the thermal boundary layer. Note that the skin depths for viscosity and thermal conduction are related by the Prandtl number,

$$\delta/\delta_\kappa = \sqrt{\mathrm{Pr}} \tag{8.9.13}$$

Calculation of the acoustic energy loss in the boundary layer from thermal conduction proceeds from (8.4.5) as before, leading to (8.4.7). In this case, however, the relevant volume of integration is a pipe-like sheath one wavelength long, of circumference $2\pi a$, and with laminar thickness z large enough to encompass the boundary layer. Because the boundary layer is so strongly damped in the z direction, the limit of integration over z can be extended to infinity. This gives us

$$\int_V \frac{\Delta q}{\Delta t}\, dV = 2\pi a \frac{\kappa}{T_{eq}} \int_0^\lambda \int_0^\infty \left(\frac{\partial T}{\partial z}\right)^2 dz\, dx \tag{8.9.14}$$

Note that T must be the real part of $\mathbf{T}$. The result, after some fairly messy manipulation, gives the change of the acoustic energy per unit time in this length λ of the pipe,

$$\frac{dE}{dt} = -\int_V \frac{\Delta q}{\Delta t}\, dV = -\pi a\lambda \frac{\kappa}{T_{eq}\delta_\kappa}\left(T_{eq}(\gamma - 1)\frac{U}{c}\right)^2 \tag{8.9.15}$$

The absorption coefficient is found from $(dE/dt)/E = -2\alpha_{w\kappa}c$. For this traveling wave, the acoustic energy in this length is $E = \frac{1}{2}\rho_0 U^2 \pi a^2 \lambda$. Furthermore, recall that for a perfect gas $T_{eq}(\gamma - 1) = c^2/c_{\mathscr{P}}$, so that the absorption coefficient for thermal conduction losses at the pipe wall is

$$\alpha_{w\kappa} = \frac{1}{ac}(\gamma - 1)\left(\frac{\kappa\omega}{2\rho_0 c}\right)^{1/2} = \frac{1}{ac}\frac{(\gamma - 1)}{\sqrt{Pr}}\left(\frac{\eta\omega}{2\rho_0}\right)^{1/2} \tag{8.9.16}$$

Comparison with (8.9.8) shows that the two coefficients are proportional,

$$\alpha_{w\kappa}/\alpha_{w\eta} = (\gamma - 1)/\sqrt{Pr} \tag{8.9.17}$$

A little thought and review of the integration over wavelength shows that $\alpha_{w\kappa}$ will have the same value for both standing and traveling waves. Because Q is inversely proportional to the absorption coefficient, it is straightforward to show that the contribution to the quality factor resulting from thermal conduction losses is

$$Q_{w\eta} = a\left(\frac{\rho_0\omega}{2\eta}\right)^{1/2}\frac{\sqrt{Pr}}{\gamma - 1} \tag{8.9.18}$$

(c) The Combined Absorption Coefficient

The combined absorption coefficient for wall losses is

$$\alpha_w = \alpha_{w\eta} + \alpha_{w\kappa} = \frac{1}{ac}\left(\frac{\eta\omega}{2\rho_0}\right)^{1/2}\left(1 + \frac{\gamma - 1}{\sqrt{Pr}}\right) \tag{8.9.19}$$

The presence of the viscous boundary layer also modifies the phase speed of the acoustic wave. Reference to (8.9.3) shows that not only does the viscosity at the wall introduce a resistance, but also the reactance is changed so that the apparent mass of the fluid disk is slightly larger. This is equivalent to the fluid having a slightly greater density $\rho_\eta = \rho_0(1 + \delta/a)$. This affects the phase speed c_p. For a given adiabatic compressibility, the speed of sound for a fluid depends inversely on the square root of the density. This means that the phase speed corrected for the viscous boundary layer is $c_{p\eta}/c = \sqrt{\rho_0/\rho_\eta} = 1 - \frac{1}{2}\delta/a = 1 - \alpha_{w\eta}/k$. If these arguments are applied to the thermal boundary layer, similar conclusions are reached. The temperature fluctuations in the boundary layer are proportional to the condensation, so calculating the effective density exactly matches the calculation of the effective magnitude of the particle velocity. The attendant correction to the density from the thermal boundary layer is $\rho_\kappa = \rho_0(1 + \delta_\kappa/a)$, and this adds its own correction to the phase speed. Through $O(\alpha/k)$, the total corrected phase speed c_p becomes

$$c_p/c = 1 - \frac{1}{2}(\delta + \delta_\kappa)/a = 1 - \alpha_w/k \tag{8.9.20}$$

Since α_w increases as $\sqrt{\omega}$ but k increases as ω, the phase speed approaches the free field value asymptotically from below.

As an example, consider a traveling sawtooth wave. Because this wave has a spectrum of frequencies, it will distort as it propagates, with higher frequencies tending to travel faster than lower frequencies. The absorption is greater for the higher frequencies, so the

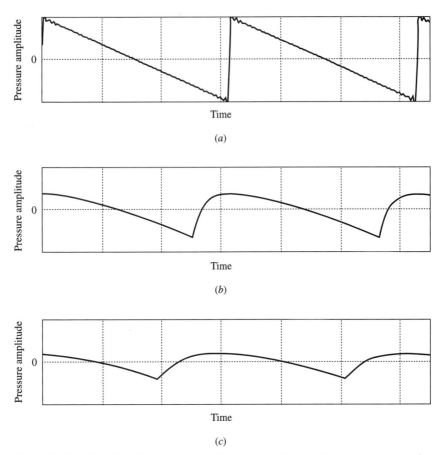

Figure 8.9.1   The time history of an initially sawtooth traveling wave at several positions down a pipe with wall losses predominating: (*a*) $\alpha x = 0$, (*b*) $\alpha x = 0.5$, and (*c*) $\alpha x = 1.5$, where α is the spatial decay coefficient of the fundamental. (One hundred harmonics are included in the calculation.)

waveform will have its peaks rounded off, but the maxima and minima of the harmonics migrate forward, leading to a sharper rise and gentler fall to the waveform. See Fig. 8.9.1.

*8.10   ATTENUATION IN SUSPENSIONS

When a fluid contains inhomogeneities such as fog droplets, suspended particles, bubbles, thermal microcells, or regions of turbulence, acoustic energy is lost from a sound beam faster than in the homogeneous medium. The excess attenuation in suspensions arises from (1) absorption—the conversion of acoustic energy into thermal energy—and (2) scattering—the reradiation of incident acoustic energy out of the incident beam. In this section, we will restrict discussion to the cases of liquid droplets in a gas and air bubbles in water, but the analysis will be general enough to cover many similar cases.

The combined effects of scattering and absorption can be described by an *extinction cross section* σ, where σ is an *effective* area whose product with the incident intensity is equal to the power lost from the sound beam. Assume that the particle density is N particles per unit volume, that they all have an extinction cross section σ, and that they do not "shadow" each other. In each incremental distance dx along the path of the incident sound beam, there will be a total cross section of $N\sigma\,dx$ for each unit cross-sectional area of the beam. The fractional intensity that is "intercepted" by the particles is $dI/I = N\sigma\,dx$ so that

$$I = I_0 e^{-N\sigma x} \tag{8.10.1}$$

For the cases of interest, the intensity is proportional to the square of the pressure amplitude P, so $P = P_0\exp(-\alpha x)$, where $\alpha = N\sigma/2$ Np/m or $a = 4.35N\sigma$ dB/m. If, instead of only one population of particles all with the same cross section, we had N particles per unit volume, each with its own cross section σ_i, then $N\sigma$ would be replaced with the summation of the individual scatterers, each of cross section σ_i, within the volume. The absorption coefficient is generalized to

$$\alpha = \frac{1}{2}\sum_{i=1}^{N}\sigma_i \tag{8.10.2}$$

Depending on frequency, each of these cross sections may be either equal to, less than, or greater than the *geometric cross section* πa_i^2, where a_i is the radius of the ith particle.

(a) Fogs

Fog and smoke particles can significantly affect acoustic propagation through the atmosphere. In the immediate neighborhood of suspended particles, thermoviscous losses occur in addition to those in the homogeneous fluid. The equilibrium between each droplet and the surrounding gas is altered by the sound wave and adjustment lags behind the perturbation. This leads to relaxation effects. We will consider two important relaxation effects that contribute to the absorption of a sound wave by a fog: viscous damping of the gross motion of the droplet and thermal damping from heat conduction into and out of the droplet.

Assume a fog consists of identical liquid droplets in a gas through which a plane wave travels in the $+x$ direction. At a coordinate x let the gas have complex particle speed $\mathbf{u}(t) = U\exp(j\omega t)$. Let the gas have density ρ_0, speed of sound c, and shear viscosity η. Let there be N droplets of water per unit volume of gas, each droplet with radius a, density ρ_d, and mass m_d. As the gas oscillates back and forth, viscous forces attempt to carry the droplet along with a speed $d\xi/dt$. In the limit of small *Reynolds number* Re, the viscous force is given by the *Stokes relation* $6\pi\eta aV$, where $V = [d\xi/dt - u(t)]$ is the relative speed of the droplet with respect to the gas.[9] [The Reynolds number of a flow is defined by the dimensionless ratio $VL/(\eta/\rho)$, where V and L are the characteristic speed and length of the object in the flow and η/ρ is the *kinematic viscosity* of the fluid surrounding the object.] The equation of motion for the displacement ξ of the droplet when the Stokes relation holds is then

$$m_d\frac{d^2\xi}{dt^2} + 6\pi\eta a\left(\frac{d\xi}{dt} - Ue^{j\omega t}\right) = 0 \tag{8.10.3}$$

Solution for the velocity of the droplet in the x direction is immediate,

$$\frac{d\xi}{dt} = \frac{1}{1 + j\omega\tau_{f\eta}}Ue^{j\omega t}$$

$$\tau_{f\eta} = \tfrac{2}{9}\rho_d a^2/\eta \tag{8.10.4}$$

where $\tau_{f\eta}$ is the relaxation time for viscous damping. The absorption coefficient for viscous losses is found by calculating the work done per unit time by the viscous forces on the N

---

[9]Stokes, *Trans. Cambridge Philos. Soc.*, **9**(2), 8 (1851). Or, see any good book on fluid dynamics.

droplets in each unit volume divided by the energy density of the acoustic wave,

$$2\alpha_{f\eta}c = \frac{\langle 6\pi\eta aN\left[(d\xi/dt) - u(t)\right]^2 \rangle_t}{\frac{1}{2}\rho_0 U^2} = \frac{6\pi\eta aN}{\rho_0}\frac{(\omega\tau_{f\eta})^2}{1 + (\omega\tau_{f\eta})^2} \tag{8.10.5}$$

If we define r_m as the ratio of the total mass of water droplets per unit volume of gas to the density of the gas, and recognize that in all fogs of practical interest the density ρ_f of the fog is very similar to that of the gas alone, then

$$r_m = \tfrac{4}{3}\pi a^3 \rho_d N/\rho_0 \approx \rho_f/\rho_0 - 1 \tag{8.10.6}$$

With the approximation $\omega/c \approx k$,

$$\alpha_{f\eta}/k = \tfrac{1}{2}r_m\omega\tau_{f\eta}/[1 + (\omega\tau_{f\eta})^2] \tag{8.10.7}$$

which has the same dependence on $\omega\tau$ as has α_M/k in (8.6.14).

As the temperature of the gas fluctuates with the density fluctuations, the droplets heat and cool, disturbing the temperature field in the surrounding gas. The thermal conductivity of the liquid in the droplets is much greater than that of the gas, so we can assume that each droplet has uniform temperature that is a function of time. Let the ambient temperature of the gas at a large distance from a droplet be T_∞ and the internal temperature of the droplet be T_d. The change $\Delta q_d/\Delta t$ in the internal energy per unit volume of the droplet is related to the change of its internal temperature by (8.4.3),

$$\frac{\Delta q_d}{\Delta t} = c_{\mathcal{P}d}\rho_d\frac{dT_d}{dt} \tag{8.10.8}$$

where $c_{\mathcal{P}d}$ is the specific heat at constant pressure for the droplet. This rate of energy loss from the surrounding gas must determine its local temperature field T by (8.4.4),

$$c_{\mathcal{P}d}\rho_d\frac{dT_d}{dt} = \kappa\nabla^2 T \tag{8.10.9}$$

with κ the thermal conductivity of the gas. This is *Poisson's equation*. The left side is the equivalent of a charge density distributed uniformly throughout the volume of the droplet and the temperature T is the equivalent of the electrical potential. By analogy, solution for the temperature throughout the volume exterior to the droplet with the boundary condition that the temperature approach T_∞ at large distances is

$$T(r) = T_\infty - \frac{c_{\mathcal{P}d}\rho_d a^3}{3\kappa r}\frac{dT_d}{dt} \qquad (r \geq a) \tag{8.10.10}$$

(The mass of the droplet has been expressed in terms of its density and volume.) The second boundary condition is $T(a) = T_d$ and this yields the differential equation for the droplet temperature,

$$\frac{dT_d}{dt} = \frac{3\kappa}{c_{\mathcal{P}d}\rho_d a^2}(T_\infty - T_d) \tag{8.10.11}$$

In the presence of a plane traveling wave of particle velocity amplitude U and angular frequency ω, the ambient temperature is

$$\mathbf{T}_\infty = T_{eq} + T_{eq}(\gamma - 1)(U/c)e^{j\omega t} \tag{8.10.12}$$

and comparison of (8.10.11) and (8.10.12) with (8.6.3), (8.6.6), and (8.6.7) shows that the droplet temperature is

$$\mathbf{T}_d = T_{eq} + \frac{T_{eq}(\gamma - 1)U/c}{1 + j\omega\tau_{f\kappa}}e^{j\omega t}$$

$$\tau_{f\kappa} = \tfrac{1}{3}c_{\mathscr{P}d}\rho_d a^2/\kappa$$

(8.10.13)

where $\tau_{f\kappa}$ is the relaxation time for the thermal losses. Calculation of the absorption coefficient measuring the thermal losses resulting from the droplets proceeds in much the same way as the calculation for bulk thermal losses in Section 8.4. The complex temperature field external to the droplet is found from substituting (8.10.12) and (8.10.13) into (8.10.10),

$$\mathbf{T}(r) = T_\infty - T_{eq}(\gamma - 1)\frac{U}{c}\frac{a}{r}\frac{j\omega\tau_{f\kappa}}{1 + j\omega\tau_{f\kappa}}e^{j\omega t}$$

(8.10.14)

The gradient of the real temperature $T(r)$ is substituted into (8.4.7) and the volume taken over all space *exterior* to the droplet and averaged over one period of the motion. This, multiplied by N, the number of droplets per unit volume, and then divided by the energy density of the plane wave, $\tfrac{1}{2}\rho_0 U^2$, gives

$$\frac{1}{\mathscr{E}}\frac{d\mathscr{E}}{dt} = -2\alpha_{f\kappa}c = -\frac{4\pi aN}{\rho_0}(\gamma - 1)\frac{\kappa}{c_{\mathscr{P}}}\frac{(\omega\tau_{f\kappa})^2}{1 + (\omega\tau_{f\kappa})^2}$$

(8.10.15)

Use of (8.10.6), the relaxation time from (8.10.13), and $T_{eq}(\gamma - 1) = c^2/c_{\mathscr{P}}$ for a perfect gas yields

$$\frac{\alpha_{f\kappa}}{k} = \tfrac{1}{2}r_m r_c(\gamma - 1)\frac{\omega\tau_{f\kappa}}{1 + (\omega\tau_{f\kappa})^2}$$

$$r_c = c_{\mathscr{P}d}/c_{\mathscr{P}}$$

(8.10.16)

The total absorption arising from these two mechanisms is simply the sum,

$$\alpha_f = \alpha_{f\eta} + \alpha_{f\kappa}$$

(8.10.17)

Calculation of the phase speed for frequencies well below the relaxation values can be accomplished fairly simply by finding the modified adiabat for the fog resulting from the effective density ρ_f and effective ratio of specific heats γ_f of the fog. Then, the phase speed is found by using $c^2 = \gamma\mathscr{P}_0/\rho$ and $c_p^2 = \gamma_f\mathscr{P}_0/\rho_f$. For low frequencies, the droplets are carried along with the motion and can be considered as simply a more massive constituent of the gas. Under this assumption, the effective density is simply the static value $\rho_f = \rho_0(1 + r_m)$. The modified ratio of specific heats can be obtained similarly. At frequencies much less than the relaxation values, the liquid in the droplets can be considered to be essentially in thermal equilibrium with the gas. With these two approximations, the effective specific heats of the fog are found from

$$\rho_f c_{\mathscr{P}f} = \rho_0 c_{\mathscr{P}} + \rho_0 r_m c_{\mathscr{P}d}$$

$$\rho_f c_{Vf} = \rho_0 c_V + \rho_0 r_m c_{Vd}$$

(8.10.18)

In liquids the ratio of specific heats is nearly unity, so $c_{\mathscr{P}d}/c_{Vd} \approx 1$, and use of $r_m r_c \ll 1$ gives

$$\gamma_f/\gamma = 1 - (\gamma - 1)r_m r_c$$

(8.10.19)

for low frequencies. Then, the phase speed c_p is

$$c_p/c = [(\rho_0/\rho_f)(\gamma_f/\gamma)]^{1/2} \approx 1 - \tfrac{1}{2}r_m[1 + (\gamma - 1)r_c] \qquad (\omega \ll \omega_\eta \text{ and } \omega_\kappa) \quad (8.10.20)$$

A more general derivation,[10] of substantially greater difficulty, shows that the phase speed is given by

$$\frac{c_p}{c} = 1 - \frac{1}{2}r_m\left[\frac{1}{1 + (\omega\tau_{f\eta})^2} + \frac{(\gamma - 1)r_c}{1 + (\omega\tau_{f\kappa})^2}\right] \qquad (8.10.21)$$

The phase speed c_p is less than c at low frequencies, having the asymptotic value obtained from (8.10.20). It increases relatively rapidly for frequencies around the relaxation values, until at higher frequencies it becomes asymptotic to c from below. This is consistent with simple physical arguments. For frequencies much less than the relaxation values the above discussion is valid. For frequencies above the relaxation values, (1) there is less thermal energy exchange between the droplets and the gas so that $\gamma_f \to \gamma$, and (2) the droplets cannot drift along with the surrounding gas and are effectively frozen out of the motion so that the effective density of the moving medium reduces to that of the gas alone. Thus, for frequencies far above resonance the behavior of the phase speed should approach that of the gas alone.

(b) Resonant Bubbles in Water

High attenuations are also produced in water containing air bubbles when the ensonifying acoustic wave causes the bubbles to resonate. A resonating bubble vibrates at such a large amplitude that an appreciable amount of energy is reradiated out of the sound beam and the temperature variations are sufficiently great to cause significant energy losses. The effects of viscosity are of much less importance. The presence of air bubbles also affects the density and compressibility of the medium, changing its speed of sound and resulting in a considerable amount of acoustic energy being reflected and refracted away from the direction of the initial sound beam. Thus, a beam of sound can be attenuated by reflection, refraction, absorption, and scattering as it enters water containing a high concentration of air bubbles. Although air bubbles do not occur in large numbers in the main body of the ocean, high concentrations are found in the wakes of ships, at shallow depths when waves are breaking at the surface, and at various depths where there are regions of intense biological activity.

When ensonified with an incident pressure p of frequency f, a bubble of radius a at depth z can be driven into radial oscillations. Let the *outward* displacement of the surface of the bubble be ξ. The overpressure within the bubble is given by the adiabat (5.5.6), $p = -\rho_b c_b^2\,\Delta V/V$, where $\Delta V/V$ is the condensation within the bubble and ρ_b and c_b are for the enclosed air. For a perfect gas, (5.6.3) gives $\rho_b c_b^2 = \gamma\mathscr{P}_b$ with $\mathscr{P}_b$ the total hydrostatic pressure within the bubble. (For very small bubbles, surface tension increases the hydrostatic pressure, but for the sizes of interest here this effect can be ignored without introducing much error.) The condensation within the bubble is $\Delta V/V$ and the compressive force on the surface of the bubble is $f = -4\pi a^2 p$. Combination of these equations yields $f = -12\pi a\gamma\mathscr{P}_b\xi$, so that the effective stiffness s of the bubble is

$$s = 12\pi a\gamma\mathscr{P}_b \qquad (8.10.22)$$

For bubbles of specified radius a, the bubble stiffness depends on the depth. The hydrostatic pressure in seawater increases by nearly 1 atm for each 10 m of depth, so we can write the

---

[10]Temkin and Dobbins, *J. Acoust. Soc. Am.*, **40**, 317 (1966).

equilibrium pressure within the bubble as

$$\mathscr{P}_b = \mathscr{P}_0(1 + z/10) \tag{8.10.23}$$

for z in meters.

When the bubble is oscillating, it acts as a small pulsating sphere and radiates sound uniformly in all directions. It therefore has the radiation impedance for a pulsating sphere in the low-frequency limit. As seen in Chapter 7, this is

$$\mathbf{Z}_r = R_r + j\omega m_r$$
$$R_r = 4\pi a^2 \rho_0 c (ka)^2 \tag{8.10.24}$$
$$m_r = 4\pi a^3 \rho_0$$

with R_r and m_r the radiation resistance and mass, respectively, and ρ_0, c, and k the density, speed of sound, and propagation constant for water. It is clear that the mass of the bubble can be ignored since it is much less than the radiation mass. Thus, a bubble of radius a at depth z acts like a harmonic oscillator with resonance angular frequency $\omega_0 = \sqrt{s/m_r}$, which gives

$$\omega_0 = \frac{1}{a}\left(\frac{3\gamma\mathscr{P}_b}{\rho_0}\right)^{1/2} \tag{8.10.25}$$

Evaluating (8.10.25) at the surface gives $k_0(0)\,a = 0.0136$, so the radius of the resonant bubble is clearly much less than the acoustical wavelength at resonance. In this frequency range, we see that $R_r \ll \omega m_r$.

In addition to the loss by radiation, there is also absorption of energy by the bubble. Because of the high thermal conductivity of water and the smallness of the bubble, the condensation of the air is not adiabatic. Analysis is involved,[11] but the result is that the bubble can be characterized by an additional (mechanical) resistance R_m given by

$$R_m/\omega m_r = 1.6 \times 10^{-4}\sqrt{\omega} \tag{8.10.26}$$

This is accurate within about 10% for bubbles resonant below about 50 kHz. Losses from viscosity are negligible in this frequency range. Thus, the total input mechanical impedance $\mathbf{Z}$ of the bubble is

$$\mathbf{Z} = (R_m + R_r) + j(\omega m_r - s/\omega) \tag{8.10.27}$$

This describes the bubble as a simple harmonic oscillator with two sources of energy loss, one arising from the radiation of acoustic energy and the other the result of thermal conduction between the bubble and the surrounding water. The driving force is $4\pi a^2 \mathbf{p}$, where $\mathbf{p}$ is the complex acoustic pressure provided by the incident plane traveling wave, and the particle speed $\mathbf{u}_r(a, t)$ describes the radial motion of the bubble surface. The quality factor of the bubble resonance is

$$Q = \frac{\omega_0 m_r}{R_m + R_r} = \frac{1}{k_0 a + 1.6 \times 10^{-4}\sqrt{\omega_0}} \tag{8.10.28}$$

For a bubble of radius $a = 0.065$ cm just beneath the ocean surface, the resonance frequency is 5 kHz and $Q = 24$.

[11]Devin, *J. Acoust. Soc. Am.*, **31**, 1654 (1959).

The intensity of the incident wave is $I = |\mathbf{p}|^2/2\rho_0 c$. The power dissipated by the bubble in radiation is

$$\Pi_r = \tfrac{1}{2}|\mathbf{u}_r(a)|^2 R_r \tag{8.10.29}$$

and the total power dissipated in both radiation and thermal losses is

$$\Pi = \tfrac{1}{2}|\mathbf{u}_r(a)|^2(R_r + R_m) \tag{8.10.30}$$

The radial complex particle speed can be related (1) to the incident pressure and the mechanical impedance of the bubble, and (2) to the radiation pressure and the radiation impedance,

$$\mathbf{u}_r(a, t) = 4\pi a^2 \mathbf{p}/Z$$
$$= 4\pi a^2 \mathbf{p}_r(a)/Z_r \tag{8.10.31}$$

Substitution and simplification give the scattering cross section σ_S and the extinction cross section σ,

$$\sigma_S = \Pi_r/I = 4\pi a^2 (\omega m_r)^2 /Z^2$$
$$\sigma = \Pi/I = 4\pi a^2 \left[(R_m + R_r)/Z^2\right](4\pi a^2 \rho_0 c) \tag{8.10.32}$$

The scattering and extinction cross sections are simply related,

$$\sigma_S/\sigma = (\omega/\omega_0)^2 Q k_0 a \tag{8.10.33}$$

This expression suggests the upper limit of validity of this model, since we must always have $\sigma_S < \sigma$. Since the quality factor is fairly high, the resonance occurs over a narrow range of frequencies. With the approximation of a high Q, the scattering cross section can be written as

$$\sigma_S \approx \frac{4\pi a^2}{(1/Q)^2 + [1 - (\omega_0/\omega)^2]^2} \tag{8.10.34}$$

At resonance, the cross sections become

$$\sigma_S = 4\pi a^2 Q^2$$
$$\sigma = 4\pi a^2 Q/k_0 a \tag{8.10.35}$$

For the 0.065 cm radius bubble mentioned earlier, these results show that at resonance the scattering cross section is about 0.33 times the extinction cross section, and the extinction cross section is about 7000 times greater than the geometrical cross section πa^2. This demonstrates in a striking way the efficiency of a resonant bubble as an attenuator of an incident sound beam.

When the frequency of the sound wave is somewhat higher than the resonance value, the extinction cross section approaches $4\pi a^2$. See Fig. 8.10.1. At still higher frequencies, where the model fails and the radiation is no longer spherically symmetric, the extinction and geometric cross sections become nearly equal. For frequencies well below resonance, Z is dominated by the stiffness and the scattering cross section becomes

$$\sigma_S = 4\pi a^2(\omega/\omega_0)^4 \tag{8.10.36}$$

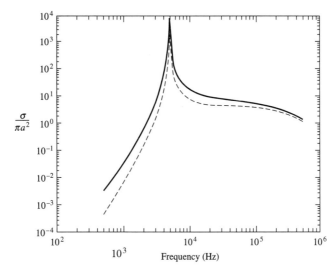

Figure 8.10.1   Acoustic cross sections for an air bubble with radius $a = 6.5 \times 10^{-4}$ m in water at 1 atm. Solid line is the total cross section and the dashed line is the scattering cross section.

This is the Rayleigh law of scattering, which in optics predicts that the sky is blue. (Since blue light has higher frequency than red light, blue light is scattered more strongly than red and more of it reaches the earth's surface from high atmospheric scattering.)

The concentration of bubbles occurring naturally in the main body of the ocean is so small that any attenuation resulting from this source is negligible compared to that caused by viscous forces and other relaxation phenomena. However, the agitation of the ocean surface produces bubbles of various sizes that may influence the propagation of sound near the surface. Also, because of scattering from their internal air sacs, fish and other marine critters, in large schools, will produce measurable attenuation at the resonance frequency of the sac. Another situation where bubbles may be important is in the wakes of ships. For example, observed attenuation coefficients for the relatively fresh wake existing 500 m astern of a destroyer making 15 kt (equivalent to 30 km/h = 8.3 m/s), range from 0.8 dB/m at 8 kHz to 1.8 dB/m at 40 kHz.[12]

Problems

8.2.1. Using the data in the appendix, (*a*) compute the viscous relaxation time for glycerin at 20°C. (*b*) For what frequency f_r is $\omega \tau_S = 1$? (*c*) Calculate the low-frequency value of α_S / f^2 and compare to the value given in Table 8.5.1.

8.2.2. (*a*) Assume the coefficient of shear viscosity in air is independent of pressure and is the dominant loss mechanism. What is τ_S in air at 20°C and 0.1 atm? (*b*) At what frequency will $\omega \tau_S = 1$? (*c*) Calculate α_S and c_p at this frequency. (*d*) What are they in air at 1.0 atm pressure at this same frequency?

8.2.3. For the set of equations (8.2.2)–(8.2.4), assume a damped plane, progressive wave $\mathbf{p} = P \exp(-\alpha x) \exp[j(\omega t - kx)]$. (*a*) Determine the phase angles between the pressure, condensation, and particle speed. Which pairs of variables are exactly in phase? (*b*) From the definition (5.9.1), determine the acoustic intensity through first order terms in α / k.

[12]NDRC, *Physics of Sound in the Sea*, U.S. Government Printing Office (1969).

8.2.4. Assume air is a perfect gas with a molecular weight of 29. At standard temperature and pressure (20°C and 1 atm) calculate (*a*) the number density of molecules, (*b*) the average speed of an air molecule, (*c*) the mean free path between collisions, and (*d*) the mean time between collisions.

8.2.5. A thin flat plate with one surface in contact with an incompressible, viscous fluid extending to infinity is oscillating parallel to its surface with speed $\mathbf{u}_0 = U_0 \exp(j\omega t)$. (*a*) Starting with the Navier–Stokes equation, show that the differential equation controlling the motion of the fluid is $\partial^2 u/\partial z^2 = (\rho_0/\eta)\, \partial u/\partial t$, where u is the particle speed parallel to the surface of the plate, z the distance from the plate, η the shear viscosity of the fluid, and ρ_0 the density of the fluid. (*b*) Solve the equation in (*a*) with the given boundary condition. (*c*) Find expressions for the phase speed and absorption coefficient. (*d*) Find the thickness of the layer of fluid in which 1 Np of absorption takes place and evaluate this thickness for air at 10 Hz and 1 kHz.

8.2.6C. For the wave of Problem 8.2.5, plot the particle speed as a function of distance from the plate for one cycle of the motion. Determine the phase speed and skin depth from the graph and compare to the input values.

8.3.1. A modified pressure–condensation relationship studied by Stokes is

$$p = \rho_0 c^2 (1 + \tau\, \partial/\partial t)s$$

(*a*) Obtain $s(t)$ if the acoustic pressure is suddenly increased from 0 to P at time $t = 0$ and identify the relaxation time of the process. (*b*) Assuming steady-state excitation $\exp(j\omega t)$, find the expression for the complex sound speed $\mathbf{c}$ in the *dynamic* equation $\mathbf{p} = \rho_0 \mathbf{c}^2 \mathbf{s}$. (*c*) Show for this steady-state motion that combination of $\mathbf{p} = \rho_0 \mathbf{c}^2 \mathbf{s}$ with Euler's equation and the equation of continuity gives the lossy Helmholtz equation (8.2.5).

8.3.2. Assume the Stokes relationship $p = \rho_0 c^2 (1 + \tau\, \partial/\partial t)s$ and combine this with the equation of continuity and Euler's equation to obtain a wave equation for pressure. What choice of τ gives (8.2.4)?

8.3.3. (*a*) Show by solving (8.3.2) in spherical coordinates with radial symmetry that in a medium with acoustic losses, $\mathbf{p}(r, t) = (A/r)\exp(-\alpha r)\exp[j(\omega t - kr)]$, where $\omega/k = c_p$. (*b*) Obtain a formula for the change in intensity level $IL(1) - IL(r)$ as a function of r and $a = 8.7\alpha$. (*c*) Find the upper limit of α for the approximation $(20 \log r + ar)$ to be within 0.1 dB of the formula in (*b*).

8.3.4. Show that the intensity of a traveling wave $\mathbf{p} = P\exp(-\alpha x)\exp[j(\omega t - kx)]$ satisfies the equation $(1/I)(dI/dx) = -2\alpha$.

8.3.5C. As a function of α/k, plot the magnitude and phase of the exact specific acoustic impedance of a damped plane wave (8.3.3) and compare to the approximate equation (8.3.4). For what value of α/k are the magnitudes within 10% and what is the phase difference at the value?

8.3.6. Derive (8.3.2) by following the steps indicated in the text after (8.3.1).

8.4.1. Derive (8.4.1) from the equation of state and the adiabat for a perfect gas.

8.4.2. Determine the energy density $\mathcal{E}$ for a plane traveling wave in terms of the magnitude of the temperature fluctuation, the equilibrium pressure, and γ.

8.4.3. Calculate the relaxation time for heat conduction in air at standard conditions (20°C and 1 atm). Compare with the corresponding value for viscosity.

8.4.4. Verify the thermodynamic relationship $T_0(\gamma - 1) = c^2/c_{\mathscr{P}}$ with the help of Appendix A9.

8.4.5. Verify the operator identity of (8.4.6).

8.4.6. Derive the diffusion equation (8.4.4). *Hint:* The rate of energy flow per unit area is proportional to the temperature gradient and the proportionality constant is the thermal conductivity.

8.5.1. Show that (8.5.2) follows from (8.5.1).

8.5.2. Evaluate the Prandtl number for dry air and helium at standard conditions (20°C and 1 atm).

8.6.1. From (8.6.14) find the frequency for which α_M/k is maximized.

8.6.2. (*a*) Use μ_{max} determined from Fig. 8.6.1 to calculate the fraction of carbon dioxide molecules in the excited state. (*b*) Use the low-frequency phase speed for carbon dioxide at 0°C tabulated in the appendix to calculate the low-frequency phase speed at 20°C and compare to the results in Fig. 8.6.1. (*c*) Use these results to calculate the high-frequency phase speed at 20°C and compare to the results plotted in Fig. 8.6.1.

8.6.3. Assume that air of 13% relative humidity has its maximum excess molecular absorption per wavelength at 5 kHz and assume a single relaxation involving oxygen and carbon dioxide. (*a*) What is the relaxation time? (*b*) If the measured excess absorption at 5 kHz is 0.14 dB/m, compute and plot the excess molecular absorption per meter over the frequency range 1 to 10 kHz. (*c*) Combine these results with the classical absorption coefficient for dry air to obtain a predicted total absorption. (*d*) By comparing these results with the values obtained from Fig. 8.6.2, comment on the importance of the nitrogen–carbon dioxide relaxation mechanism for this relative humidity and range of frequencies.

8.6.4. A siren is to operate at 500 Hz in air and at a small height above the ground. Assuming hemispherical divergence and no absorption by the ground, what is the absorption coefficient and what must be the acoustic output of the siren in watts if it is to produce an intensity level of 60 dB *re* 20 μPa at a distance of 1000 ft for each of the following conditions: (*a*) no absorption by the air, (*b*) according to the classical absorption coefficient, (*c*) completely dry air, and (*d*) air of very high relative humidity.

8.6.5. Given that the ratio of heat capacities for CO_2 gas is $\gamma = 1.31$ and that its gas constant is $r = 189$ J/(kg · K), use the data of Fig. 8.6.1 and the equations of Section 8.6 to compute values for $C_{\mathcal{P}}, C_V, C_e$, and C.

8.6.6. The population function is found from statistical mechanics to be given by

$$H(T_K) - u^2 e^{-u}/(1 - e^{-u})^2$$

where $u = T_D/T_K$ and the Debye temperature is $T_D = \Delta E/k_B$ with ΔE the *excitation energy* of the state and $k_B = 1.3807 \times 10^{-23}$ J/K the *Boltzmann constant*. A certain internal vibrational energy state is 10% filled at room temperature. Calculate the Debye temperature and the excitation energy for that state.

8.7.1. (*a*) Show that the attenuation constant α as given by (8.7.1) has the dimensions of a reciprocal length. What is the predicted absorption in decibels for sounds of 40 kHz in traversing a path length of 4 km (*b*) in freshwater at 5°C and (*c*) in seawater at 5°C? Assume zero depth and $S = 35$ ppt.

8.7.2. A 1 kHz plane wave traverses freshwater at 15°C. (*a*) In what distance will it be attenuated by 10 dB? (*b*) Work the same problem for 20 kHz. (*c*) What are the corresponding distances in seawater at 15°C? (*d*) In dry air at 20°C? (*e*) In air of 10% relative humidity at 20°C?

8.7.3C. (*a*) Plot the absorption coefficient in dB/m for seawater (35 ppt salinity) at 1 atm for temperatures of 5°C, 15°C, and 30°C for frequencies between 100 Hz and 1000 Hz. (*b*) Repeat for seawater (35 ppt salinity) at 5°C for depths of 0 m, 1000 m, and 4000 m. (*c*) Repeat for seawater at 5°C at 0 m for salinities 0 ppt and 35 ppt.

8.7.4C. Plot the attenuation per wavelength for seawater (35 ppt salinity), at 5°C and 1 atm for frequencies between 100 Hz and 1000 Hz.

8.8.1. With the primary and secondary particle velocity fields as defined in the discussion following (8.8.1), show that this vector equation has as components the three equations of (8.8.3).

8.8.2. With the solution (8.8.6) for $\mathbf{u}'$, show that $|\partial p'/\partial x|/|\rho_0\, \partial \mathbf{u}'/\partial t| = \frac{1}{2}(k\delta)^2$ so that the approximation resulting in (8.8.4) is justified.

8.9.1C. (*a*) Plot, as function of frequency from 1 kHz to 100 kHz, the absorption coefficients in dB/m in dry air for plane traveling waves in pipes of 1.0 cm, 10 cm, and 100 cm radius. (*b*) Calculate the ratios of the viscous and thermal skin depths to radius for the 100 cm pipe at 1 kHz and 100 kHz.

8.9.2. Calculate a value for the absorption in dB/m at 20 kHz (*a*) in freshwater contained in a pipe of 1.0 cm radius, (*b*) in a large body of freshwater at 15°C, and (*c*) in a large body of seawater at 15°C.

8.9.3. (*a*) Calculate the thermoviscous absorption in air at 1 atm and 20°C at normal humidities for a plane traveling wave of 200 Hz. (*b*) Calculate the absorption from wall losses including both viscous and thermal effects in a pipe of 1 cm radius. (*c*) What is the total absorption coefficient in Np/m? (*d*) What is the change in intensity level produced in 2 m in this pipe?

8.9.4. An *effective viscosity coefficient* η_e utilized by Rayleigh,

$$\eta_e - \eta \left[1 + (\gamma - 1)/\sqrt{\text{Pr}} \right]^2$$

is often used to include thermal conductivity in the absorption coefficients for wall losses. Show that the substitution of η_e for η in α_{wn} converts it into α_w.

8.9.5. Calculate and compare the absorption coefficient in dB/m in dry air for plane waves in a pipe of 1.0 cm radius with that for plane waves in the unbounded medium at frequencies of 1 kHz, 10 kHz, and 100 kHz.

8.9.6. (*a*) Calculate the value of $\sqrt{2\eta/\rho_0\omega}$ for acoustic waves of 200 Hz in air. (*b*) Including the effects of thermoviscous losses at the walls of the pipe, calculate the phase speed of 200 Hz plane waves through air in a pipe of 1 cm radius. (*c*) What is the corresponding absorption coefficient in Np/m? (*d*) What is the attenuation in dB produced in a 2 m length of this pipe?

8.10.1. A fog has 400 droplets/cm³, each having an average radius of 6×10^{-4} cm. (*a*) Find the attenuation caused by the fog at 1 kHz. Compare this value with that at the same frequency for (*b*) very dry air and (*c*) moist air.

8.10.2. (*a*) Find an expression for the extinction cross section of a water droplet of radius a in air. (*b*) If the water droplet has a radius of $a = 4 \times 10^{-6}$ m, find the relaxation frequencies. Calculate the extinction cross section at (*c*) 100 Hz, (*d*) 1 kHz, and (*e*) 10 kHz.

8.10.3. Evaluate the Reynolds number for a representative water droplet in a fog ensonified with a sound wave having $IL = 120$ dB *re* 10^{-12} W/m².

8.10.4. Show that $r_m \approx \rho_f/\rho_0 - 1$ when the total droplet volume per unit volume of the gas is small.

8.10.5. Show that under the approximation $\omega/c = k$ the expressions for $\alpha_M/k, \alpha_{f_\eta}/k$, and α_{f_κ}/k all have the same dependence on $\omega\tau$, where τ is the appropriate relaxation time.

8.10.6. For the resonant bubble, find (*a*) the quality factor Q_r if the only loss were the acoustic radiation and (*b*) the quality factor Q_κ if the only loss were thermal. (*c*) Show that these results are consistent with (8.5.2).

8.10.7. (*a*) What is the resonance frequency for an air bubble of 0.01 cm radius in water at a depth of 10 m? (*b*) Find the extinction and scattering cross sections at resonance. (*c*) How many bubbles per cubic meter will be required to produce an attenuation of 0.01 dB/m? (*d*) How does this attenuation compare with that of bubble-free seawater at the same frequency? Assume a temperature of 15°C.

8.10.8C. For a bubble of 0.065 cm radius just below the ocean surface, plot the extinction cross section (divided by the geometrical cross section) against the frequency (divided by the resonance frequency) for frequencies from 0.1 to 10 times the resonance frequency. From the graph, determine the value of Q.

CAVITIES
AND WAVEGUIDES

9.1 INTRODUCTION

In this and the next chapter we concentrate on the confinement of acoustic energy to closed or partly closed regions of space. In completely enclosed spaces, two- and three-dimensional standing waves can be stimulated. The normal modes associated with these standing waves determine the acoustic behavior of rooms, auditoriums, and concert halls. If the space is open in one or two dimensions, it can form a waveguide. Applications of waveguides include surface-wave delay lines, high-frequency electronic systems, folded-horn loudspeakers, and propagation of sound in the oceans and the atmosphere.

9.2 THE RECTANGULAR CAVITY

Consider a rectangular cavity of dimensions L_x, L_y, L_z, as indicated in Fig. 9.2.1. This box could represent a living room or auditorium, a simple model of a concert hall, or any other right-hexahedral space that has few windows or other openings and fairly rigid walls. Such applications will be encountered in Chapter 12. Assume that all surfaces of the cavity are perfectly rigid so that the normal component of the particle velocity vanishes at all boundaries,

$$\left(\frac{\partial p}{\partial x}\right)_{x=0} = \left(\frac{\partial p}{\partial x}\right)_{x=L_x} = 0$$

$$\left(\frac{\partial p}{\partial y}\right)_{y=0} = \left(\frac{\partial p}{\partial y}\right)_{y=L_y} = 0 \qquad (9.2.1)$$

$$\left(\frac{\partial p}{\partial z}\right)_{z=0} = \left(\frac{\partial p}{\partial z}\right)_{z=L_z} = 0$$

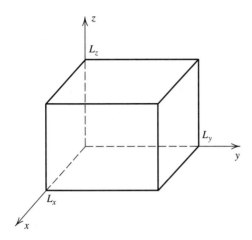

Figure 9.2.1 The rectangular cavity with dimensions L_x, L_y, and L_z.

Since acoustic energy cannot escape from a closed cavity with rigid boundaries, appropriate solutions of the wave equation are standing waves. Substitution of

$$\mathbf{p}(x, y, z, t) = \mathbf{X}(x)\mathbf{Y}(y)\mathbf{Z}(z)e^{j\omega t} \tag{9.2.2}$$

into the wave equation and separation of variables (as performed in Chapter 4) results in the set of equations

$$\left(\frac{d^2}{dx^2} + k_x^2\right)\mathbf{X} = 0$$

$$\left(\frac{d^2}{dy^2} + k_y^2\right)\mathbf{Y} = 0 \tag{9.2.3}$$

$$\left(\frac{d^2}{dz^2} + k_z^2\right)\mathbf{Z} = 0$$

where the angular frequency must be given by

$$(\omega/c)^2 = k^2 = k_x^2 + k_y^2 + k_z^2 \tag{9.2.4}$$

Application of the boundary conditions (9.2.1) shows that cosines are appropriate solutions, and (9.2.2) becomes

$$\mathbf{p}_{lmn} = \mathbf{A}_{lmn} \cos k_{xl}x \cos k_{ym}y \cos k_{zn}z \, e^{j\omega_{lmn}t} \tag{9.2.5}$$

where the components of k are

$$
\begin{aligned}
k_{xl} &= l\pi/L_x & l &= 0, 1, 2, \ldots \\
k_{ym} &= m\pi/L_y & m &= 0, 1, 2, \ldots \\
k_{zn} &= n\pi/L_z & n &= 0, 1, 2, \ldots
\end{aligned} \tag{9.2.6}
$$

Thus, the allowed angular frequencies of vibration are quantized,

$$\omega_{lmn} = c[(l\pi/L_x)^2 + (m\pi/L_y)^2 + (n\pi/L_z)^2]^{1/2} \tag{9.2.7}$$

Each standing wave given by (9.2.5) has its own angular frequency (9.2.7) and can be specified by the ordered integers (l,m,n).

The form (9.2.5) gives three-dimensional standing waves in the cavity with nodal planes parallel to the walls. Between these nodal planes the pressure varies sinusoidally, with the pressure within a given loop in phase, and with adjacent loops 180° out of phase. Comparison of the mathematical developments of this section with those for the rectangular membrane with fixed rim of Section 4.3 reveals similarities and analogs:

1. If only those modes for which $n = 0$ are considered, the z component of the propagation vector vanishes, and the resulting standing wave patterns become two-dimensional, like those for the rectangular membrane.

2. A rigid boundary for a pressure wave in a fluid is analogous to a free boundary for a membrane displacement wave in that both correspond to respective antinodes. The distribution of nodes and antinodes of these respective pressure and displacement waves in planes perpendicular to any axis will be identical for the same dimensions and modal numbers. Similarly, a pressure release boundary for a fluid is analogous to a fixed boundary for a membrane, both requiring nodes in pressure and displacement, respectively.

If a pressure source is located anywhere on a nodal surface of a normal mode of pressure, that mode will not be excited. The closer a source is to an antinode of the mode, the greater the excitation of that mode. Similarly, a pressure-sensitive receiver will have greatest output if it is placed at an antinode of the mode. These effects are used to either emphasize or suppress selected modes or families of modes. For example, if it is desired to excite and detect all the modes of a rectangular room, the source and receiver must be placed in the corners (junctions of three surfaces). (If, in a hard-walled room like a shower, one hums at an eigenfrequency and moves around in the enclosure, strong fluctuations in loudness will be heard, with maxima when the head is close to a corner or any other pressure antinode. In contrast, the hummer will experience difficulty in trying to drive a mode at a pressure node.)

If two or more modes have the same eigenfrequency, they are called *degenerate*. Degenerate modes can be isolated by judicious placement of the source and receiver. A receiver placed on a nodal plane of one of a set of degenerate modes will not respond to that mode. Similarly, a source located at a node of one of the degenerate modes cannot excite that mode.

Just as a standing wave on a string could be considered as two traveling waves moving in opposite directions, the standing waves in the rectangular cavity can be decomposed into traveling plane waves. If the solutions (9.2.5) are represented in complex exponential form and expanded as a sum of products, it is seen that

$$\mathbf{p}_{lmn} = \tfrac{1}{8}\mathbf{A}_{lmn} \sum_{\pm} e^{j(\omega_{lmn}t \pm k_{xl}x \pm k_{ym}y \pm k_{zn}z)} \tag{9.2.8}$$

where the summation is taken over all permutations of plus and minus signs. Each of these eight terms represents a plane wave traveling in the direction of its propagation vector k_{lmn} whose projections on the coordinate axes are $\pm k_{xl}$, $\pm k_{ym}$, and $\pm k_{zn}$. Thus, the standing wave solution can be viewed as a superposition of eight traveling waves (one into each octant) whose directions of propagation are fixed by the boundary conditions.

*9.3 THE CYLINDRICAL CAVITY

Figure 9.3.1 shows a rigid-walled, right circular cavity with radius a and height L. In cylindrical coordinates (Appendix A7), the Helmholtz equation $\nabla^2 \mathbf{p} + k^2 \mathbf{p} = 0$ with $\mathbf{p} = P \exp(j\omega t)$ becomes

$$\frac{\partial^2 P}{\partial r^2} + \frac{1}{r}\frac{\partial P}{\partial r} + \frac{1}{r^2}\frac{\partial^2 P}{\partial \theta^2} + \frac{\partial^2 P}{\partial z^2} + k^2 P = 0 \tag{9.3.1}$$

and the boundary conditions at the rigid walls are

$$\left(\frac{\partial P}{\partial z}\right)_{z=0} = \left(\frac{\partial P}{\partial z}\right)_{z=L} = \left(\frac{\partial P}{\partial r}\right)_{r=a} = 0 \tag{9.3.2}$$

If a solution of the form

$$P(r,\theta,z) = R(r)\Theta(\theta)Z(z) \tag{9.3.3}$$

is assumed, separation of variables results in three equations:

$$\frac{d^2 Z}{dz^2} = -k_{zl}^2 Z$$

$$\frac{d^2 \Theta}{d\theta^2} = -m^2 \Theta \tag{9.3.4}$$

$$r^2 \frac{d^2 R}{dr^2} + r\frac{dR}{dr} + (k_{mn}^2 r^2 - m^2)R = 0$$

where

$$k^2 = k_{mn}^2 + k_{zl}^2 \tag{9.3.5}$$

These equations have solutions

$$Z = \cos k_{zl}z$$

$$\Theta = \cos(m\theta + \gamma_{lmn}) \tag{9.3.6}$$

$$R = J_m(k_{mn}r)$$

with $m = 0, 1, 2, \ldots$ (since Θ must be single valued), $k_{zl}L = l\pi$, where $l = 0, 1, 2, \ldots$, and $k_{mn}a = j'_{mn}$, where j'_{mn} is the nth extremum of the mth Bessel function of the first kind. The normal modes are designated by the three integers (l,m,n), which denote the number of null surfaces in the z, θ, and r directions, respectively. The pressure of the (l,m,n) mode is

$$\mathbf{p}_{lmn} = A_{lmn}J_m(k_{mn}r)\cos(m\theta + \gamma_{lmn})\cos k_{zl}z\, e^{j\omega_{lmn}t} \tag{9.3.7}$$

where the angular frequencies are determined from

$$(\omega_{lmn}/c)^2 = k_{lmn}^2 = k_{mn}^2 + k_{zl}^2 \tag{9.3.8}$$

Comparison with the discussion of the circular membrane with fixed rim of Section 4.4 shows that the introduction of the third spatial dimension z has, as in the rectangular case, simply introduced a new component of the propagation vector.

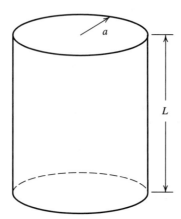

Figure 9.3.1   The right circular cylindrical cavity
with height L and radius a.

Just as for the circular membrane, if there were an inner boundary, i.e., a perfectly
reflecting cylinder with radius $r = b < a$ and no acoustic field for $r < b$, then the Bessel
functions $Y_m(k_{mn}r)$ would also be acceptable solutions and the boundary conditions at $r = a$
and $r = b$ would have to be satisfied by some combination $A_{lmn}J_m(k_{mn}r) + B_{lmn}Y_m(k_{mn}r)$.

As with the rectangular cavity, the standing waves in the cylindrical cavity can be
expressed as traveling waves. Expand $\cos(k_z z)$ in terms of exponentials, use $2J_n = H_n^{(1)} + H_n^{(2)}$,
and then, purely for ease of interpretation, expand the Hankel functions in their asymptotic
approximations. The resulting eight terms in each $\mathbf{p}_{lmn}$ have the forms

$$(2/\pi k_{mn}r)^{1/2}e^{j\left(\omega_{lmn}t \pm m\theta \pm k_{mn}r \pm k_{zl}z\right)} \tag{9.3.9}$$

with all permutations of the $+$ and $-$ signs. We have suppressed amplitude factors and
the γ's. (See Problem 9.3.1.) These describe eight conical traveling waves whose phases
are shaded according to the polar angle θ. In general, the surfaces of constant phase are
conical spirals. The intersection of a surface of constant phase with the z plane forms a spiral
that propagates with a radial speed ω_{lmn}/k_{mn} outward (or inward), appearing to emanate
from (or disappear into) the origin. The propagation vectors have angles of elevation and
depression given by $\pm \tan^{-1}(k_{zl}/k_{mn})$.

*9.4 THE SPHERICAL CAVITY

The Helmholtz equation in spherical coordinates (Appendix A7) is

$$\frac{\partial}{\partial r}\left(r^2 \frac{\partial P}{\partial r}\right) + \frac{1}{\sin\theta}\frac{\partial}{\partial\theta}\left(\sin\theta\frac{\partial P}{\partial\theta}\right) + \frac{1}{\sin^2\theta}\frac{\partial^2 P}{\partial\phi^2} + k^2 r^2 P = 0 \tag{9.4.1}$$

and the boundary condition for a rigid-walled sphere of radius a is

$$\left(\frac{\partial P}{\partial r}\right)_{r=a} = 0 \tag{9.4.2}$$

For a solution of the form

$$P = R(r)\Theta(\theta)\Phi(\phi) \tag{9.4.3}$$

separation of variables gives

$$\frac{d^2\Phi}{d\phi^2} + m^2\Phi = 0$$

$$\frac{1}{\sin\theta}\frac{d}{d\theta}\left(\sin\theta\frac{d\Theta}{d\theta}\right) + \left(\eta^2 - \frac{m^2}{\sin^2\theta}\right)\Theta = 0 \qquad (9.4.4)$$

$$\frac{d}{dr}\left(r^2\frac{dR}{dr}\right) + (k^2r^2 - \eta^2)R = 0$$

where m and η are the separation constants.

Solutions for the Φ dependence are

$$\Phi_m = A\cos(m\phi + \gamma_{lmn}) \qquad (9.4.5)$$

Since Φ must be single valued, m must be integral. As discussed in Section 4.4, each phase angle γ_{lmn} must be determined by the initial conditions. If there is no condition to determine γ_{lmn}, then except for $m = 0$ each Φ_m must be considered a pair of degenerate modes. (They can be made orthogonal if, for example, one is chosen to be $\cos m\phi$ and the other $\sin m\phi$.)

The equation for Θ is related to the *Legendre equation*. Solutions to this equation that are continuous, single valued, and finite must have $\eta^2 = l(l + 1)$, where $l = 0, 1, 2, \ldots$, and must also have $m \leq l$. These solutions are the *associated Legendre functions of the first kind of order l and degree m*, denoted by $P_l^m(\cos\theta)$. (See Appendix A4 for more details and properties of these functions.)

The equation for the radial dependence can be rewritten as

$$r^2\frac{d^2R}{dr^2} + 2r\frac{dR}{dr} + \left[k^2r^2 - l(l + 1)\right]R = 0 \qquad (9.4.6)$$

The solutions to this equation that are finite at the origin of r are the *spherical Bessel functions of order l*:

$$R = j_l(k_{ln}r) \qquad (9.4.7)$$

[If the cavity is the space between two perfectly reflecting concentric boundaries of radii a and b, then the spherical Bessel functions of the second kind, $y_l(k_{ln}r)$, are also admissible solutions.] For a rigid-walled cavity, $k_{ln}a = \zeta'_{ln}$ where ζ'_{ln} are the extrema of j_{ln}.

The pressure amplitude in the cavity is then

$$\mathbf{p}_{lmn} = \mathbf{A}_{lmn}j_l(k_{ln}r)P_l^m(\cos\theta)\cos(m\phi + \gamma_{lmn})e^{j\omega_{ln}t} \qquad (9.4.8)$$

and the angular frequencies are given by $\omega_{ln} = ck_{ln}$. The lack of any dependence of the propagation constant on m means that all modes having the same values of l and n but different values of m are degenerate.

A study of the spherical Bessel functions shows that the lowest eigenfrequency, found from $k_{11}a = 2.08$, is shared by the single (1, 0, 1) mode and the pair (1, 1, 1). Together they constitute a threefold degeneracy. The spatial pressure distributions of the three are

$$P_{101} = A_{101}\left(\frac{\sin k_{11}r}{(k_{11}r)^2} - \frac{\cos k_{11}r}{k_{11}r}\right)\cos\theta$$

$$P_{111}^{(1)} = A_{111}^{(1)}\left(\frac{\sin k_{11}r}{(k_{11}r)^2} - \frac{\cos k_{11}r}{k_{11}r}\right)\sin\theta\cos\phi \qquad (9.4.9)$$

$$P_{111}^{(2)} = A_{111}^{(2)}\left(\frac{\sin k_{11}r}{(k_{11}r)^2} - \frac{\cos k_{11}r}{k_{11}r}\right)\sin\theta\sin\phi$$

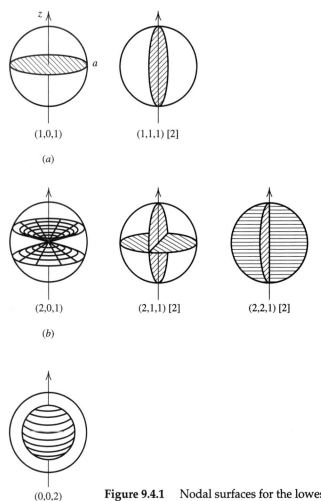

Figure 9.4.1   Nodal surfaces for the lowest three sets of normal modes in a rigid-walled spherical cavity of radius $r = a$. (a) $\omega_{101} = \omega_{111} = 2.08\, c/a$. (b) $\omega_{201} = \omega_{211} = \omega_{221} = 3.34\, c/a$. (c) $\omega_{002} = 4.49\, c/a$.

The next set of modes, forming a fivefold degeneracy, have $k_{21}a = 3.34$. The radial dependence is $j_2(k_{21}r)$ and the angular dependences are one mode with $(l, m, n) = (2, 0, 1)$, two modes with $(2, 1, 1)$, and two with $(2, 2, 1)$. The third set, for which $k_{02}a = 4.49$, has a single member with radial dependence $j_0(k_{02}r) = (\sin k_{02}r)/(k_{02}r)$ and no angular dependence since $P_0(\cos\theta) = 1$. Nodal surfaces for the lowest three sets of normal modes are shown in Fig. 9.4.1.

9.5 THE WAVEGUIDE OF CONSTANT CROSS SECTION

Waveguides having different, but uniform, cross sections and the same boundary conditions will display similar behaviors. We will develop the properties for a waveguide with a rectangular cross section, as shown in Fig. 9.5.1, and then generalize the results to other cross-sectional geometries. Assume the side walls to be rigid and the boundary at $z = 0$ to be a source of acoustic energy. The absence of another boundary on the z axis allows energy to propagate down the waveguide.

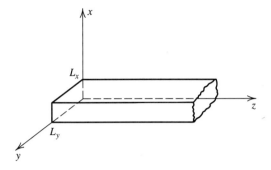

Figure 9.5.1 The rectangular waveguide with dimensions L_x and L_y.

This suggests a wave consisting of standing waves in the transverse directions (x and y) and a traveling wave in the z direction.

Since the cross section is rectangular and the boundaries are rigid, acceptable solutions are

$$
\begin{aligned}
\mathbf{p}_{lm} &= \mathbf{A}_{lm} \cos k_{xl}x \cos k_{ym}y \, e^{j(\omega t - k_z z)} \\
k_z &= [(\omega/c)^2 - (k_{xl}^2 + k_{ym}^2)]^{1/2} \\
k_{xl} &= l\pi/L_x \qquad l = 0, 1, 2, \ldots \\
k_{ym} &= m\pi/L_y \qquad m = 0, 1, 2, \ldots
\end{aligned}
\tag{9.5.1}
$$

Since ω can have any value, k_z is not fixed.

It is convenient to define k_{lm} as the *transverse component* of the propagation vector. For a rectangular cross section,

$$
k_{lm} = (k_{xl}^2 + k_{ym}^2)^{1/2}
\tag{9.5.2}
$$

and the required value of k_z can be written more succinctly as

$$
k_z = [(\omega/c)^2 - k_{lm}^2]^{1/2}
\tag{9.5.3}
$$

When $\omega/c > k_{lm}$, then k_z is real. The wave advances in the $+z$ direction and is called a *propagating mode*. The limiting value of ω/c for which k_{lm} remains real is given by $\omega/c = k_{lm}$, and this defines the *cutoff angular frequency*

$$
\boxed{\omega_{lm} = ck_{lm}}
\tag{9.5.4}
$$

for the (l, m) mode. If the input frequency is lowered below cutoff, the argument of the square root in (9.5.3) becomes negative and k_z must be pure imaginary

$$
k_z = \pm j[k_{lm}^2 - (\omega/c)^2]^{1/2}
\tag{9.5.5}
$$

The minus sign must be taken on physical grounds so that $\mathbf{p} \to 0$ as $z \to \infty$, and (9.5.1) has the form

$$
\mathbf{p}_{lm} = \mathbf{A}_{lm} \cos k_{xl}x \cos k_{ym}y \exp\{-[k_{lm}^2 - (\omega/c)^2]^{1/2} z\} e^{j\omega t}
\tag{9.5.6}
$$

This is an *evanescent* standing wave that attenuates exponentially with z. No energy propagates down the waveguide. If the waveguide is excited with a frequency just below the cutoff frequency of some particular mode, then this and higher modes are evanescent and not important at appreciable distances from the source. All modes having cutoff frequencies below the driving frequency may propagate energy and may be detected at large distances.

In a rigid-walled waveguide, only plane waves propagate if the frequency of the sound is sufficiently low. For a waveguide of rectangular cross section of greater dimension L, this frequency is easily shown to be $f = c/2L$.

The *phase speed* of a mode is

$$c_p = \omega/k_z = c/[1 - (k_{lm}/k)^2]^{1/2} = c/[1 - (\omega_{lm}/\omega)^2]^{1/2} \qquad (9.5.7)$$

and is greater than c. An understanding of this is obtained by writing the cosines in (9.5.1) in complex exponential form. The solution then consists of the sum

$$\mathbf{p}_{lm} = \tfrac{1}{4}\mathbf{A}_{lm} \sum_{\pm} e^{j(\omega t \pm k_{xl}x \pm k_{ym}y - k_z z)} \qquad (9.5.8)$$

(Note that only the minus sign appears before k_z.) The propagation vector $\vec{k}$ for each of the four traveling waves makes an angle θ with the z axis given by

$$\cos\theta = k_z/k = [1 - (\omega_{lm}/\omega)^2]^{1/2} \qquad (9.5.9)$$

so that the phase speed (9.5.7) is

$$c_p = c/\cos\theta \qquad (9.5.10)$$

This is simply the speed with which a surface of constant phase appears to propagate along the z axis. (See Problems. 9.2.3 and 9.3.2.)

Figure 9.5.2 gives the surfaces of constant phase for the two component waves that represent the (0,1) mode of a rigid-walled rectangular waveguide. The waves exactly cancel each other for $y = L_y/2$, so that there is a nodal plane midway between the walls. At the upper and lower walls the waves are always in phase so that the pressure amplitude is maximized at these (rigid) boundaries. The *apparent wavelength* λ_z measured in the z direction is $\lambda_z = \lambda/(\cos\theta)$.

The lowest mode for a rigid-walled waveguide is the (0, 0) mode. For this case, $k_z = k$ and the four component waves collapse into a single plane wave that travels down the axis of the waveguide with phase speed c. For all other modes, the propagation vectors of the component waves can be at angles to the waveguide axis, one pointing into each of the four forward octants. From (9.5.9) and (9.5.10), at frequencies far above the cutoff of the (l, m) mode, we have $\omega \gg \omega_{lm}$ so that θ tends to zero and the waves are traveling almost straight down the waveguide with $c_p \approx c$. As the input frequency is decreased toward cutoff, the angle θ increases so that the component waves travel in increasingly transverse directions. If we imagine that each component wave carries energy down the waveguide by a process of continual reflection from the walls (much like a bullet ricocheting down a hard-walled corridor), and remember that the energy of a wave is propagated with speed c in the direction of $\hat{k} = \vec{k}/k$, then we see that the speed with which energy moves in the z direction is given by the *group speed* $c_g = c\hat{k} \cdot \hat{z}$, the projection of the component wave velocity along the waveguide axis,

$$c_g = c\cos\theta = c[1 - (\omega_{lm}/\omega)^2]^{1/2} \qquad (9.5.11)$$

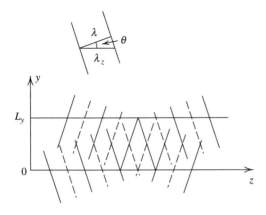

Figure 9.5.2 Component plane waves for the (0,1) mode in a rigid-walled, rectangular cavity. These waves travel with speed c in directions that make angles $\pm\theta$ with the z axis of the waveguide.

For a given angular frequency ω, each modal wave with $\omega_{lm} < \omega$ has its own individual values of c_p and c_g. The behaviors of the group and phase speeds as functions of frequency for three modes in a rigid-walled waveguide are shown in Fig. 9.5.3.

It is straightforward to generalize the above discussion and derive the behavior of a rigid-walled waveguide with a circular cross section of radius $r = a$. Separation of variables and solution results in

$$\mathbf{p}_{ml} = \mathbf{A}_{ml}J_m(k_{ml}r)\cos m\theta\, e^{j(\omega t - k_z z)}$$
$$k_z = [(\omega/c)^2 - k_{ml}^2]^{1/2}$$

(9.5.12)

where r, θ, and z are the cylindrical coordinates, J_m is the mth order Bessel function, and the allowed k_{ml} are determined by the boundary condition for the rigid wall,

$$k_{ml} = j'_{ml}/a$$

(9.5.13)

where j'_{ml} are the extrema of $J_m(z)$. These values are tabulated in Appendix A5. Once the values of k_{ml} are found, all the salient results developed for rectangular waveguides can be applied simply by substituting the values of k_{ml} for a circular waveguide. For example, the (0, 0) mode is a plane wave that propagates with $c_p = c$ for all $\omega > 0$. The nonplanar mode with the lowest cutoff frequency is the (1, 1) mode (the first "sloshing" mode) with cutoff frequency $\omega_{11} = 1.84\, c/a$ or $f_{11} = 100/a$ for air. It is of great practical importance that for frequencies below f_{11} only plane waves can propagate in a rigid-walled, circular waveguide.

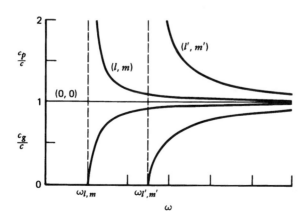

Figure 9.5.3 Group and phase speeds for the lowest three normal modes in a rigid-walled waveguide.

*9.6 SOURCES AND TRANSIENTS IN CAVITIES AND WAVEGUIDES

Up to this point we have not dealt with the acoustic source. If we know the pressure or velocity distribution of the source, then these can be related to the behavior of the pressure or the velocity of the total acoustic field as was done for membranes in Section 4.10. In what follows we will sketch the development of a few special cases to demonstrate the basics.

Assume that a rigid-walled rectangular enclosure is excited by a point impulsive pressure source (like a cap or starter pistol shot). This is the three-dimensional extension and analog of the impulsive point excitation of a rectangular membrane. The source can be described by an initial condition at $t = 0$ of

$$\mathbf{p}(x, y, z, 0) = \delta(x - x_0)\,\delta(y - y_0)\,\delta(z - z_0) \tag{9.6.1}$$

and this must be matched to (9.2.5). Since the space must be quiescent before the pressure impulse, the particle velocity field throughout the enclosure must be zero at $t = 0$. This requires $\mathbf{A}_{lmn} = A_{lmn}$ so that the real pressure standing waves are cosinusoidal in time and we must have

$$\delta(x - x_0)\,\delta(y - y_0)\,\delta(z - z_0) = \sum_{l,m,n} A_{lmn} \cos k_{xl}x \cos k_{ym}y \cos k_{zn}z \tag{9.6.2}$$

Inversion and use of orthogonality to solve for the coefficients provides the resultant pressure field,

$$p(x, y, z, t) = \frac{8}{L_x L_y L_z} \sum_{l,m,n} \cos k_{xl}x_0 \cos k_{ym}y_0 \cos k_{zn}z_0 \cos k_{xl}x \cos k_{ym}y \cos k_{zn}z \cos \omega_{lmn}t$$
$$\tag{9.6.3}$$

This result is simply an extension of what has been done before. Application to cylindrical enclosures proceeds similarly with no surprises. If there are losses, each standing wave will decay as $\exp(-\beta_{lmn}t)$.

Excitation of the enclosure by a monofrequency source presents a few more difficulties: losses must be included, and these require introduction of the frequency dependence of the amplitude of each of the driven lossy standing waves. Excitation of the cavity with losses by a monofrequency source is deferred until Section 12.9.

In the case of excitation of a waveguide of uniform cross section on the plane $z = 0$, assume that the source distribution is

$$\mathbf{p}(x, y, 0, t) = \mathbf{P}(x, y)e^{j\omega t} \tag{9.6.4}$$

Again, $\mathbf{p}$ can be written as a superposition of the normal modes of the waveguide as in Section 4.10. For a waveguide with rectangular cross section and rigid walls, we have

$$\mathbf{p}(x, y, z, t) = \sum_{l,m} \mathbf{A}_{lm} \cos k_{xl}x \cos k_{ym}y \, e^{j(\omega t - k_z z)} \tag{9.6.5}$$

Evaluation at $z = 0$ and use of (9.6.4) gives

$$\mathbf{P}(x, y) = \sum_{l,m} \mathbf{A}_{lm} \cos k_{xl}x \cos k_{ym}y \tag{9.6.6}$$

from which we can determine the required values of $\mathbf{A}_{lm}$.

The existence of three speeds c_p, c_g, and c in the description of each traveling wave in a waveguide serves to elucidate the propagation behavior of transient signals. First, we will develop some general results based on the *method of stationary phase* and then examine more exactly the behavior of a particular transient. Consider a well-defined pulse generated at the source and propagating down the waveguide. Recalling the elements of Fourier superposition stated in Section 1.15, we can write the dependence of the pulse on distance and time in the form of a weighted superposition of monofrequency components. The spectral density $\mathbf{g}(\omega)$ can be found from the behavior of the source at $z = 0$. If, instead of $\exp(j\omega t)$, the source generates a known signature $f(t)$, then

$$f(t) = \int_{-\infty}^{\infty} \mathbf{g}(\omega)e^{j\omega t}\,d\omega$$

$$\mathbf{g}(\omega) = \frac{1}{2\pi}\int_{-\infty}^{\infty} f(t)e^{-j\omega t}\,dt$$

(9.6.7)

The extension of (9.6.5) to a transient excitation becomes

$$\mathbf{p}(x,y,z,t) = \sum_{l,m}\left[\mathbf{A}_{lm}\cos k_{xl}x\cos k_{ym}y\int_{-\infty}^{\infty} \mathbf{g}(\omega)e^{j(\omega t - k_z z)}\,d\omega\right]$$

(9.6.8)

(Recall that k_z is a function of ω and is different for each nondegenerate wave.) Because of the distance-dependent phase in the integrand, it is clear that the pulse will evolve in shape as it travels along the z axis. If the pulse is initially well defined, then $\mathbf{g}(\omega)$ is a smoothly varying function of frequency and strong over a broad bandwidth. In this case, the portion of the integrand that contributes the most to the pulse is that for which the phase is nearly *stationary* (constant) as a function of frequency. For other frequencies the phase of the integrand is rapidly varying so that adjacent cycles of the integrand tend to cancel. Thus, the major portion of the pulse will begin near the time for which the phase is stationary, and for each mode this time is found from

$$\frac{d}{d\omega}(\omega t - k_z z) = 0$$

$$t = \frac{dk_z}{d\omega}z$$

(9.6.9)

The speed with which this major portion of the pulse travels down the waveguide is the group speed,

$$\boxed{c_g = \frac{d\omega}{dk_z}}$$

(9.6.10)

[It is straightforward to show that this is identical with (9.5.11) for the waveguide of rectangular cross section, but (9.6.10) is more general and can be applied to any lossless dispersive medium.] The phase speed c_p of each frequency component of the signal is, of course, still given by

$$\boxed{c_p = \omega/k_z}$$

(9.6.11)

Now, let us analyze a simple transient signal exciting a single mode of the waveguide. Write the pressure p_{lm} and the z component u_{zlm} of the particle velocity $\vec{u}_{lm}$ associated with the (l,m) normal mode in the forms

$$p_{lm}(x,y,z,t) = P_{lm}(x,y)f(z,t)$$

$$u_{zlm}(x,y,z,t) = P_{lm}(x,y)v(z,t)$$

(9.6.12)

If at the source location, $z = 0$, the function $v(0, t)$ is taken to be $1(t)/\rho_0 c$, where $1(t)$ is the unit step function, then with the help of Table 1.15.1 and standard acoustic relations,

$$v(z, t) = \frac{1}{\rho_0 c} \frac{1}{2\pi} \int_{-\infty}^{\infty} \frac{1}{j\omega} e^{j(\omega t - k_z z)} \, d\omega$$

$$f(z, t) = \frac{1}{2\pi c} \int_{-\infty}^{\infty} \frac{1}{jk_z} e^{j(\omega t - k_z z)} \, d\omega$$

(9.6.13)

Evaluation of $f(z, t)$ from a more general table of Fourier transforms or use of Problems 9.6.5–9.6.7 gives

$$f(z, t) = J_0\left(\omega_{lm} \sqrt{t^2 - T^2}\right) \cdot 1(t - T) \qquad T = z/c$$

(9.6.14)

where $T = z/c$ is the time of flight of the leading edge of the signal, which travels with the free field speed of sound c. [The basic mechanism for sound propagation, collisions between molecules, is not changed by the presence of boundaries. Consequently, the first information that the source has been turned on must arrive at z by the shortest path (directly down the z axis) with speed c.]

Recognizing that the Bessel function behaves very much like a cosinusoidal function of the same argument (and with a slightly shifted phase), write the argument of J_0 in the form appropriate for a traveling wave with instantaneous angular frequency ω and propagation constant k_z,

$$\omega_{lm}[t^2 - (z/c)^2]^{1/2} = \omega t - k_z z$$

(9.6.15)

Differentiating with respect to t gives ω as a function of z and t, and differentiation with respect to z does the same for k_z,

$$\omega = \omega_{lm} t \, / \, [t^2 - (z/c)^2]^{1/2}$$

$$k_z = \omega_{lm} z \, / \, \{c^2[t^2 - (z/c)^2]^{1/2}\}$$

(9.6.16)

Examination of the first of (9.6.16) shows that when t is just slightly larger than T, corresponding to the earliest portions of the signal arriving at location z, ω is very much larger than ω_{lm}. For very long elapsed times, $t \gg T$, the cutoff angular frequency appears at z. Higher frequencies arrive much faster than do lower frequencies, and none less than the cutoff value propagate down the waveguide. If this first equation is solved for z/t in terms of ω_{lm}/ω, we get

$$z/t = c[1 - (\omega_{lm}/\omega)^2]^{1/2}$$

(9.6.17)

This gives the time t at which the portion of the signal with angular frequency ω will appear at z. Thus z/t is the group speed c_g for energy associated with angular frequency ω,

$$c_g/c = [1 - (\omega_{lm}/\omega)^2]^{1/2}$$

(9.6.18)

Taking the ratio of the two equations in (9.6.16) and then eliminating z/t with (9.6.17) gives the phase speed c_p associated with the angular frequency ω,

$$c_p/c = 1 \, / \, [1 - (\omega_{lm}/\omega)^2]^{1/2}$$

(9.6.19)

These are identical with the earlier results.

*9.7   THE LAYER AS A WAVEGUIDE

Another important case of waveguide propagation is encountered when a source radiates into a horizontally stratified fluid contained between two horizontal planes. This and the following sections will provide a simplified introduction to the normal-mode approach to this subject. For further information and more mathematically sophisticated methods of analysis, start with the references.[1]

In cylindrical coordinates, assume a point source with time dependence $\exp(j\omega t)$ and unit pressure amplitude at a distance of 1 m is located at a depth $z = z_0$ on the axis ($r = 0$) within a layer of fluid that is bounded at two depths by perfectly reflecting planes (Fig. 9.7.1) at $z = 0$ and $z = H$. The speed of sound within the layer of fluid can be a function of depth z, but not of range r. If the pressure field is written as $\mathbf{p}(r,z,t) = \mathbf{P}(r,z)\exp(j\omega t)$, then the appropriate Helmholtz equation is found from (5.16.5) to be

$$\left[\frac{1}{r}\frac{\partial}{\partial r}\left(r\frac{\partial}{\partial r}\right) + \frac{\partial^2}{\partial z^2} + \left(\frac{\omega}{c}\right)^2\right]\mathbf{P}(r,z) = -\frac{2}{r}\delta(r)\,\delta(z-z_0) \tag{9.7.1}$$

where $\delta(\vec{r} - \vec{r}_0)$ has been expressed in cylindrical coordinates with the help of Problem 5.16.3. Since this is a case of waveguide propagation, we can assume a solution of the form

$$\mathbf{p}(r,z,t) = e^{j\omega t}\sum_n \mathbf{R}_n(r)Z_n(z) \tag{9.7.2}$$

where Z_n satisfies the one-dimensional Helmholtz equation solution

$$\frac{d^2 Z_n}{dz^2} + \left[\left(\frac{\omega}{c}\right)^2 - \kappa_n^2\right]Z_n = 0 \tag{9.7.3}$$

and κ_n is the separation constant. With appropriate normalization, the Z_n form an *orthonormal* set of eigenfunctions,

$$\int_0^H Z_n(z)Z_m(z)\,dz = \delta_{nm} \tag{9.7.4}$$

Substitution of (9.7.2) and (9.7.3) into (9.7.1) yields

$$\sum_n\left[Z_n\frac{1}{r}\frac{d}{dr}\left(r\frac{d\mathbf{R}_n}{dr}\right) + \kappa_n^2 Z_n\mathbf{R}_n\right] = -\frac{2}{r}\delta(r)\,\delta(z-z_0) \tag{9.7.5}$$

Multiplication by Z_m, integration over the depth, and use of orthonormality gives an inhomogeneous Helmholtz equation for $\mathbf{R}_n$,

$$\frac{1}{r}\frac{d}{dr}\left(r\frac{d\mathbf{R}_n}{dr}\right) + \kappa_n^2\mathbf{R}_n = -\frac{2}{r}\delta(r)Z_n(z_0) \tag{9.7.6}$$

Figure 9.7.1   The fluid layer with a source at depth z_0 between two perfectly reflecting parallel planes.

[1]Officer, *Sound Transmission*, McGraw-Hill (1958). Stephen (ed.), *Underwater Acoustics*, Wiley (1970). Frisk, *Ocean and Seabed Acoustics*, Prentice Hall (1994).

The solution of this equation corresponding to outgoing waves and valid for all r (including the origin) is

$$\mathbf{R_n}(r) = -j\pi Z_n(z_0)H_0^{(2)}(\kappa_n r) \tag{9.7.7}$$

so that the complex pressure field is given by

$$\mathbf{p}(r,z,t) = -j\pi e^{j\omega t}\sum_n Z_n(z_0)Z_n(z)H_0^{(2)}(\kappa_n r) \tag{9.7.8}$$

The allowed values of the κ_n and the form of the orthonormal functions $Z_n(z)$ are found by solving (9.7.3) with the desired speed of sound profile and appropriate boundary conditions.

In many circumstances the separation constants are not all discrete but may also be *continuous* over some interval of κ. The solutions to (9.7.3) for these continuous values of κ form a set of continuous eigenfunctions. Fortunately, these continuous eigenfunctions are associated with untrapped energy or evanescent modes and generate waves significant only close to the source. They can therefore be neglected for our purposes.

At sufficiently large distances the Hankel functions can be replaced with their asymptotic forms and (9.7.8) becomes

$$\mathbf{p}(r,z,t) = -j\sum_n (2\pi/\kappa_n r)^{1/2}Z_n(z_0)Z_n(z)e^{j(\omega t - \kappa_n r + \pi/4)} \tag{9.7.9}$$

Thus, each term of (9.7.8) is a propagating cylindrical wave with phase speed $c_p = \omega/\kappa_n$. The values of the discrete κ_n are fixed, but the magnitude of the propagation vector $k = \omega/c$ can be a function of depth. The angle θ of elevation or depression of the local direction of propagation of the traveling wave is found from $\cos\theta = \kappa_n/k(z)$. Thus, each traveling wave corresponds to a collection of rays traveling in the fluid whose local directions of propagation at each depth z are given by the angles $\pm\theta(z)$.

A convenient analogy can be used to provide further insight for readers having some acquaintance with quantum mechanics. Write the *minimum value* of the speed of sound as c_{min}. Then with the definitions

$$\begin{aligned} E_n &= (\omega/c_{min})^2 - \kappa_n^2 \\ U(z) &= (\omega/c_{min})^2 - (\omega/c)^2 \end{aligned} \tag{9.7.10}$$

the Helmholtz equation (9.7.3) takes on the form

$$\frac{d^2 Z_n}{dz^2} + [E_n - U(z)]Z_n = 0 \tag{9.7.11}$$

This is the one-dimensional time-independent Schroedinger equation with $\hbar^2/2m = 1$. The definition of the minimum speed c_{min} ensures in this analog that $U(z)$ is the potential energy well (with zero minimum value) and E_n is the energy level of the wave function $Z_n(z)$. Now the argument about continuous and discrete values of κ_n can be couched in quantum mechanical terms. If the potential energy $U(z)$ has a finite maximum value, then quantum states having energies E_n large enough that the wave function extends to infinity in either or both directions along the z axis form a continuous set of eigenfunctions so that E_n and therefore κ_n take on continuous values. Thus, unbound quantum states correspond to the untrapped and evanescent modes. When the energy levels lie within the potential well, each wave function has two turning points, E_n and κ_n have discrete values, and these states correspond to the modes trapped in a channel. For a given speed of sound profile, $U(z)$ depends on ω^2. The well becomes more deeply notched with higher walls as frequency increases above cutoff. This means that for a given normal mode the vertical "spread" of the function over depth will tend to be greatest for frequencies close to cutoff and diminish as frequency increases.

For all but a handful of profiles (9.7.3) must be solved by numerical computation. Among those that can be solved analytically, there are a couple of simple cases that provide some physical insight.

*9.8 AN ISOSPEED CHANNEL

Assume that a layer of fluid has constant speed of sound c_0 throughout and is contained by a pressure release surface at $z = 0$ and a rigid bottom at $z = H$. The boundary conditions are $Z(0) = 0$ and $\partial Z/\partial z = 0$ at $z = H$. Solution of (9.7.3) is straightforward,

$$Z_n(z) = \sqrt{2/H} \sin k_{zn} z \qquad k_{zn} = (n - \tfrac{1}{2})\pi/H \tag{9.8.1}$$

and the values of the separation constants κ_n are determined by

$$\kappa_n = [(\omega/c_0)^2 - k_{zn}^2]^{1/2} \tag{9.8.2}$$

For values of k_{zn} exceeding ω/c_0 the associated κ_n values must be imaginary. This yields waves that do not propagate, but decay exponentially with range. Thus, all waves with indices n exceeding the integer N given by

$$N \le (H/\pi)(\omega/c_0) + \tfrac{1}{2} \tag{9.8.3}$$

are evanescent and important only near $r = 0$. At larger distances, the solution is well approximated by

$$\mathbf{p}(r, z, t) \approx -j\frac{2}{H} \sum_{n=1}^{N} \left(\frac{2\pi}{\kappa_n r}\right)^{1/2} \sin k_{zn} z_0 \sin k_{zn} z \, e^{j(\omega t - \kappa_n r + \pi/4)} \tag{9.8.4}$$

The phase speed c_p associated with each mode is given by (9.5.7) with ω_n replacing ω_{lm}:

$$c_p/c = 1/[1 - (\omega_n/\omega)^2]^{1/2} \tag{9.8.5}$$

*9.9 A TWO-FLUID CHANNEL

Let a fluid layer of constant density ρ_1 and sound speed c_1 overlie a fluid bottom of constant density ρ_2 and sound speed $c_2 > c_1$. Let the surface of fluid 1 be a pressure release boundary at $z = 0$ and let the interface between the two fluids be at a depth $z = H$. Figure 9.9.1 shows the geometry. Because the fluid bottom has a greater speed of sound, reflection in fluid 1 from the interface at $z = H$ will be total for grazing angles of incidence less than the grazing critical angle given by $\cos \theta_c = c_1/c_2$.

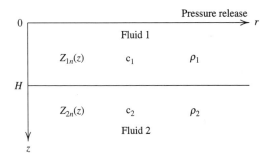

Figure 9.9.1   A channel consisting of a fluid layer of depth H with sound speed c_1 and density ρ_1 overlaying a fluid bottom of infinite depth with sound speed c_2 and density ρ_2.

While c is a function of depth, it is a constant within each layer but changes discontinuously across the interface. We therefore separate the Helmholtz equation

$$\left\{ \frac{d^2}{dz^2} + \left[\left(\frac{\omega}{c}\right)^2 - \kappa_n^2 \right] \right\} Z_n(z) = 0 \tag{9.9.1}$$

into two equations, one for each region,

$$\left\{ \frac{d^2}{dz^2} + \left[\left(\frac{\omega}{c_1}\right)^2 - \kappa_n^2 \right] \right\} Z_{1n}(z) = 0 \qquad 0 \le r \le H$$

$$\left\{ \frac{d^2}{dz^2} + \left[\left(\frac{\omega}{c_2}\right)^2 - \kappa_n^2 \right] \right\} Z_{2n}(z) = 0 \qquad H \le r \le \infty \tag{9.9.2}$$

The boundary conditions are (1) $p_1 = 0$ at $z = 0$, (2) $p_1 = p_2$ and $u_{z1} = u_{z2}$ at $z = H$, and (3) $p_2 \to 0$ as $z \to \infty$. These give us

$$Z_{1n}(0) = 0$$

$$Z_{1n}(H) = Z_{2n}(H)$$

$$\frac{1}{\rho_1}\left(\frac{dZ_{1n}}{dz}\right)_H = \frac{1}{\rho_2}\left(\frac{dZ_{2n}}{dz}\right)_H \tag{9.9.3}$$

$$\lim_{z \to \infty} Z_{2n}(z) = 0$$

Solutions that satisfy the boundary conditions at the surface, at the interface, and at infinite depth are

$$Z_{1n}(z) = \sin k_{zn}z \qquad\qquad 0 \le z \le H$$

$$Z_{2n}(z) = \sin k_{zn}H \, e^{-\beta_n(z-H)} \qquad H \le z \le \infty$$

$$k_{zn}^2 = (\omega/c_1)^2 - \kappa_n^2 \tag{9.9.4}$$

$$\beta_n^2 = \kappa_n^2 - (\omega/c_2)^2$$

Both k_{zn} and β_n must be real for trapped normal modes. This restricts κ_n to the interval $\omega/c_2 \le \kappa_n \le \omega/c_1$ and is equivalent to

$$c_1 \le c_{pn} \le c_2 \tag{9.9.5}$$

Manipulation of the boundary conditions at $z = H$ provides a transcendental equation for the allowed values of k_{zn} (and therefore κ_n) at each angular frequency,

$$\tan k_{zn}H = (\rho_2/\rho_1)(k_{zn}/\beta_n) \tag{9.9.6}$$

Definition of

$$y = k_{zn}H \qquad b = \rho_2/\rho_1 \qquad a = \omega H \sqrt{1/c_1^2 - 1/c_2^2} = (\omega/c_1)H \sin \theta_c \tag{9.9.7}$$

allows (9.9.6) to be expressed in a form amenable to graphical or numerical analysis,

$$\tan y = -by/(a^2 - y^2)^{1/2} \tag{9.9.8}$$

See Fig. 9.9.2. The tangent curves have been numbered to designate the associated normal mode. Since a is proportional to frequency, the tangent curves will be intersected

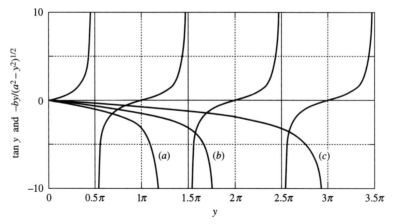

Figure 9.9.2 Graphical solutions for the lower modes of propagation at various frequencies in a shallow-water channel with a fast fluid bottom. The top layer is water with $c_1 = 1500$ m/s, $\rho_1 = 1000$ kg/m^3, and thickness $H = 30$ m. The bottom is quartz sand with $c_2 = 1730$ m/s, $\rho_2 = 2070$ kg/m^3, and infinite thickness. The driving frequencies are (a) 60 Hz, (b) 90 Hz, and (c) 150 Hz.

at different points as the frequency is changed. This is suggested in the figure by the three curves (a), (b) and (c). Since each curve is asymptotic to the appropriate value of $a(\omega)$, it is clear that as a increases with ω the line $y = a$ moves to the right and more normal modes can be excited. The nth normal mode cannot be excited until $a \geq (n - \frac{1}{2})\pi$, and substitution of this into (9.9.7) gives the cutoff angular frequencies,

$$\frac{\omega_n}{c_1} = (n - \tfrac{1}{2})\frac{\pi}{H}\frac{1}{\sin\theta_c} \tag{9.9.9}$$

Once the κ_{zn} have been obtained, κ_n and β_n can be found from (9.9.4). Combination of (9.9.7) and (9.9.9) shows that a is closely related to the ratio of input frequency to the cutoff value,

$$\frac{\omega}{\omega_n} = \frac{a}{(n - \frac{1}{2})\pi} \tag{9.9.10}$$

Figure 9.9.3 reveals the depth dependence of a mode. As the frequency is increased above the cutoff frequency for the nth mode, the value of $k_{zn}H$ increases from $(n - \frac{1}{2})\pi$ at cutoff to $n\pi$ as $\omega \to \infty$. The pressure has an antinode at the interface $z = H$ at cutoff and approaches having a node there at high frequencies. Thus, the interface appears to be rigid at cutoff and pressure release at high frequencies. Evaluation of κ_n at cutoff gives $\kappa_n = \omega_n/c_2$, and (9.9.4) shows that $\beta_n = 0$ so that the normal mode has an extended tail down to infinite depths. As frequency increases above cutoff, κ_n increases, β_n becomes positive real, and the tail decays more rapidly with depth. As frequency becomes arbitrarily large, the tail disappears. This is consistent with the discussion after (9.7.11) about the diminishing vertical extent of the normal mode with increasing frequency.

Trying to form an orthonormal set from Z_n by assuming $Z_n = A_n Z_{1n}$ in the layer and $Z_n = A_n Z_{2n}$ below the layer will not work because of the discontinuity in slope across the boundary at $z = H$. Applying the orthonormality condition to (9.9.1) gives

$$\int_0^\infty \frac{d}{dz}\left(Z_m \frac{dZ_n}{dz} - Z_n \frac{dZ_m}{dz}\right)dz = 0 \tag{9.9.11}$$

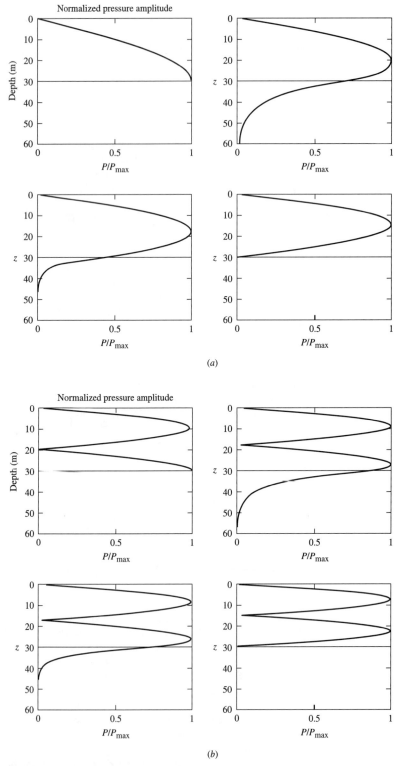

Figure 9.9.3 The depth dependence of the pressure amplitude for various driving frequencies in the shallow-water channel for Figure 9.9.2. (*a*) First propagating mode. (*b*) Second propagating mode. For each mode, the driving frequency increases from just above cutoff (upper left), where the bottom behaves like a rigid surface, to a high frequency (lower right), where the behavior of the bottom approaches that of a free surface.

Integrating and applying the boundary conditions results in

$$\left(Z_m \frac{dZ_n}{dz} - Z_n \frac{dZ_m}{dz}\right)\Bigg|_{H+}^{H-} = 0 \tag{9.9.12}$$

where the upper evaluation is accomplished by approaching the boundary from fluid 1 (z increasing to H) and the lower by approaching the boundary from fluid 2. Direct substitution of the boundary conditions (9.9.3) into (9.9.12) shows that the equation cannot be satisfied by the above assumption. However, the slightly more complicated choice

$$Z_n(z) = \begin{cases} A_n \sin k_{zn}z & 0 \le z \le H \\ A_n(\rho_1/\rho_2)^{1/2} \sin k_{zn}H\, e^{-\beta_n(z-H)} & H \le z \le \infty \end{cases} \tag{9.9.13}$$

does satisfy (9.9.12). Thus, (9.9.13) forms a set of orthogonal eigenfunctions with respect to the *weighting function* $\sqrt{\rho_1/\rho(z)}$. Normalization of the set provides the required values of the A_n for orthonormality,

$$\frac{1}{A_n^2} = \int_0^H Z_{1n}^2(z)\, dz + \frac{\rho_1}{\rho_2} \int_H^\infty Z_{2n}^2(z)\, dz \tag{9.9.14}$$

$$= (1/2k_{zn})[k_{zn}H - \cos k_{zn}H \sin k_{zn}H - (\rho_1/\rho_2)^2 \sin^2 k_{zn}H \tan k_{zn}H]$$

The acoustic pressures in fluid 1 and fluid 2 are found by substituting (9.9.13) into (9.7.8),

$$\mathbf{p}_1(r, z, t) = -j\pi \sum_n A_n^2 \sin k_{zn}z_0 \sin k_{zn}z\, H_0^{(2)}(\kappa_n r)e^{j\omega t}$$

$$\rightarrow -j\sum_n (2\pi/\kappa_n r)^{1/2} A_n^2 \sin k_{zn}z_0 \sin k_{zn}z\, e^{j(\omega t - \kappa_n r + \pi/4)}$$

$$\mathbf{p}_2(r, z, t) = -j\pi \sum_n A_n^2 \sin k_{zn}z_0 \sin k_{zn}z\, e^{-\beta_n(z-H)} H_0^{(2)}(\kappa_n r)e^{j\omega t} \tag{9.9.15}$$

$$\rightarrow -j\sum_n (2\pi/\kappa_n r)^{1/2} A_n^2 \sin k_{zn}z_0 \sin k_{zn}z\, e^{-\beta_n(z-H)} e^{j(\omega t - \kappa_n r + \pi/4)}$$

Calculation of the group and phase speeds is a little tricky and the details are treated in Problem 9.9.8. Results can be expressed in implicit forms,

$$\left(\frac{c_1}{c_{pn}}\right)^2 = 1 - \frac{\sin^2 \theta_c}{1 + (b \cot y)^2} \tag{9.9.16}$$

$$\frac{c_1}{c_{gn}}\frac{c_1}{c_{pn}} = 1 - \frac{(\sin \theta_c \sin y)^2}{\sin^2 y + b^2(\cos^2 y - y \cot y)}$$

where $(n - \frac{1}{2})\pi \le y \le n\pi$. Since y increases monotonically with frequency, certain properties of the group and phase speeds can be determined. (1) At cutoff $\cos y = 0$ and (9.9.16) shows that $c_{pn} = c_{gn} = c_2$. With increasing frequency, (2) the phase speed falls monotonically toward an asymptotic value of c_1 and (3) the group speed also approaches the value c_1, but from below, so that (4) the group speed has a minimum value that is less than c_1 at some intermediate frequency. See Fig. 9.9.4 and Problem 9.9.14C.

The fact that for each mode the group speed has a minimum and approaches c_2 for frequencies near cutoff leads to a complicated waveform for a transient excitation. The following general features, sketched in Fig. 9.9.5, can be identified with propagation in each mode.

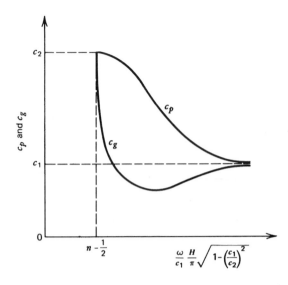

Figure 9.9.4   Group and phase speeds for a normal mode propagating in an isospeed shallow-water channel of depth H with a fast fluid bottom. The speed of sound in the channel is c_1 and that in the bottom is c_2.

1. The *first arrival* reaches the receiver at a time $t = r/c_2$. It consists of Fourier components of the transient having frequencies very close to the cutoff value for the mode and propagating with the group speeds near cutoff. As time increases, slightly higher frequencies traveling with lower group speeds will arrive. This portion of the signal is the *ground wave* and corresponds to energy propagating along the boundary in fluid 2 and radiating back into the layer.

2. At a later time $t = r/c_1$, the highest frequency components arrive with group speeds at and slightly below c_1 and are superimposed on the trailing portion of the ground wave. This high-frequency portion is the *water wave* and corresponds to the high-frequency energy that is propagated radially outward in the channel with angles of elevation and depression very close to zero.

3. For still later times the increasing frequencies in the ground wave and the decreasing frequencies in the water wave become similar and merge into a signal traveling at group speeds slightly above the minimum group speed for the mode. This *Airy phase* comes to a relatively abrupt termination when the energy traveling at the minimum group speed arrives.

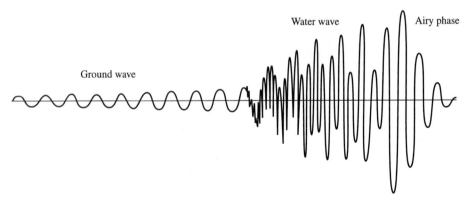

Figure 9.9.5   Sketch of the signal received from a transient propagated in a shallow-water channel with a fast fluid bottom. (After Ewing, Jardetzky, and Press, *Elastic Waves in Layered Media*, McGraw-Hill, 1957.)

PROBLEMS

Unless otherwise indicated, the fluid is air. If not specified, assume that $c = 343$ m/s for air and $c = 1500$ m/s for water.

9.2.1. Calculate the frequencies for the ten lowest normal modes of a rigid-walled room of dimensions 2.59 m $\times$ 2.42 m $\times$ 2.82 m.

9.2.2. Calculate the five lowest normal-mode frequencies of a water-filled cubical cavity (L on a side), which has five rigid sides and one pressure release side. Sketch the pressure distributions associated with these modes. Which of these modes, if any, are degenerate?

9.2.3. For the $(l, m, 0)$ modes in a rigid-walled rectangular cavity, show that (a) the surfaces of constant phase of the traveling plane waves are perpendicular to the z axis, (b) the propagation velocity of the surface has magnitude ω/k, lies in the x-y plane, and makes an angle $\tan^{-1}(k_y/k_x)$ measured from the x axis, and (c) the intersection of the surface with the y axis travels along that axis with a speed ω/k_y.

9.2.4. (a) Calculate the frequencies for the lowest ten normal modes of a rigid-walled, cubical room L on a side. (b) Sketch the nodal patterns of these modes. (c) Which modes are degenerate? (d) For each set of degenerate modes, where should a source be located to excite just one of the modes?

9.2.5. A rigid-walled, cubical room L on a side has a sound source in one corner and a receiver in the furthest opposite corner. For each of the lowest ten normal modes, what is the phase between the source and receiver?

9.2.6C. For the room in Problem 9.2.1, assume that the amplitude of the nth mode is $A_n = f_n/f_1$ and the quality factor is $Q_n = 10 f_n/f_1$. Plot the pressure at the receiver for frequencies covering the lowest ten normal modes. Comment on what effect the absorption can have on determining the normal modes of a room experimentally.

9.3.1. Starting from (9.3.7), work out the details to get the complex amplitudes and phases for the cylindrical traveling waves represented by (9.3.9).

9.3.2. For the $(l, 0, n)$ modes of a cylindrical standing wave, show that (a) the surfaces of constant phase of the traveling plane waves are cones with vertices lying on the z axis, (b) the propagation velocity of each surface has magnitude ω/k, lies in the plane $\theta = constant$, and has an angle $\tan^{-1}(k_z/k_{0n})$ measured from the $z = 0$ plane, (c) the vertices travel along the z axis with a speed ω/k_z, and (d) the circles marking the intersection of the surfaces with the plane $z = constant$ travel inward or outward with speeds ω/k_{0n}.

9.3.3. (a) Calculate the frequencies for the five lowest normal modes for a cylindrical room of 10 m diameter and 3 m height. (b) Sketch the nodal pattern for each of these modes. (c) Which of these modes, if any, are degenerate?

9.3.4. A cylindrical water tank of 1 m radius and 1 m depth has walls and bottom that are rigid and a pressure release top. (a) Calculate frequencies for the five lowest normal modes. (b) Calculate the locations of all nodes for each of these modes.

9.3.5C. Calculate the fundamental frequency of an air-filled, rigid-walled cavity between concentric cylinders of inner radius 10 cm, outer radius 50 cm, and height 10 cm.

9.4.1. (a) Calculate the frequencies for the lowest three sets of normal modes of an air-filled spherical cavity of 20 cm radius if the walls are rigid. (b) Calculate the radii of all spherical nodal and antinodal surfaces for each of these modes. (c) Which modes can be excited by a source at the center of the sphere?

9.4.2. Repeat Problem 9.4.1 for the same sized and shaped cavity, but filled with water and having pressure release walls. Use the speed of sound given by (5.6.8) and assume $T = 4°C$.

9.4.3C. For the cavity of Problem 9.4.2, plot the change in the fundamental frequency as a function of temperature for $0 < T < 10°C$. The coefficient of volume expansion for water in this temperature range is $\beta \approx 15(T - 4) \times 10^{-6}$ per C° and the speed of sound is given by (5.6.8).

9.5.1. An open irrigation canal measures 30 m wide and 10 m deep. If it is completely filled with water, calculate the cutoff frequencies for the five lowest modes of propagation. Assume the sides and bottom are perfectly rigid.

9.5.2. Consider a fluid in an infinitely long rigid-walled pipe of square cross section of dimension L. For the lowest five sets of modes, (a) find the cutoff frequencies and (b) sketch the pressure distribution across the pipe.

9.5.3. Calculate the cutoff frequencies for the lowest five modes in a water-filled circular waveguide of 10 cm radius and with pressure release walls.

9.5.4. Show from (9.6.10) that $c_g/c = \sqrt{1 - (\omega_{lm}/\omega)^2}$ for a rigid-walled waveguide.

9.5.5. A water-filled waveguide has a square cross section 10 cm on a side and pressure release walls. (a) Calculate the cutoff frequencies for the lowest five modes. (b) Which of these modes, if any, have the same cutoff frequencies? (c) If a point source were placed in the center of the cross section, which of these modes would be excited? (d) Is it possible to position a point source so that, of these five modes, only one is excited? If so, where?

9.5.6C. A water-filled waveguide of square cross section 10 cm on a side has pressure release walls. (a) Plot the phase and group speeds for excitation of the propagating mode with the lowest cutoff frequency. (b) For frequencies below the lowest cutoff frequency, plot the distance along the axis of the waveguide at which the amplitude of the evanescent mode is reduced to $1/e$ against frequency.

9.6.1. A point impulse pressure source is located at $(x, y, z) = (0, 0, 1.41)$ m in a rigid-walled room with $L_x = 2.59$ m, $L_y = 2.42$ m, and $L_z = 2.82$ m. Find the ratio of the maximum pressure amplitudes of the lowest two excited normal modes.

9.6.2. A point impulse pressure source is located at $(0, L/2, L/4)$ in a cubical, rigid-walled room measuring L on each side. Compare the maximum pressure amplitude of each mode in the lowest two sets of degenerate normal modes to that of the $(1, 0, 0)$ mode.

9.6.3. Use $c_p = \omega/k_z = c/\cos\theta$, where $\cos\theta = \sqrt{1 - (\omega_n/\omega)^2}$ and $c_g = d\omega/dk_z$ to show that group speed in a waveguide of uniform cross section is $c_g = c\cos\theta$.

9.6.4. A point pressure source, located at the center of the cross section of a square waveguide with pressure release sides, is driven at a frequency just above three times that of the cutoff for the lowest propagating mode. (a) Which modes have cutoff frequencies that would allow them to propagate? (b) Of these, which are excited by the source? (c) What are the relative amplitudes of the excited modes compared to the lowest propagating mode?

9.6.5. Obtain the velocity potential Φ_{lm} from the pressure in (9.6.12) by direct integration over time. *Hint:* Note that all acoustic quantities must vanish for $t < 0$. (b) Generate u_{zlm} from Φ_{lm} and obtain thereby an equation relating the integral over t of df/dz and the integral expression for $v(z, t)$. (c) Differentiate the equation in (b) with respect to t and then integrate with respect to z to obtain $f(z, t)$ in (9.6.13).

9.6.6. (*a*) Show that the wave equation can be separated so that the differential equation for the fragment f in (9.6.12) for the (z, t) dependence is

$$\frac{\partial^2 f}{\partial z^2} - \frac{1}{c^2}\frac{\partial^2 f}{\partial t^2} - \left(\frac{\omega_{lm}}{c}\right)^2 f = 0$$

(*b*) If $f(z, t)$ satisfies this differential equation, show that $f(z, t) \cdot 1(t - z/c)$ also does if the condition

$$\frac{\partial f}{\partial z} = -\frac{1}{c}\frac{\partial f}{\partial t}\Big|_{t=z/c}$$

is satisfied. (*c*) Show that if $f(z, t)$ has argument $[t^2 - (z/c)^2]$ then the condition in (*b*) is satisfied. (*d*) Show that the differential equation is solved by $J_0(u) \cdot 1(t - T)$, where $u = \omega_{lm}\sqrt{t^2 - T^2}$ and $T = z/c$.

9.6.7. Show by generating Φ from p and then obtaining u_z that the pressure associated with (9.6.14) is consistent with the boundary condition for $v(0, t)$.

9.6.8C. (*a*) On one set of axes, graph the solution (9.6.14) at two large but not too different distances z_1 and z_2. Plot the waves as functions of *elapsed* time $\tau = t - z_1/c$ at z_1 and $\tau = t - z_2/c$ at z_2 so that both functions start at $\tau = 0$. Use an interval long enough that the dispersion is clear and an increment small enough that all the cycles of the motion are reproduced. (*b*) Select an axis crossing the same number of cycles behind the fronts of the signals and see how it appears further forward in the wave packet at the larger distance. Determine from the graphs the phase speed of the local frequency around that axis crossing and compare with theory. (Your accuracy will depend on your selections of distances and the axis crossing, so some experimentation with the parameters may be necessary, and it will be helpful if you can "zoom" in on your graphs.) (*c*) Repeat (*a*) but for two distances whose ratio is about 2 or more. (*d*) In the graphs of (*c*), find the locations of two regions of equal frequency many cycles removed from the front of the wave packet, determine the group speed of that frequency, and compare with the theoretical prediction.

9.7.1. Verify that (9.7.7) is a solution of (9.7.6) for all r including $r = 0$. *Hint:* Adapt the approach of Problem 5.16.4, but for a monofrequency wave in cylindrical coordinates.

9.7.2. Show that (9.7.8) exhibits pure cylindrical spreading plus a depth-dependent correction term accounting for the ability of the source to excite the individual normal modes and the receiver to sense them.

9.8.1. A layer of fluid has constant sound speed c and is contained between rigid walls at $z = 0$ and $z = H$. (*a*) Find the depth dependences of the propagating modes. (*b*) Find the phase speeds of the propagating modes. (*c*) For a given driving angular frequency ω, find the number of the highest mode that will propagate. (*d*) Write the equation for $p(r, z, t)$ applicable at great distance from a source at the origin.

9.8.2. A 12 m layer of isospeed water with a pressure release top overlies a rigid bottom. A unit pressure source at a depth of 8 m is driven at a frequency four times that of the cutoff frequency of the lowest propagating mode. (*a*) How many modes propagate? (*b*) Find $P(r, z)$ at great distances from the source.

9.8.3. Assume the channel of Problem 9.8.2. A unit pressure source is at a depth of 4 m and is driven at a frequency four times that of the cutoff frequency of the lowest propagating mode. (*a*) How many modes propagate? (*b*) If the receiver is at a depth of 8 m, find the received $P(r)$ at great distances from the source.

9.8.4. Assume the channel of Problem 9.8.2. A unit pressure source is at a depth of 4 m and is driven at a frequency four times that of the cutoff frequency of the lowest propagating mode. The receiver is also at a depth of 4 m. (*a*) Find $P(r)$ at great distances from the source. (*b*) Find the distance over which the lowest two modes will go through a relative phase shift of 180°.

9.8.5C. For the conditions of Problem 9.8.4, plot the pressure amplitude for a range of distances at least twice the distance calculated in Problem 9.8.4*b*.

9.8.6C. Assume the channel of Problem 9.8.2. A unit pressure source is at a depth of 4 m and is driven at a frequency four times that of the cutoff frequency of the lowest propagating mode. Plot equal contours of pressure amplitude for all depths and for large ranges.

9.8.7C. Because of tides, the depth of the water channel described in Problem 9.8.2 varies from 8 m to 12 m. A unit pressure source on the bottom is driven at a frequency four times that of the cutoff frequency of the lowest propagating mode when the water depth is 12 m. The receiver is also on the bottom. Plot the pressure amplitude at the receiver as a function of water depth.

9.8.8. For an isospeed layer with perfectly reflecting surfaces, show that conservation of energy yields $P(1)/P(r) = \sqrt{rH/2}$.

9.8.9. (*a*) Show that the isospeed channel with pressure release surface and rigid bottom can be modeled by a square potential well that is zero between $z = -H$ to $z = H$ and has infinite height. (*b*) Show that the boundary conditions at the surface and bottom are satisfied by retaining only the antisymmetric wave functions and requiring that they have antinodes at $z = \pm H$. (*c*) Find the allowed values of k_{zn} and κ_n and compare with (9.8.1) and (9.8.2).

9.9.1. Verify that the solution and equalities of (9.9.4) satisfy the Helmholtz equation and the boundary conditions for both fluids.

9.9.2. (*a*) If β is allowed to be imaginary, find solutions to (9.9.2) for all depths (these solutions do not decay to zero at infinite depth). (*b*) Verify that for these solutions k_z can have any value greater than $(\omega/c_1)\sin\theta_c$ and is not quantized. (*c*) Find the values of κ allowed for these solutions. (*d*) Show that these solutions correspond to rays whose angles of incidence on the bottom are less grazing than critical and therefore are not trapped in the layer.

9.9.3. (*a*) Show that the two-fluid channel can be modeled as a square potential well that is zero for $|z| < H$ and has height $(\omega/c_1)^2\sin^2\theta_c$ for $|z| > H$. (*b*) Show that the pressure release surface at $z = 0$ is equivalent to retaining only the antisymmetric quantum wave functions. (*c*) Find the values of k_z and κ corresponding to the untrapped waves.

9.9.4. Show that the choice $Z_n = A_n Z_{1n}$ in the layer and $Z_n = A_n Z_{2n}$ in the bottom will not yield Z_n as an orthogonal set of eigenfunctions. *Hint:* Show this choice does not satisfy (9.9.12).

9.9.5. Show that for $Z_n = A_n Z_{1n}$ in the layer and $Z_n = A_n\sqrt{\rho_1/\rho_2}\,Z_{2n}$ in the bottom the Z_n form an orthogonal set of eigenfunctions.

9.9.6. Fill in the steps to evaluate A_n in (9.9.14).

9.9.7. Obtain (9.9.6) from the boundary conditions and show that it generates (9.9.8).

9.9.8. Derive the expressions in (9.9.16) for the group and phase speeds. For the phase speed, solve (9.9.8) for $(y/a)^2$ and then use a in (9.9.7) and κ_n from (9.9.4). For the group speed, define $x = \kappa_n H$, show $da/dx = (c_{gn}/c_1)\sin\theta_c$, then take differentials of $(y/a)^2$ and κ_n to get equations in da, dx, and dy, and eliminate dy.

9.9.9. For the two-fluid channel, the cutoff frequencies are found from (9.9.9). Derive this equation from ray concepts utilizing the grazing critical angle and that $k_{zn}H = (n-\frac{1}{2})\pi$ at cutoff.

9.9.10. An isospeed layer of water 12 m deep ($c_1 = 1500$ m/s, $\rho_1 = 10^3$ kg/m^3) overlies a layer of infinite depth ($c_2 = 1600$ m/s, $\rho_2 = 1.25 \times 10^3$ kg/m^3). (a) Calculate the critical grazing angle at the interface. (b) Calculate the cutoff frequencies for the lowest three propagating modes. (c) For the lowest two propagating modes, find k_{zn}, κ_n, and A_n for frequencies 1, 2, 3, and 4 times the cutoff frequency of the lowest propagating mode.

9.9.11C. Assume the layer and frequencies of Problem 9.9.10. For a source at a depth of 6 m and a receiver at a range of 100 m, plot the pressure amplitude as a function of depth in both the water and in the bottom.

9.9.12C. Assume the layer and frequencies of Problem 9.9.10. For source and receiver both at a depth of 6 m, plot the pressure amplitude as a function of range.

9.9.13C. Assume the layer and frequencies of Problem 9.9.10. For a source at a depth of 6 m, plot contours of equal pressure amplitude.

9.9.14C. For the layer of Problem 9.9.10, plot c_p and c_g as functions of f.

9.9.15. An isospeed layer of sea water ($c_1 = 1.5 \times 10^3$ m/s) of depth 100 m overlies a fluid bottom with $\rho_2 = 1.25\rho_1$ and $c_2 = 1.6 \times 10^3$ m/s. An omnidirectional 3.5 kHz source has a source level $SL(re\ 1\ \mu Pa) = 200$ dB and is at a depth of 50 m below the sea surface (a) Evaluate the intensity level at a range of 20 km and the same depth, assuming that just the lowest mode is important at this distance. (b) Comment on the results for other choices of depths. (c) Find the distance over which the lowest two normal modes will go through a relative phase shift of 180°.

PIPES, RESONATORS, AND FILTERS

10.1 INTRODUCTION

The behavior of sound in a rigid-walled pipe depends strongly on the properties of the driver, the length of the pipe, the behavior of its cross section as a function of distance, the presence of any perforations of its wall, and the boundary conditions describing any termination. If the wavelength of the sound is sufficiently large, the wave motion can be considered to be well approximated by a collimated plane wave; this affords great simplification. Applications include measuring the absorptive and reflective properties of materials, predicting the behavior of wind instruments (brasses, woodwinds, organ pipes, etc.), and determining the design of ventilation ducts.

Segments of pipes having all dimensions sufficiently small compared to the relevant wavelengths can be considered as *lumped acoustic elements* whose behaviors resemble those of simple oscillators. These lumped elements find application as convenient models for more complicated systems at low frequencies, allowing straightforward design of the noise transmission characteristics of pipes, ducts, mufflers, and so forth, without materially affecting any required steady flow of fluid through the system.

10.2 RESONANCE IN PIPES

Assume that the fluid in a pipe of cross-sectional area S and length L is driven by a piston at $x = 0$ and that the pipe is terminated at $x = L$ in a mechanical impedance $\mathbf{Z}_{mL}$. If the piston vibrates at frequencies for which only plane waves propagate, the wave in the pipe will be of the form

$$\mathbf{p} = \mathbf{A}e^{j[\omega t + k(L-x)]} + \mathbf{B}e^{j[\omega t - k(L-x)]} \tag{10.2.1}$$

where $\mathbf{A}$ and $\mathbf{B}$ are determined by the boundary conditions at $x = 0$ and $x = L$.

The continuities of force and particle speed require that the mechanical impedance of the wave at $x = L$ equals the mechanical impedance $\mathbf{Z}_{mL}$ of the termination. Since the force of the fluid on the termination is $\mathbf{p}(L, t)S$ and the particle speed is $\mathbf{u}(L, t) = -(1/\rho_0) \int (\partial \mathbf{p}/\partial x) \, dt$,

$$\mathbf{Z}_{mL} = \rho_0 c S \frac{\mathbf{A} + \mathbf{B}}{\mathbf{A} - \mathbf{B}} \tag{10.2.2}$$

The input mechanical impedance $\mathbf{Z}_{m0}$ at $x = 0$ is

$$\mathbf{Z}_{m0} = \rho_0 c S \frac{\mathbf{A}e^{jkL} + \mathbf{B}e^{-jkL}}{\mathbf{A}e^{jkL} - \mathbf{B}e^{-jkL}} \tag{10.2.3}$$

Combining these equations to eliminate $\mathbf{A}$ and $\mathbf{B}$, we obtain

$$\frac{\mathbf{Z}_{m0}}{\rho_0 c S} = \frac{(\mathbf{Z}_{mL}/\rho_0 c S) + j \tan kL}{1 + j(\mathbf{Z}_{mL}/\rho_0 c S) \tan kL} \tag{10.2.4}$$

which is identical to (3.7.3) with the replacement of $\rho_L c$ with $\rho_0 c S$, the *characteristic mechanical impedance* of the fluid. The substitution

$$\mathbf{Z}_{mL}/\rho_0 c S = r + jx \tag{10.2.5}$$

leads directly to (3.7.5). Recalling the discussion following that equation, the frequencies of resonance and antiresonance are determined by the vanishing of the input mechanical reactance, which requires

$$x \tan^2 kL + (r^2 + x^2 - 1) \tan kL - x = 0 \tag{10.2.6}$$

The solution identified with *small* input resistance denotes *resonance*, and that identified with *large* input resistance denotes *antiresonance*. (In the limiting case $r = 0$, there is only one solution, corresponding to resonance.)

Let the pipe be driven at $x = 0$ and *closed* at $x = L$ by a rigid cap. To obtain the condition of resonance most simply, let $|\mathbf{Z}_{mL}/\rho_0 c S| \to \infty$ in (10.2.4). This yields

$$\mathbf{Z}_{m0}/\rho_0 c S = -j \cot kL \tag{10.2.7}$$

The reactance is zero and resonance occurs when $\cot kL = 0$,

$$k_n L = (2n - 1)\pi/2 \qquad n = 1, 2, 3, \ldots \tag{10.2.8}$$

This is identical to (2.9.9) for the forced, fixed string. The resonance frequencies are the odd harmonics of the fundamental. The driven, closed pipe has a pressure antinode at $x = L$ and a pressure node at $x = 0$. Note that this requires that the driver presents a vanishing mechanical impedance to the pipe. The implication of this, and the effects of the mechanical properties of the driver on the behavior of the driver–pipe system, will be discussed in Section 10.6.

Now, consider a pipe driven at $x = 0$ and *open-ended* at $x = L$. On first examination, it might be thought that this will lead to $\mathbf{Z}_{mL} = 0$ for which

$Z_{m0}/\rho_0 cS = j \tan kL$ with resonances occurring at $f_n = (n/2)c/L$ for $n = 1, 2, 3, \ldots$. However, this is *not* the case, most elementary physics textbooks notwithstanding. The condition at $x = L$ is not $Z_{mL} = 0$ since the open end of the pipe radiates sound into the surrounding medium. The appropriate value for Z_{mL} is therefore

$$Z_{mL} = Z_r \qquad (10.2.9)$$

where Z_r is the radiation impedance of the open end of the pipe.

For example, assume that the open end of a circular pipe of radius a is surrounded by a *flange* large with respect to the wavelength of the sound. Consistent with the assumption that the wavelength is large compared to the transverse dimensions of the pipe ($\lambda \gg a$), the opening resembles a baffled piston in the low-frequency limit. We have, then, from (7.5.11) *et seq.*

$$Z_{mL}/\rho_0 cS = \tfrac{1}{2}(ka)^2 + j(8/3\pi)ka \qquad \text{(flanged)} \qquad (10.2.10)$$

where both $r = (ka)^2/2$ and $x = 8ka/3\pi$ are much less than unity. Solution of (10.2.6) under these conditions gives $\tan kL = -x$ for the resonance frequencies. Since $x \ll 1$, this yields

$$\tan(n\pi - k_n L) = (8/3\pi)ka \approx \tan(8ka/3\pi) \qquad n = 1, 2, 3, \ldots \quad (10.2.11)$$

Therefore,

$$n\pi = k_n L + (8/3\pi)k_n a \qquad (10.2.12)$$

and the resonance frequencies are

$$f_n = \frac{n}{2} \frac{c}{L + (8/3\pi)a} \qquad (10.2.13)$$

These resonance frequencies are all harmonics of the fundamental, and the *effective length* L_{eff} of such a pipe is not L but rather $L + 8a/3\pi$. This predicted *end correction* is in reasonable agreement with measured values of around $0.85a$.

For an *unflanged* open pipe, both experiments and theory indicate that the radiation impedance is approximately

$$Z_{mL}/\rho_0 cS = \tfrac{1}{4}(ka)^2 + j\, 0.6\, ka \qquad \text{(unflanged)} \qquad (10.2.14)$$

so the effective length of an unflanged open pipe is $L_{eff} = L + 0.6a$.

In both cases, the end corrections are independent of frequency. The resonance frequencies of flanged and unflanged open pipes are harmonics of the fundamental (as long as $\lambda_n \gg a$). This result has been obtained only for pipes of constant cross section. The presence of *flare* in the pipe, as found in many wind instruments and some organ pipes, modifies these results. In particular, the resonance frequencies may no longer be harmonics of the fundamental. Designing the flare is very important in emphasizing or reducing certain of the harmonics present in the forcing function, thereby controlling the *timbre* of the sound radiated by the pipe.

10.3 POWER RADIATION FROM OPEN-ENDED PIPES

Solution of (10.2.2) for $\mathbf{B}/\mathbf{A}$ yields

$$\frac{\mathbf{B}}{\mathbf{A}} = \frac{\mathbf{Z}_{mL}/\rho_0 c S - 1}{\mathbf{Z}_{mL}/\rho_0 c S + 1} \tag{10.3.1}$$

and the power transmission coefficient can be found from

$$T_\Pi = 1 - |\mathbf{B}/\mathbf{A}|^2 \tag{10.3.2}$$

once the termination impedance $\mathbf{Z}_{mL}$ is known.

For an open-ended pipe terminated in a flange, $\mathbf{Z}_{mL}$ is given by (10.2.10), and (10.3.1) becomes

$$\frac{\mathbf{B}}{\mathbf{A}} = -\frac{\left[1 - \frac{1}{2}(ka)^2\right] - j(8/3\pi)ka}{\left[1 + \frac{1}{2}(ka)^2\right] + j(8/3\pi)ka} \tag{10.3.3}$$

For $ka \ll 1$ this shows that $\mathbf{B}/\mathbf{A}$ is very nearly -1. The transmission coefficient is extremely small and can be further simplified,

$$T_\Pi \approx 2(ka)^2 \qquad \text{(flanged)} \tag{10.3.4}$$

The pressure amplitude of the reflected wave is only slightly less than that of the incident wave and (10.2.1) shows that the two are nearly 180° out of phase at the open end. A condensation is reflected as a rarefaction. In contrast, the incident and reflected particle speeds are nearly in phase so that this position is approximately an antinode of particle speed. Thus, even though the amplitude of the particle speed at the orifice is almost twice that of the incident wave, only a small fraction of the incident power is transmitted out of the pipe. As seen previously, small ($ka \ll 1$) sources are inefficient radiators.

For an unflanged pipe, $\mathbf{Z}_{mL}$ is given by (10.2.14), and the transmission coefficient becomes

$$T_\Pi \approx (ka)^2 \qquad \text{(unflanged)} \tag{10.3.5}$$

A wide flange at the end of a pipe approximately doubles the radiated acoustic power at low frequencies. (When a pipe is terminated in a gradual flare the low-frequency power transmission is increased still further.)

In the vicinity of resonance we can write $\omega = \omega_n + \Delta\omega$, and the input impedance of the unflanged pipe (10.2.4) with (10.2.14) is then well approximated by

$$Z_{m0}/\rho_0 c S \approx \tfrac{1}{4}\left(k_n a\right)^2 + j\,\Delta\omega\,L/c \tag{10.3.6}$$

The half-power points are

$$\omega_{u,l} = \omega_n \pm \tfrac{1}{4}\left(k_n a\right)^2 c/L \tag{10.3.7}$$

and the Q for the nth resonance is

$$Q_n = \frac{\omega_n}{\omega_u - \omega_l} = \frac{2}{n\pi}\frac{L}{a}\frac{L + 0.6a}{a} \tag{10.3.8}$$

The radiated power, $\Pi = F^2 R_{m0}/2Z_{m0}^2$, where $R_{m0} = Re\{\mathbf{Z}_{m0}\}$ and F is the force amplitude, has the value

$$\Pi_n = \frac{F^2}{\rho_0 cS}\frac{2}{(k_n a)^2} = \frac{2}{(n\pi)^2}\frac{F^2}{\rho_0 cS}\left(\frac{L + 0.6a}{a}\right)^2 \tag{10.3.9}$$

Thus, we see that, for constant applied force amplitude, the Q values of the resonances decrease as $1/n$ and the power radiated at resonance decreases as $1/n^2$ in the low-frequency region.

10.4 STANDING WAVE PATTERNS

The phase interference between the transmitted and reflected waves in a terminated pipe results in a standing wave pattern whose properties can be used to determine the load impedance. Let us choose to write

$$\mathbf{A} = A \qquad \mathbf{B} = Be^{j\theta} \tag{10.4.1}$$

where A and B are real and positive. Substituting (10.4.1) into (10.2.2) gives

$$\frac{\mathbf{Z}_{mL}}{\rho_0 cS} = \frac{1 + (B/A)e^{j\theta}}{1 - (B/A)e^{j\theta}} \tag{10.4.2}$$

Given B/A and θ, $\mathbf{Z}_{mL}$ can be determined. By substituting (10.4.1) into (10.2.1) and solving for the amplitude $P = |\mathbf{p}|$ of the wave, we obtain

$$P = \{(A + B)^2 \cos^2[k(L - x) - \theta/2] + (A - B)^2 \sin^2[k(L - x) - \theta/2]\}^{1/2} \tag{10.4.3}$$

The amplitude at a pressure antinode is $A + B$, and the amplitude at a node is $A - B$. The ratio of pressure amplitude at an antinode to that at a node is the *standing wave ratio*

$$SWR = \frac{A + B}{A - B} \tag{10.4.4}$$

which can be rearranged to provide

$$\frac{B}{A} = \frac{SWR - 1}{SWR + 1} \tag{10.4.5}$$

Thus, measurement of the SWR by probing the pressure field in the pipe with a small microphone yields a value for B/A. The phase angle θ can be evaluated from the distance of the first node from the end at $x = L$. From (10.4.3), these nodes are located at $k(L - x_n) - \theta/2 = (n - \frac{1}{2})\pi$, so for the first node

$$\theta = 2k(L - x_1) - \pi \tag{10.4.6}$$

For example, let us assume that in some terminated pipe the standing wave ratio is $SWR = 2$ and the first node is $3/8$ of a wavelength from the end. Then $L - x_1 = 3\lambda/8$, and $\theta = 2(2\pi/\lambda)(3\lambda/8) - \pi = \pi/2$. Furthermore, $B/A = (2 - 1)/(2 + 1) = 1/3$ and

$$\mathbf{Z}_{mL}/\rho_0 cS = \left(1 + \tfrac{1}{3}e^{j\pi/2}\right)\Big/\left(1 - \tfrac{1}{3}e^{j\pi/2}\right) \approx 0.80 + j0.60 \tag{10.4.7}$$

Since the mechanical impedance of the termination can be a complicated function of frequency, it may be necessary to repeat the above measurements over the frequency range of interest.

The reflective and absorptive properties, at normal incidence, of such materials as acoustic tiles and other sound control materials can be determined by mounting a small section of the material at the end of a standing wave tube and making the measurements and calculations described in this section. A *Smith chart* can be used to facilitate the calculations.[1]

10.5 ABSORPTION OF SOUND IN PIPES

If absorptive processes within the fluid and at the walls of the pipe are considered, all we have to do is replace k in the lossless solutions obtained above with the complex propagation constant $\mathbf{k} = k - j\alpha$, just as in Chapter 8.

As an example, for a rigid termination at $x = L$ the pressure is

$$\mathbf{p}(x, t) = \frac{F}{S} \frac{\cos[\mathbf{k}(L - x)]}{\cos \mathbf{k}L} e^{j\omega t} \tag{10.5.1}$$

and the input impedance (10.2.7) is

$$\mathbf{Z}_{m0} = -j\rho_0(\omega/\mathbf{k})S \cot \mathbf{k}L \tag{10.5.2}$$

With the help of the expansions of sines and cosines of complex argument, this becomes

$$\frac{\mathbf{Z}_{m0}}{\rho_0 cS} = -j \frac{1 + j\alpha/k}{1 + (\alpha/k)^2} \frac{\cos kL \sin kL + j \sinh \alpha L \cosh \alpha L}{\sin^2 kL \cosh^2 \alpha L + \cos^2 kL \sinh^2 \alpha L} \tag{10.5.3}$$

The terms in α/k can be ignored with no loss of accuracy when $\alpha/k \ll 1$. Furthermore, if $\alpha L \ll 1$, then the input impedance assumes a simpler form

$$\frac{\mathbf{Z}_{m0}}{\rho_0 cS} = \frac{\alpha L - j \cos kL \sin kL}{\sin^2 kL + (\alpha L)^2 \cos^2 kL} \tag{10.5.4}$$

[1]Beranek, *Acoustic Measurements*, p. 317, Wiley (1949).

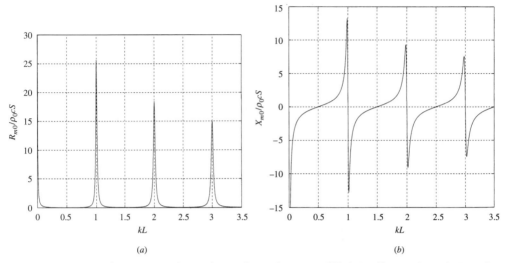

Figure 10.5.1   The input mechanical impedance for an air-filled rigidly terminated pipe of 1 m length and 1 cm radius. (*a*) Input mechanical resistance R_{m0} and (*b*) input mechanical reactance X_{m0}.

In the lossless limit, the resistance vanishes and the reactance is proportional to $\cot kL$. The major effects of absorption are to introduce a small resistance, which maximizes when $k_n L = n\pi$, and to alter the behavior of the reactance in this same region so that it no longer becomes infinite in magnitude but rather remains bounded and changes from positive to negative value very rapidly. These effects are shown in Fig. 10.5.1.

The power dissipated within the pipe is delivered by the source, $\Pi = F^2 R_{m0}/2Z_{m0}^2$, where $\mathbf{Z}_{m0} = R_{m0} + jX_{m0}$. This becomes

$$\Pi = \frac{1}{2}\frac{F^2}{\rho_0 cS}\alpha L \frac{\sin^2 kL + (\alpha L)^2 \cos^2 kL}{(\alpha L)^2 + \cos^2 kL \sin^2 kL} \tag{10.5.5}$$

At mechanical *resonance* $\cos kL = 0$, and the power consumption is

$$\Pi_r = \frac{1}{2}\frac{F^2}{\rho_0 cS}\frac{1}{\alpha L} \tag{10.5.6}$$

whereas at *antiresonance* $\sin kL = 0$ and we have

$$\Pi_a = \frac{1}{2}\frac{F^2}{\rho_0 cS}\alpha L \tag{10.5.7}$$

Note that the frequencies of resonance and antiresonance are close to the natural frequencies of the undamped open, rigid pipe and the undamped rigid, rigid pipe, respectively.

If we define the deviation from resonance by the incremental angular frequency $\Delta\omega$, then

$$kL = (\omega_n + \Delta\omega)L/c = (2n - 1)\pi/2 + \Delta\omega\, L/c$$
$$\mathbf{Z}_{m0}/\rho_0 cS = \alpha L + j\,\Delta\omega\, L/c \tag{10.5.8}$$

The resistance is constant over this range, and the half-power points are determined from the frequencies for which reactance equals resistance, $\Delta\omega = \pm\alpha c$. The frequency interval between upper and lower half-power points is $2\alpha c$, so that $Q_n = \omega_n/2\alpha c$ or

$$Q_n = \frac{1}{2}\frac{1}{\alpha/k_n} \tag{10.5.9}$$

Laboratory measurements of acoustic absorption in fluids are frequently made on fluids contained within cylindrical pipes. In one method, a probe microphone is used to measure pressure amplitudes of a plane progressive wave at two or more positions along the length of the pipe. If P_1 is the pressure amplitude at x_1 and P_2 that at x_2, then the attenuation constant may be determined from

$$P_2 = P_1 e^{-\alpha(x_2-x_1)} \tag{10.5.10}$$

When this equation is used, steps must be taken to eliminate reflected waves either through the use of a nonreflecting termination at the end of the pipe or through the use of short pulses and long pipes so that measurements at x_1 and x_2 may be made before a reflected pulse is returned.

A second method utilizes the standing wave (Fig. 10.5.2). Let us assume that the termination at $x = L$ is rigid. Then at $x = L$ the amplitude P_L of the reflected wave will equal that of the incident wave. The resulting pressure amplitude P at any position along the pipe may be shown from (10.5.1) to be

$$P = 2P_L\{\cos^2[k(L-x)]\cosh^2[\alpha(L-x)] + \sin^2[k(L-x)]\sinh^2[\alpha(L-x)]\}^{1/2}$$

$$\tag{10.5.11}$$

The nodes occur at

$$k(L-x) = (2n-1)\pi/2 \qquad n = 1,2,3,\dots \tag{10.5.12}$$

and have relative amplitudes

$$P_{min}/P_L = 2\sinh[\alpha(L-x)] \approx 2\alpha(L-x) \tag{10.5.13}$$

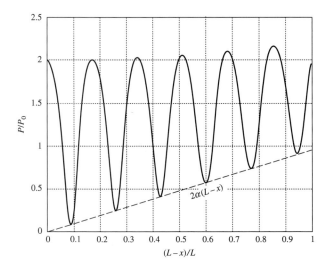

Figure 10.5.2 Spatial dependence of the pressure amplitude of a 1 kHz damped standing wave ($\alpha/k = 0.0253$) in a pipe 1 m long with a radius of 0.20 cm, rigidly terminated at $x = L$.

The pressure amplitudes at successive nodes can be measured with a probe microphone. The value of α may then be determined by drawing a smooth curve through these points, as indicated in Fig. 10.5.2. The antinodes occur at $k(L - x) = n\pi$, where $n = 0, 1, 2, \ldots$, and give maximum pressure amplitudes

$$P_{max}/P_L = 2\cosh[\alpha(L - x)] \approx 2 + [\alpha(L - x)]^2 \qquad (10.5.14)$$

Acoustic absorption determined by either of the above methods is always higher than that measured in large volumes of the fluid because of the losses taking place at the walls of the pipe. See Chapter 8 and Section 8.9. The total absorption coefficient will be the sum of the absorption coefficients for all the individual loss mechanisms.

10.6 BEHAVIOR OF THE COMBINED DRIVER–PIPE SYSTEM

Up to this point we have considered the resonance properties of the pipe. A more realistic investigation of resonating pipes must account for the properties of the mechanical driver, which itself is driven by the applied force. The driver has its own mechanical impedance, so that when a force is applied to the *driver–pipe system* the mechanical resonances of the combined system involve the mechanical behaviors of both driver and pipe.

For example, let the driver be a damped harmonic oscillator, as suggested in Fig. 10.6.1, excited with the externally applied force $\mathbf{f} = F\exp(j\omega t)$. Newton's second law for the motion of the mass is

$$m\frac{d^2\boldsymbol{\xi}}{dt^2} = -R_m\frac{d\boldsymbol{\xi}}{dt} - s\boldsymbol{\xi} - S\mathbf{p}(0, t) + \mathbf{f} \qquad (10.6.1)$$

where $\boldsymbol{\xi}$ is the displacement of the mass to the right and $\mathbf{p}(0, t)$ is the pressure in the pipe at $x = 0$. The complex speed of the mass is $\mathbf{u}(0, t) = d\boldsymbol{\xi}/dt$, the particle speed of the fluid in the pipe at $x = 0$, so that (10.6.1) becomes

$$\left[R_m + j\left(\omega m - \frac{s}{\omega}\right) + \frac{S\mathbf{p}(0, t)}{\mathbf{u}(0, t)}\right]\mathbf{u}(0, t) = \mathbf{f} \qquad (10.6.2)$$

The input mechanical impedance $\mathbf{Z}_{md}$ of the driver is

$$\mathbf{Z}_{md} = R_m + j(\omega m - s/\omega) \qquad (10.6.3)$$

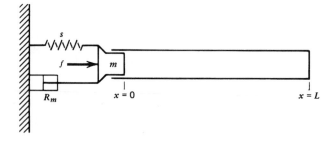

Figure 10.6.1 Schematic representation of a driver–pipe system. The pipe, of length L, is driven at $x = 0$ by a simple oscillator of mass m, mechanical resistance R_m, and stiffness s.

and the input mechanical impedance of the pipe is

$$\mathbf{Z}_{m0} = \frac{S\mathbf{p}(0,t)}{\mathbf{u}(0,t)} \tag{10.6.4}$$

Thus, (10.6.2) shows that the input mechanical impedance $\mathbf{Z}_m$ of this system is the series combination of $\mathbf{Z}_{md}$ and $\mathbf{Z}_{m0}$, so that

$$\mathbf{f} = \mathbf{Z}_m\mathbf{u}(0,t) = \left(\mathbf{Z}_{md} + \mathbf{Z}_{m0}\right)\mathbf{u}(0,t) \tag{10.6.5}$$

The driver alone would resonate when its reactance vanishes, $\omega_0 = \sqrt{s/m}$, and the pipe alone would resonate when $\text{Im}\{\mathbf{Z}_{m0}\} = 0$. When the combined system is driven, however, the input impedance seen by the applied force is the sum of the impedances of the driver and pipe, so that the frequencies of mechanical resonance of the system are found from

$$\text{Im}\left\{\mathbf{Z}_{md} + \mathbf{Z}_{m0}\right\} = 0 \tag{10.6.6}$$

Assume the driven pipe has a rigid termination at $x = L$. Using (10.5.4) and (10.6.3), we can express (10.6.6) as

$$\omega m - \frac{s}{\omega} - \frac{S\rho_0 c \cos kL \sin kL}{\sin^2 kL + (\alpha L)^2 \cos^2 kL} = 0 \tag{10.6.7}$$

Use of $\omega/k = c$ and rearrangement yields

$$\frac{\cos kL \sin kL}{\sin^2 kL + (\alpha L)^2 \cos^2 kL} = akL - \frac{b}{kL} \tag{10.6.8}$$

$$a = m/S\rho_0 L \quad \text{and} \quad b = sL/S\rho_0 c^2$$

Note that a is the ratio of the mass of the driver to the mass of the fluid in the pipe and b is the ratio of the stiffness of the driver suspension to the stiffness of the fluid filling the pipe. Plotting both sides of (10.6.8) against kL on the same set of axes provides the frequencies of mechanical resonance from the values of kL for which the two curves intersect. Two examples are given in Fig. 10.6.2. The examples illustrate the effects of two different driver conditions. (1) For a light, flexible driver with small values of a and b (Fig.10.6.2a), the curves tend to intersect for $k_n L \sim (2n - 1)\pi/2$, so that $L \sim (2n - 1)\lambda/4$ and there is nearly a pressure node at $x = 0$. (2) If the driver is heavy and stiff so that a and b are large, Fig. 10.6.2b shows that most of the resonances occur for $k_n L \sim n\pi$; there is nearly a pressure antinode at $x = 0$. However, in the vicinity of the driver resonance, $kL = 3.6\pi$ in Fig. 10.6.2b, the system resonances tend toward values of $k_n L$ corresponding to a pressure node at $x = 0$.

Since there will always be a pressure antinode at $x = L$, if we obtain the pressure amplitude at this end in terms of the applied force and the mechanical impedance, then we can determine the behavior of the antinodal pressure amplitude as a function of frequency. From (10.5.1) we have

$$\mathbf{p}(x,t) = \mathbf{p}(0,t)\frac{\cos\left[\mathbf{k}(L-x)\right]}{\cos kL} \tag{10.6.9}$$

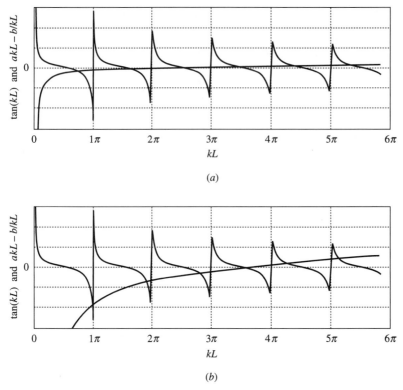

Figure 10.6.2 Graphical solution for the resonance frequencies of a rigidly terminated pipe of 1 m length and 1 cm radius driven by: (*a*) a light, flexible driver with $a = 0.04$ and $b = 2.57$; and (*b*) a heavy, stiff driver with $a = 0.25$ and $b = 32$.

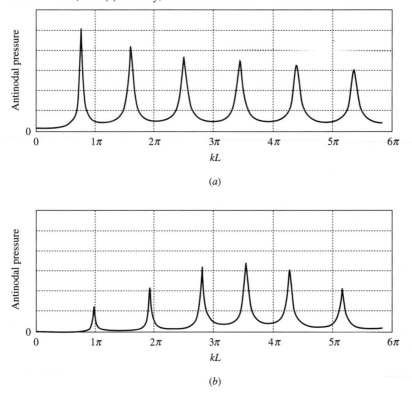

Figure 10.6.3 The antinodal pressure amplitude for a rigidly terminated driver–pipe system excited by a force of constant amplitude: (*a*) for the light, flexible driver of Fig. 10.6.2*a*; and (*b*) for the heavy, stiff driver of Fig. 10.6.2*b*. For both drivers, $R_m / S\rho_0 c = 0.0715$.

Evaluation of the above equation at $x = L$ and then use of (10.6.4) and (10.6.5) result in

$$\mathbf{p}(L, t) = \frac{F}{S} \frac{\mathbf{Z}_{m0}}{\mathbf{Z}_m \cos kL} e^{j\omega t} \tag{10.6.10}$$

For $\alpha L \ll 1$, the pressure amplitude $P(L)$ at the rigid end is

$$P(L) = \rho_0 c \frac{F}{Z_m} \frac{1}{[\sin^2 kL + (\alpha L)^2 \cos^2 kL]^{1/2}} \tag{10.6.11}$$

Figure 10.6.3 displays the pressure amplitude at the rigid end for the same driver–pipe systems illustrated in Fig.10.6.2. (1) The system with the light, flexible driver displays resonances of nearly constant spacing in frequency and with nearly equal maximum pressure amplitudes. (2) The system with the heavy, stiff driver produces pressure amplitudes at the resonances that are much stronger for frequencies close to the resonance frequency of the driver. Furthermore, the driver resonance introduces an "extra" resonance for kL between 3π and 4π. Finally, note that for sufficiently large b the two curves in Fig. 10.6.2b may not intersect near $kL = \pi$. Even though this would not be a true resonance according to our definition, since the reactance of the system would not vanish, there would be a relative minimum in the reactance so that the response $P(L)$ of the driver–pipe system would peak at this frequency.

This interaction of the driver with the pipe in determining the resonances of the system is displayed prominently in many musical instruments. In the brasses, for example, the player can, by altering the *embouchure* (the tension and position of the lips and tongue), influence the reactance of the driver and, therefore, the resonance frequencies of the system. The player may thus be able to "lip" the desired note about a semitone away from the pertinent resonance frequency of the instrument. (In the even-tempered musical scale, two frequencies f_2 and f_1 a semitone apart are related by $f_2/f_1 = 2^{1/12}$.)

10.7 THE LONG WAVELENGTH LIMIT

Analysis of many acoustic devices becomes simple if the wavelength in the fluid is much longer than the dimensions of the device. For example, we have shown for a rigid-walled waveguide that there is a frequency below which the only propagating waveform that can exist is a plane wave traveling straight down the waveguide with phase speed $c_p = c$. Propagation in such a waveguide is extremely simple if the wavelength is sufficiently long compared to the cross-sectional dimensions of the waveguide. If the wavelength is greater than all dimensions, further simplifications are possible: all acoustic variables are constant over the dimensions of the device. Thus, spatial coordinates can be ignored in the equation of motion, and the device behaves as if it were a harmonic oscillator with two degrees of freedom. In such cases, all the results of Chapter 1 can be applied. Acoustic devices in this long wavelength limit are termed *lumped acoustic elements*.

10.8 THE HELMHOLTZ RESONATOR

A simple example of a lumped acoustic system is the Helmholtz resonator shown in Fig. 10.8.1. This device consists of a rigid-walled cavity of volume V with a neck of area S and length L. For the frequencies of interest, assume that $\lambda \gg L$, $\lambda \gg V^{1/3}$, and $\lambda \gg S^{1/2}$. The open end of the neck radiates sound, providing radiation resistance and a radiation mass. The fluid in the neck, moving as a unit, provides another mass element and the thermoviscous losses at the neck walls provide additional resistance. The compression of the fluid in the cavity provides stiffness.

It was seen in Section 10.2 that at low frequencies a circular opening of radius a is loaded with a radiation mass equal to that of the fluid contained in a cylinder of area πa^2 and length $0.85a$ if terminated in a wide flange and $0.6a$ if unflanged. This accounts for the outer opening of the neck. If the inner opening of the neck is assumed to be equivalent to a flanged termination, then combination gives the total effective mass

$$m = \rho_0 SL' \tag{10.8.1}$$

where L', the effective length of the neck, is

$$
\begin{aligned}
L' &= L + (0.85 + 0.85)a = L + 1.7a &&\text{(outer end flanged)} \\
L' &= L + (0.85 + 0.6)a = L + 1.4a &&\text{(outer end unflanged)}
\end{aligned}
\tag{10.8.2}
$$

(An opening consisting of a circular hole in the thin wall of a resonator will have an effective length of about $1.6a$.)

To determine the stiffness of the system, consider the neck to be fitted with an airtight piston. When this piston is pushed in a distance ξ, the volume of the cavity is changed by $\Delta V = -S\xi$, resulting in a condensation $\Delta \rho / \rho = -\Delta V / V = S\xi / V$. The pressure increase (in the acoustic approximation) is

$$p = \rho_0 c^2 \Delta\rho/\rho = \rho_0 c^2 S\xi / V \tag{10.8.3}$$

The force f required to maintain the displacement is $f = Sp = s\xi$ and the effective stiffness s is

$$s = \rho_0 c^2 S^2 / V \tag{10.8.4}$$

The total resistance is the sum of the radiation resistance R_r and the resistance R_w arising from the wall losses. The instantaneous complex driving force produced by a pressure wave of amplitude P impinging on the resonator opening is

$$\mathbf{f} = SPe^{j\omega t} \tag{10.8.5}$$

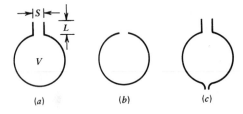

(a)   (b)   (c)

Figure 10.8.1   Three examples of simple Helmholtz resonators.

The resulting differential equation for the inward displacement ξ of the fluid in the neck is

$$m\frac{d^2\xi}{dt^2} + (R_r + R_w)\frac{d\xi}{dt} + s\xi = SPe^{j\omega t} \tag{10.8.6}$$

The input mechanical impedance of the resonator is

$$\mathbf{Z}_m = (R_r + R_w) + j(\omega m - s/\omega) \tag{10.8.7}$$

and resonance occurs when the reactance goes to zero,

$$\omega_0 = c(S/L'V)^{1/2} \tag{10.8.8}$$

In deriving this equation, no assumption has been made restricting the shape of the resonator. As long as all dimensions of the cavity are considerably less than a wavelength and the opening is not too large with reasonably circular cross section, the resonant frequency depends only on the value of $S/L'V$. Resonators having this same value, but very different shapes, are essentially identical.

If it is assumed that the moving fluid in the neck radiates sound into the surrounding medium in the same manner as an open-ended pipe, the radiation resistance is as given in Section 10.2:

$$\begin{aligned} R_r &= \rho_0 c k^2 S^2/2\pi \quad \text{(flanged)} \\ R_r &= \rho_0 c k^2 S^2/4\pi \quad \text{(unflanged)} \end{aligned} \tag{10.8.9}$$

The thermoviscous resistance R_w is found by calculating the absorption coefficient for wall losses α_w from (8.9.19) and then using $Q_w = k/2\alpha_w = \omega_0 m/R_w$. This gives

$$R_w = \omega_0 m/Q_w = 2mc\alpha_w \tag{10.8.10}$$

The quality factor Q of the driven Helmholtz resonator is

$$\begin{aligned} Q &= \omega_0 m/R_m \\ R_m &= R_r + R_w \end{aligned} \tag{10.8.11}$$

For a *flanged* oscillator with negligible thermoviscous losses, the Q has particularly simple form,

$$Q = 2\pi[V(L'/S)^3]^{1/2} \quad \text{(flanged, } R_w \ll R_r) \tag{10.8.12}$$

If the resonator is *unflanged*, multiply the right side by 2.

The *pressure amplification factor* of the resonator is the ratio of the acoustic pressure amplitude P_c within the cavity to the external driving pressure amplitude P of the incident wave. The pressure amplitude P_c is obtained from (10.8.3). Then, from the mechanical impedance $\mathbf{Z}_m = F/(d\xi/dt)$, we have at resonance $|\xi| = PS/\omega_0 R_m$, which when combined with (10.8.8) yields

$$P_c/P = Q \quad \text{(resonance)} \tag{10.8.13}$$

Thus, at resonance, the Helmholtz resonator acts like an amplifier of gain Q.

Helmholtz used a series of graduated resonators as acoustic filters in investigating the harmonic content of complex musical waveforms. Their amplification factors allowed their individual resonances to the harmonics present in the waveform to be detected aurally by coupling a small nipple opposite the neck of the resonator to the ear.

When a loudspeaker is mounted in a closed cabinet, the combined system may be treated as a Helmholtz resonator, in which both the reactance of the air and the mass of the speaker cone contribute to the effective mass of the system. Similarly, both the stiffness of the cavity and that of the speaker suspension contribute to the effective stiffness. The effective resistance is the sum of the radiation resistance, the mechanical resistance of the speaker cone, and thermoviscous resistance from losses within the cabinet.

10.9 ACOUSTIC IMPEDANCE

All simple acoustic systems that can be converted into analogous mechanical systems can also be represented by electrical circuits in which the motion of the fluid is equivalent to the electrical current and the electrical analog of the pressure difference across an acoustic element is the voltage across the corresponding part of the electric circuit. One acoustic analog of current at some point in the circuit is the *volume velocity* U of the fluid in the corresponding acoustic element. (Strictly speaking, since U is not a vector, it should not be termed a "velocity," but this is the accepted convention.) The *acoustic impedance* Z of a fluid acting on a surface of area S is the quotient of the complex acoustic pressure at the surface divided by the complex volume velocity at the surface:

$$\mathbf{Z} = \mathbf{p}/\mathbf{U} \qquad\qquad (10.9.1)$$

We have now encountered three kinds of impedances, an inexcusable redundancy if it were not for the fact that these different impedances are useful in different kinds of calculations.

1. The *specific acoustic impedance* z (pressure/particle speed) is a characteristic property of the medium and of the type of wave that is being propagated. It is useful in calculations involving the transmission of acoustic waves from one medium to another.

2. The *acoustic impedance* Z (pressure/volume velocity) is useful in discussing acoustic radiation from vibrating surfaces, and the transmission of this radiation through lumped acoustic elements or through pipes and horns. The acoustic impedance is related to the specific acoustic impedance at a surface by

$$\mathbf{Z} = \mathbf{z}/S \qquad\qquad (10.9.2)$$

3. The *radiation impedance* $\mathbf{Z}_r$ (force/particle speed) is used in calculating the coupling between acoustic waves and a driving source or driven load. It is part of the *mechanical impedance* $\mathbf{Z}_m$ of a vibrating system associated with the radiation of sound. Radiation impedance is related to specific acoustic impedance at a surface by

$$\mathbf{Z}_r = S\mathbf{z} \tag{10.9.3}$$

(a) Lumped Acoustic Impedance

When concentrated rather than distributed impedances are considered, the impedance of a portion of the acoustic system is defined as the complex ratio of the pressure difference $\mathbf{p}$ that is driving that portion to the resultant volume velocity $\mathbf{U}$. The acoustic impedance unit is the Pa·s/m^3, often termed an *acoustic ohm*.

One example of a lumped acoustic system is the Helmholtz resonator. To cast (10.8.7) into the form of acoustic impedance, divide by S and note that $\mathbf{U} = (d\boldsymbol{\xi}/dt)S$. The result is

$$\mathbf{Z} = R + j(\omega M - 1/\omega C) \tag{10.9.4}$$

with

$$R = R_m/S^2$$
$$M = m/S^2 \tag{10.9.5}$$
$$C = S^2/s$$

where R, M, and C are the (acoustic) *resistance*, *inertance*, and *compliance* of the equivalent mechanical system. These are calculated from R_m, m, and s as given by (10.8.1), (10.8.4), and (10.8.9)–(10.8.11).

The *inertance M* of an acoustic system is represented by the fluid contained in a constriction that is short enough that all particles may be assumed to move as a unit. The *compliance C* of the system is represented by an enclosed volume, with its associated stiffness. Although *resistance* of an acoustic system can arise from a number of different factors, irrespective of its origin it is conventionally represented by narrow slits in a pipe. These three equivalent mechanical elements for the Helmholtz resonator are represented schematically in Fig. 10.9.1a. Figure 10.9.1b shows the equivalent electrical circuit.

(b) Distributed Acoustic Impedance

When one or more dimensions of an acoustic system are not small compared to a wavelength, it may no longer be possible to treat the system as having

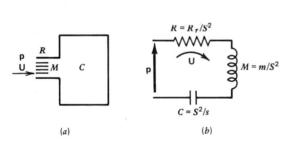

(a) (b)

Figure 10.9.1 Schematic representation of a Helmholtz resonator. (a) Acoustic analog with inertance M, resistance R, and compliance C. The oscillator is driven by an incident pressure $\mathbf{p}$ and the air in the neck moves with volume velocity $\mathbf{U}$. (b) Electrical analog with inductance M, resistance R, and capacitance C. This series circuit is driven by voltage $\mathbf{p}$ and carries current $\mathbf{U}$.

lumped elements, and it then must be considered as having *distributed elements.* The simplest system of this type is one in which a low-frequency plane wave travels through a pipe. If the wave is propagating in the $+x$ direction, the ratio of acoustic pressure to particle speed is given by the characteristic impedance $\rho_0 c$ of the medium, and hence the acoustic impedance at any cross section S in the pipe is

$$\mathbf{Z} = \mathbf{p}/\mathbf{U} = \rho_0 c / S \qquad (10.9.6)$$

The propagation of plane waves in such a pipe is analogous to high-frequency currents along a transmission line. If such a transmission line possesses inductance per unit length L_1 and capacitance per unit length C_1, it can be shown that the input electrical impedance is $\sqrt{L_1/C_1}$. (See any standard textbook on electrical circuits.)

The medium in the pipe may be considered to possess a (*distributed*) *acoustic inertance per unit length* M_1 and a (*distributed*) *acoustic compliance per unit length* C_1. A fluid element of length l and cross-sectional area S has an acoustic inertance $M = \rho_0 l / S$, so the acoustic inertance per unit length is

$$M_1 = \rho_0 / S \qquad (10.9.7)$$

If the length l of the fluid is compressed by ξ, then $p = \rho_0 c^2 \xi / l$. The applied force is pS, the stiffness is $s = \rho_0 c^2 S / l$, so the acoustic compliance is $Sl/\rho_0 c^2$ and the acoustic compliance per unit length C_1 is

$$C_1 = S/\rho_0 c^2 \qquad (10.9.8)$$

The acoustic impedance of a pipe is then, by analogy,

$$\mathbf{Z} = \sqrt{M_1/C_1} = \rho_0 c / S \qquad (10.9.9)$$

which is in agreement with (10.9.6).

10.10 REFLECTION AND TRANSMISSION OF WAVES IN A PIPE

Assume that at some point $x = 0$ along a pipe filled with a single fluid, the acoustic impedance changes from $\rho_0 c / S$ to $\mathbf{Z}_0 = R_0 + jX_0$. This can be accomplished by a change in cross-sectional area, a branch, or a side port.

Our boundary conditions must be continuity of pressure $\mathbf{p}$ and continuity of volume velocity $\mathbf{U}$. (For subsonic flow there is no change in density across the interface. Conservation of mass then requires that the volume velocity be conserved.) Paralleling the development in Section 6.2, but using the volume velocity $\mathbf{U}$ instead of $\mathbf{u}$ in (6.2.5), the incident and reflected waves are represented, respectively, by

$$\mathbf{p}_i = \mathbf{P}_i e^{j(\omega t - kx)}$$
$$\mathbf{p}_r = \mathbf{P}_r e^{j(\omega t + kx)} \qquad (10.10.1)$$

The transmitted wave $\mathbf{p}_t$ can be a standing wave, traveling wave, or any combination thereof. The continuities of pressure and volume velocity give us

$$\frac{\mathbf{P}_i + \mathbf{P}_r}{\mathbf{U}_i + \mathbf{U}_r} = \frac{\mathbf{P}_t}{\mathbf{U}_t} \qquad x = 0 \qquad (10.10.2)$$

The acoustic impedance $\mathbf{Z}$ for $x < 0$ is

$$\mathbf{Z} = R + jX = \frac{\mathbf{P}_i + \mathbf{P}_r}{\mathbf{U}_i + \mathbf{U}_r} = \frac{\rho_0 c}{S} \frac{\mathbf{P}_i e^{-jkx} + \mathbf{P}_r e^{jkx}}{\mathbf{P}_i e^{-jkx} - \mathbf{P}_r e^{jkx}} \qquad (10.10.3)$$

In general, the flow near $x = 0$ will be complicated, but in the long wavelength limit the acoustic impedance in the vicinity of $x = 0$ is well approximated by evaluating (10.10.3) at $x = 0$:

$$\mathbf{Z}_0 = \frac{\rho_0 c}{S} \frac{\mathbf{P}_i + \mathbf{P}_r}{\mathbf{P}_i - \mathbf{P}_r} \qquad (10.10.4)$$

Solving for the pressure reflection coefficient $\mathbf{P}_r/\mathbf{P}_i$ gives

$$\frac{\mathbf{P}_r}{\mathbf{P}_i} = \frac{\mathbf{Z}_0 - \rho_0 c/S}{\mathbf{Z}_0 + \rho_0 c/S} \qquad (10.10.5)$$

The *power reflection coefficient* $R_\Pi = |\mathbf{P}_r/\mathbf{P}_i|^2$ is

$$R_\Pi = \frac{(R_0 - \rho_0 c/S)^2 + X_0^2}{(R_0 + \rho_0 c/S)^2 + X_0^2} \qquad (10.10.6)$$

and the *power transmission coefficient* $T_\Pi = 1 - R_\Pi$ is

$$T_\Pi = \frac{4R_0 \rho_0 c/S}{(R_0 + \rho_0 c/S)^2 + X_0^2} \qquad (10.10.7)$$

These last two equations are identical in form to those developed for normal reflection from a plane interface between two fluids if $\mathbf{Z}_0 = \rho_2 c_2/S$, where ρ_2 and c_2 characterize the system for $x \geq 0$. (See Section 6.2.)

Let us apply these equations to a plane wave in a pipe of cross-sectional area S_1 as it enters a second pipe of area S_2, as shown in Fig. 10.10.1. If the second pipe is either infinitely long or terminated so that there is no reflected wave, the acoustic impedance seen by the wave incident on the junction is $\mathbf{Z}_0 = \rho_0 c/S_2$. Substituting this into (10.10.6) and (10.10.7) gives

$$R_\Pi = \frac{(S_1 - S_2)^2}{(S_1 + S_2)^2}$$

$$T_\Pi = \frac{4S_1 S_2}{(S_1 + S_2)^2} \qquad (10.10.8)$$

If the end of the pipe is closed, $S_2 = 0$. Then $\mathbf{Z}_0 = \infty$ and $R_\Pi = 1$, as expected for a rigidly terminated pipe. If the end of the pipe is open, $\mathbf{Z}_0 = \mathbf{Z}_{mL}/S^2$, where $\mathbf{Z}_{mL}$ is given by (10.2.10) or (10.2.14).

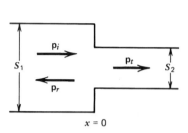

Figure 10.10.1   Transmission and reflection of a plane wave in the vicinity of a junction between two pipes where the cross-sectional area changes from S_1 to S_2.

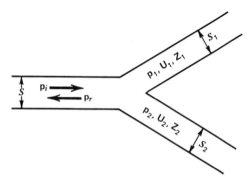

Figure 10.10.2   Conditions in the vicinity of a branch. The branches have cross-sectional areas S_1 and S_2 and input acoustic impedances z_1 and z_2.

As a second example, consider a pipe that branches into two pipes each with arbitrary input impedance, as indicated in Fig. 10.10.2. If the junction is at $x = 0$, the pressures in the three pipes very close to $x = 0$ are

$$\mathbf{p}_i = \mathbf{P}_i e^{j\omega t} \qquad \mathbf{p}_r = \mathbf{P}_r e^{j\omega t}$$
$$\mathbf{p}_1 = \mathbf{Z}_1 \mathbf{U}_1 e^{j\omega t} \qquad \mathbf{p}_2 = \mathbf{Z}_2 \mathbf{U}_2 e^{j\omega t} \tag{10.10.9}$$

where $\mathbf{Z}_1$, $\mathbf{Z}_2$, and $\mathbf{U}_1$, $\mathbf{U}_2$ are the input impedances and complex volume velocity amplitudes in the two branches. In the long wavelength approximation, continuity of pressure at the junction requires

$$\mathbf{p}_i + \mathbf{p}_r = \mathbf{p}_1 = \mathbf{p}_2 \tag{10.10.10}$$

and continuity of volume velocity requires

$$\mathbf{U}_i + \mathbf{U}_r = \mathbf{U}_1 + \mathbf{U}_2 \tag{10.10.11}$$

Dividing (10.10.11) by (10.10.10) gives

$$(\mathbf{U}_i + \mathbf{U}_r)/(\mathbf{p}_i + \mathbf{p}_r) = \mathbf{U}_1/\mathbf{p}_1 + \mathbf{U}_2/\mathbf{p}_2 \tag{10.10.12}$$

which may be written as

$$1/\mathbf{Z}_0 = 1/\mathbf{Z}_1 + 1/\mathbf{Z}_2 \tag{10.10.13}$$

Thus, the combined *acoustic admittance* $1/\mathbf{Z}_0$ associated with the incident and reflected waves equals the sum of the acoustic admittances $1/\mathbf{Z}_1$ and $1/\mathbf{Z}_2$ of the two branches.

As a special case of branching, consider a side branch of arbitrary acoustic impedance $\mathbf{Z}_1 = \mathbf{Z}_b$ connected at $x = 0$ to an infinitely long pipe of cross-sectional area S with $\mathbf{Z}_2 = \rho_0 c/S$. Solving for the pressure reflection coefficient gives

$$\frac{\mathbf{P}_r}{\mathbf{P}_i} = -\frac{\rho_0 c/2S}{\rho_0 c/2S + \mathbf{Z}_b} \tag{10.10.14}$$

Condition (10.10.10) shows that $\mathbf{P}_2/\mathbf{P}_i = 1 + \mathbf{P}_r/\mathbf{P}_i$, so the pressure transmission coefficient is

$$\frac{\mathbf{P}_t}{\mathbf{P}_i} = \frac{\mathbf{Z}_b}{\rho_0 c/2S + \mathbf{Z}_b} \tag{10.10.15}$$

where $\mathbf{P}_t = \mathbf{P}_2$. Denoting the acoustic impedance of the branch by $\mathbf{Z}_b = R_b + jX_b$ and solving for the power reflection and transmission coefficients gives

$$R_\Pi = \frac{(\rho_0 c/2S)^2}{(\rho_0 c/2S + R_b)^2 + X_b^2} \tag{10.10.16}$$

$$T_\Pi = \frac{R_b^2 + X_b^2}{(\rho_0 c/2S + R_b)^2 + X_b^2}$$

The power transmission coefficient for the side branch is

$$T_{\Pi b} = \frac{(\rho_0 c/S)R_b}{(\rho_0 c/2S + R_b)^2 + X_b^2} \tag{10.10.17}$$

The power transmitted past the junction and along the main pipe is zero when $\mathbf{Z}_b = 0$. In this case, all the power is reflected back to the source. If R_b is greater than zero but not infinite, some power is consumed in the side branch and some transmitted beyond the junction, irrespective of the value of X_b. If either R_b or X_b is very large compared to $\rho_0 c/S$, almost all the incident power is transmitted past the side branch. In the limit $R_b = X_b = \infty$, which corresponds to no side branch, $T_\Pi = 1$.

10.11 ACOUSTIC FILTERS

The ability of a side branch to attenuate the sound energy transmitted in a pipe is the basis of a class of acoustic filters. Depending on the input acoustic impedance of the side branch, such systems can act as low-pass, high-pass, or band-stop filters. One example of each type will be considered.

(a) Low-Pass Filters

Assume we insert an *enlarged* section of pipe of *total* cross-sectional area S_1 and length L in an infinitely long (or suitably terminated) pipe of cross section S, as shown in Fig. 10.11.1. At low frequencies such that $kL \ll 1$, this section acts like a side branch of approximate volume $V = (S_1 - S)L$ and has an acoustic compliance $C = V/\rho_0 c^2$. The acoustic impedance of this chamber is, therefore,

$$\mathbf{Z}_b \approx -j\rho_0 c^2/[\omega(S_1 - S)L] \tag{10.11.1}$$

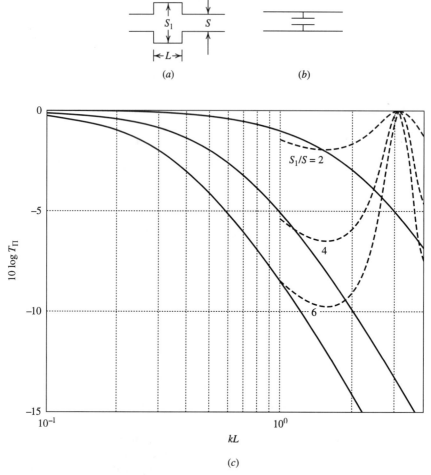

Figure 10.11.1   A simple low-pass acoustic filter consists of an enlarged section of cross-sectional area S_1 and length L in a pipe of cross-sectional area S. (a) Schematic. (b) Analogous electrical filter. (c) Attenuation for several values of S_1/S. Solid lines are from (10.11.2) for $kL \ll 1$. Dashed lines are from Problem 10.11.6 for $kL \gg 1$.

and, according to (10.10.13), is in parallel with the impedance $\rho_0 c/S$ of the continuation of the pipe. Substituting (10.11.1) into (10.10.16) yields a power transmission coefficient of

$$T_\Pi \approx \frac{1}{1 + \left(\dfrac{S_1 - S}{2S}kL\right)^2} \qquad (10.11.2)$$

This equation shows that at low frequencies the power transmission is total and gradually decreases with increasing frequency. T_Π is 0.50 when $kL = 2S/(S_1 - S)$. This type of acoustic filter is analogous to the low-pass electrical filter produced by shunting a capacitor across a transmission line, as shown in Fig. 10.11.1b, but only when $kL < 1$. The equation fails when $kL > 1$. Figure 10.11.1c shows the approximate power transmission coefficient in decibels as a function of kL for several values of S_1/S. Note that T_Π does not drop to 0.5 at $kL = 1$ until $S_1/S > 3$.

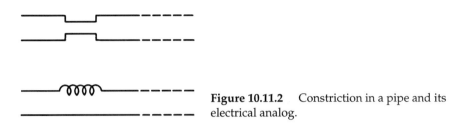

Figure 10.11.2 Constriction in a pipe and its electrical analog.

A more general equation valid for $kL > 1$ can be derived by considering the various incident, reflected, and transmitted waves present in the three sections of pipe. When the wavelength is large compared with the radii of all sections, the appropriate boundary conditions remain continuity of pressure and continuity of volume velocity at the two ends of the expanded pipe. The transmission coefficient can then be found as in Section 6.3, but using the volume velocity instead of the particle velocity. The result is identical to (6.3.9) when r_2/r_1 is replaced with S/S_1. (See Problem 10.11.6.)

Another type of low-pass filter is produced by a *constriction* in a pipe, as shown in Fig. 10.11.2. This can be modeled as an inertance in *series* with the continuation of the pipe. However, as is the case with an enlarged section, the analogy between the electric and the acoustic cases is valid only for low frequencies. The same general equation [(6.3.9) with r_2/r_1 replaced by S/S_1] can be used for calculating the transmission coefficient for this type of filter, since its derivation is independent of whether S or S_1 is the larger.

There exist practical limitations, other than those of frequency, which must be taken into consideration in designing low-pass acoustic filters. The equations are not applicable when there is an extreme difference between the cross sections of the filter and the original pipe. In spite of all these limitations, filters of this type are basic to the design of simple automobile mufflers, gun silencers, and sound-absorbing plenum chambers installed in ventilating systems.

(b) High-Pass Filters

Next, consider the effect of a short length of unflanged pipe as a branch. If both the radius a and the length L of this pipe are small compared to a wavelength, the impedance of the branch is, from (10.8.2) and (10.8.9),

$$\mathbf{Z}_b = \rho_0 c k^2/4\pi + j\omega(\rho_0 L'/\pi a^2) \tag{10.11.3}$$

where $L' = L + 1.4a$. The first term results from the radiation of sound through the branch into the external medium, and the second from the inertance of the gas in the orifice.

The ratio of the acoustic resistance of the branch to its acoustic reactance is

$$R_b/X_b = ka^2/4L' \tag{10.11.4}$$

Since the radius of the branch is assumed small compared to the wavelength, $ka \ll 1$, the acoustic resistance can be neglected compared with the acoustic reactance in calculating the transmission coefficient. Thus, (10.10.16) becomes

$$T_{\Pi} = \frac{1}{1 + (\pi a^2/2SL'k)^2} \tag{10.11.5}$$

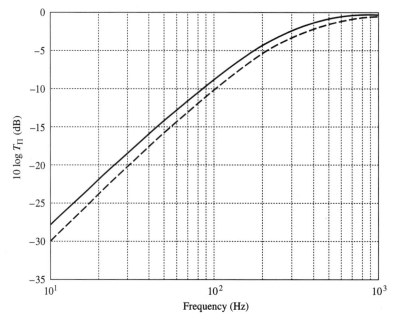

Figure 10.11.3   Attenuation for a high-pass filter in a pipe of cross-sectional area $S = 28$ cm^2. The side branch has a 1.55 cm radius. The solid curve is for $L = 0.6$ cm and the dashed curve for $L = 0$.

This transmission coefficient is very nearly zero at low frequencies and rises to nearly unity at higher frequencies, as is indicated in Fig. 10.11.3, where the power transmission coefficient is shown as a function of kL for two cases—$L \gg a$ and $L = 0$ (a hole in the main pipe). The transmission coefficient is 0.5 when

$$k = \pi a^2 / 2SL' \tag{10.11.6}$$

The presence of a single branch converts a pipe into a high-pass filter. As the radius of the branch is increased, the power transmitted at low frequencies is decreased and the frequency for which $T_\Pi = 0.5$ is increased. If the pipe has several branches, located near enough to one another so that they may be considered as being at a single point (separated by a small fraction of a wavelength), the action of the group is that of their equivalent parallel impedance. If, however, the distance between orifices is an appreciable fraction of the wavelength, the waves reflected from the various orifices are not in phase with one another and the power transmission coefficient must be determined by methods analogous to those for reflections in multilayered media. In general, the low-frequency attenuation by a number of suitably spaced branches can be made much greater than that of a single branch of equal total area.

The power transmission coefficient into a single branch is approximately

$$T_{\Pi b} = \frac{2}{\pi} \frac{k^2 S}{1 + (2SkL' / \pi a^2)^2} \tag{10.11.7}$$

At the frequency corresponding to $T_\Pi = 0.5$ this expression reduces to $T_{\Pi b} = k^2 S / \pi$. For the example of Fig. 10.11.3 (solid curve) this is only 0.015 at 225 Hz. It is therefore quite apparent that the filtering action of an orifice does not result from the transmission of acoustic energy out of the pipe, but rather from the reflection of energy back toward the source.

The influence of an orifice may be used to explain qualitatively the action of a wind instrument, such as a flute or clarinet. When these instruments are played in their fundamental registers the player opens all (or nearly all) the orifices lying beyond some particular distance from the mouthpiece. Since the diameters of the orifices are almost as large as the bore of the tube, this effectively shortens the length of the instrument, and the acoustic energy reflected back from the first open orifice sets up a pattern of standing waves between this orifice and the mouthpiece. In a flute, which acts essentially like an open, open pipe, the wavelength is approximately equal to twice the distance from the opening in the mouthpiece to the first open orifice. In a clarinet, however, the action of the vibrating reed causes the conditions at the mouthpiece to approximate those at the closed end of a tube, and hence the wavelength is nearly four times the distance from the reed to the first open orifice. In both instruments there are also a number of harmonic overtones, those of the clarinet being predominately the odd harmonics, as is to be expected from an open, closed pipe. When either instrument is played in a high register, the fingering is more complex; some orifices beyond the first open orifice are left closed and others opened, the purpose being to emphasize the desired standing wave pattern.

(c) Band-Stop Filters

If the side branch possesses both inertance and compliance, then it will act as a band-stop filter. One such side branch would be a long pipe rigidly capped on the far end. Another example is the Helmholtz resonator for Fig. 10.11.4. If we neglect thermoviscous losses, there is no net dissipation of energy so $R_b = 0$. If the area of the opening into the resonator is $S_b = \pi a^2$, the length of its neck is L, and its volume is V, then the acoustic reactance of the branch is

$$X_b = \rho_0(\omega L'/S_b - c^2/\omega V) \tag{10.11.8}$$

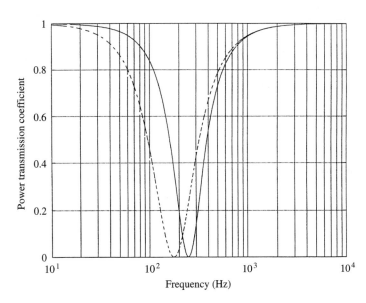

Figure 10.11.4 The power transmission coefficient for a band-stop filter consisting of a Helmholtz resonator. The resonator has a neck of length 0.6 cm and radius 1.55 cm. The pipe has a cross-sectional area 28 cm^2. The solid line is for a resonator volume of 1120 cm^3. The dashed line is for a resonator volume of 2240 cm^3.

where $L' = L + 1.7a$. Substitution into (10.10.16) leads to a transmission coefficient of

$$T_\Pi = \frac{1}{1 + \left(\dfrac{c/2S}{\omega L'/S_b - c^2/\omega V}\right)^2} \qquad (10.11.9)$$

This transmission coefficient is zero when

$$\omega = c(S_b/L'V)^{1/2} \qquad (10.11.10)$$

which is the resonance frequency of the Helmholtz resonator. At this frequency large volume velocity amplitudes exist in the neck of the resonator, but all acoustic energy that is transmitted into the resonator cavity from the incident wave is returned to the main pipe with a phase relationship causing reflection back toward the source.

Calculated values of the transmission coefficient as a function of frequency are plotted in Fig. 10.11.4 for two representative resonators under the assumption of no losses. One striking characteristic of the curves is that each indicates a significant reduction in the transmission over a frequency range extending for more than an octave on either side of the resonance frequency.

PROBLEMS

Unless otherwise specified, the fluid in the pipe is air at 20°C and 1 atm.

10.2.1. Find the shortest length pipe for which the input mechanical resistance equals the input mechanical reactance at 500 Hz when the terminating load impedance is three times the characteristic mechanical impedance $\rho_0 cS$ of the air in the pipe.

10.2.2. Show that for $kL \ll 1$ the input impedance of the forced, rigid pipe is spring-like. Explain the physical significance of this stiffness for a perfect gas.

10.2.3C. The diaphragm of a condenser microphone is stretched across one end of a pipe of 0.02 m radius and 0.01 m length, open at the other end. Compute and plot the ratio of the pressure at the diaphragm, considered rigid, and the pressure at the open end as a function of frequency from 100 to 2000 Hz.

10.2.4. A cylindrical pipe of 1 m length and cross-sectional radius 1 cm is driven at one end and open to the air at the other end, which is flanged. (a) Make graphs of the input mechanical resistance and reactance as functions of kL. (b) From the graphs, find the percent differences between the first three resonance frequencies and those for an integral number of half-waves in the pipe.

10.3.1. The air in a pipe of 0.05 m radius and 1.0 m length is being driven by a piston of 0.015 kg mass and 0.05 m radius inserted in one end of the pipe. The other end of the pipe is terminated in an infinite baffle. (a) At 150 Hz, what is the mechanical impedance of the piston, including the loading effect of the air in the pipe? (b) What is the amplitude of the force required to drive the piston with a displacement amplitude of 0.005 m at this frequency? (c) How much acoustic power in watts will be radiated from the open end of the pipe?

10.3.2. The air in a pipe of 1 m length and 0.05 m radius is being driven at one end by a piston of negligible mass. The far end of the pipe is open and has a large flange attached to it. (*a*) What is the fundamental resonance frequency of the system? (*b*) If the peak displacement amplitude of the piston is 0.01 m when driven at the above frequency, what acoustic power is being transmitted by the plane waves moving toward the open end of the pipe? (*c*) What acoustic power is being transmitted out through the open end of the pipe?

10.3.3C. (*a*) For an open-ended, flanged pipe, plot the power transmission coefficient as a function of *ka* for *ka* < 0.2 and compare to the approximate values given by (10.3.5). (*b*) Find the value of *ka* above which the approximate result deviates from the exact result by 10%.

10.3.4C. For a pipe driven at one end, with an unflanged opening at the other end, and with a ratio of length to radius of 100, (*a*) plot the radiated power for frequencies in the neighborhood of the first resonance. (*b*) From this graph, determine Q_1 and compare with the value obtained using (10.3.9).

10.4.1. A 1480 Hz plane wave in water is normally incident on a flat concrete wall that absorbs all the acoustic energy transmitted into it. Assume the speed of sound in the water is 1480 m/s. The standing wave has a peak pressure amplitude of 15 Pa at the wall and a pressure amplitude of 5 Pa at the nearest pressure node at a distance of 0.25 m from the wall. (*a*) What is the ratio of the intensity of the reflected wave to that of the incident wave? (*b*) What is the specific acoustic impedance of the wall?

10.4.2. A 500 Hz sound wave in air having a pressure level of 60 dB *re* 0.0002 μbar is normally incident on a boundary between the air and a second medium having a characteristic impedance of 830 Pa · s/m. What is the effective (rms) pressure amplitude of (*a*) the reflected wave and (*b*) the transmitted wave? (*c*) At what distance from the boundary is the pressure amplitude in the standing wave pattern equal to that of the incident wave?

10.4.3. Given that a 200 Hz plane wave in air is normally incident on an acoustic tile panel having a normal specific acoustic impedance of $1000 - j2000$ Pa · s/m, (*a*) what is the standing wave ratio in the resulting pattern? (*b*) Where are the first two nodes located?

10.4.4. A plane wave at 1 kHz in water is normally incident on a concrete wall that can be considered as infinitely thick. (*a*) What is the standing wave ratio? (*b*) To what difference in pressure levels is this equivalent? (*c*) Where are the first three nodes located?

10.4.5. A 200 Hz plane wave in air is normally incident on an acoustic tile panel. The resulting standing wave has a $SWR = 10$ and the node closest to the tile is 50 cm from the tile. Find the normal specific acoustic impedance of the tile.

10.4.6C. For the tile in Problem 10.4.5, plot the pressure amplitude as a function of distance from the tile.

10.5.1. Plane waves at 6 kHz are being propagated in an air-filled pipe of 0.1 m radius by means of a loudspeaker fitted into one end of the pipe. The far end of the pipe is closed by a rigid cap. The measured standing wave ratio at one position in the pipe is 8. At a second position 0.5 m further down the pipe, the measured standing wave ratio is 9. (*a*) Derive an equation, involving these ratios and the distance between them, that may be used in calculating the absorption constant for the wave. Simplify your equation for $\alpha \ll 1$ Np/m. (*b*) What is the numerical value of α corresponding to the above data? (*c*) Calculate the absorption constant to be anticipated if there were only thermoviscous losses at the walls of the pipe. (*d*) Assuming the remainder of the measured absorption constant to result from the presence of water vapor, use Fig. 8.6.2 to estimate the relative humidity of the air in the pipe.

10.5.2C. An air-filled pipe of length 1 m and radius 1 cm is driven at one end and rigidly capped at the other end. The attenuation in the pipe is given by (10.5.9) with $\alpha = 2.93 \times 10^{-5} f^{1/2}/a$ Np/m. Plot the input mechanical resistance and reactance as functions of frequency using (a) the exact expression (10.5.3) and (b) the approximate expression (10.5.4). (c) Determine the frequency at which the approximate solution begins to fail and find the values of α/k and αL at this frequency.

10.5.3C. For the pipe in Problem 10.5.2C, (a) plot the power delivered by the source for frequencies around the first resonance. (b) Use this graph to determine Q and compare to the value calculated from (10.5.9).

10.5.4C. Plot the spatial dependence of the pressure amplitude of a damped standing wave in a driven pipe rigidly terminated at $x = L$ with $\alpha/k = 0.04$. From this graph, determine the slope and intercept of a straight line drawn through the nodes and compare to the predictions of (10.5.13).

10.6.1. Verify that the frequencies of resonance and those near antiresonance of the forced, open pipe with damping correspond to the frequencies of maximum and minimum power dissipation, respectively. Assume $ka \ll 1$.

10.6.2. A piston of mass m and radius a is mounted in one end of a pipe of length L and radius a, where $a \ll L$. The other end of the pipe opens into an infinite plane flange. (a) Derive an approximate equation giving the acoustic power radiated out the open end of the pipe when the piston is driven by a force $F \cos \omega t$ at such high frequencies that $ka \gg 1$. (Assume plane waves.) (b) Derive an approximate equation that is valid when both $ka \ll 1$ and $kL \ll 1$.

10.6.3. For the driver–pipe systems illustrated in Fig. 10.6.2, find the values of f and kL at which the driver is resonant. Are your results consistent with the graphs?

10.6.4C. An air-filled pipe of length 1 m and radius 1 cm is driven at one end and rigidly capped at the other end. The attenuation is $\alpha = 2.93 \times 10^{-5} f^{1/2}/a$ Np/m. (a) Find the first six resonance frequencies if the mass of the driver is 1 g and its stiffness is 10^4 N/m. (b) Plot the pressure at the rigid end for a constant driving force.

10.8.1. The sphere of a Helmholtz resonator has a diameter of 0.1 m. (a) What diameter hole should be drilled in the sphere if it is to resonate in air at 320 Hz? (b) What must be the pressure amplitude of an incident acoustic plane wave at 320 Hz if it is to produce an internal pressure amplitude of 20 μbar? (c) What will be the resonance frequency if a hole of twice the cross-sectional area as that in part (a) is drilled in the sphere? (d) What will be the resonant frequency if two *independent* holes having the diameter of part (a) are drilled in the sphere?

10.8.2. A rigid-walled enclosed loudspeaker cabinet has inside dimensions of 0.3 m×0.5 m× 0.4 m. The front panel of the cabinet is 0.03 m thick and has a 0.2 m diameter hole cut in it for mounting a loudspeaker. (a) What is the fundamental frequency of the cabinet (considered as a Helmholtz resonator)? (b) If a loudspeaker having a cone of 0.2 m diameter, a mass of 0.01 kg, and a suspension of 1000 N/m stiffness is mounted in this cabinet, what is the resonance frequency of the cone? Assume that the effective mass of the system is the sum of that of the cone and the radiation mass and that the effective stiffness is the sum of that of the cone and that of the volume of air within the cabinet. (c) What would be the resonance frequency of the cone if it were not mounted in the cabinet and had no air loading? (d) What is the acoustic power radiated if the cone is driven at the frequency of part (b) with an amplitude of 0.002 m? (e) Under these same conditions, what is the excess pressure amplitude inside the cabinet and what is the amplitude of the associated force acting on one of the 0.4 m × 0.5 m panels?

10.8.3. Show that (10.8.8) can be put in the form $k_0 L' = [(\text{effective neck volume})/(\text{cavity volume})]^{1/2}$, where $k_0 = \omega_0/c$.

10.9.1. A loudspeaker has cone mass m, suspension stiffness s, and radiation impedance $R_r + jX_r$. If the loudspeaker is activated by a force $F \exp(j\omega t)$, (a) obtain the equivalent circuit in terms of the acoustic impedances of the elements of the source. (b) Find the velocity amplitude of the cone. (c) Does this agree with the analysis of Section 7.5?

10.9.2. A rectangular room has internal dimensions of 2.5 m × 4.0 m × 4.0 m and walls of 0.1 m thickness. A door opening into this room has dimensions of 0.8 m × 2.0 m. (a) Assuming the inertance of the door opening to be equal to that of a circular opening of equal area, calculate the resonance frequency of the room (considered as a Helmholtz resonator). (b) What is the acoustic compliance of the room and the acoustic inertance of the door opening? (c) Considering only the compliance of the room and the inertance of the door opening, what acoustic impedance is presented by the room to a sound source within the room at 20 Hz?

10.10.1. A pipe of cross section S_1 is connected to a pipe of cross section S_2. (a) Derive a general expression for the ratio of the intensity of the waves transmitted into the second pipe to that of the incident waves. (b) Under what conditions is the transmitted intensity greater than the incident intensity? Explain. (c) Derive a general expression for the standing wave ratio produced in pipe S_1 in terms of the areas S_1 and S_2.

10.10.2. A plane wave is traveling in a pipe of area S_1 containing a fluid of characteristic impedance $\rho_1 c_1$. A second pipe of area S_2 and infinite length containing a fluid of characteristic impedance $\rho_2 c_2$ is attached to the end of this pipe. The two fluids are separated by means of a thin rubber diaphragm. (a) Derive the power transmission coefficient. (b) What is the condition for complete transmission?

10.10.3. A plane wave of pressure amplitude P is traveling to the right in a pipe of cross-sectional area S_1. A narrower pipe of area S_2 and infinite length is attached to the right end of the pipe. (a) What must be the ratio of the two areas if the transmitted pressure amplitude in the second pipe is to be 1.5 that of the incident wave in the first pipe? (b) If the smaller pipe is cut off at a distance of a quarter-wavelength from the junction of the two pipes and then covered with a rigid cap, derive an expression giving the pressure amplitude at the cap in terms of that of the incident wave in the large pipe. (c) What is the ratio of these two pressures for the area ratio determined in (a)?

10.10.4. The side branch from an infinitely long main pipe of area S is another infinitely long pipe of area S_b. The main pipe is transmitting plane waves at a frequency with a wavelength large compared to the diameters of both pipes. (a) Derive an equation for T_Π. (b) Derive an equation for $T_{\Pi b}$. (c) If the area of the main pipe is twice that of the branch, calculate values for the transmission coefficient into each pipe. (d) Is the sum of these two coefficients equal to unity? If not, where has the remaining power gone? Support your explanation by computation.

10.10.5. A square ventilating duct measures 0.3 m on each side. A Helmholtz resonator band filter is constructed by drilling a hole of 0.08 m radius in one wall of the duct leading into a surrounding closed chamber of volume V. (a) What volume V is required to filter sounds most effectively at 30 Hz? (b) What will be the power transmission coefficient of the filter at 60 Hz?

10.10.6. (*a*) If a hole of 1 cm radius is drilled in a thin-walled pipe of 2 cm radius, what is T_{Π} at 500 Hz? *Hint:* Use (10.8.2) and (10.8.9) to estimate $\mathbf{Z}_r$ and then obtain $\mathbf{Z}_b$. (*b*) What will T_{Π} be at this frequency if a second hole of 1 cm radius is drilled in the pipe directly opposite the first hole?

10.10.7. Show that the radius a of the hole that must be drilled into a thin-walled pipe of radius a_0 to yield $T_{\Pi} = 0.5$ at a frequency f is $a = \frac{64}{3} f a_0^2 / c$.

10.10.8. A 300 Hz plane wave of 0.1 W power is traveling through an infinitely long pipe of 2.0 cm radius. What will be the power reflected, the power transmitted along the pipe, and the power transmitted out through a simple orifice of 0.50 cm radius?

10.10.9. A ventilating duct has a radius of 0.6 m. It is desired to prevent the transmission of low-frequency sound by introducing an appropriately shaped piston of mass m and radius 0.5 m into the wall of the duct. (*a*) Obtain a plot of the power transmission coefficient as a function of $\omega m / \rho_0 cS$. (*b*) If T_{Π} is not to exceed 0.5 at 60 Hz, what is the largest allowed piston mass?

10.10.10. A water pipeline has a diameter of 0.04 m. It is desired to filter out plane waves traveling in the water by use of a rigidly-capped side branch pipe of 0.02 m diameter. Calculate the minimum length of the branch that will most effectively filter out 900 Hz.

10.10.11. Show that the attachment of a side branch consisting of a short length of pipe of volume V and a hole of radius a drilled in its cap is equivalent to shunting the main pipe with an inertance $M = 1.7\rho_0 / \pi a$ in parallel with a compliance $C = V / \rho_0 c^2$. (*a*) What kind of filtering will result? (*b*) Given that the main pipe has $S = 0.005$ m^2 and that $a = 0.02$ m, what must V be if the filter is to produce a minimum in transmission of waves along the main pipe at 400 Hz? (*c*) What will T_{Π} be at 200 Hz?

10.10.12C. An air-filled pipe of cross-sectional area S_1 terminates in an infinitely long pipe of cross-sectional area S. (*a*) Plot the power transmission coefficient as a function of $\log(S_2 / S_1)$ for $0 < S_2 / S_1 < 10$. (*b*) Calculate the values of S_2 / S_1 for which T_{Π} is at least 0.5.

10.10.13. An air-filled pipe of cross-sectional area S has a side branch of cross-sectional area S_b. If the length of the side branch is infinite, calculate the value of S_b / S for which $T_{\Pi} = 0.5$.

10.10.14C. An air-filled pipe with a cross-sectional area of 25 cm branches into two pipes each with a radius of 2 cm and lengths 50 cm and 100 cm. The ends of both branches are open and unflanged. Plot the input resistance and reactance for $0 < f < 1$ kHz and determine the resonance frequencies.

10.11.1. Derive (10.11.2) from (10.11.1) and (10.10.16).

10.11.2. Approximate (10.11.2) for $[(S_1 - S)/S]kL \ll 1$ and compare with the approximate power transmission coefficient for the fluid layer of Section 6.3 under the conditions $r_1 = r_3 \gg r_2$ and $k_2 L \ll 1$.

10.11.3C. Plot the power transmission coefficient versus $\log f$ for a low-pass filter inserted into an infinitely long air-filled pipe of 1 cm radius. The filter is a cylindrical section of 2 cm radius and 5 cm length. The frequencies considered should not exceed those for which the wavelength becomes smaller than about ten times the larger dimension of the filter.

10.11.4. (*a*) A low-pass filter consists of three identical cavities connected by two identical narrow pipes, all physically in series. Show that this is a low-pass filter with the cutoff frequency given by $f_c = (1/\pi)(c^2 S/LV)^{1/2}$, where S is the cross-sectional area of each of the narrow pipes and L its effective length, and V is the excess volume of each of the cavities. (*b*) If an air-conditioning system has a fan that produces noise above 60 Hz, design this filter using reasonable dimensions so that, when fitted into a duct of cross-sectional area 1 m^2, it will filter out most of the noise.

10.11.5. (*a*) Design a low-pass filter to be fitted into an air-filled pipe of 0.1 m^2 cross-sectional area that achieves $T_\Pi = 0.5$ at 1 kHz. (*b*) At what frequency will $T_\Pi = 0.9$?

10.11.6. A low-pass filter consists of a cylindrical expansion chamber of cross-sectional area S_1 and length L inserted into a pipe of cross-sectional area S. (*a*) Derive an expression for T_Π for all kL for which $\lambda > \sqrt{S_1}$. *Hint:* Follow the procedure in Section 6.3. (*b*) Show that your result is identical with (6.3.9) if S/S_1 is replaced with r_2/r_1. (*c*) Find the lowest value of kL for which T_Π has a minimum value, and write that minimum in a form displaying its symmetrical dependence on $(S_1 - S)/S$ and its reciprocal.

10.11.7C. An air-filled pipe with a cross-sectional area of 28 cm^2 has a high-pass filter consisting of an unflanged side branch with a 7.5 cm^2 circular cross-sectional area and length L. Plot the power transmission coefficient as a function of $\log f$ for $f < 2.5$ kHz and for $0 < L < 0.6$ cm.

10.11.8C. An air-filled pipe with a cross-sectional area of 7.5 cm^2 is fitted with a band filter consisting of a Helmholtz resonator with a volume of 1120 cm^3 and a cylindrical neck of 1.55 cm radius and 0.6 cm length. (*a*) Plot the power transmission coefficient as a function of $\log f$ for $f < 2.5$ kHz. (*b*) What parameters can be varied to change the width of the band-stop without changing the frequency for which the power transmission coefficient is zero?

10.11.9C. (*a*) For the pipe of Problem 10.11.8C, design a band-pass filter consisting of a rigidly capped side branch of length L with the same frequency at which $T_\Pi = 0$. (*b*) Plot the power transmission coefficient for this filter and compare with the curve obtained for the Helmholtz resonator.

Chapter 11

NOISE, SIGNAL DETECTION, HEARING, AND SPEECH

11.1 INTRODUCTION

The physical characteristics of speech, music, and noise can be measured with precision by standard acoustic instruments (microphones, filters, spectrum analyzers, oscilloscopes, etc.) and the results expressed quantitatively in terms of physical parameters. By contrast, the *interpretative* characteristics of hearing are expressible in terms of *subjective* parameters that lead to statistical predictions of the judgments of an average listener under assumed or known conditions. For example, judgments of the relative loudnesses of two sounds of different frequencies allow relating the subjective *loudness* to the physical parameters of *intensity* and *frequency*. Similar investigations have determined statistical values for the onset of pain, the *just noticeable differences* for frequency and intensity, and so on. Such experiments often suffer from uncertainty as to whether or not all pertinent variables, including any bias or attitude on the part of the subject toward the experiment, are controlled. Consequently, it must be borne in mind that the data from experiments based on subjective impressions were obtained by investigators using particular stimuli, presented to selected subjects under specific conditions. Other experimenters attempting to repeat a given experiment may obtain different results unless great care is taken to duplicate all factors involved in the initial experiment. For a wealth of information beyond that given here, the reader is advised to consult the references given below[1] and in subsequent sections.

11.2 NOISE, SPECTRUM LEVEL, AND BAND LEVEL

Until now, emphasis has been on monofrequency acoustic signals (tones). This has been a serious restriction. Whether listening to a string quartet or trying to ignore a jet aircraft, we usually hear sounds containing a great number of frequencies.

---

[1]Fletcher, *Sound and Hearing in Communication*, Van Nostrand (1953). *Handbook of Noise Control*, ed. Harris, McGraw-Hill (1979). *Encyclopedia of Acoustics*, ed. Crocker, Vols. 3 and 4, Wiley (1997).

The acoustic intensity for most sounds is nonuniformly distributed over frequency and time. It is convenient to describe the distribution by the *spectral density*

$$\mathcal{I} = \Delta I / \Delta f \tag{11.2.1}$$

where $\mathcal{I}$ is the intensity within the frequency interval $\Delta f = 1\,\text{Hz}$. The total intensity I contained within a band with upper and lower frequencies f_2 and f_1 is

$$I = \int_{f_1}^{f_2} \mathcal{I}\, df \tag{11.2.2}$$

The interval $w = f_2 - f_1$ is the bandwidth.

For almost all noise, the *instantaneous* spectral density $\mathcal{I}(t)$ is a time-varying quantity and $\mathcal{I}$ in (11.2.2) is an average over some suitable time interval τ, $\mathcal{I} = \langle \mathcal{I}(t)\rangle_\tau$. If $\mathcal{I}$ is constant for each frequency regardless of when the average is performed, the noise is termed *stationary*.

Many conventional acoustic filters and meters have both *fast* ($\frac{1}{8}$ s) and *slow* (1 s) integration times. The slow integration time is particularly useful in smoothing fluctuations. In general, fluctuations in I will decrease with increasing integration time and with increasing bandwidth.

The decibel measure of $\mathcal{I}$ is the *intensity spectrum level (ISL)*,

$$ISL = 10\log(\mathcal{I} \cdot 1\,\text{Hz}/I_{ref}) \tag{11.2.3}$$

where for air $I_{ref} = 10^{-12}\,\text{W/m}^2$. Obtaining the intensity level $IL = 10\log(I/I_{ref})$ over a desired bandwidth is straightforward if the ISL is essentially constant over the interval, but becomes involved if it depends strongly on frequency.

1. If the ISL is constant over the bandwidth w, the total intensity calculated from (11.2.2) has the simple form $I = \mathcal{I}w$ and the IL for the band is

$$IL = 10\log(\mathcal{I} \cdot 1\,\text{Hz}/I_{ref}) + 10\log(w/1\,\text{Hz}) \tag{11.2.4}$$

The first term on the right is the ISL, so this becomes

$$\boxed{IL = ISL + 10\log w} \tag{11.2.5}$$

Although incorrect, the 1 Hz divisor of w is conventionally suppressed for notational simplicity.

2. If the ISL is a function of frequency, the problem can be handled by subdividing the total bandwidth into small intervals within each of which the ISL changes by no more than a couple dB. The intensity level IL in each is calculated from (11.2.5) and converted to intensity I. These intensities are summed and then converted to a total intensity level,

$$IL = 10\log\left[\left(\sum_i I_i\right)\bigg/ I_{ref}\right] \tag{11.2.6}$$

This calculation will be simplified in the next section.

If the reference intensity and reference pressure are compatible, $I_{ref} = P_{ref}^2/\rho_0 c$, then the intensity level IL and the sound pressure level SPL are usually equal for progressive waves and (11.2.5) can be represented by

$$SPL = PSL + 10\log w \qquad (11.2.7)$$

where PSL, the *pressure spectrum level*, is equivalent to the intensity spectrum level ISL. The sound pressure level SPL and the intensity level IL of a frequency band are each referred to as the *band level*. The meaning of the pressure spectrum level for sounds other than tones may give the reader pause. However, this is the acoustic analog of the total effective voltage generated across a resistance R by several independent voltages of different frequencies. Each effective voltage amplitude V_i acting alone dissipates power V_i^2/R in the resistance. The total dissipated power is $\sum(V_i^2/R)$ and this yields a total effective voltage $V = \left(\sum V_i^2\right)^{1/2}$ across the resistance that dissipates the same power.

Instruments used to analyze noise may be either *constant bandwidth* or *proportional bandwidth* devices. The constant bandwidth instrument is essentially a tunable narrowband filter with (constant) bandwidth $w = f_u - f_l$, where f_u and f_l are the upper and lower half-power frequencies. The center frequency f_c of the band, defined by

$$f_c = \sqrt{f_u f_l} \qquad (11.2.8)$$

is usually continuously variable so that the filter can be swept over the desired frequency range. Depending on the instrument and the band, bandwidths may range from a few kHz to less than 0.01 Hz.

The proportional bandwidth instrument consists of a series of relatively broadband filters with upper and lower half-power frequencies satisfying the relationship $f_u/f_l = constant$. Each bandwidth, being proportional to the center frequency, increases linearly with increasing center frequency. Most of these filters have fixed center frequencies and contiguous bands. Common instruments of this type are the octave-band filter with $f_u/f_l = 2$, the 1/3-octave-band filter with $f_u/f_l = 2^{1/3}$, and the 1/10-octave-band filter with $f_u/f_l = 2^{1/10}$. (See the homework problems for some additional properties.) The preferred center frequencies and the corresponding logarithmic bandwidths for modern octave-band and 1/3-octave-band filters are shown in Table 11.2.1. Octave-band and 1/3-octave-band filters are used most often to analyze relatively smoothly varying spectra. If tonals are present, a 1/10-octave or a narrowband filter should be used.

In theory, the PSL of a noise can be measured by detecting the sound with a calibrated microphone and passing the resulting voltage through a filter of 1 Hz bandwidth tunable over the desired frequency range. In practice, it is more convenient to use a filter of wider bandwidth w, determine the SPL of the band, and then invert (11.2.7) to obtain a "smoothed" spectrum level,

$$\langle PSL\rangle_w = SPL - 10\log w \qquad (11.2.9)$$

If w is not too great and if the spectrum does not contain strong tones, then the $\langle PSL\rangle_w$ will be a relatively good estimate of the PSL. If there is significant

Table 11.2.1 Center frequencies and bandwidths for the preferred octave and 1/3-octave bands

Center Frequency (Hz)		10 log(Bandwidth)	
Octave	*1/3-Octave*	*Octave*	*1/3-Octave*
	10		3.6
	12.5		4.6
16	16	10.5	5.7
	20		6.6
	25		7.6
31.5	31.5	13.4	8.6
	40		9.7
	50		10.6
63	63	16.5	11.6
	80		12.7
	100		13.6
125	125	19.5	14.6
	160		15.7
	200		16.7
250	250	22.5	17.6
	315		18.6
	400		19.7
500	500	25.5	20.6
	630		21.6
	800		22.7
1000	1000	28.5	23.6
	1250		24.6
	1600		25.7
2000	2000	31.5	26.7
	2500		27.6
	3150		28.6
4000	4000	34.5	29.7
	5000		30.6
	6300		31.6
8000	8000	37.5	32.7

structure, however, the $\langle PSL \rangle_w$ will be a smoothed version of the *PSL*. If tones are important, then (11.2.9) may yield misleading results and measurements should be augmented with the help of a narrowband filter. A representative spectrum for noise containing both broadband noise and tones is given in Fig. 11.2.1. Although the $\langle PSL \rangle_w$ allows construction of a bar graph from which the overall band level can be calculated with relative ease, the effects of important tones can be obscured.

If a broadband noise has a pressure spectrum level that is independent of frequency, the noise is termed *white noise*. This noise sounds rather "hissing" and sharp. Another type of noise is *pink noise*. This contains equal powers within equal fractions of an octave. The spectrum level decreases uniformly with increasing frequency with a slope of −3 dB/octave. This noise sounds more "hushing" and is less irritating than white noise.

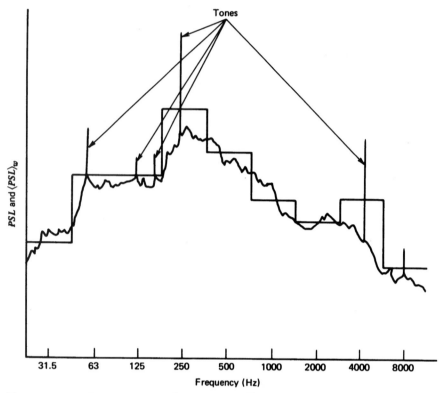

Figure 11.2.1   Representation of a spectrum of broadband noise with embedded tones. The $\langle PSL \rangle_w$ was determined over the preferred octave bands.

11.3 COMBINING BAND LEVELS AND TONES

Calculation of the overall band level can be simplified with the help of a nomogram (Fig. 11.3.1) that avoids the necessity of taking antilogs and logs as discussed in developing (11.2.6). As an example of its use, Table 11.3.1 shows the calculation of the overall level for a noise analyzed with an octave-band filter. If a more exact calculation is desired, (11.2.6) can be rewritten in a form convenient for use on pocket or desktop computers,

$$IL = 10 \log \left(\sum_i 10^{IL_i/10} \right) \tag{11.3.1}$$

Because the *PSL* is defined over a 1 Hz interval, the *SPL* of a tone is identical with its *PSL*. If a spectrum contains background noise and tones, the band level of the continuous spectrum is combined with the *PSL*s of the tones. For example, consider the spectrum of Fig 11.3.2 (for calculational ease, the background noise

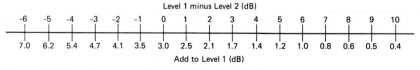

Figure 11.3.1   Nomogram for combining levels.

Table 11.3.1 Sample calculation of an overall band level from octave-band values

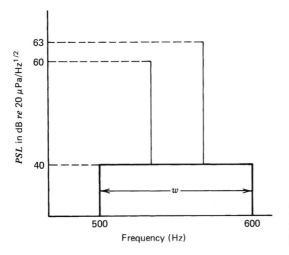

Octave-Band Center Frequency	Octave-Band Level dB re 20 μPa	Calculation of Overall Band Level
31.5	70	
		76.2
63	75	
		79.6
125	75	
		77.1
250	73	
		85.1
500	76	
		80.1
1000	78	
		83.7
2000	80	85.1 + ≈ 85 dB re 20 μPa
		81.2
4000	75	
8000	65 — 65 — 65 — 65	

Figure 11.3.2 The spectrum in a narrow bandwidth w with broadband noise and two tones.

is constant over the bandwidth). From (11.2.7), the band level of the background alone is 60 dB. When this is combined with the tones of levels 60 dB and 63 dB, the overall band level is 66 dB.

*11.4 DETECTING SIGNALS IN NOISE

The detection of a signal in the presence of noise ultimately reduces to a subjective judgment.[2] Whether trying to focus on one voice at a noisy party or to locate a submarine underwater, a listener is attempting to isolate desired information, *signal,* from undesired information, *noise.* In humans, the complicated chain of bioacoustic, neurological, and psychoacoustic processes leading from the mechanical stimulation of the eardrum to

[2]Green and Swets, *Signal Detection Theory and Psychophysics,* Wiley (1966).

the perception of sound in the brain constitutes a signal processing system of stunning sophistication contained within an impressively small volume. Many of its functions and attributes can be described quantitatively with the aid of *detection theory*.

Consider a signal that is *detected, filtered* over bandwidth w, *processed* for a time τ, and finally presented as an *output*. This output may be the time-averaged rectified voltage from a microphone, the subjective loudness of a sample of noise that may contain a tone, the instantaneous voltage developed by a thermosensitive bimetallic strip, or the perceived pitches of two successive samples of frequency-limited noise. Whatever the detection system, the output will be designated as A. If the output A_i of the processor for each time $t_i = t_{i-1} + \tau$ is recorded for a long period of time, the frequency of occurrence of each value can be determined and a histogram constructed giving the probability of any particular value A being observed (Fig. 11.4.1). Figure 11.4.1a represents the *probability density function* ρ_N if only *noise* is present, and Fig. 11.4.1b the probability density function $\rho_{S,N}$ for *signal with noise*. Each has its own mean A_N and $A_{S,N}$ and standard deviation σ_N and $\sigma_{S,N}$. The probability of *any* value of A_i occurring must be unity, so that

$$\int_0^\infty \rho_N \, dA = \int_0^\infty \rho_{S,N} \, dA = 1 \tag{11.4.1}$$

For any one time interval τ, the only way to decide if a signal is present is to select a *threshold criterion* A_T and assume that if $A_i > A_T$ there is a signal present and if $A_i < A_T$ there is no signal. This yields an independent decision for each time interval, with each decision having a certain probability of being correct or incorrect. From Fig. 11.4.1c the area under the *signal and noise* curve to the right of A_T gives the *probability of a true detection P(D)* and the area under the *noise* curve to the right of A_T gives the *probability of a false alarm P(FA)*,

$$P(D) = \int_{A_T}^\infty \rho_{S,N} \, dA \qquad P(FA) = \int_{A_T}^\infty \rho_S \, dA \tag{11.4.2}$$

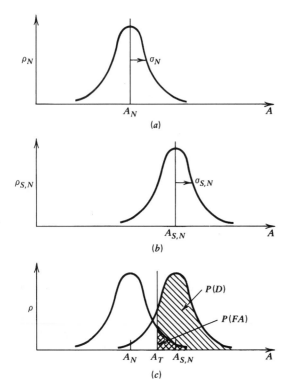

Figure 11.4.1   Probability density functions for (*a*) noise and (*b*) signal with noise. (*c*) The probability of detection and the probability of a false alarm for a given threshold criterion.

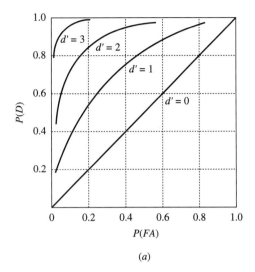

(a)

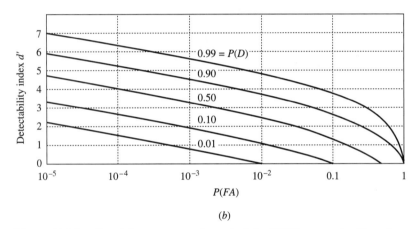

(b)

Figure 11.4.2 Receiving operator characteristic (*ROC*) curves for Gaussian distributions with equal standard deviations.

Thus, the decision whether or not a signal is present during each τ is statistical, and choosing a threshold A_T amounts to finding a balance between the probability of a detection and the probability of a false alarm.

As A_T increases from zero to an arbitrarily large value, $P(D)$ and $P(FA)$ decrease from unity to zero (Fig. 11.4.2a). Which one of these *receiver operating characteristic (ROC)* curves will be followed depends on the separation between the two probability functions. For example, if the value of A_T is held constant and the signal is allowed to increase, the probability density curve for signal with noise will move to the right and $P(D)$ will increase. This shifts the operating point of the system vertically upward to a higher curve in the figure. Each individual detection system (an echo-locating bat, a submarine detection system with its operator, or a smoke detector) has its own set of *ROC* curves. Each *ROC* curve of a set is labeled by a *detectability index* that has larger values for the higher curves.

A very important special case is that for which ρ_N and $\rho_{S,N}$ are both Gaussian with the same standard deviation σ. Under these circumstances, the detectability index is given by

$$d' = (A_{S,N} - A_N)/\sigma \tag{11.4.3}$$

and measures the number of standard deviations separating the means. If the standard deviations σ_N and $\sigma_{S,N}$ are different, replace σ in (11.4.3) with $\sqrt{\sigma_S^2 + \sigma_{S,N}^2}$.

While any particular detection problem may have its own specific probability density functions, for many processes the assumption of Gaussian distributions with equal σ values is adequate. In a real situation, deviation of experimentally determined ROC curves from the shapes shown in Fig. 11.4.2a demonstrates that this assumption is not always satisfied.

Determining the ROC curves for some detection process is an extremely laborious task. It takes hundreds of trials to obtain good estimates of $P(D)$ and $P(FA)$ for a single point on a ROC curve and at least several points to determine each of the family of curves. Three methodologies of analysis are summarized below.

1. **Yes–No Task.**   The subject is presented during each trial with either noise or signal with noise. Usually the signal strength is kept constant (d' fixed) so that the results will fall on a single ROC curve. The particular criterion is specified by the instructions to the observer (e.g., "say yes only when you are sure the signal is present" sets a higher value of A_T than does "say yes when there is even a hint of the signal"). By forcing the subject to change the criterion between sets of trials, several points on a single ROC curve can be obtained. The procedure can then be repeated for different signal strengths to obtain curves for different values of d'. The decision process used as the basis of the discussion in this section, and to which the ROC curves of Fig. 11.4.2 apply, is a yes–no task.

2. **Rating Task.**   The subject is presented with either noise or signal with noise and is asked to rate (e.g., on a scale from 1 to 4) the chance that the signal is present. This amounts to applying four different values of d' to each task. If the observer remains reasonably consistent, the resulting ROC curves should be the same as those obtained from the yes–no task.

3. ***n*-Alternative Forced-Choice (*n*AFC) Task.** The subject is presented with n samples in each trial, only one of which has signal present, and must choose which sample is most likely to contain the signal. If the distributions of noise and signal with noise are Gaussian with $\sigma_N = \sigma_{S,N}$, then the ROC curves for 2AFC will be the same as for the yes–no task, but the value of d' for each curve will be $\sqrt{2}$ times that indicated in Fig. 11.4.2.

*11.5   DETECTION THRESHOLD

Up to now we have dealt with the detection process at the *output* of the detection system. A most important task is to relate the above considerations to the *input* of the detection system. For example, the ear receives acoustic signals plus acoustic noise. There is then a succession of mechanical, neurological, and psychological transmutations of the combined signal and noise that results in a decision as to whether or not the signal is present.

The receiver (ear) has as input the *signal power S* and the *noise power N* within the bandwidth w of the system. The minimum ratio S/N that, after all the processing of the input is completed, will guarantee a given $P(D)$ for a specified $P(FA)$ is specified by the *detection threshold* over the bandwidth w,

$$\boxed{DT = 10\log(S/N) = DT_1 - 10\log w}$$ (11.5.1)

where DT_1 is the detection threshold over a 1 Hz bandwidth,

$$DT_1 = 10\log[S/(N/w)]$$ (11.5.2)

(Various texts use either DT or DT_1, or both, in discussing detection processes. The reader must be careful to determine which is being referred to, since notation is often not clear.)

Relating this to the properties observed at the output (perception) is nontrivial. The calculations lie beyond the scope of this book, but we can quote two results[3] for Gaussian probability densities with equal σ values that correspond to two extremes.

[3]Peterson, Birdsall, and Fox, *Trans. IRE*, **PGIT4,** 171 (1954).

(a) Correlation Detection

If the signal is known exactly, then optimal signal processing is based on *cross correlation*, which searches for a facsimile of the known signal in the received signal with noise. Assume the signal $s(t)$ has duration τ and the noise is given by $n(t)$. The signal with noise is $r(t) = s(t - T') + n(t)$, where T' is some unspecified delay time (i.e., the time of flight of the signal from the source to the input of the detection system). Cross correlation amounts to generating and studying the function

$$F(T) = \int_{-\infty}^{\infty} s(t)r(t + T)\, dt \tag{11.5.3}$$

where T is an adjustable time delay in the correlator. Inserting the value for $r(t)$, we have

$$F(T) = \int_{-\infty}^{\infty} s(t)s(t - T' + T)\, dt + \int_{-\infty}^{\infty} s(t)n(t + T)\, dt \tag{11.5.4}$$

On the right side the first integral will be maximized when $T = T'$. The second integral will, on the average, contribute very little for any value of T since the integrand is the product of independent oscillatory functions of time. Thus, $F(T)$, which peaks when $T = T'$, plays the role of the detection system output A. The detectability index is

$$d' = (2w\tau S/N)^{1/2} \qquad \text{(correlation)} \tag{11.5.5}$$

Such correlation may play a role in the ability of the experienced listener to concentrate on some particular instrument in an orchestra and follow its performance even during *tutti* passages.

(b) Energy Detection

If the salient features of the signal are unknown, *square-law* processing is best. A detection system receives acoustic pressure, squares it to determine the acoustic energy in each observation interval, and presents that as the output A. In the process of hearing, there is evidence that this may occur for the detection of a pure tone in noise. In the absence of transients, associated with musical instruments as one example, or stored information on overtone prominence, the hearing process may fall back on pure energy detection. Under the restrictions $S/N \ll 1$ and $w\tau \gg 1$, the detectability index is found to be

$$d' = (w\tau)^{1/2}S/N \qquad \text{(square-law)} \tag{11.5.6}$$

The detection thresholds for these two processing schemes are

$$DT = \begin{cases} 10\log\left[(d')^2/2w\tau\right] & \text{(correlation)} \\ 5\log\left[(d')^2/w\tau\right] & \text{(square-law)} \end{cases} \tag{11.5.7}$$

For example, assume that the system is to detect an unknown signal with $P(D) = 0.5$ and $P(FA) = 0.02$. From Fig. 11.4.2 we have $d' = 2$. Let the bandwidth of the system be 100 Hz and the processing time be $\tau = 0.5$ s. (1) With square-law processing $DT = -5.5$ dB and the required ratio of signal power to noise power is 0.28. If the signal were a tone, its *SPL* would be 14 dB above the *PSL* of the noise (assuming the noise has constant *PSL* over the bandwidth). (2) If the signal is known exactly, then with correlation processing $DT = -14$ dB and the required $S/N = 0.04$. The *SPL* of the pure tone can be only 6 dB above the *PSL* of the noise and still satisfy the required $P(D)$ and $P(FA)$.

In real detection systems (perhaps including some aspects of hearing), there is often a *postdetection filter* that averages the output A over a time interval T_s to reduce output fluctuations. It can be shown that the effect on the detection threshold is to increase it by $|5\log(\tau/T_s)|$. When such a device is present, the resultant detection threshold DT' of the combined system of processor and postdetection filter is then increased above the DT of the processor alone,

$$DT' = DT + |5\log(\tau/T_s)| \tag{11.5.8}$$

Postdetection filtering can never decrease DT, but if $\tau \approx T_s$ it has little deleterious effect.

*11.6 THE EAR

The human ear can respond to frequencies from approximately 20 Hz to 20 kHz. At 1 kHz, sounds that displace the eardrum only one-tenth the diameter of the hydrogen molecule can be detected. However, it is much more than a sensitive, broadband receiver. In conjunction with the nervous system, it acts as a frequency analyzer of impressive selectivity. In this book we can give only a limited introduction to the ear and hearing. For more information, consult the references.[4]

The human ear (Fig. 11.6.1) is one of the most intricate and delicate mechanical structures in the human body. It consists of three main parts: the outer, middle, and inner ears.

The *pinna* of the outer ear serves as a horn collecting sound into the *auditory canal*. In humans the pinna is a relatively ineffective device, but in some animals it supplies an appreciable gain over certain frequency ranges. The auditory canal is an approximately straight tube, about 0.8 cm in diameter and 2.8 cm long, closed at its inner end by the *tympanic membrane* (eardrum). The lowest resonance of this tube is broadly peaked around 3 kHz and affords appreciable gains from about 2 to 6 kHz. When exposed to an incident directional sound field, and including the diffractive effects of the head, the maximum *SPL* at the eardrum can be about 7 to 20 dB higher than in the incident field, depending on the direction of the sound.

The eardrum is a flattened cone lying obliquely across the auditory canal with its apex facing inward. It is quite flexible in the center and attached around its edges to the end of the canal. This membrane is the entrance to the middle ear, an air-filled cavity of about 2 cm^3 volume that contains three *ossicles* (bones). The eardrum is connected to the first of these, the *malleus* (hammer), which communicates to the *stapes* (stirrup) through the *incus* (anvil). The stapes ties directly to the *oval window* of the inner ear. There is a collection of supporting muscles and ligaments that control the lever ratio of the system. This linkage of bones, in combination with the area ratio of about 30:1 between the eardrum and the oval window, forms a broadly resonant coupler to the liquid of the inner ear. The resonance frequency lies around 3 kHz. The mechanical behavior of this coupler appears to account for the overall shape of the minimum audibility curve shown in Fig. 11.7.1. The coupling also varies with the intensity of the sound. For high intensities, the muscles controlling the motion of the ossicles change their tension to reduce the amplitude of motion of the stapes, thereby protecting the inner ear from damage. This is the *acoustic reflex*. Since for loud sounds it takes about 20 to 40 ms for the reflex to activate (shorter times for higher frequencies), it offers no protection from sudden impulsive sounds such as gunshots, explosions, and so forth. The cavity of the middle ear is connected to the throat through the *Eustachian tube* (normally closed, but sometimes opening during swallowing or yawning to reduce any pressure gradient across the eardrum).

The inner ear (*labyrinth*) has three parts: the *vestibule* (entrance chamber), *semicircular canals,* and *cochlea.* The vestibule connects with the middle ear through two openings, the *oval window* and the *round window*. Both are sealed to prevent the escape of the liquid filling

---

[4]Fletcher, *ibid.* Gelard, *The Human Senses,* 2nd ed., Wiley (1972). Jerger, *Modern Developments in Audiology,* 2nd ed., Academic Press (1973). Gelfand, *Hearing,* Dekker (1981).

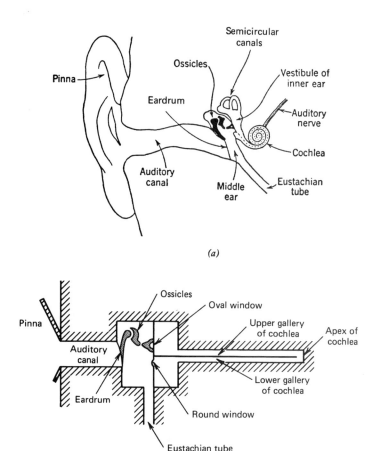

Figure 11.6.1   Sketch of the ear.

the inner ear: the former by the stapes and its support, and the latter by a thin membrane. With these two exceptions, the entire inner ear is surrounded by bone. The semicircular canals play no part in hearing but provide us with a sense of balance. The cochlea contains a tube of roughly circular cross section, wound in the shape of a snail shell about a bony core and forming a conical structure with a base diameter of about 0.9 cm and a height of about 0.5 cm. The tube has about 2.7 turns, a total length of about 3.5 cm, and a volume of about 0.05 cm³.

The tube of the cochlea is divided by the *cochlear partition* into an *upper gallery* (*scala vestibuli*), the *cochlear duct* (*scala media*), and a *lower gallery* (*scala tympani*). The galleries are joined at the *helicotrema*, a small opening at the apex of the cochlea. The other ends of the upper and lower galleries connect with the oval and round windows, respectively. A cross section of one of the turns of the cochlea is shown in Fig. 11.6.2. The *bony ledge* projects into the tube from the bony core and carries the *auditory nerve*. At the termination of the ledge the nerve fibers enter the *basilar membrane,* which continues across the tube to the farther side, where it is attached to the *spiral ligament*. This membrane is about 3.2 cm in length and is around 0.05 cm wide near the apex of the cochlea, narrowing to about 0.01 cm at the base, where its thickness is greatest. Lying above the basilar membrane is the *tectorial membrane,* attached along one edge to the bony ledge by the *spiral limbus* with its opposite edge projecting into the cochlear duct and joining the basilar membrane through the *reticular lamina*. Running diagonally across the cochlear canal from the bony ledge to the

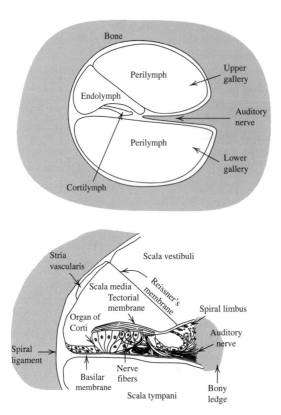

Figure 11.6.2   Cross section of the cochlear duct.

opposite wall is *Reissner's membrane,* just two cells thick. This membrane, with the basilar membrane, isolates the upper and lower galleries from the cochlear duct. The duct is filled with *endolymph* and the galleries are filled with *perilymph*. (The endolymph, potassium rich, is closely related to the intracellular fluid found throughout the body. The perilymph, sodium rich, is similar to spinal fluid.)

Attached to the top of the basilar membrane and beneath the tectorial membrane is the *organ of Corti*. It contains a third fluid, the *cortilymph* (high in sodium), and is protected from the endolymph by the reticular lamina. The organ contains nominally four rows of hair cells (about 16×10^3 cells in all) spanning the entire length of the cochlea. The 3500 hair cells in the inner row are less vulnerable to damage than are the cells in the outer three rows. Several dozen small hairs (*cilia*) extend from each hair cell through the reticular lamina toward the under surface of the tectorial membrane. The taller cilia of each outer cell are embedded in the tectorial membrane and the others of the cell appear to serve as nonlinear stiffening elements.

When the ear is exposed to a pure tone, the motion of the eardrum is transmitted by the bones of the middle ear to the oval window, creating a fluid disturbance that travels in the upper gallery toward the apex, through the helicotrema into the lower gallery, and then propagates in the lower gallery to the round window, which acts as a pressure release termination. The detailed properties of this disturbance and its role in the mechanism of hearing were elucidated in a Nobel Prize-winning series of investigations by Bekesy. His experiments demonstrated that the basilar membrane is driven into highly damped motion with a peak amplitude that increases slowly with distance away from the stapes, reaches a maximum, and then diminishes rapidly toward the apex. The peak amplitude maximizes closer to the apex for lower frequencies. Two examples are

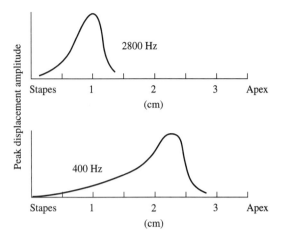

Figure 11.6.3 Peak displacement amplitude of the basilar membrane for a pure tone.

sketched in Fig. 11.6.3. The relation between frequency f in Hz and distance z in cm from the stapes of the maximum peak amplitude can be estimated from a simple empirical fit to observed data,

$$f = 10^{4 - 1.5 \tan(z/3)} \tag{11.6.1}$$

These motions of the basilar membrane occur whether the mechanical excitation is produced by airborne sounds (through the eardrum) or by sounds conducted through the skull.

Since the organ of Corti is attached to the basilar membrane and the tectorial membrane is attached to the bony ledge, the relative shearing motion between them flexes the hairs. Those cilia of the outer hair cells embedded in the tectorial membrane respond to the relative shearing displacement between the membranes, while the cilia of the inner hair cells respond to fluid drag and, therefore, are velocity sensitive. These responses, which depend on the vector nature of the excitation, cause the nerve endings attached to the hair cells to fire electrical impulses. These nerves do not necessarily fire at the frequency with which they are excited, but quasi-randomly when they are stressed beyond certain limits, usually firing more often when highly stressed. These pulses form the information communicated from the cochlea to the brain. From each ear, information routes go to a number of interlinked processing centers within the brain. There is considerable communication between processors on different sides of the brain so that the signals from both ears are mixed. There are, in addition, sets of nerves that send information from the brain to the hair cells. (For example, the left ear transmits information to the left superior olivary complex, which in turn has nerves going directly to the right ear.) For further information, the reference[5] may be a suitable start.

11.7 SOME FUNDAMENTAL PROPERTIES OF HEARING

This section deals with certain properties of behavior exhibited by the human aural apparatus: thresholds of audibility and frequency discrimination, masking, and some nonlinear effects. The subjective evaluations of loudness and pitch will be discussed separately.

---

[5]Carteratte and Friedman, *Handbook of Perception*, Vol. 2, Academic Press (1974).

(a) Thresholds

The symbol L_I is recommended by the International Organization for Standard-ization (ISO) for *IL re* 10^{-12} W/m^2. The *threshold of audibility* is the minimum perceptible L_I of a tone that can be detected at each frequency over the entire range of the ear. The tone (presented to one or both ears over headphones, in a diffuse sound field, or generated in an anechoic chamber by a source placed at a particular location with respect to the head of the subject) should have a duration of about 1 s. For tones shorter than about 0.1 s the apparent loudness increases with increasing tone duration, very similar to the ear being sensitive to the total energy of the tone burst. For tones longer than a few seconds, a reduction of sensitivity sets in and the apparent strength slowly diminishes with time, corresponding to an apparent decrease in signal level of about 30 dB over periods of about 5 min.

A representative threshold of audibility for a young undamaged ear is shown as the lowest curve in Fig. 11.7.1. The frequency of maximum sensitivity is near 4 kHz. Below this, the threshold rises with decreasing frequency, the minimum power required to produce an audible sound at 30 Hz being nearly a million times as great as at 4 kHz. For high frequencies, the threshold also rises rapidly to a

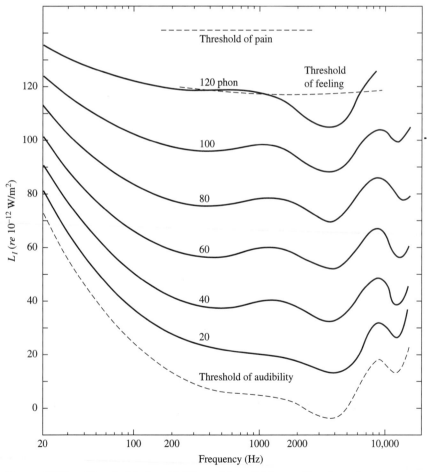

Figure 11.7.1   Threshold of audibility and free field, equal loudness level contours[6] for pure tones with subject facing the source.

cutoff. It is in this higher frequency region that the greatest variability is observed among different listeners, particularly if they are over 30 years of age. The cutoff frequency for a young person may be as high as 20 kHz or even 25 kHz, but people 40 or 50 years of age with typical hearing (and living in Western cultures) can seldom hear frequencies near or above 15 kHz. In the range below 1 kHz, the threshold is usually independent of the age of the listener.

As the intensity of the incident acoustic wave is increased, the sound grows louder and eventually produces a tickling sensation. This occurs at an intensity level of about 120 dB and is called the *threshold of feeling*. As with the lower threshold, it varies somewhat from individual to individual, but not to so great an extent. As the intensity is increased still further, the tickling sensation becomes one of pain at about 140 dB.

Since the ear responds, relatively slowly, to loud sounds by reducing the lever action of the middle ear (the acoustic reflex mentioned earlier), the threshold of audibility shifts upward under exposure, the amount of shift depending on the intensity and duration of the sound. After the sound is removed, the threshold of hearing will begin to reduce and, if the ear fully recovers its original threshold, it has experienced *temporary threshold shift (TTS)*. (The communication between right and left ears evinces itself here in that strong stimulation of one ear results in a small threshold shift in the other ear.) The amount of time required for a complete recovery increases with increasing intensity and duration of the sound. If the exposure is long enough or the intensity high enough, the recovery of the ear is not complete. The threshold never returns to its original value and *permanent threshold shift (PTS)* has occurred. It is important to realize that the damage leading to *PTS* occurs in the inner ear. The hair cells are damaged. Once thought to be irreversible, there is now evidence that in certain animals the damage can be repaired to some extent. This possibility that with treatment hair cells can be regrown is being actively investigated.

Also of importance are *differential thresholds*, one of which is the differential threshold for intensity determination. If two tones of almost identical frequency are sounded together, one tone much weaker than the other, the resultant signal is indistinguishable from a single frequency whose amplitude fluctuates slightly and sinusoidally (the "beat" phenomenon discussed in Section 1.13). The amount of fluctuation that the ear can just barely detect, when converted into the difference in intensity between the stronger and weaker portions, determines the differential threshold. As might be expected, values depend on frequency, number of beats per second, and intensity level. Generally, the greatest sensitivity to intensity changes is found for about 3 beats per second. Sensitivity decreases at the frequency extremes, particularly for low frequencies, but this effect diminishes with increasing sound level. For sounds more than about 40 dB above threshold, the ear is sensitive to intensity level fluctuations of less than 2 dB at the frequency extremes, and less than about 1 dB between 100 and 1000 Hz.

Other differential thresholds involve the ability to discriminate between two sequential signals of nearly the same frequency. The frequency difference required to make the discrimination is termed the *difference limen*. Older methods of measuring the difference limen consist of exposing the ear to a frequency modulated tone with controlled amount and rate of modulation. The difference limen depends on the intensity of the signal, its center frequency, and its rate of modulation. Above about 200 Hz, fractional changes of frequency $\Delta f / f$ clustering around about 0.005 within a factor of 2 (with higher values at lower frequencies, and vice versa) can be detected. More recent experiments use carefully shaped tone bursts of each frequency. The two methods show order of magnitude agreement, but the difference limens for pulses tend to be up to about five times those for frequency modulation above about 2 kHz.

(b) Equal Loudness Level Contours[6]

Experiments in which listeners gauge when two tones of different frequencies, sounded alternately, are equally loud provide contours as functions of frequency. As seen in Fig. 11.7.1, high- and low-frequency tones require greater values of L_I to sound as loud as those in the mid-frequency range. The curves resulting from such comparisons are labeled by the L_I they have at 1 kHz. Each curve is an *equal loudness level contour* and expresses the loudness level L_N in *phon*, which is assigned to all tones whose L_I fall on the contour. Thus, $L_N = L_I$ for a 1 kHz tone, regardless of its level. However, a 40 Hz tone with $L_I = 90$ dB has a loudness level $L_N = 70$ phon, as does a 4 kHz tone with $L_I = 61$ dB. The curves become straighter at higher loudness levels and L_N and L_I become more similar at all frequencies.

The flattening of the equal loudness level contours for higher loudness levels explains why increasing the loudness of a hi-fi system causes the sound to have disproportionately more bass and treble, while decreasing the loudness removes both "body" and "brilliance," the sound taking on a "thin" and "tinny" quality. This also reveals a problem in choosing between loudspeakers of roughly equivalent frequency responses. The more efficient one will sound louder than the other (for equal signal input) and may therefore appear to have a broader frequency response.

(c) Critical Bandwidth

If a subject listens to a sample of noise with a tone present, the tone cannot be detected until its L_I exceeds a value that depends on the amount of noise present. In a pivotal set of experiments, Fletcher and Munson[7] found that the masking of a tone by broadband noise is independent of the noise bandwidth until the bandwidth becomes smaller than some critical value that depends on the frequency of the tone. In this task the ear appears to act like a collection of *parallel filters*, each with its own bandwidth, and the detection of a tone requires that its level exceed the noise level in its particular band by some detection threshold.

In early experiments it was assumed that the signal must equal the noise for detection to occur ($DT = 0$). On this basis, and assuming that the sensitivity of the ear is constant across each bandwidth w_{cr}, it follows that $w_{cr} = S/N_1$, where S is the signal power and N_1 the noise power per hertz. The bandwidths measured this way are now termed the *critical ratios* (see Fig. 11.7.2).

Later experiments based on the perceived loudness of noise have yielded *critical bandwidths* w_{cb} larger than the critical ratios. In some of these experiments, the loudness of a band of noise is observed as a function of bandwidth while the overall noise level is held constant. For noise bandwidths less than critical, the loudness will be constant but when the bandwidth exceeds the critical bandwidth,

---

[6]Robinson and Dadson, *Br. J. Appl. Phys.*, **7**, 166 (1956). International Organization for Standardization ISO R226-1961. For more recent comments, see Yost and Killion, *Encyclopedia of Acoustics*, ed. Crocker, Chap. 23, Wiley (1997).
[7]Fletcher and Munson, *J. Acoust. Soc. Am.*, **9**, 1 (1937).

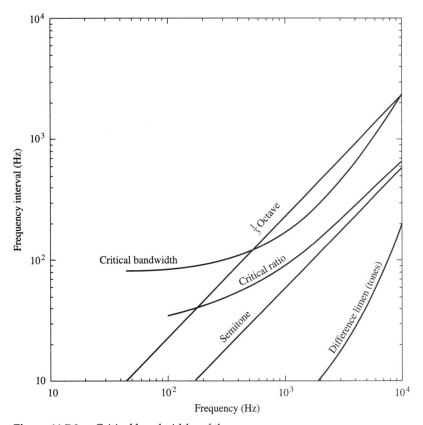

Figure 11.7.2 Critical bandwidths of the ear.

the loudness will increase (see Section 11.8). A representative curve[8] of this critical bandwidth is shown in Fig. 11.7.2. The critical bandwidth is nearly 1/3 octave for frequencies above about 400 Hz. The critical bandwidth is about two to four times the critical ratio. The curves of critical bandwidth and critical ratio would overlap fairly closely if the DT for the latter were taken to be around -4 dB.

For frequencies below about 2 kHz, the difference limen obtained by tone comparison shows that the ear can perceive frequency changes of less than about 0.03 of the critical bandwidth. The mechanism for this frequency selectivity cannot be explained by the simple filtering action mentioned earlier. Suggestions of mechanisms causing such a heightened frequency discrimination have included signal processing within the brain causing an effective narrowing of the region of cochlear response. Another postulation was that the ear may recover frequency information from the long-term combined rate at which the nerves fire. Even though each nerve usually fires with a repetition rate far below the detected frequency, if there are many nerves firing at lower rates, the collection of all the nerve firings will yield "volleys" generated with a repetition rate equal to the frequency. More recent findings point to electromechanical effects within the cochlea itself.[9] Modern measurements of the membrane vibrations show that the

[8]Zwicker, Flottop, and Stevens, *J. Acoust. Soc. Am.*, **29**, 548 (1957). Buus, *Encyclopedia of Acoustics,* Chap. 115, Wiley (1997).
[9]Zwislocki, *Am. Sci.*, **69**, 184 (1981).

frequency response at any location on the basilar membrane is much sharper than observed by Bekesy (whose work was confined to postmortem samples). It appears that the flexings of the cilia embedded in the tectorial membrane lead to biochemical processes that change the outer hair cell lengths and thus the geometry and mechanical behavior of the links between the tectorial and basilar membranes. This constitutes strongly nonlinear reactive behavior that provides significant positive feedback. Studies of the resulting nonlinear alterations in the mechanical properties of this rather complicated coupled motion support this sharpening of response. (The Q's appear to be on the order of about 20. This corresponds to a half-power bandwidth of about $0.05f$, somewhat more than about a quarter tone, which is consistent with practical observations.)

This relatively recent finding that the cochlea is not merely a passive receiver, but actively interacts with the mechanical motion generated by the external stimulus, thereby forming a strongly nonlinear process, is further supported by observation: the ear generates acoustic signatures (otoacoustic emissions). These are probably connected with the reactive deformations of the outer hair cells and can be observed at distinct time intervals after removal of a stimulus.

(d) Masking

This is the increase of the level of audibility in the presence of noise. First consider the masking of one pure tone by another. The subject is exposed to a single tone of fixed frequency and L_I, and then asked to detect another tone of different frequency and level. Analysis yields the *threshold shift*, the increase in L_I of the masked tone above its value for the threshold of audibility before it can be detected. Figure 11.7.3 gives representative results for masking frequencies of 400 and 2000 Hz. The frequency range over which there is appreciable masking increases with the L_I of the masker, the increase being greater for frequencies above that of the masker. This is to be expected because the region of the basilar membrane excited into appreciable motion at moderate values of L_I extends from the maximum further toward the stapes than the apex. For stronger excitation of the membrane both regions grow, the region toward the stapes more significantly. It is this region that covers the frequencies higher than that of the masker (low frequencies mask high frequencies). The notches in the curves will be explained later.

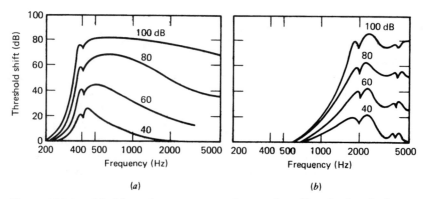

Figure 11.7.3   Masking of one pure tone by another. The abscissa is the frequency of the masked tone and the curves are labeled with the L_I of the masking tone. The frequency of the masking tone is 400 Hz in (a) and 2000 Hz in (b).

Masking of pure tones by a band of noise narrower than w_{cb} is essentially the same as that of an equally intense pure tone having the same frequency as the center of the band. Consequently, when the spectrum level is relatively constant, the intensity of a narrow band of noise is directly proportional to the bandwidth $w < w_{cb}$ and the masking (in dB) increases as $10 \log w$. Ultimately, the bandwidth will equal the critical bandwidth, and beyond that any further increase in the bandwidth of the noise has little influence on the amount of masking of a pure tone at the center of the band.

(e) Beats, Combination Tones, and Aural Harmonics

Let two tones of similar frequency f_1 and f_2 and of equal L_I be presented to one ear (or both ears). When the two frequencies are very close together, the ear perceives a tone of single frequency $f_c = (f_1 + f_2)/2$ fluctuating in intensity at the beat frequency $f_B = |f_1 - f_2|$ (see Section 1.13). As the frequency interval between the two tones increases, the sensation of beating changes to throbbing and then to roughness. As the frequency interval increases further, the roughness gradually diminishes and the sound becomes smoother, finally resolving into two separate tones. For frequencies falling in the midrange of hearing, the transition from beats to throbbing occurs at about 5 to 10 beats per second and this turns into roughness at about 15 to 30 beats per second. These transitions occur for higher beat frequencies as the frequencies of the primary tones are increased. Transition to separate tones occurs when the frequency interval has increased to about the critical bandwidth. None of this occurs if each tone is presented to a different ear. When each ear is exposed to a separate tone, the combined sound does not exhibit intensity fluctuations as discussed above; this kind of beating is absent. This suggests that the beats arise because the two tones generate overlapping regions of excitation on the basilar membrane, and it is not until these regions become separated by the distance corresponding to the critical bandwidth that they can be separately sensed by the ear. When the tones are presented one in each ear, each basilar membrane is separately excited and these effects do not occur.

If the two tones (both presented together in one ear or both ears) are separated far enough and are of sufficient loudness, combination tones can be detected. These combination tones are not present in the original sound but are manufactured by the ear. There is a collection of possible combination tones whose frequencies are various sums and differences of the original frequencies f_1 and f_2,

$$f_{nm} = |mf_2 \pm nf_1| \qquad n, m = 1, 2, 3, \dots \tag{11.7.1}$$

Only a few of these frequencies will be sensed. One of the easiest to detect is the difference frequency $|f_2 - f_1|$.

A linear combination of two tones at different frequencies does not result in any combination tones, but if two frequencies are supplied to a nonlinear system, the response will contain not only the original frequencies but also various sum and difference frequencies. Nonlinear electrical circuits are common in radio receivers and transmitters for generating such combination tones. The cochlea is the origin of the combination tones. (The middle ear remains remarkably linear up to the threshold of pain.) One means of studying this nonlinearity is through measurement of cochlear potentials. When the basilar membrane is set into motion and the hair cells stressed, slight potential differences are generated that can be

322

detected by an electrical probe inserted into the cochlea. The amplitudes and waveforms of the cochlear potentials appear to represent quite accurately those of the sounds heard. For two pure tones of low intensity, the cochlear potential has a waveform identical with that of the received sound. As the incident intensity is increased, however, the wave becomes distorted, indicating the generation of combination tones. In general, difference tones appear at lower intensity levels than do summation tones, but both are observed.

The fact that there are nonlinear mechanisms operating in the cochlea leads to the formation of various combination tones when an ear is exposed to two pure tones. Furthermore, there is nonlinear distortion when only one tone is received. If m is set equal to zero in (11.7.1), which removes the tone of frequency f_2, there are still nonlinearly generated tones with frequencies nf_1, the *aural harmonics*. Tones with frequencies above 500 Hz and loudness levels below 40 phon do not generate aural harmonics of appreciable magnitude. For frequencies around 100 Hz, the loudness level at which distortion first appears is about 20 phon. With increasing loudness the aural harmonics appear in the order of the harmonic. In general, the loudnesses of the aural harmonics are less than that of the fundamental, and they decrease progressively with increasing order. (However, for an intense low-pitched sound, such as a 60 Hz tone at 100 dB, the second harmonic may be louder than the fundamental and a number of the higher harmonics may also approach the fundamental in loudness.) The strongest evidence supporting their existence is from cochlear potentials.

Another method introduces a second tone whose frequency is brought close to a frequency of an aural harmonic so that beats are generated between the second tone and the harmonic. (This is less convincing because the vibratory motion of the basilar membrane is perturbed by the introduction of the second tone.) This second method, however, demonstrates the effectiveness of beats in enhancing detection. The notches in the masking curves of Fig. 11.7.3 show that these beats lower the threshold at the frequencies of the aural harmonics.

(f) Consonance and the Restored Fundamental

In addition to the effects described earlier, there are other more subtle nonlinear effects that have great importance for music.

One involves the perception of mistuned intervals. If two tones almost an octave apart are sounded simultaneously and tuned toward the octave, there will be a sensation similar to beating that slows and dies away as the interval of the octave is approached. This sensation is not like that for the amplitude modulation previously discussed but is apparently a response to the slowly time-varying nature of the waveform (Fig. 11.7.4). It is not necessary that both tones be presented to the same ear. If one tone is presented to one ear and the second tone to the other ear, the sensation is still produced. Thus, this interference is not the result of nonlinearities of the cochlea but arises in the brain. Besides the octave, this effect is also observed when the ratio of the two frequencies is nearly that of two integer numbers, $f_2/f_1 = 1/1, 3/2, 4/3, \ldots$, the effect becoming more subtle as the integers get larger. When the frequencies are exactly aligned to these ratios, the beating vanishes and the sensation is of *consonance*. It is no accident that these ratios form the foundation of our musical scales.[10]

---

[10]Roederer, *Introduction to the Physics and Psychophysics of Music*, Springer-Verlag (1973). Strong and Plitnik, *Music, Speech, and High Fidelity*, Brigham Young University Publications (1977).

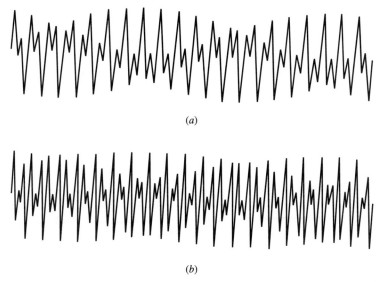

(a)

(b)

Figure 11.7.4   Waveforms for mistuned consonances: (*a*) mistuned octave and (*b*) mistuned fifth. (The mistuning has been exaggerated greatly for visual clarity.)

A second effect is the generation of the fundamental when two or more consonances are presented. If a signal consisting of tones of 1000, 1200, and 1400 Hz (e.g., a 1200 Hz carrier amplitude modulated at 200 Hz) are presented to the ears, a frequency of 200 Hz can be perceived. This tone is the fundamental of which the others are adjacent harmonics. (Recall the discussion of the kettledrum in Section 4.7.) Unlike the difference tone discussed earlier, which can be masked by noise, this fundamental can be perceived even when (1) it should be masked, (2) the signals are too weak to generate a detectable difference frequency, and (3) the tones are fed into different ears. Another aspect of this restoration is observed if the frequencies are shifted. For example, if in the above example the carrier is shifted from 1200 to 1236 Hz, tones of 1036, 1236, and 1436 Hz are generated, and the ear hears a restored pitch with a frequency around 206 Hz. (There are other possible pitches that can be perceived; see Problem 11.7.6.) This is referred to as *pitch shift*. One possible explanation suggests a temporal analysis of the total signal wherein the ear and mind "lock onto" the time separating a specific peak of the waveform in successive beats; the above shifted frequencies give rise to a peak that shifts forward in the amplitude envelope of successive beats, leading to a higher observed fundamental. Another possibility requires a spectral analysis of the signal: the mind searches for the harmonic series most closely fitting the observed frequency components. In the above example, the three shifted frequencies provide a close match to the 5th, 6th, and 7th harmonics. Both explanations present difficulties, and some current research attempts to find a viable combination of the two possibilities.[11]

This generation of the missing fundamental has practical application in the design of inexpensive, small radios. To eliminate the cost of electrical filters to remove the 60 Hz line frequency and the 120 Hz harmonic generated in the rectifier, manufacturers deliberately limit the low-frequency response of such

---

[11]Moore, *Encyclopedia of Acoustics*, Chap. 116, Wiley (1997).

radios to eliminate frequencies below about 150 Hz. (Higher harmonics of the line frequency are filtered out quite inexpensively.) Thus, while there is no significant output below about 150 Hz, the nonlinear processing in the brain restores the fundamentals of the bass notes from the higher harmonics, which are still present.

11.8 LOUDNESS LEVEL AND LOUDNESS

Although two sounds having the same loudness level are judged to be equally loud, this does *not* mean that the subjective *loudness* N is proportional to *loudness level* L_N. A tone of $L_N = 60$ phon will not sound twice as loud as one of 30 phon. The unit of loudness is the *sone,* and a loudness $N = 1$ sone is equal by definition to a loudness level of 40 phon, independent of frequency. A loudness of 16 sone is twice as loud as one of 8 sone and four times as loud as one of 4 sone.

Loudness is not easy to measure, and its determination requires ingenuity. The results of a great many experiments at different frequencies have been summarized by Fletcher (*op. cit.*) and a graph representing the relationship between loudness and loudness level is given in Fig. 11.8.1. While the lower portion of the graph is noticeably curved, the portion in excess of 1 sone is straight. In the linear portion (corresponding to sounds from comfortably audible to unpleasantly loud), an increase in the loudness level of 9 phon is approximately equivalent to doubling the loudness. An empirical formula relating loudness and loudness level over the linear portion is[12]

$$N = 0.046 \times 10^{L_N/30} \tag{11.8.1}$$

There is no dependence on frequency.

Loudness and intensity can now be related. By definition, at 1 kHz the loudness level L_N is equal to the intensity level L_I. Substitution of this into (11.8.1) and use of $L_I = 10\log(I/10^{-12})$ results in

$$N(1 \text{ kHz}) = 460\,I^{1/3} \tag{11.8.2}$$

where I is the intensity (W/m^2) of the 1 kHz tone. For any other frequency, refer to Fig. 11.7.1 and make a plot of L_N versus L_I. The resulting curve will be very close to a straight line with a slope of $+1$. Fitting this curve with a straight line of this slope over the loudness levels of interest will yield

$$L_N \approx L_I + 30\log \mathscr{F} \tag{11.8.3}$$

where $\mathscr{F}$ is an empirically determined parameter depending only on frequency. Substitution into (11.8.1) gives

$$N \approx 460\mathscr{F}I^{1/3} \tag{11.8.4}$$

Equation (11.8.4) is quite accurate for the range 500 Hz to 5 kHz and moderate loudness levels. This relation is one example of a general psychophysical "power law" of sensation postulated by Stevens.[13] For our purposes, this can be

[12]For a standardized engineering approximation, see Problem 11.8.6.
[13]Stevens, *Science,* **133,** 80 (1961).

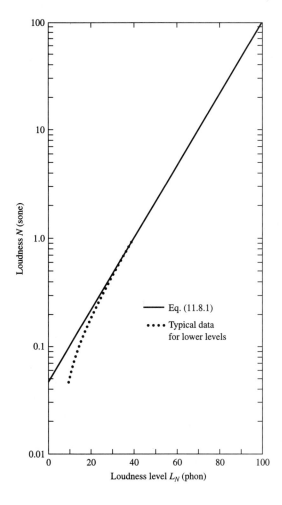

Figure 11.8.1 Loudness versus loudness level.

expressed as

$$\Omega = \begin{cases} C(S - S_T)^E & S \geq S_T \\ 0 & S < S_T \end{cases} \qquad (11.8.5)$$

where Ω is the subjective sensation, S the physical stimulus, S_T the threshold value, and C and E are constants depending on the quantities represented by Ω and S.

If two or more tones are sounded simultaneously, the total loudness depends on whether they lie within a critical band.

1. Tones lying within one critical bandwidth are sensed according to the overall power, so their intensities add and the loudness is given by

$$N = 460\mathscr{F}\left(\sum_i I_i\right)^{1/3} \qquad (11.8.6)$$

2. Tones differing by more than the relevant critical bands are sensed at well-separated regions on the basilar membrane and the loudnesses add,

Table 11.8.1   Sample calculation of loudness and loudness level

Frequency f (Hz)	Intensity Level L_I (dB)	Loudness Level L_N (phon)	Loudness N (sone)
125	60	55	3.2
250	60	62	5.4
500	60	63	5.9
1000	60	60	4.7
2000	60	62	5.4
4000	60	69	9.3
Total			33.9

$$N = \sum_i N_i \tag{11.8.7}$$

3. If the tones are considerably different in their loudness, or widely separated in frequency, the evaluation of loudness becomes difficult, often tending to be based on the loudest of the tones.

As an example of the second case, let us determine the composite loudness of six pure tones of frequencies 125, 250, 500, 1000, 2000, and 4000 Hz and each having an L_I of 60 dB. The computation is summarized in Table 11.8.1. The result of 34 sone is equivalent to about 84 phon. This corresponds to an intensity level of 84 dB for a 1 kHz tone. Since the sum of the intensities of the six tones corresponds to an L_I of only 68 dB, it is apparent that the sound energy appears to be louder when it is distributed over several critical bandwidths.

Calculations of the overall loudness of broadband noise (or even more complicated combinations of tones) becomes quite complex. The loudness becomes a function of the masking interactions among the sounds in the various critical bandwidths. The procedure has been systematized by Stevens through a number of permutations, and the methodology along with the associated tables necessary for the calculation is contained in the reference.[14]

11.9 PITCH AND FREQUENCY

Another subjective descriptor of sound is *pitch*. Like loudness, this is a complex characteristic and is dependent on various physical quantities as well as on the observer. While determined primarily by frequency, intensity and waveform are also influential.

For any particular loudness it is possible to assign numbers to the perceived pitches of pure tones describing how "high" they sound. This establishes, for that loudness, a relationship between pitch and frequency. The reference frequency is usually 1 kHz, and the tone corresponding to this frequency is assigned a pitch of 1000 mel. A tone whose pitch is 500 mel sounds half as high, and a tone of 2000 mel sounds twice as high.

For some but not all observers, increasing loudness decreases the pitch for tones below about 500 Hz and increases the pitch for tones above about 3 kHz.

---

[14]Stevens, *J. Acoust. Soc. Am.*, **51,** 575 (1972).

Tones in between show little change in pitch. When the loudnesses of tones below about 200 Hz or above about 6 kHz are increased appreciably, observers sensitive to the effect may perceive a pitch change up to nearly a whole tone. (Since the fundamental frequencies of adjacent semitones in the even-tempered musical scale are related by $f_1/f_2 = 2^{1/12} = 1.059$, a whole tone corresponds to about a 12% change in pitch.)

Complex sounds of differing loudness generate much smaller deviations between pitch and frequency. Such sounds are rich in harmonics, some of which may have amplitudes exceeding that of the fundamental. Even if the fundamental lies in the frequency range where a pure tone can show a decrease in pitch with loudness, the harmonics will have frequencies for which the pitch changes very little, so that the ear, aided by the consonances of all the harmonics, judges the sound as remaining at essentially the same pitch. The pitch of a complex sound is thus determined principally by the harmonics.

*11.10 THE VOICE

The acoustic energy associated with speech originates in the chest, diaphragm, and stomach muscles, which, by contraction, force air from the lungs up through the various components of the vocal mechanism (Fig. 11.10.1). This steady stream of air may be looked on as a carrier of energy that must be modulated in its velocity and corresponding pressure to produce sounds. The requisite modulation is accomplished in two basic ways, leading to voiced and unvoiced sounds.

Voiced sounds include the vowels of ordinary speech as well as tones characteristic of the singing voice. The primary modulating agent for voiced sounds is the *larynx*, across which are stretched the *vocal cords* (also known as *folds*). The vocal cords are two membrane-like bands controlled by sets of muscles that can change both the tension and the separation of the folds. They form a diaphragm with a slit-like opening that, as it opens and shuts, modulates the airstream. This produces pressure waveforms rich in harmonics. The length of this opening (about 2 cm in males and 1 cm in females) and the tensions to which the folds are stretched determine the fundamental frequency of this modulation. There appear to be three basic modes of vibration: (1) *pulse* in which the folds are thick and loose and produce groups of three or more pressure pulses separated by relatively long time intervals,

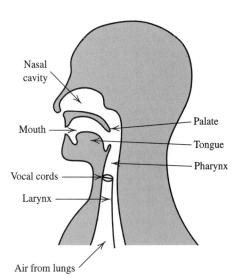

Nasal cavity

Mouth

Vocal cords

Larynx

Palate

Tongue

Pharynx

Air from lungs

Figure 11.10.1   Sectional view of the head showing the important elements of the voice mechanism.

(2) *modal* (chest) in which the folds are under greater tension and pulses arrive evenly spaced at time intervals approximating the pulse duration, and (3) *falsetto* (head) in which the folds are quite thin and stretched and do not completely close between cycles of vibration. The shorter and lighter vocal cords of most females vibrate almost twice as rapidly as those of most males. For singers, the ranges of fundamentals are approximately 70 to 300 Hz (basses), 100 to 300 Hz (tenors), 200 to 700 Hz (altos), and 250 to 1300 Hz (sopranos), with some notable exceptions.

The resonating cavities and orifices of the nose, mouth, and airways above and below the larynx form a curved tube about 15 cm in length and 70 cm³ in volume with quite variable cross sections, extending from 0 to 10 cm² and averaging about 3.5 cm². These various bulges constitute an acoustic filtering network altering the relative amounts of the harmonics. There appear to be three or four prominent resonances that impose themselves as fairly broadband pass filters, or *formants*. The lowest three are centered around $F_1 \sim$ 500 Hz, $F_2 \sim$ 1.6 kHz, and $F_3 \sim$ 3.0 kHz for a typical female and range about 20% higher for children and 20% lower for males. They are controllable within limits by changing the tensions and separation of the vocal cords, the position of the tongue, configuration of the oral cavity, shape of the lips, and so on. The formants F_1 and F_2 appear to be quite variable, each being controlled over a range of around an octave and a half. The considerable flexibility of all these elements allows production of a wide variety of voiced vowel sounds with adjustable pitch, volume, and timbre ranging from cavernous to nasal, whining to belligerent, and melodious to grating.

The voice mechanism also produces unvoiced sounds. These sounds arise from turbulence generated by air rushing through the vocal tract and include unvoiced *fricative consonants* like *f*, *h*, *s*, and *sh* as well as unvoiced *stop consonants* like *k*, *p*, and *t*, and voiceless vowels. As with voiced sounds, modulating the airstream with the lips, teeth, and tongue alters the formants, producing a wide variety of whispered sounds so that recognizable speech can be generated. Analysis of the unvoiced sounds reveals a nearly continuous spectral density over the upper portion of the audible range.

Typical average speech spectra are shown in Fig. 11.10.2. The average L_I at 1 m is about 65 dB. The L_I for female speakers is typically about 5 to 6 dB less than for male speakers. Male singers can generate maximum levels at 1 m of from about 75 dB at low pitches to 90 dB at high pitches. Female singers can produce about 85 dB at 1 m across their ranges. The acoustic power, averaged over 2 to 4 s, generated by a speaker at

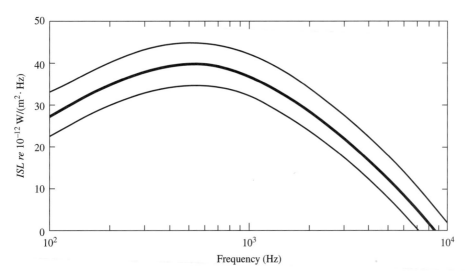

Figure 11.10.2   Representative average *ISL* for conversational speech at 1 m from the mouth.

conversational level is of order 10 μW. Very loud talking generates on the order of 100 μW and shouting in excess of about 1000 μW.

PROBLEMS

11.2.1. A proportional bandwidth filter has $f_u/f_l = r = 2^{1/n}$. Find n and r for a filter designed to have a bandwidth of (a) 1/3 octave, (b) 1/2 octave, and (c) 1/12 octave.

11.2.2. The even-tempered musical scale is designed so that adjacent semitones have $f_{i+1}/f_i = 2^{1/12}$. (a) How many semitones are there in an octave? (b) For an octave with lowest note (first semitone) tuned to 440 Hz, determine the frequencies of the remaining semitones in the octave. (c) Calculate the ratio of frequencies of 8th to 1st semitones. How close is that value to 3/2? (d) Repeat for the 6th and 1st semitones and compare with 4/3.

11.2.3. The proportional bandwidth filter has $f_u/f_l = r = 2^{1/n}$. (a) Show that $f_u = f_c\sqrt{r}$ and $f_l = f_c/\sqrt{r}$. (b) Show that for the bandwidth w of each band, $w = f_c(\sqrt{r} - 1/\sqrt{r})$. (c) Show that the center frequencies of ith and $(i + 1)$th contiguous bands are related by $f_{c(i+1)}/f_{ci} = r$.

11.2.4. Show that for a proportional bandwidth filter the upper and lower frequencies of a band are each given by the geometric mean of the adjacent center frequencies.

11.3.1. An acoustic signal consists of three tones, each of different frequency and different effective pressure amplitude: $P_1 = 5 \times 10^{-2}$, $P_2 = 7 \times 10^{-2}$, $P_3 = 0.1$ Pa, and $f_1 = 104$, $f_2 = 190$, $f_3 = 237$ Hz. Find the intensity in each of the following bands: (a) 100 to 110 Hz, (b) 100 to 150 Hz, and (c) 150 to 300 Hz.

11.3.2. The results of a noise analysis are as follows:

Filter #	f_l	f_u	V = Effective Voltage Output
1	100	200	7.1 mV
2	200	400	6.3
3	400	800	11.2
4	800	1600	8.9
5	1600	3200	11.2
6	3200	6400	7.9

(a) If the sensitivity of the receiver is 5×10^{-2} V/Pa, find the effective pressure of the sound within the bandwidth of each filter. (b) Find the intensity of the sound within the bandwidth of each filter. (c) Find the band level (re 20 μPa) for each filter. (d) Find the pressure spectrum level (re 20 μPa/Hz$^{1/2}$) for the bandwidth of each filter. (e) Use the intensities obtained in (b) to find the band level between 100 and 6400 Hz. (f) Use the band levels from (c) to find the total band level between 100 and 6400 Hz. Compare this to the answer for (e).

11.3.3. (a) A tone with a sound pressure level of 140 dB re 1 μPa is superimposed on a background noise with a constant pressure spectrum level of 150 dB re 1 Pa/Hz$^{1/2}$. Calculate the band levels obtained when the tone and the background are combined in filters of bandwidths 1, 10, or 100 Hz. Find the averaged pressure spectrum level in each of these filters. (b) Repeat for a tone with a sound pressure level of 150 dB re 1 μPa. (c) Repeat for a tone with a sound pressure level of 160 dB re 1 μPa. (d) Comment on the effect of increasing the sound pressure level of the tone.

11.3.4. A noise has a spectrum such that the intensity I_1 in each 1 Hz band is given by $I_1 = (10^{-6}/f)$ W/m^2 with f the center frequency of the band in Hz. (a) Compute the intensity spectrum level at 100, 500, and 1000 Hz. (b) What is the intensity level within the band 0.1 to 1 kHz.

11.3.5. A noise is represented by an rms acoustic pressure $P_1 = (500/f)$ μbar, where P_1 is the pressure in a 1 Hz band centered on the frequency f in Hz. (a) Derive a general expression for the pressure spectrum level PSL of this sound. (b) How does the pressure spectrum level change with frequency in dB/octave? (c) What is the band level of this noise in a band 50 Hz wide centered on 2500 Hz?

11.3.6. If $\mathcal{I} = Af^n$, show that $ISL = 10\log(A/I_{ref}) + 10n\log f$. Obtain the intensity levels of (a) pink noise and (b) white noise over the frequency intervals 0.1 to 1 kHz, 0.5 to 2 kHz, and 1 to 10 kHz if for both signals $ISL = 35$ dB re 10^{-12} W/m^2 at 1 kHz.

11.4.1. In a certain detection system the probability of a false alarm is to be reduced by an order of magnitude from 0.005 with no change in $P(D)$. How must the probability density function of signal with noise shift with respect to that for noise alone?

11.4.2. A subject is presented with samples of noise and samples of noise with a signal. In each of the following cases, qualitatively explain what happens to $P(D)$, $P(FA)$, and d'. (a) The mean amplitude of the signal gets weaker and $P(FA)$ does not change. (b) The subject is instructed to say there is a signal present only "when he is sure this is true" rather than "when it may be true." (c) The mean amplitude of the noise gradually gets smaller. (d) The subject is instructed "don't give so many false alarms."

11.4.3. Two experiments are designed to detect the presence of a signal in noise. One is a yes–no task and the other is a 2AFC task, and both use the same samples of noise and signal with noise. In the first experiment $P(FA) = 0.002$ when $P(D) = 0.5$. For the same $P(D)$, what $P(FA)$ will be expected in the second experiment?

11.4.4. In a yes–no task with specific instructions to each of the subjects for making the choice, three subjects listen to samples of Gaussian noise with or without a signal of constant amplitude such that $d' = 1$. Subject 1 detected the signal 10% of the time when it was present, subject 2 detected the signal 70% of the time it was present, and subject 3 detected the signal 40% of the time it was present. (a) Estimate the respective $P(FA)$ for each, and (b) list the subjects in order of increasing desire to avoid a false detection.

11.5.1. A detection system has fixed bandwidth w and a postdetection filter of fixed integration time T_s, but the processing time τ is adjustable. For samples of noise and signal with noise with fixed d', qualitatively explain how the detection threshold changes for square-law detection.

11.5.2. A signal consists of a tone burst of duration $\tau = 1$ s, of frequency 200 Hz, and with effective pressure amplitude of 0.02 Pa. The noise with which it is combined has a mean pressure amplitude of 2.83×10^{-2} Pa, a bandwidth of 100 Hz, and a standard deviation of 0.41×10^{-2} Pa. (a) Calculate the sound pressure levels re 20 μPa of the signal, noise, and signal with noise. (b) Calculate d' for this case assuming a yes–no task. (c) Find $P(FA)$ for $P(D) = 0.5$. (d) If the detector is a square-law processor with a postdetection filter with $T_s = \tau$, calculate the detection threshold for the specified $P(FA)$ and $P(D)$. (e) Repeat (d) for $T_s = 500$ ms and for $T_s = 2$ s.

11.5.3. Two receiving systems have the same bandwidth and the same processing time. One is a square-law and the other a correlation detector. If the two systems are operated for the same probability of false alarm and also the same probability of detection, for what value of signal-to-noise will the detection thresholds be the same, and what will be the value of the detection thresholds?

11.5.4. Assume two Gaussian distributions, both with equal standard deviations σ but with different means A_N and $A_{S,N}$. For $d' = 1, 2$, and 3, find $P(D)$ and $P(FA)$ for a number of A_T and plot $P(D)$ versus $P(FA)$. Compare to the ROC curves in the text. *Hint:* For a Gaussian distribution with mean x, the probability of a measurement greater than $x + a\sigma$ is $0.5 - F(a)$ where

$$F(a) = \frac{1}{\sqrt{2\pi}} \int_0^a e^{-z^2/2}\, dz$$

11.6.1. Invert (11.6.1) to obtain z as a function of f.

11.6.2. Find the distances on the basilar membrane measured from the oval window of the peak amplitudes for 30, 100, 300 Hz and 1, 3, 10 kHz.

11.7.1. Use Figs. 11.7.1 and 11.7.3 to answer the following questions. (*a*) What must be the sound pressure level of a 1 kHz tone if it is to be heard in the presence of a 2 kHz tone of sound pressure level 80 dB *re* 20 μPa? (*b*) What must be the sound pressure level of a 5 kHz tone for it to be heard in the presence of the 2 kHz tone?

11.7.2. Assume that the ear can be modeled by a collection of parallel square-law detectors, each of which has a bandwidth given by w_{cb} and is followed by a postdetection filter that integrates over time T_s. In three separate experiments, using 200 Hz signals of fixed duration τ in Gaussian noise, the following relationships were found to result in $P(D) = 0.6$ and $P(FA) = 0.05$:

Experiment	τ	$A_{S,N}$	A_N	σ
1	4 s	3.44×10^{-2} Pa	3×10^{-2} Pa	0.22×10^{-2} Pa
2	1	3.44×10^{-2}	3×10^{-2}	0.22×10^{-2}
3	0.1	4.51×10^{-2}	3×10^{-2}	0.22×10^{-2}

(*a*) Find the behavior of the detection threshold DT' as a function of τ for fixed d'. (*b*) Determine the detection threshold of the processor with filter for each experiment. (*c*) Determine the integration time T_s. (*d*) Assuming that T_s is independent of frequency, calculate the resultant DT' for tones of 1 s duration at 100, 200, 500, 1000, 2000, and 5000 Hz for the same $P(D)$ and $P(FA)$.

11.7.3. What must be the sound pressure level of a 200 Hz pure tone if it is to be audible in a factory noise spectrum level of 73 dB *re* 20 μPa/Hz$^{1/2}$ between 100 and 300 Hz? Assume that the critical ratio is the operative bandwidth.

11.7.4. (*a*) Let the pressure on the eardrum exerted by an incident sound be $p = P\cos\omega t$. Assuming that the subjective response r may be represented by $r = a_1 p + a_2 p^2$, where the a's are constants, prove that the response contains a constant term and terms having angular frequencies ω and 2ω. Compute the amplitude of each term as a function of P and the constants. (*b*) If the incident sound consists of frequencies ω_1 and ω_2 and produces a pressure $p = P_1 \cos\omega_1 t + P_2 \cos\omega_2 t$, determine the frequencies and amplitudes of the response.

11.7.5. If two tones of frequencies nf and $(n + 1)f$ are both shifted in frequency by the same very small amount $\Delta f \ll f$, show that the time interval T' between maxima of successive constructive interferences is consistent with a repetition rate around $f + \Delta f/n$. *Hint:* Note that successive beats of the combined signal will occur at time intervals $T = 1/f$ very slightly greater than T'.

11.7.6. For the example of shifted frequencies in Section 11.7, show that shifted pitches with frequencies around 177 and 247 Hz might also be perceived.

11.8.1. (*a*) Determine the loudness level and loudness of a 100 Hz tone with an intensity level of 60 dB *re* 10^{-12} W/m^2. (*b*) To what intensity level must this tone be reduced to lower its loudness to one-tenth of the value obtained in (*a*)? (*c*) To what intensity level must it be increased to raise its loudness to ten times the value determined in (*a*)?

11.8.2. Six pure tones have the following frequencies and intensity levels (*re* 10^{-12} W/m^2): 50 Hz at 85 dB, 100 Hz at 80 dB, 200 Hz at 75 dB, 500 Hz at 80 dB, 1 kHz at 75 dB, and 10 kHz at 70 dB. (*a*) Determine the loudness level for each of the tones. (*b*) Assume the intensity level of each of the tones is decreased by 30 dB. Determine the new loudness levels of the tones.

11.8.3. (*a*) Compute the overall intensity level for the six tones of Problem 11.8.2*a*. (*b*) What is the total loudness in sone of these six tones? (*c*) What is the intensity level of a single 1 kHz tone having the loudness of part (*b*)?

11.8.4. Three pure tones have the following frequencies and values of L_I: 100 Hz at 60 dB, 200 Hz at 60 dB, and 500 Hz at 55 dB. (*a*) Which tone is the loudest? (*b*) What is the overall sound pressure level of these three tones when sounded simultaneously? (*c*) What is their total loudness level in phon?

11.8.5. An approximation is that a 10 dB increase in loudness level is equivalent to twice the loudness. Plot this on Fig. 11.8.1 normalized so that 1 sone and 40 phon are equivalent, and compare with the plot of (11.8.1).

11.8.6. For engineering applications, ISO R131-1959 recommends

$$N = 0.0625 \times 10^{0.03 L_N}$$

Plot this on Fig. 11.8.1, and compare with the results of Problem 11.8.5.

11.8.7. Assume in (11.8.5) that Ω is the loudness N and S is the intensity I. (*a*) At 1 kHz, estimate the threshold value of S from Fig. 11.7.1. (*b*) Assuming (11.8.4) is an appropriate fit to the data for the larger loudness levels, find a modification following Stevens's law to extend the equation to smaller levels. (*c*) Calculate a corrected theoretical curve for Fig. 11.8.1 and compare it with the data presented in the figure. (*d*) Now, assume that S is the effective pressure amplitude P and repeat (*a*)–(*c*). (*e*) Compare the two corrected curves with the data and comment.

Chapter *12*

ARCHITECTURAL ACOUSTICS

12.1 SOUND IN ENCLOSURES

Around the turn of the 19th century, Wallace Sabine[1] (1868–1919) arrived at an empirical relation among the reverberation characteristics of a room, its size, and the amount of absorbing material present. The Sabine equation

$$T \propto V/A \qquad (12.1.1)$$

relates the *reverberation time T* of a room, its volume *V*, and its *total sound absorption A* (also known as the *absorption area*). This equation is based on a ray model. Sound is assumed to travel outward from the source along diverging rays. At each encounter with the boundaries of the room, the rays are partially absorbed and reflected as described in earlier chapters. After a large number of reflections the sound in the room is assumed to have become *diffuse*. In a diffuse sound field, the energy density $\mathscr{E}$ is the same throughout the space and all directions of propagation are equally probable. This model oversimplifies the actual behavior of sound in a room, particularly for lower frequencies and higher absorption, because it requires a large number of reflections before appreciable attenuation accumulates and neglects the existence of standing waves, the distribution of absorptive materials, and the shape of the room. Nevertheless, with properly chosen values of *A*, (12.1.1) leads to valid conclusions.

When a source of constant acoustic power output is turned on, the acoustic energy density in the enclosure reaches higher values (often greater than tenfold) than would exist if the source were operated in open air. If the source is shut off, reception of sound by the direct path ceases after a time interval $t = r/c$, where r is the distance from the source to the receiver and c the speed of sound in air. The reflected waves continue to be received as a succession of arrivals of decreasing intensity. The presence of this *reverberant* acoustic energy tends to mask

[1]Sabine, *Collected Papers on Acoustics,* Harvard University Press (1922); republished, Acoustical Society of America (1993).

the immediate recognition of any new sound until sufficient time has elapsed. Since both loudness and masking increase with longer reverberation times, the choice of the best reverberation time for a particular purpose in an enclosure must strike a balance between these two effects.

For more information about architectural acoustics than presented here, several sources are recommended as starting points. [2]

12.2 A SIMPLE MODEL FOR THE GROWTH OF SOUND IN A ROOM

If a source of sound is operated continuously in an enclosure, absorption in the air and at the surrounding surfaces prevents the acoustic pressure amplitude from becoming infinitely large. In smaller enclosures, absorption in air is negligible so that both the rate at which amplitude increases and its ultimate value are controlled by surface absorption. If the total sound absorption is large, the pressure amplitude quickly reaches an ultimate value only slightly in excess of that produced by the direct wave alone. Rooms with very high absorption are called *dead* or *anechoic*. By contrast, if the absorption is small, considerable time will elapse before an ultimately higher amplitude is reached. Rooms of this type are *live, reverberant,* or *echoic*.

When a source of sound is started in a live room, reflections at the walls produce a sound energy distribution that becomes more and more uniform with increasing time. Ultimately, except close to the source or to the absorbing surfaces, this energy distribution may be assumed to be completely diffuse. This lends itself to ray acoustic descriptions.

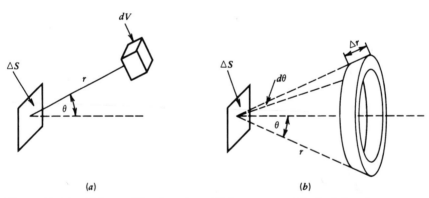

Figure 12.2.1   Volume dV and surface ΔS elements used in deriving the expression for the intensity of diffuse sound fields.

From Fig. 12.2.1, let ΔS be an element of a boundary and dV an element of volume in the air at a distance r from ΔS, where r makes an angle θ with the normal to ΔS. Let the acoustic energy density $\mathscr{E}$ be uniform throughout the region

[2]Knudsen and Harris, *Acoustical Designing in Architecture*, Wiley (1950); republished, Acoustical Society of America (1980). Doelle, *Environmental Acoustics*, McGraw-Hill (1972). Rettinger, *Acoustic Design and Noise Control, Vol. I*, Chemical Publishing Co. (1977). Beranek, *Music, Acoustics, and Architecture*, Wiley (1962); *Concert and Opera Halls: How They Sound*, Acoustical Society of America (1996).

so that the acoustic energy present in dV is $\mathscr{E}\,dV$. The amount of this energy that will strike ΔS by direct transmission is $\mathscr{E}\,dV/4\pi r^2$ multiplied by the projection of ΔS on the sphere of radius r centered on dV,

$$(\mathscr{E}\,dV/4\pi r^2)\Delta S \cos\theta \tag{12.2.1}$$

Now let dV be part of a hemispherical shell of thickness Δr and radius r centered on ΔS. The acoustic energy ΔE contributed to ΔS by this entire shell is obtained by assuming that energy arrives from any direction with equal probability. Integrating over the hemisphere with $dV = 2\pi r \sin\theta r\,\Delta r\,d\theta$ yields

$$\Delta E = \frac{\mathscr{E}\Delta S\,\Delta r}{2}\int_0^{\pi/2}\sin\theta\cos\theta\,d\theta = \frac{\mathscr{E}\Delta S\,\Delta r}{4} \tag{12.2.2}$$

This energy arrives during a time interval $\Delta t = \Delta r/c$, so that (12.2.2) can be rewritten $\Delta E/\Delta t = \mathscr{E}c\,\Delta S/4$. Thus, the rate dE/dt at which energy falls on a unit area of the wall is

$$\frac{dE}{dt} = \frac{\mathscr{E}c}{4} \tag{12.2.3}$$

Assume that at any point within the room (1) energy is arriving and departing along individual ray paths and (2) the rays have random phases. The energy density $\mathscr{E}$ at the point is then the sum of the energy densities $\mathscr{E}_j$ of each of the rays. From (5.8.10), if the jth ray has effective pressure amplitude P_{ej}, we have $\mathscr{E} = \sum\mathscr{E}_j = \sum(P_{ej}^2/\rho_0c^2)$ and thus

$$\mathscr{E} = P_r^2/\rho_0c^2 \tag{12.2.4}$$

where $P_r = (\sum P_{ej}^2)^{1/2}$ is the *effective pressure amplitude of the reverberant sound field.*

If the total sound absorption by the surfaces of and within the room is A, then from (12.2.3) the rate at which energy is being absorbed is $A\mathscr{E}c/4$. The sound absorption A has the dimensions of area and is expressed as either *metric sabin* (m²) or *English sabin* (ft²). This rate of absorption of energy by the surfaces plus the rate $V\,d\mathscr{E}/dt$ at which it increases in the volume V of the room must equal the input power Π. The differential equation governing the growth of sound energy in a live room is therefore

$$V\frac{d\mathscr{E}}{dt} + \frac{Ac}{4}\mathscr{E} = \Pi \tag{12.2.5}$$

If the sound source is started at $t = 0$, solution gives

$$\mathscr{E} = (4\Pi/Ac)(1 - e^{-t/\tau_E}) \\ \tau_E = 4V/Ac \tag{12.2.6}$$

where τ_E is the time constant. It is clear that $1/\tau_E = 2\beta$ with β the temporal absorption coefficient. If the space has large volume and small total absorption, then τ_E is large and a relatively long time will be required for the energy density to approach its limiting value. The final equilibrium energy density is

$$\mathcal{E}(\infty) = P_r^2(\infty)/\rho_0 c^2 = 4\Pi/Ac \tag{12.2.7}$$

For the same input power, the smaller A is, the larger $\mathcal{E}(\infty)$ will become.

Since these results are based on a diffuse field, there are limitations. For instance, (12.2.5) may not be used until the energy traveling along each ray path has had enough time to accumulate several reflections at the boundaries. This time can range from around 50 ms for a small room to above 1 s for an auditorium. Equation (12.2.7) indicates that the final energy density and effective pressure amplitude are independent of the volume and shape of the room, are the same at all points in the room, and depend only on the source strength and the total absorption. This is not true for a room that has sound focusing surfaces or deep recesses, or that is coupled to another space by an opening. These equations may also be invalid if some large surfaces of the room are abnormally absorptive, since the energy density near such surfaces may be lower than elsewhere.

12.3 REVERBERATION TIME—SABINE

Assume that in a live room a sound source has been turned on long enough to establish a steady-state energy density $\mathcal{E}_0$ and is then turned off at time $t = 0$. Solution of (12.2.5) provides the energy density at any later time $t > 0$,

$$\mathcal{E} = \mathcal{E}_0 e^{-t/\tau_E} \tag{12.3.1}$$

With the help of (12.2.4) the sound pressure level is seen to reduce with time as $\Delta SPL = 4.34t/\tau_E$. The *reverberation time* T, defined as the time required for the level of the sound to drop by 60 dB, is $T = 13.82\tau_E = 55.3V/Ac$. Expressing V in cubic meters, A in metric sabin, and with $c = 343$ m/s, we obtain the metric form of the *Sabine reverberation formula*,

$$T = 0.161V/A \tag{12.3.2}$$

(In English units, if V is in cubic feet, A in English sabin, and $c = 1125$ ft/s, then $T = 0.049V/A$.)

If the total surface area in the room is S, the *average Sabine absorptivity* $\bar{a}$ is defined by

$$\boxed{\bar{a} = A/S} \tag{12.3.3}$$

With this definition, (12.3.2) becomes

$$\boxed{T = \frac{0.161V}{S\bar{a}}} \tag{12.3.4}$$

If the reverberation time T is known, the total sound absorption A and the average Sabine absorptivity $\bar{a}$ can be calculated. However, the goal is to reverse the argument and *predict* the reverberation time *given* the acoustic properties of the room. While it is clear that A must depend on the areas and absorptive properties of all the diverse materials within the room, the form of this dependence is subject to a variety of simplifying assumptions. Sabine assumed that the total sound absorption is the sum of the sound absorptions A_i of the individual surfaces,

$$A = \sum_i A_i = \sum_i S_i a_i \qquad (12.3.5)$$

where a_i is the *Sabine absorptivity* for the ith surface of area S_i. With this assumption, the average Sabine absorptivity $\bar{a}$ is the area-weighted average of the individual absorptivities a_i,

$$\bar{a} = \frac{1}{S} \sum_i S_i a_i \qquad (12.3.6)$$

[In architectural acoustics a is called the *energy absorption coefficient* or *absorption coefficient* of the item and is conventionally written as α. Note that it is not an absorption coefficient like those in Chapter 9 (which are designated by α or β and have units m^{-1} or s^{-1}). It is a *dimensionless* quantity based, as will be seen below, on the power reflection coefficient R_Π of Chapter 6. To avoid confusion in this book, we use a rather than α and *absorptivity* rather than *absorption coefficient*.]

Each a for a surface or object is evaluated from standardized measurements on a sample of the material in a *reverberation chamber*. (This is an enclosure with very small $\bar{a}$ and reasonable dimensions, the longest edge being less than about three times the shortest.) When the chamber is empty, the reverberation time T will be given by (12.3.4). If the sample of surface area S_s and unknown absorptivity a_s replaces a surface of the chamber with the same area but original absorptivity a_0, the new reverberation time T_s will be

$$T_s = \frac{0.161V}{S\bar{a} + S_s(a_s - a_0)} \qquad (12.3.7)$$

Combination of (12.3.4) and (12.3.7) yields the desired evaluation,

$$a_s = a_0 + \frac{0.161V}{S_s}\left(\frac{1}{T_s} - \frac{1}{T}\right) \qquad (12.3.8)$$

In practice, it is found that the measured a_s for a sample depends somewhat on its surface area and its location within the room.

Difficulties encountered in reverberation measurements include the existence of local anomalies resulting from the formation of standing wave patterns. Sabine's method of overcoming this was to place near the center of the chamber a number of large, highly reflective surfaces that were rotated while measurements were being made. The variations in the standing wave patterns averaged out the local anomalies. Another approach was to make measurements at a large number of different points in the chamber. Current methods include the use of (1) a warble oscillator (whose varying frequency continuously changes the standing wave patterns) or (2) 1/3-octave bands of noise. Another difficulty in measurement arises from diffraction effects at the edges of the sample. When the sample is highly absorptive (like an open window), then absorptivities in excess of unity can be obtained. (It is a convention in architectural acoustics to specify $a = 1.00$ as the maximum value for any absorptivity that measures in excess of 1.0.)

Sabine's initial studies of reverberation were limited to 512 Hz. His later experiments included measurements in octave steps between 64 and 4096 Hz. Custom has attached so much importance to the 512 Hz frequency that, when the reverberation time is given without specification of any particular frequency, it is generally understood to refer to this frequency (or more recently 500 Hz). Because of the frequency dependence of the absorptivity of a surface, it is necessary to specify the reverberation time for representative frequencies covering the entire range important to speech and music. The frequencies usually chosen are 125, 250, 500, 1000, 2000, and 4000 Hz.

Consider a rectangular room measuring 3 m × 5 m × 9 m whose interior boundaries have an average Sabine absorptivity $\bar{a} = 0.1$. Then $A = 17.4$ m^2 and $T = 1.25$ s, which corresponds to a fairly live room. Equation (12.2.7) gives the ultimate effective pressure amplitude developed by a 10 μW source as 0.031 Pa, which corresponds to a sound pressure level of 64 dB re 20 μPa. For comparison, the sound pressure level produced by direct transmission at a distance of 5 m from the source in free space is 45 dB re 20 μPa.

In the development of the reverberation time (12.3.4), acoustic losses in the volume of the air were neglected. These increase the total absorption and decrease the reverberation time. The total temporal absorption coefficient is simply the sum of the individual coefficients. We have seen that the pressure amplitude of a standing wave experiencing only absorption in the air decreases as $P = P_0 \exp(-\alpha ct)$. Thus, $\mathcal{E} = \mathcal{E}_0 \exp(-mct)$, where $m = 2\alpha$. (In architectural acoustics, m rather than 2α is used.) Consequently, (12.3.1) may be rewritten

$$\mathcal{E} = \mathcal{E}_0 e^{-(A/4V+m)t} \tag{12.3.9}$$

and the expression for the reverberation time becomes

$$\boxed{T = \frac{0.161V}{S\bar{a} + 4mV}} \tag{12.3.10}$$

The importance of absorption in the air is determined by the ratio $4mV/S\bar{a}$. Since m increases with f whereas $\bar{a}$ tends to decrease above 1 kHz, absorption in the air can be significant at higher frequencies in large volumes. Also, in highly reverberant spaces the major portion of the sound absorption may occur in the air rather than at the surfaces. For *relative humidities h* in percent between about 20 and 70 and frequencies in the range 1.5 to 10 kHz, a sufficiently accurate approximation for most architectural applications is

$$m = 5.5 \times 10^{-4}(50/h)(f/1000)^{1.7} \tag{12.3.11}$$

12.4 REVERBERATION TIME—EYRING AND NORRIS

One of other proposed reverberation-time equations is that attributed to Eyring[3] and Norris. It is based on the mean free path between reflections. It can be shown[4] that the mean distance traveled by a ray between successive reflections from the

[3]Eyring, *J. Acoust. Soc. Am.*, **1**, 217 (1930).
[4]Bate and Pillow, *Proc. Phys. Soc.*, **59**, 535 (1947).

walls of a rectangular enclosure is $L_M = 4V/S$, so the number of reflections per second is $N = cS/4V$. With each reflection, the sound is reduced in energy (on the average) by the factor $(1 - \bar{a}_E)$, where $\bar{a}_E$ is the *area-averaged random-incidence energy absorption coefficient,*

$$\bar{a}_E = \frac{1}{S} \sum_i S_i a_{Ei} \qquad (12.4.1)$$

and a_{Ei} is the random-incidence energy absorption coefficient of the ith surface. The total attenuation of the energy over a time interval of one reverberation time T must therefore be $(1 - \bar{a}_E)^{NT}$. This must correspond to a reduction of 60 dB in the sound pressure level, so that $10 \log(1 - \bar{a}_E)^{NT} = -60$. Solution for T and use of the above expression for N yields the *Eyring–Norris reverberation formula,*

$$T = \frac{0.161V}{-S \ln(1 - \bar{a}_E)} \qquad (12.4.2)$$

Expanding $\ln(1 - \bar{a}_E)$ for small $\bar{a}_E$ and comparing the Sabine and Eyring formulas show that for highly reverberant rooms ($a \ll 1$) we can make the identification

$$a = a_E \qquad (12.4.3)$$

The Sabine absorptivity and the random-incidence energy absorption coefficient for a specific surface can be taken as the same. The Eyring and Sabine predictions for T are identical for small $\bar{a}$, but for large $\bar{a}$ the Eyring formula predicts a smaller value.

If the power reflection coefficient R_{Π} is known for each angle of incidence θ, then the relationship between $R_{\Pi}(\theta)$ and $a_E(\theta)$ is

$$a_E(\theta) = 1 - R_{\Pi}(\theta) \qquad (12.4.4)$$

So that, from (6.1.7), $a_E(\theta)$ is the power transmission coefficient T_{Π}. With the help of (12.2.2) it is straightforward to see that for a diffuse incident field

$$a_E = 2 \int_0^{\pi/2} a_E(\theta) \sin\theta \cos\theta \, d\theta \qquad (12.4.5)$$

While it is usual to assume a diffuse sound field, there is no guarantee that this is the case. If the angular dependence of the power reflection coefficient is known, the diffuse field absorptivity can be evaluated from (12.4.5). If the distribution of rays is not diffuse the integral can be modified appropriately. Otherwise, the absorptivity can be measured according to Sabine's method in a reverberant chamber. The results may not be in agreement. In particular, as mentioned earlier, the reverberation chamber measurement can lead to absorptivities in excess of unity. There is still discussion about the relative merits of the different methods of determining the effective a_E for a surface, but many investigators recommend the use of Sabine's equation and the experimentally determined Sabine absorptivities for most practical applications.[5]

[5]Young, *J. Acoust. Soc. Am.,* **31,** 912 (1959). *Noise and Vibration Control,* ed. Beranek, McGraw-Hill (1971).

Further discussions of these and other formulas for reverberation times are available.[6]

12.5 SOUND ABSORPTION MATERIALS

Table 12.5.1 provides absorptivities and absorptions for various materials. For more information, refer to the bulletin "Performance Data: Architectural—Acoustic Materials," published annually by the Acoustical and Insulating Materials Association, and to the sources cited in footnote 2.

Sound absorbers important in acoustic design can be loosely classified as (1) porous materials, (2) panel absorbers, (3) cavity resonators, and (4) individual people and items of furniture.

1. *Porous materials,* such as acoustic tiles and plasters, mineral wools (fiberglass), carpets, and draperies are networks of interconnected pores within which viscous losses convert acoustic energy into heat. The absorptivities of such materials are strong functions of frequency, being relatively small at the lower frequencies and increasing to relatively high values above about 500 Hz. The absorptivities increase with increasing material thickness. Low-frequency absorption can be increased by mounting the material away from the wall. Painting acoustic plasters and tiles will invariably result in a substantial reduction in effectiveness.

2. A *nonporous panel* mounted away from a solid backing vibrates under the influence of an incident sound, and the dissipative mechanisms in the panel convert some of the incident acoustic energy into heat. Such absorbers (gypsum sheetrock, plywood, thin wooden paneling, etc.) are quite effective at low frequencies. The addition of a porous absorber in the space between the panel and the wall will further increase the efficiency of the low-frequency absorption.

3. A *cavity resonator* consists of a confined volume of air connected to the room by a narrow opening. It acts like a Helmholtz resonator, absorbing acoustic energy most efficiently in a narrow band of frequencies near its resonance. These absorbers may be in the form of individual elements, such as concrete blocks with slotted cavities. Other forms consist of perforated panels and wood lattices spaced away from a solid backing with absorption blankets in between. Besides allowing for free architectural expression, these provide useful absorption over a wider frequency range than is possible with individual cavity elements.

4. Table 12.5.1 also includes the *sound absorption per item* for clothed people, upholstered seats, and wooden furniture. Wooden furniture includes chairs with very little upholstery, school desks, and tables (a table providing work space for five people counts as five tables). For widely dispersed audiences with wooden desks, tables, or chairs (as are found in sparsely filled classrooms and many lecture halls), it may be more appropriate to use the absorption per body and per article of furniture rather than the audience absorptivity.

[6]A good starting point is the *Encyclopedia of Acoustics,* ed. Crocker, Wiley (1997): in particular, the chapters by Tohyama (Chap. 77), Kuttruff (Chap. 91), and Bies and Hansen (Chap. 92).

Table 12.5.1   Representative Sabine absorptivities and absorptions

Description	Frequency (Hz)					
	125	250	500	1000	2000	4000
	Sabine Absorptivity a					
Occupied audience, orchestra, chorus	0.40	0.55	0.80	0.95	0.90	0.85
Upholstered seats, cloth-covered, perforated bottoms	0.20	0.35	0.55	0.65	0.60	0.60
Upholstered seats, leather-covered	0.15	0.25	0.35	0.40	0.35	0.35
Carpet, heavy on undercarpet (1.35 kg/m^2 felt or foam rubber)	0.08	0.25	0.55	0.70	0.70	0.75
Carpet, heavy on concrete	0.02	0.06	0.14	0.35	0.60	0.65
Acoustic plaster (approximate)	0.07	0.17	0.40	0.55	0.65	0.65
Acoustic tile on rigid surface	0.10	0.25	0.55	0.65	0.65	0.60
Acoustic tile, suspended (false ceiling)	0.40	0.50	0.60	0.75	0.70	0.60
Curtains, 0.48 kg/m^2 velour, draped to half area	0.07	0.30	0.50	0.75	0.70	0.60
Wooden platform with airspace	0.40	0.30	0.20	0.17	0.15	0.10
Wood paneling, 3/8–1/2 in. over 2–4 in. airspace	0.30	0.25	0.20	0.17	0.15	0.10
Plywood, 1/4 in. on studs, fiberglass backing	0.60	0.30	0.10	0.09	0.09	0.09
Wooden walls, 2 in.	0.14	0.10	0.07	0.05	0.05	0.05
Floor, wooden	0.15	0.11	0.10	0.07	0.06	0.07
Floor, linoleum, flexible tile, on concrete	0.02	0.03	0.03	0.03	0.03	0.02
Floor, linoleum, flexible tile, on subfloor	0.02	0.04	0.05	0.05	0.10	0.05
Floor, terrazzo	0.01	0.01	0.02	0.02	0.02	0.02
Concrete (poured, unpainted)	0.01	0.01	0.02	0.02	0.02	0.02
Gypsum, 1/2 in. on studs	0.30	0.10	0.05	0.04	0.07	0.09
Plaster, smooth on lath	0.14	0.10	0.06	0.04	0.04	0.03
Plaster, smooth on lath on studs	0.30	0.15	0.10	0.05	0.04	0.05
Plaster, 1 in. damped on concrete block, brick, lath	0.14	0.10	0.07	0.05	0.05	0.05
Glass, heavy plate	0.18	0.06	0.04	0.03	0.02	0.02
Glass, windowpane	0.35	0.25	0.18	0.12	0.07	0.04
Brick, unglazed, no paint	0.03	0.03	0.03	0.04	0.05	0.07
Brick, smooth plaster finish	0.01	0.02	0.02	0.03	0.04	0.05
Concrete block, no paint	0.35	0.45	0.30	0.30	0.40	0.25
Concrete block, painted	0.10	0.05	0.06	0.07	0.09	0.08
Concrete block, smooth plaster finish	0.12	0.09	0.07	0.05	0.05	0.04
Concrete block, slotted two-well	0.10	0.90	0.50	0.45	0.45	0.40
Perforated panel over isolation blanket, 10% open area	0.20	0.90	0.90	0.90	0.85	0.85
Fiberglass, 1 in. on rigid backing	0.08	0.25	0.45	0.75	0.75	0.65
Fiberglass, 2 in. on rigid backing	0.21	0.50	0.75	0.90	0.85	0.80
Fiberglass, 2 in. on rigid backing, 1 in. airspace	0.35	0.65	0.80	0.90	0.85	0.80
Fiberglass, 4 in. on rigid backing	0.45	0.90	0.95	1.00	0.95	0.85
	Sound Absorption A in m^2					
Single person or heavily upholstered seat ($\pm$0.10 m^2)	0.40	0.70	0.85	0.95	0.90	0.80
Wooden chair, table, furnishing, for one person	0.02	0.03	0.05	0.08	0.08	0.05

   Proper choice of the amounts and distributions of these classes of absorbers can tailor the behavior of the reverberation time with frequency to obtain almost any desired acoustic environment. Since the optimum reverberation time depends on the use of the room, it is possible to design multipurpose rooms with sliding or rotating panels that expose surfaces of different absorption properties. However, artificial reverberation introduced electronically can be a less expensive and more flexible solution, especially in large rooms.

12.6 MEASUREMENT OF THE ACOUSTIC OUTPUT OF SOUND SOURCES IN LIVE ROOMS

The most accurate methods of measuring the acoustic output of sound sources require an anechoic chamber with walls as completely absorptive as possible. This is the closest approximation of an unbounded, homogeneous space that can be obtained in the laboratory. However, source outputs can be measured with acceptable accuracies in reverberant rooms. When the sound energy in such a room is completely diffuse, the acoustic power output is given by (12.2.7). If $P_r(\infty)$ were truly uniform throughout the room, one measurement of its magnitude would be sufficient. When it is not, either a number of measurements can be made and averaged or the microphone can be rotated on an arm (over a distance of at least a quarter-wavelength) to measure an averaged pressure. The only other unknown parameter in (12.2.7) is the total sound absorption A of the room. If the absorptivities of the walls of the room are known, A may be computed from the equations presented earlier. If not, it may be determined from (12.3.2) by measuring the reverberation time T of the room. Combination of (12.2.7) with (12.3.2) to eliminate A yields

$$\Pi = 13.9(P_r^2/\rho_0 c^2)V/T = 9.7 \times 10^{-5}P_r^2 V/T \qquad (12.6.1)$$

(If P_r is in μbar instead of Pa, replace 13.9 with 0.139 and the exponent -5 with -7.)

12.7 DIRECT AND REVERBERANT SOUND

Whenever a continuous source of sound is present in a room, two sound fields are produced. One, the *direct sound field,* is the direct arrival from the source. The other, the *reverberant sound field,* is produced by the reflections. The energy density $\mathcal{E}_d$ produced by the direct sound field of an omnidirectional source is given by

$$\mathcal{E}_d = (\Pi/c)/4\pi r^2 \qquad (12.7.1)$$

where r is the radial distance from the effective center of the sound source and Π is the acoustic power output of the source. The energy density $\mathcal{E}(\infty)$ of the reverberant field is obtained from (12.2.7) and the total field is $\mathcal{E}_d + \mathcal{E}(\infty)$. The ratio of reverberant to direct energy densities is

$$\mathcal{E}(\infty)/\mathcal{E}_d = (r/r_d)^2 \qquad (12.7.2)$$

where $r_d = \frac{1}{4}\sqrt{A/\pi}$ is the distance at which the direct field has fallen to the same value as the reverberant field. This equation shows that for locations very close to the source ($r \ll r_d$) the shape or acoustic treatment of the room will have little influence on measured sound pressure levels. By contrast, at distances for which $r \gg r_d$, the sound pressure level will be reduced by 3 dB for each doubling of the total sound absorption A.

For example, a worker near a noisy machine will receive little benefit from increasing the total absorption of the room. However, the acoustic exposure of other workers at some distance from the machine will be reduced by such treatment. As another example, when two people are alone in a quiet room and relatively close together, the acoustic characteristics of the surroundings have negligible influence on their ability to carry on a conversation. However, if many talkers are present in the room, the reverberant sound pressure level will increase by $10 \log N$, where N is the number of talkers, and will make conversation difficult unless the total sound absorption A is large. This is why it is often hard to converse in a large ballroom or dining hall when many people are talking. Assume 100 talkers, each with an acoustic output of 100 μW, to be present in a room 5 m $\times$ 20 m $\times$ 40 m with $T = 3$ s. Substitution into (12.3.2) leads to $A = 215$ m^2. The reverberant sound pressure (12.6.1) is then $P_r = 0.28$ Pa, which corresponds to a sound pressure level of 83 dB *re* 20 μPa. This background level is much too high for a normal conversation: substitution into (12.7.1) shows that this same pressure is reached in the direct field from an individual talker at a distance of 0.2 m. If all the talkers were to reduce their acoustic output to 10 μW, the background level of the reverberant sound would be reduced to 73 dB *re* 20 μPa without altering the intelligibility of the conversation. Unfortunately, when a large number of talkers are present in a room, the "cocktail party effect"[7] becomes evident as each talker raises his or her individual acoustic output to be heard. On the average, this does not increase the intelligibility, but merely increases the background level to an unpleasantly high value.

12.8 ACOUSTIC FACTORS IN ARCHITECTURAL DESIGN

Whether designing a music room, conference room, lecture hall, concert hall, or large auditorium, the acoustic consultant must consider several different factors whose relative importances depend on the purpose of the enclosure.

(a) The Direct Arrival

In any enclosure, there should be a direct and clear line of sight between the audience and the source of sound. Not only is this psychologically important, but it also guarantees that there will be a well-defined direct arrival of sound. This is very important in giving an aural sense of the direction from which the sound comes. Generally, in large spaces this requires that seating areas, including balconies, be raked from front to back. This also helps avoid the low-frequency attenuation that occurs for sounds crossing the audience at near-grazing incidence. Raking the stage or having risers on its rear sections will also enhance direct arrivals and reduce grazing incidence problems. If there is electronic reinforcement of the sound (often found in multipurpose enclosures), sufficient time delay must be inserted in these signals so that the direct arrival precedes the arrivals from other directions by an interval of about 10 to 30 ms. This *precedence effect* is necessary for a good sense of acoustic direction.

(b) Reverberation at 500 Hz

There must be an appropriate balance between the direct arrival and the reverberant sound field. Combination of (12.7.2) and (12.3.2) yields

$$\mathscr{E}(\infty)/\mathscr{E}_d = 312\, r^2 T/V \tag{12.8.1}$$

Since the energy of the direct arrival falls off as the square of the distance from the source, it is impossible to have a constant ratio throughout the enclosure. However,

---

[7]MacLean, *J. Acoust. Soc. Am.*, **31,** 79 (1959).

for equivalent locations in spaces of similar geometry r^3 will be proportional to the volume V. This means that if it is desired to maintain the same value of $\mathscr{E}(\infty)/\mathscr{E}_d$ at the same relative location in enclosures of similar shape but different volumes, we must have

$$T = RV^{1/3} \tag{12.8.2}$$

where R is a constant that depends on the purpose of the enclosure. Despite the approximate nature of this relationship, it is a useful empirical estimate. Table 12.8.1 gives the results of fitting this formula to conventionally accepted values of T and V for enclosures used for various purposes. These formulas allow the estimation of certain limits in the design criteria. Conditions for a lecture by an accomplished speaker in a quiet hall will be very good if the reverberation time does not exceed about 0.8 s, which suggests a maximum volume for a lecture hall of about 2.4×10^3 m^3. For music in a concert hall, the reverberation time should not exceed about 2 s, which yields an estimated maximum volume of about 2.4×10^4 m^3. It must be emphasized that these predicted values are only estimates and are untrustworthy for V lying outside the typically encountered range of values. For volumes approaching or exceeding the indicated upper limits, it is found in practice that the value of R may decrease toward its indicated lower limit. The basic trend, however, is obvious. For example, Carnegie Hall is a poor place to hold a lecture without electronic enhancement, and a symphony orchestra or rock band will be overpowering in a lecture hall unless output is deliberately and severely limited.

The observed reverberation may exhibit different values for T at different times during the decay. Evidence indicates that the initial apparent reverberation time is of greatest importance to the listener. This means that the earlier delayed arrivals are quite important in establishing a satisfactory impression of reverberation. This can present problems in fan-shaped spaces with diverging side walls.

The problem of specifying an optimum reverberation time for an enclosure designed primarily for the playing or reproduction of music is more complex than for speech, because the optimum time varies with the type of music and the desired effect. Music rooms should be more reverberant than lecture halls or conference rooms of similar size. The optimum reverberation time is found to range from about 0.5 s in living rooms, through about 1.0 s in small rooms used for soloists or chamber music, up to about 2.5 s for organ music or oratorios in large cathedrals.

Table 12.8.1   Approximate values of $R = T/V^{1/3}$ for rooms used for various purposes

Purpose	$R \pm 10\%$ (s/m)	Range of Volumes Conventionally Encountered (m^3)[a]
Concert hall	0.07	$10 \times 10^3 < V < 25 \times 10^3$
Opera house	0.06	$7 \times 10^3 < V < 20 \times 10^3$
Motion picture theater	0.05	$V < 10 \times 10^3$
Auditorium Legitimate theater Lecture hall Conference room	0.06	$V < 4 \times 10^3$
Recording studio Broadcasting studio	0.04	$V < 1 \times 10^3$

[a]For conversion to British units, 1 m^3 = 35.3 ft^3.

Classical and Baroque music benefit from reverberation times of about 1.0 to 1.4 s, whereas 19th century orchestral music may require about 2.0 s of reverberation for the best effect. These values are also subject to individual tastes and cultural attitudes.

In the design of a recording studio for nonclassical music (especially rock, pop rock, or country and western) or a television broadcast studio, the reverberation time should be short, particularly since these uses are usually accompanied by considerable electronic manipulation, including amplitude limiting, artificial reverberation and echo, and frequency contouring.

A factor that must be considered in designing an auditorium is the effect of an audience on reverberation. Variations in the size of the audience may produce large changes in reverberation times at all frequencies, particularly above about 250 Hz. The change is most apparent when an empty concert hall with unupholstered seats is used for rehearsals; inexperienced conductors can suffer a rude awakening when the public performance occurs. The influence of variations in the size of an audience can be materially reduced by using seats that are well upholstered and have perforated bottoms, since the resulting absorption is nearly the same whether the seat is empty or occupied. If economy requires that inexpensive (lightly upholstered or bare) seating be used, then a reasonable rule of thumb is to design the hall to have the desired reverberation time when it is about two-thirds occupied.

For representative concert halls, Table 12.8.1 gives a ratio of direct energy to reverberant energy of about 0.07. This means that for sounds lasting long enough to allow the full reverberation of the hall to develop, the reverberant sound energy is roughly 15 times that of the direct arrival. Subjectively, the combined sound field has a loudness somewhat more than twice that for the direct sound alone. This is what generates the sense of power or grandeur associated with music of the Romantic Period. (The absence of reverberation is frequently what makes the unamplified sound of outdoor concerts so unsatisfying.)

As a result of an extensive survey of European and New World concert halls and opera houses, an empirical relationship between reverberation time and the parameters of the enclosure has been developed by Beranek (*op. cit.,* footnote 2):

$$1/T = 0.1 + 5.4 S_T/V \tag{12.8.3}$$

where S_T is the total floor area of audience, orchestra, and chorus. This appears to describe above-average concert halls better than (12.8.2). The absorptivity attributed to S_T is obtained from the first entry in Table 12.5.1. A selection of concert halls and opera houses judged "very good" to "excellent" is listed in Table 12.8.2, and the inverse reverberation times as functions of S_T/V are plotted in Fig. 12.8.1. The close fit to the curve of (12.8.3) is apparent, particularly for the larger reverberation times. A plot of T against $V^{1/3}$ for the same halls and comparison with (12.8.2) shows much poorer correlation.

(c) Warmth

The *warmth* of the enclosure depends on the comparison between low-frequency and midfrequency reverberation times. The desired behaviors of the reverberation times as functions of frequency for the limiting cases of speech and music are sketched in Fig. 12.8.2. The cross-hatched regions indicate desired values, with Baroque music tending toward the speech curve and Romantic tending toward the music curve. Study of Table 12.5.1 with these curves in mind will reveal that the use of thin paneling or other lightweight wall material may result in undesirably large absorption

Table 12.8.2 Acoustic environments of selected concert halls and opera houses

Hall	$V/10^3$ (m³)	$S_T/10^3$ (m²)	Reverberation Time (s) at Various Frequencies (Hz)						Delayed Arrival Time (ms)	Seats
			125	250	500	1000	2000	4000		
J	Jerusalem, Binyanei Ha'oomah	24.7	2.4	2.2	2.0	1.75	1.75	1.65	1.5	13–26	3100
N	New York, Carnegie Hall (before renovation)	24.3	2.0	1.8	1.8	1.8	1.6	1.6	1.4	16–23	2800
Bo	Boston, Symphony Hall	18.7	1.6	2.2	2.0	1.8	1.8	1.7	1.5	7–15	2600
A	Amsterdam, Concertgebouw	18.7	1.3	2.2	2.2	2.1	1.9	1.8	1.6	9–21	2200
Gl	Glasgow, St. Andrew's Hall	16.1	1.4	1.8	1.8	1.9	1.9	1.8	1.5	8–20	2100
P	Philadelphia, Academy of Music	15.7	1.7	1.4	1.7	1.45	1.35	1.25	1.15	10–19	3000
V	Vienna, Grosser Musikvereinsaal	15.0	1.1	2.4	2.2	2.1	2.0	1.9	1.6	9–12	1700
Bri	Bristol, Colston Hall	13.5	1.3	1.85	1.7	1.7	1.7	1.6	1.35	6–14	2200
Bru	Brussels, Palais des Beaux Arts	12.5	1.5	1.9	1.75	1.5	1.35	1.25	1.1	4–23	2200
Go	Gothenburg, Konserthus	11.9	1.0	1.9	1.7	1.7	1.7	1.55	1.45	22–33	1400
L	Leipzig, Neues Gewandhaus	10.6	1.0	1.5	1.6	1.55	1.55	1.35	1.2	6–8	1600
Ba	Basel, Stadt-Casino	10.5	0.9	2.2	2.0	1.8	1.6	1.5	1.4	6–16	1400
C	Cambridge, Mass., Kresge Auditorium	10.0	1.0	1.65	1.55	1.5	1.45	1.35	1.25	10–15	1200
(Bu)	Buenos Aires, Teatro Colon	20.6	2.1	—	—	1.7	—	—	—	13–19	2800
(NM)	New York, Metropolitan Opera	19.5	2.6	1.8	1.5	1.3	1.1	1.0	0.9	18–22	2800
(M)	Milan, Teatro alla Scala	11.2	1.6	1.5	1.4	1.3	1.2	1.0	0.9	12–15	2500

Source: Adapted from Beranek, *op. cit.*

346

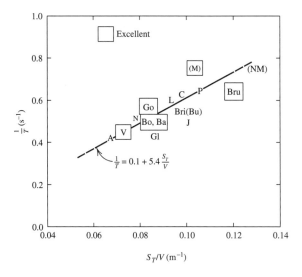

Figure 12.8.1 Reverberation times for "good" to "excellent" concert halls and opera houses. For identification of buildings, see Table 12.8.2.

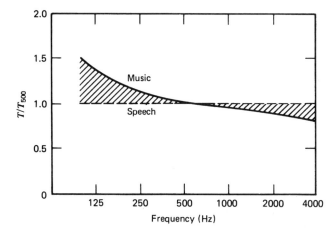

Figure 12.8.2 Relative reverberation-time limits for music and speech.

at low frequencies. For large concert halls, in which the audience is the dominant source of absorption, thick well-sealed wooden surfaces are an excellent choice.

(d) Intimacy

The perceived *intimacy* of the sound is of great importance for speech and of even greater importance for music. This quality depends on the reception of reflected arrivals immediately after the direct arrival from the source. These early delayed arrivals should be many and evenly distributed with time, beginning no later than about 20 ms (30 ms for opera) after the direct arrival and blending smoothly together to form the reverberation. (A 20 ms delay corresponds to a difference in path length of about 7 m.) Early arrivals from the sides of a hall are of greater importance than early arrivals from overhead. Modern audiences appear to prefer more of a "stereo" effect than in the past, perhaps because of greater access to well-designed home sound systems. Obtaining the proper balance of early laterally reflected arrivals can be a problem in large or fan-shaped enclosures, particularly close to the source. Careful design of the stage, a heavy and rigid shell, judiciously positioned ceiling and wall reflectors, and suspended reflectors can enhance these early arrivals throughout the audience. Reflectors should be made of rigidly braced plywood at least 1.9 cm ($\frac{3}{4}$ in.) thick, or of some reasonable equivalent. Care must be taken that these remedies do not introduce undesired effects such as (1) resonances of the reflectors, (2) excess absorption, (3) high-frequency reflection toward the audience but low-frequency diffraction into the space above the reflectors, with an attendant loss of low-frequency energy in the audience space, or (4) focusing of the sound.

Scale drawings or models of the enclosure with its shell and reflectors must be studied to determine, at representative positions, the earlier arrivals along reflected paths so that shadows and "hot spots" can be eliminated and early reflections distributed throughout the enclosure. If the enclosure is of complicated or irregular shape, construction of a scaled model and use of small pulsed sources and probe receivers may reveal acoustic problems that would otherwise be missed. In studying an existing facility, using a starter's pistol and recording the output of well-placed microphones can help isolate unexpected sources of reflected arrivals and echoes. The back wall in any enclosure (particularly fan-shaped ones) and any flat or concave surface can be troublesome. Large, flat, parallel surfaces can cause *flutter* (repetitive echoes). Structural irregularities on these surfaces to scatter the reflections, or randomly spaced absorptive elements to suppress reflections and provide some diffraction, are useful remedies. Balconies must be shallow and their undersurfaces designed to prevent focusing, allow early reflected arrivals, and allow the reverberant sound field to penetrate to the underlying audience. If possible, they should be steeply raked so that the listeners receive good direct paths.

For halls used primarily for unamplified speech or music, the rows of seats closest to the source should not subtend an angle measured from the source of more than about 120° unless particular attention is paid to providing early reflections from ceiling, walls, shell (if present), or well-located reflecting surfaces.

(e) Diffusion, Blend, and Ensemble

The reverberant sound field must quickly become diffuse so that there is good blend of the sound throughout the enclosure. This is important both for the audience and for the speaker or orchestra. There must be a return of reverberant sound to the stage; otherwise, the performers will feel they are projecting into a sonic void. While many lecturers are amazingly insensitive to their environment, others depend on their perception of the acoustics. For these latter, an impression of "deadness" may cause them to overcompensate and speak too rapidly or too loudly for the enclosure. Lack of reverberant return can cause an orchestra and conductor to misjudge the behavior of the hall with much the same results.

The stage should be designed also to project the sound evenly from its boundaries back into itself as well as out into the body of the hall. This allows speakers and performers to hear themselves and each other. This is very important in obtaining good *ensemble*. In a large enclosure a stage should be no wider than about twice its depth. If there are any problems in obtaining good diffusion, the walls and ceiling forming the stage, any shell, and relevant reflecting surfaces should be broken up with carefully chosen irregularities.

*12.9 STANDING WAVES AND NORMAL MODES IN ENCLOSURES

Ray acoustics does not provide a complete description of the behavior of sound in an enclosure. A more adequate approach must include wave theory. The wave equation has been solved for simple enclosures and new insights have emerged from examining the transient and steady-state behaviors of sound in such enclosures. Even in complicated enclosures for which the wave equation cannot be solved, the theory has been used to supplement and extend results predicted by ray acoustic methods.

(a) The Rectangular Enclosure

As was seen in Section 9.2, solution of the wave equation in a lossless, rigid-walled, rectangular cavity of dimensions L_x, L_y, and L_z results in the standing waves

$$\mathbf{p}_{lmn} = \mathbf{P}_{lmn}(x, y, z)e^{j\omega_{lmn}t}$$
$$\mathbf{P}_{lmn}(x, y, z) = \mathbf{A}_{lmn}\cos k_{xl}x \cos k_{ym}y \cos k_{zn}z$$

(12.9.1)

The components of k are given by (9.2.6), the associated natural frequencies by (9.2.7), and the modes are labeled by the integer set (l, m, n). If no integer is zero, the mode is termed *oblique*. If one of the integers is zero, the mode is termed *tangential* because the propagation vector is parallel to one pair of surfaces. If two integers are zero, the mode is termed *axial* because the propagation vector is parallel to one of the axes. In what follows, we will begin by investigating the behavior of individual normal modes, and so suppress the subscripts l, m, and n for notational convenience. The *natural* angular frequency of a damped standing wave will be designated as ω_d and any *forcing* angular frequency as ω.

(b) Damped Normal Modes

Because the walls of the enclosure are not perfectly rigid, there are losses of acoustic energy from the system and modifications of the above normal modes. First, since there are energy losses at the walls, a normal mode will no longer have normal particle velocities that vanish on the walls. Each standing wave will decay with its own temporal absorption coefficient β and each normal mode will have spatial absorption coefficients α_x, α_y, and α_z associated with each of the components k_x, k_y, and k_z of its propagation vector. These damped waves must satisfy the lossless wave equation (absorption in the air will be considered later), so

$$\mathbf{p}^D = \mathbf{A}\cos(\mathbf{k}_x x + \phi_x)\cos(\mathbf{k}_y y + \phi_y)\cos(\mathbf{k}_z z + \phi_z)e^{j\omega_d t}$$
$$\boldsymbol{\omega}_d = \omega_d + j\beta$$
$$\mathbf{k}_x^2 + \mathbf{k}_y^2 + \mathbf{k}_z^2 = \boldsymbol{\omega}_d^2/c^2$$
$$\mathbf{k}_i = k_i + j\alpha_i \qquad i = x, y, z$$

(12.9.2)

The values of the k's, ϕ's, and α's must be found by application of the boundary conditions describing the lossy walls. These in turn will determine the temporal absorption coefficient β for each of the standing waves. This is no more than the three-dimensional generalization of the analysis of the fixed, resistance-loaded string in Section 2.11(b).

The normal modes and their natural frequencies depend on the shape and size of the enclosure, whereas their rates of damping depend on the specific values of the normal specific acoustic impedances of the walls. This division is fortunate, for it enables us to use the simplest possible boundary condition (perfectly rigid walls with no damping) to derive the normal modes and their natural frequencies. The effect of energy absorption by the walls on the damping of the normal modes can then be considered as a perturbation of these simple conditions.

Since independent loss mechanisms are additive in the total absorption coefficient, we will first restrict attention to an enclosure having only one lossy wall at $x = L_x$, all other surfaces being perfectly rigid. Assumption that the absorbing characteristics of the wall are determined by its normal specific acoustic impedance $\mathbf{z}_x$ (Section 6.5) yields a simple boundary condition. For relatively live spaces with conventional walls, the reactive part of

$\mathbf{z}_x$ can be neglected. Thus,

$$\mathbf{z}_x = \rho_0 c v_x \qquad (12.9.3)$$

where v_x is the dimensionless ratio of the normal specific acoustic resistance of the wall to the characteristic acoustic impedance $\rho_0 c$ of the air. The remaining walls at $x = 0, y = 0, L_y$ and $z = 0, L_z$ are perfectly rigid. The normal mode in (12.9.2) must then have $\alpha_y = \alpha_z = 0$, all $\phi = 0$, $\beta = \beta_x$, and k_y and k_z given by (9.2.6). The conditions (12.9.2) take the forms

$$k^2 - \alpha_x^2 = (\omega_d/c)^2 - (\beta_x/c)^2$$

$$\alpha_x k_x = (\omega_d/c)(\beta_x/c) \qquad (12.9.4)$$

Applying the boundary condition $\mathbf{p}/\mathbf{u}_x = \rho_0 c v_x$ at $x = L_x$ results in

$$\tan[(k_x + j\alpha_x)L_x] = j\frac{1}{v_x}\frac{\omega_d/c + j\beta_x/c}{k_x + j\alpha_x} \qquad (12.9.5)$$

For all normal modes, the fact that the wall is nearly rigid suggests $k_x L_x \approx l\pi$. Expansion of the tangent yields

$$(k_x L_x - l\pi) + j\alpha_x L_x = j\frac{1}{v_x}\frac{\omega_d/c + j\beta_x/c}{k_x + j\alpha_x} \qquad l = 0, 1, 2, \ldots \qquad (12.9.6)$$

1. For the case $l \neq 0$, through lowest order terms, we have $k_x L_x = l\pi$ and $\alpha_x k_x \approx \omega_d/v_x c L_x$. Substitution into (12.9.4) gives

$$\beta_x = c/v_x L_x \qquad l \neq 0 \qquad (12.9.7a)$$

2. For the case $l = 0$, cross multiply in (12.9.6) and solve. From the imaginary terms, $\alpha_x k_x \approx \omega_d/2v_x c L_x$ and thence

$$\beta_x = \tfrac{1}{2}c/v_x L_x \qquad l = 0 \qquad (12.9.7b)$$

Note that β_x for this "grazing" mode is half that for modes with $l \neq 0$. [From the real terms, $k_x \approx \alpha_x$ and we see that for the $(0, m, n)$ modes k_x does not vanish. This is plausible on physical grounds. Because this wall is not perfectly rigid, any pressure on it must cause it to yield slightly. This means u_x cannot vanish on the wall so there must be a small but finite k_x.]

Since independent absorptive effects are additive, the total temporal absorption coefficient for the standing wave resulting from all the surfaces of the enclosure is

$$\beta = \frac{c}{V}\sum_{i=1}^{6} \varepsilon_i S_i \frac{1}{v_i} \qquad (12.9.8)$$

where ε_i is $\tfrac{1}{2}$ if the mode grazes the ith surface and is unity otherwise, S_i is the area of each surface, and $V = L_x L_y L_z$. Equation (12.9.8) can be extended to enclosures other than rectangular. Calculation of the reverberation time for the single normal mode proceeds just as in Section 12.3 (recall $2\beta = 1/\tau_E$),

$$T = \frac{0.0201V}{\sum \varepsilon_i S_i/v_i} \qquad (12.9.9)$$

and, as before, absorption in the air is included by adding $4mV$ into the denominator of (12.9.9).

Each v_x must be related to the random-incidence energy absorption coefficient a_E (or the Sabine absorptivity a) of its wall. For a wall described by (12.9.3), the form of the plane wave power transmission coefficient is given in Problem 6.6.4. This must equal the energy absorption coefficient $a_E(\theta)$ of the wall for a wave reflecting from it with angle of incidence θ measured from the normal to the surface. Thus,

$$a_E(\theta) = \frac{4v_x \cos\theta}{(v_x \cos\theta + 1)^2} \tag{12.9.10}$$

Assumption of a diffuse sound field allows us to calculate a_E. Substitution of (12.9.10) into (12.4.5) and integration yields

$$a_E = \frac{8}{v_x}\left(1 + \frac{1}{1 + v_x} - \frac{2}{v_x}\ln(1 + v_x)\right) \tag{12.9.11}$$

For $v_x > 25$, this has the approximate solution

$$v_x \to 8/a_E \tag{12.9.12}$$

The approximation improves with decreasing absorptivity. Substitution into (12.9.9) yields the Sabine reverberation formula (12.3.4) corrected by the weighting factors ε_i. Recall that (12.9.9) is for a single normal mode, and the Sabine formula assumed a diffuse sound field.

The grazing modes have smaller damping because the average mean square pressures produced by these modes at the surface they graze are only half those produced on the other walls. This illustrates an important general principle: an absorbing surface is most effective in damping a normal mode if it is located in a region of maximum mean square pressure. The most effective method of reducing any particular undesired mode is to place absorbing material on those parts of the walls where the pressures corresponding to this mode are the greatest. Also, since all normal modes have pressure maxima at the corners, absorbing material placed in the corners of a room can be nearly twice as effective as it would be if placed elsewhere.

An example of the behavior of the reverberant sound field as a function of time is presented in Fig. 12.9.1. In this highly simplified model, the space measures 10 m × 20 m × 30 m with the height being 10 m. Only the floor absorbs and is assumed to behave like an audience (first entry in Table 12.5.1). The steep initial part of the curve corresponds to the decay of the oblique modes and the final part to the decay of the grazing modes.

In conventional spaces at frequencies high enough that most of the acoustic energy resides in oblique modes, the break in the reverberation curve comes late enough that the first 20 or 30 dB of decay forms a nearly straight line. The slope of this initial part determines the reverberation time associated with the major portion of the acoustic energy. Presently, the total decay of sound in reverberant environments is often characterized by two decay times. The *early decay time* is defined as six times the duration for the first 10 dB decay in the sound field. The extended (or classical) reverberation time is twice the time taken for the reverberant field to decay from -5 to -35 dB below the initial value. This latter corresponds closely to the -60 dB criterion of Sabine.

(c) The Growth and Decay of Sound from a Source

In a space with walls that are basically rigid but lossy, we may expect pressure antinodes at the boundaries. Heeding the discussion after (12.9.2), we can express each driven standing wave in the form (12.9.1) but with the driving angular frequency ω in place of the natural angular frequency ω_d of the mode. The amplitude of each mode depends on the location of the source and also on the difference between the driving frequency and the resonance frequency for that standing wave. We can borrow directly from the analyses of the driven, damped harmonic oscillator of Chapter 1, the driven microphone diaphragm of Chapter 4, and the rectangular cavity of Chapter 9. From (1.10.7), the Q of a standing wave is

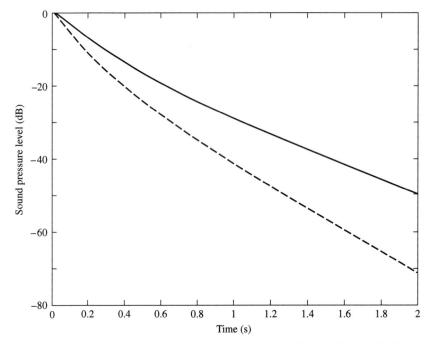

Figure 12.9.1   Reverberation decay in a room 10 m × 20 m × 30 m with all surfaces completely reflecting except the floor. The solid line is for a driving frequency of 250 Hz; the early decay time is about 1.7 s and the extended reverberation time around 2.3 s. The dashed line is for a driving frequency of 500 Hz, for which these times are 1.0 s and 1.4 s, respectively.

$Q = \omega_d/2\beta$. The resonance angular frequency of a driven standing wave is within $O(\beta)^2$ of its damped value ω_d, so the two can be assumed equal. The amplitude of each normal mode thus varies inversely as $[(\omega/\omega_d - \omega_d/\omega)^2 + 1/Q^2]^{1/2}$ so each steady-state standing wave has the form

$$\mathbf{p}^S = \mathbf{P}^S(x, y, z)e^{j\omega t}$$

$$\mathbf{P}^S(x, y, z) = \frac{\mathbf{B}\cos k_x x \cos k_y y \cos k_z z}{\left[(\omega/\omega_d - \omega_d/\omega)^2 + 1/Q^2\right]^{1/2}} \tag{12.9.13}$$

The k's are given by (9.2.6), and the $\mathbf{B}$'s are found from the location and configuration of the source. In summing over all the standing waves, just those waves for which a frequency component of the source lies near or within the bandwidth of the relevant response curve will be of significant amplitude. The homogeneous term corresponding to (12.9.13) and having the same amplitude is

$$\mathbf{p}^D = \mathbf{P}^S(x, y, z)e^{-\beta t}e^{j\omega_d t} \tag{12.9.14}$$

Thus, when a source of sound is turned on at $t = 0$, the solution for each angular frequency ω present in the Fourier spectrum of the source must be

$$\mathbf{p}_\omega = 1(t) \cdot \sum_{l,m,n} \mathbf{P}^S_{lmn}(x, y, z)(e^{j\omega t} - e^{-\beta_{lmn}t}e^{j\omega_{lmn}t}) \tag{12.9.15}$$

The subscripts designating the normal modes have been restored for clarity, $\omega_{lmn} \equiv (\omega_d)_{lmn}$, and $1(t)$ is the unit step function. Mathematically, each steady-state standing wave is canceled by the associated decaying standing wave at $t = 0$, so that the enclosure is quiescent before the source is turned on. As the decaying wave dies away the combination grows to attain the final steady-state value. For each frequency component, the growth will be relatively smooth for those modes with natural frequencies very close to it and more irregular the more dissimilar are the driving and natural frequencies. (For an exaggerated illustration, refer to Fig. 1.8.1.) Thus, an enclosure may be treated as a resonator having numerous allowed modes of vibration, each with its own natural frequency and temporal decay constant. When a sound source is started, steady-state standing waves having the frequencies of the source are set up, each with an attendant decaying standing wave. As time increases, each transient dies out at its own rate, eventually leaving only the steady-state standing waves.

Turning off the signal after a long time is mathematically equivalent to turning on another signal 180° out of phase with the original one. By superposition, the steady-state solution for the new signal annihilates the steady-state solution of the original signal, and what is left is the same collection of damped waves (but with opposite sign) for each of the original frequencies,

$$\mathbf{p}_\omega = \sum_{l,m,n} \mathbf{P}^S_{lmn}(x, y, z)e^{-\beta_{lmn}t}e^{j\omega_{lmn}t} \tag{12.9.16}$$

where $t = 0$ now designates the time at which the source is turned off. Because each normal mode has its own natural frequency, in poorly designed enclosures these reverberant components may interfere with each other and produce beats or may even sound to the listener as if they are at a pitch different from that of their excitation.

(d) Frequency Distribution of Enclosure Resonances

A knowledge of the natural frequencies of a room is essential to a complete understanding of its acoustic properties. The room will respond strongly to those sounds having frequencies in the immediate vicinity of any of these natural frequencies. It is this characteristic that affects the output of a loudspeaker measured in a reverberant room and causes the results to have limited significance as an accurate measure of the speaker's properties. Furthermore, each normal mode has its own particular spatial pattern of nodes and antinodes. In effect, every enclosure superimposes its own characteristics on those of any sound source, so that the fluctuations in sound pressure that occur as a microphone is moved from one point to another, or as the frequency of the source is varied, may obscure the true characteristics of the source. It is for this reason that the response curve of a loudspeaker should be determined either in open air or in an anechoic chamber. If the absorptivities at the walls of the anechoic chamber exceed about 0.9, the reverberant field will have little effect on measurements taken near the source and away from the walls. (This case clearly violates the approximations made above in obtaining the normal modes of an enclosure.)

Each of the individual normal modes of an enclosure can be fully excited only when the sound source is located where the normal mode has a pressure antinode. For a rectangular space, all the normal modes have pressure antinodes in the corners. If a source or microphone is in a corner of the space, it will excite or receive a mode to its fullest extent. When a source or receiver is located where a mode has a pressure node, that mode will be excited or detected only weakly, if at all. For instance, if a loudspeaker is located at the center of a rectangular room, only those modes having l, m, and n all even will be excited (about 1 mode in 10) as the driving frequency is slowly varied from low to high frequencies, and a microphone at that location will be deaf to all but those modes.

An example of the response of a microphone on the floor in one corner of a rectangular room to a source on the floor in the opposite corner is given in Fig. 12.9.2 for the room described in Table 12.9.1. The dashed line represents the output of the source measured in an anechoic chamber. The influence of the room is quite apparent.

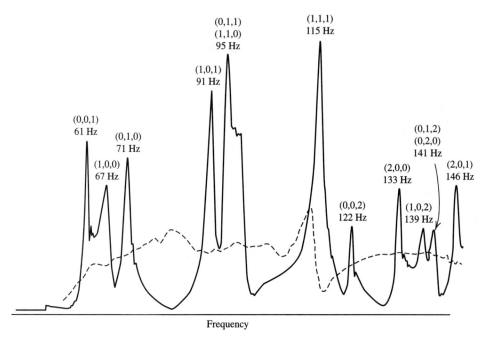

Figure 12.9.2   Experimentally determined response of the reverberant room in Table 12.9.1. The solid line is the pressure amplitude measured for source and receiver on the floor ($L_z = 0$) in opposite corners (0, 0, 0) and (L_x, L_y, 0), respectively. The dashed line is the pressure amplitude for the same source in an anechoic environment.

Table 12.9.1   The twelve lowest normal modes and their natural frequencies for a rigid-walled room measuring 2.59 m $\times$ 2.42 m $\times$ 2.82 m and for a speed of sound $c = 343.6$ m/s

Mode	Frequency (Hz)	Mode	Frequency (Hz)
(0, 0, 1)	60.9	(1, 1, 1)	114.7
(1, 0, 0)	66.3	(0, 0, 2)	121.8
(0, 1, 0)	71.0	(2, 0, 0)	132.7
(1, 0, 1)	90.1	(1, 0, 2)	138.7
(0, 1, 1)	93.6	(0, 1, 2)	141.0
(1, 1, 0)	97.2	(0, 2, 0)	142.0

Equation (9.2.7) suggests that each of the natural frequencies f may be considered a vector in frequency space with components $f_x = l(c/2L_x)$, $f_y = m(c/2L_y)$, and $f_z = n(c/2L_z)$. Each normal mode can, therefore, be represented by a point in this space. All the normal modes with natural frequency f and below are points within the octant of frequency space between the positive axes and a spherical surface of radius f. Each lattice point occupies a rectangular block in frequency space of dimensions $c/(2L_x)$, $c/(2L_y)$, and $c/(2L_z)$ and, therefore, volume $c^3/8V$, where $V = L_xL_yL_z$. The number of points N in the octant can be estimated by dividing the volume of the octant by $c^3/8V$,

$$N \sim \left(4\pi V/3c^3\right)f^3 \tag{12.9.17}$$

Differentiation with respect to f yields the number of normal modes dN with natural frequencies in a band of width df centered on f,

$$\frac{dN}{df} \sim \frac{4\pi V}{c^3} f^2 \tag{12.9.18}$$

The frequency density dN/df for the normal modes increases rapidly as the center frequency of the band (or the size of the enclosure) is increased. The more closely packed the natural frequencies are, the more the response curves of the driven standing waves will overlap, the more standing waves can be excited, and the more diffuse the combined field will be. If at a frequency f at least three standing waves can be excited within their respective half-power points, then the resulting field is fairly random. If the average bandwidth of the resonances is Δf, the required frequency density must be $dN/df > 3/\Delta f$. But $\Delta f = \beta/Q$, and $\beta = 1/2\tau_E = 6.9/T$. Combining these relations with (12.9.18) gives a criterion for the frequency above which a room of specified volume V and reverberation time T can be considered to contain fairly diffuse fields,

$$f \geq 2000 \sqrt{T/V} \tag{12.9.19}$$

(all quantities are metric). The lower limit is known as the *Schroeder frequency*.

The response of a room is observed to become less uniform as its symmetry is increased. This results from the increase in the number of degenerate modes having different (l, m, n) but the same natural frequency. If it is desired to optimize the room dimensions for greatest uniformity in the distribution of natural frequencies, Bolt[8] has shown that acceptable length ratios $1 : X : Y$ (where $1 < X < Y$) satisfy the joint conditions $2 < (X + Y) < 4$ and $\frac{3}{2}(X - 1) < (Y - 1) < 3(X - 1)$.

In spite of greater mathematical complexity, the application of wave theory to architectural acoustics contributes to understanding the behavior of sounds in enclosures. Particularly at lower frequencies, it aids in understanding the effects of the shape of the enclosure, the distribution of absorbing surfaces, and the positions of source and receiver.

PROBLEMS

As aids in conversions, note that $10 \text{ ft} = 3.05 \text{ m}$, $100 \text{ ft}^2 = 9.29 \text{ m}^2$, $1000 \text{ ft}^3 = 28.3 \text{ m}^3$, $0.0002 \mu\text{bar} = 20 \mu\text{Pa}$, $1 \text{ ft}^2 = 1$ English sabin $= 0.093 \text{ m}^2$, and $1 \text{ m}^2 = 1$ metric sabin $= 10.8 \text{ ft}^2$.

12.3.1. When steady-state conditions are reached in a live room, the sound pressure level is 74 dB *re* 20 μPa. (*a*) If the average absorptivity of the walls is 0.05, what is the rate at which sound energy is being absorbed per square meter of wall surface? (*b*) If the total area of absorbing surface in the room is 50 m², at what rate in watts is sound energy being generated in the room?

12.3.2. The surfaces of an auditorium 200 ft × 50 ft × 30 ft have an average absorptivity $\bar{a} = 0.29$. (*a*) What is the reverberation time? (*b*) What must be the power output of a source if it is to produce a steady-state sound pressure level of 65 dB *re* 20 μPa? (*c*) What average absorptivity would be required if a speaker having an acoustic output of 100 μW is to produce a steady-state level of 65 dB *re* 20 μPa? (*d*) Calculate the resulting reverberation time, and comment on its influence on the intelligibility of the speaker.

[8]Bolt, *J. Acoust. Soc. Am.* **19**, 120 (1946).

12.3.3. A room $10\text{ m} \times 10\text{ m} \times 4\text{ m}$ has an average absorptivity $\bar{a} = 0.1$. (*a*) Calculate its reverberation time. (*b*) What must be the output of a source if it is to produce a steady-state sound pressure level of 60 dB *re* 20 μPa? (*c*) At what rate in W/m^2 is sound energy incident on the walls of the room?

12.3.4. An auditorium is observed to have a reverberation time of 2.0 s. Its dimensions are $7\text{ m} \times 15\text{ m} \times 30\text{ m}$. (*a*) What acoustic power is required to produce a steady-state sound pressure level of 60 dB *re* 20 μPa? (*b*) What is the average absorptivity of the surfaces in the auditorium? (*c*) If 400 people are present, each adding 0.5 m^2 to its total absorption, what is the new reverberation time? (*d*) What will be the new sound pressure level produced by the sound source of part (*a*)?

12.3.5. Relate the average absorptivity (*a*) to an equivalent bulk temporal absorption coefficient β and (*b*) to an equivalent spatial absorption coefficient α. (*c*) Show that (12.3.6) is consistent with the principle assumed in (8.5.1).

12.3.6. (*a*) Show that m in (12.3.10) must have the units of m^{-1}. (*b*) Compute m at 3000 Hz in air of 38% relative humidity. (*c*) Compare with the value estimated with the help of Fig. 8.6.3. (*d*) With absorption only at the walls, the reverberation time at 3000 Hz in a room of 10,000 ft^3 volume is 1.2 s. What would be the reverberation time if absorption in the air is included?

12.3.7. The reverberation time in a small reverberation chamber $9\text{ ft} \times 10\text{ ft} \times 11\text{ ft}$ is 4.0 s. (*a*) What is the effective absorptivity of the surfaces of the chamber? (*b*) When 50 ft^2 of one wall is covered with acoustic tile, the reverberation time is reduced to 1.3 s. What is the effective absorptivity of the tile? (*c*) What would be the reverberation time if all surfaces of the chamber were covered with this tile?

12.4.1. Given a room $12\text{ ft} \times 18\text{ ft} \times 30\text{ ft}$, (*a*) what is the mean free path of a ray in this room? (*b*) If the average decrease in intensity level per encounter with the walls of the room is 1 dB, what is the average absorptivity of the room?

12.4.2. A room $3\text{ m} \times 6\text{ m} \times 10\text{ m}$ has walls of average absorptivity 0.05. The floor is covered with a rug ($a = 0.6$), and the ceiling is wood ($a = 0.1$). (*a*) Calculate the average absorptivity. (*b*) Calculate the reverberation time from (12.3.4). (*c*) Calculate the reverberation time from (12.4.2) and compare with the answer to (*b*).

12.4.3. (*a*) Evaluate the ratio of absorption times predicted by Eyring and by Sabine for $\bar{a} = 0.02, 0.05, 0.1, 0.2$, and 0.5. (*b*) If all the walls of the enclosure were completely absorbing of incident energy, calculate the predicted reverberation times. (*c*) Using the mean free path, calculate the mean number of reflections required to attain each predicted reverberation time. Which seems the more realistic in this *anechoic* limit?

12.4.4. Given a cubical room 10 ft on a side, (*a*) what is the mean free path of a sound ray in this room? (*b*) How many reflections per second does an average ray make with the walls? (*c*) If the loss in level per reflection is 1.5 dB, what are the reverberation time of the room and the average absorptivity for the walls?

12.4.5. The relative dimensions of most enclosures range from about 1:1:1 to 1:3:5. (*a*) Show that the mean free path length L_M between consecutive reflections along a ray path is within about 30% of the smallest dimension L of a typical enclosure. (*b*) Within the same accuracy, show that $T \sim 0.040L/\bar{a}$.

12.4.6. Assume $a_i = -\ln(1 - a_{Ei})$ for every surface in an enclosure. (*a*) Show that Sabine's formula becomes the Millington–Sette reverberation formula

$$T = \frac{0.161V}{-\sum_i S_i \ln(1 - a_{Ei})}$$

(b) Show that this is equivalent to assuming in the Eyring formula that the averaged energy reflection coefficient is the *area-weighted geometric mean* of the individual energy reflection coefficients

$$1 - \bar{a}_E = \prod_i (1 - a_{Ei})^{S_i/S}$$

(c) What happens if one of the S_i is a perfect absorber? Is this plausible?

12.5.1. A room 10 ft on a side has acoustic tile on its walls, acoustic plaster on its ceiling, and a carpeted concrete floor. Using the data presented in Section 12.5, calculate the reverberation time at (a) 125 Hz, (b) 500 Hz, and (c) 2000 Hz.

12.5.2. A narrowband noise source centered at 125 Hz has 10 μW acoustic output. (a) What ambient background sound pressure level will it generate in the room of Problem 12.5.1? (b) What will be the loudness level in phon? (c) Repeat (a) and (b) for a similar noise source at 500 Hz.

12.6.1. A motor produces a steady-state reverberant sound pressure level of 74 dB *re* 20 μPa in a room 10 ft $\times$ 20 ft $\times$ 50 ft. The measured reverberation time of the room is 2 s. (a) What is the acoustic output of the motor? (b) How much additional sound absorption in English sabin must be added to this room to lower the sound level by 10 dB? (c) What is the new reverberation time?

12.6.2. An existing concrete reverberant room has height to width to length in the proportions 2:4:5 and is to be replaced with one also of concrete and of the same height, but twice as wide and three times as long. What will be the change in the *SPL* of the reverberant field when a source of constant power output is moved from the first to the second room?

12.6.3. A small reverberation chamber is constructed with concrete walls. Its internal dimensions are 6 ft $\times$ 7 ft $\times$ 8 ft. (a) Calculate the final steady-state sound pressure level produced by a 2000 Hz source of 7.5 μW output. Assume the absorptivity of the concrete to be 0.02. (b) How many seconds will be required from the time the source is turned on until the pressure level reaches within 3 dB of the value of part (a)? (c) When an observer enters the chamber, the steady-state level is observed to decrease by 3 dB. What is the absorbing ability of the observer in English sabin?

12.7.1. A work station is 1 m in front of the acoustic center of a noisy machine, which is mounted 0.5 m away from a poured concrete wall. Assume all other surfaces are highly absorbing. The annoying sound is in the band 1 to 2 kHz. Is it better for the worker to (a) move the work station an additional 1 m away from the wall or (b) glue acoustic tile to the wall?

12.7.2. A sound source has a directivity factor of 10 and an acoustic output of 100 μW in a room of 5000 ft^3 volume with a reverberation time of 0.7 s. What is the maximum distance from the sound source at which the sound pressure level in the direct field will be 10 dB above that of the reverberant field? Assume far-field behavior.

12.8.1. A concert hall has its dimensions in the ratio 1:1:2, where the length from the stage to the rear of the audience is the long dimension. Given the rough design criterion of Table 12.8.1, evaluate the ratio of reverberant to direct intensity for a seat in the middle of the hall.

12.8.2. A concert hall has floor dimensions 20 m $\times$ 50 m and is 15 m high. The entire floor is taken up by the orchestra and audience, and there is a balcony 5 m deep across the back, 3 m deep on the sides, and extending 40 m along the sides. The sides, ceiling, and underside of the balcony are constructed of heavy, sealed wood. There is, in addition, a total of 200 m^2 of heavy plate glass windows. Calculate the reverberation time as a function of frequency when the hall is fully occupied (assume 36% humidity). How does the hall compare with Beranek's criterion (12.8.3)?

12.8.3. A concert hall with a reverberation time $T = 1.7$ s measures 25 m across, 10 m high, and 50 m long. (*a*) Estimate how acceptable this value of T is, using (12.8.3). (*b*) Evaluate the delay times of the first reflected arrivals from the rear of center stage to a seat in the middle of the hall. (Assume both points are at floor level.) Comment on choosing a seat. (*c*) Calculate the mean free path between reflections and the number of reflections during time T. (*d*) Compare the average absorptivities predicted by Sabine's formula and by that of Ewing.

12.9.1. A closed loudspeaker cabinet has internal dimensions 2 ft $\times$ 3 ft $\times$ 4 ft. Its internal surfaces are lined with an absorbing material having an effective absorptivity of 0.2. (*a*) The influence of internal resonances on the output of the loudspeaker may be assumed to be negligible when the average number of standing waves per 1 Hz band exceeds one. Above what frequency does this occur? (*b*) The duration of any transient vibrations of the loudspeaker will be influenced by the reverberation time of the cabinet. What is the reverberation time in the cabinet? (*c*) Steady-state sounds radiated into the cabinet may produce very large sound pressures within it. If 0.1 W of acoustic power is radiated into the cabinet, what is the steady-state sound pressure level within it? (*d*) What are the natural frequencies for the three lowest normal modes of the cabinet?

12.9.2. A cubical room is 5 m on a side. What are the natural frequencies of the lowest-frequency (*a*) axial wave, (*b*) tangential wave, and (*c*) oblique wave?

12.9.3. Given a room of 4 m $\times$ 6 m $\times$ 10 m, (*a*) what is the frequency of the (1,1,1) mode? (*b*) If the average absorptivity at this frequency is 0.1, what is the reverberation time for this mode? (*c*) What is the minimum frequency for which the average number of natural frequencies per 1 Hz band will exceed 5?

12.9.4. A cubical room is 5 m on a side. (*a*) Compute the natural frequencies between 30 and 70 Hz. (*b*) Locate the nodal planes for each of the associated normal modes.

12.9.5. All the walls of a rectangular room 3 m $\times$ 4 m $\times$ 7 m have a relative normal specific acoustic impedance $v_x = 20$. (*a*) What is the absorptivity of the walls for randomly incident sound waves? (*b*) What is the reverberation time of the room for oblique waves? (*c*) What is the reverberation time for axial waves parallel to the longest dimension of the room?

12.9.6. A cubical room is 5 m on a side and has $v_x = 32$ for the floor and ceiling and 160 for the walls. What are the reverberation times for (*a*) those axial waves that strike the floor and ceiling, (*b*) those tangential waves that strike all four walls, and (*c*) the oblique waves?

12.9.7. The speed of sound in concrete is 3500 m/s and its density is 2700 kg/m^3. (*a*) What is the normal specific acoustic resistance of concrete relative to water? (*b*) Applying (12.9.11), calculate the random-incidence energy absorption coefficient. (*c*) If a concrete-walled tank 3 m on a side is filled with water, what is the reverberation time for waterborne sounds? Assume the upper surface reflects all incident energy.

12.9.8. Estimate the Schroeder frequency and the number of modes that can be excited below that frequency for (*a*) the sample room given in Section 12.3 and (*b*) Boston Symphony Hall.

12.9.9C. For the room described in Table 12.9.1, plot the number of excited modes as a function of $\log f$ and show that, for frequencies greater than the Schroeder frequency, this curve is well approximated by (12.9.17) and (12.9.18).

12.9.10. Determine the allowed values of X and Y in a room of relative dimensions $1 : X : Y$ that yield acceptably uniform distributions of natural frequencies. Assume $1 < X < Y$.

ENVIRONMENTAL ACOUSTICS

13.1 INTRODUCTION

The effect of noise on human emotions ranges from negligible, through annoyance and anger, to psychologically disruptive. Physiologically, noise can range from harmless to painful and physically damaging. Noise can also exert economic factors by decreasing worker efficiency, affecting turnover, decreasing property values, and so forth.

The first step in controlling noise is to compare the existing or potential noise with appropriate rating criteria. Such a comparison not only allows specification of the degree of noise suppression necessary to attain a desired noise environment, but also provides guidance as to what aspects of the noise should be attacked, and how to provide the most cost-effective solution.

The development of noise rating procedures and criteria is complicated by the variety of spectra and time histories displayed by noise and the variability of physiological and psychological responses not only among people but for the same person at different times.

The easiest noise environments to rate are those that are steady or slowly varying both in level and in spectral content. Examples are the noise produced by machinery (such as a ventilating system), which runs at a constant rate, and distant ambient noise, which varies slowly between day and night in a community. Rating procedures for such noises can be tailored to provide accurate predictions of the impact on an "average" individual, along with statements on the percent of the population that will be affected to varying degrees. Examples of such rating procedures include *speech interference levels (SIL)* and *balanced noise criterion (NCB)* curves.

Most environmental noises are not steady. Examples of unsteady noise range from *impulses*, where the sound pressure level is 40 dB *re* 20 μPa or more for 0.5 s or less (a slammed door or a sonic boom), to *single events* of relatively long duration (an aircraft flyover or a motorcycle passby), to the widely *fluctuating* noise measured near a busy intersection.

Because of the number of variables involved, the rating of noise acceptable to a community is very difficult. No single-number measure yet devised seems capable of satisfying all situations and, instead, a witches' brew of rating systems exists, each applying to a different noise or sociological condition. However, there appears to be general agreement that analysis of instantaneous spectra gives too much information and that the A-weighted sound level (to be discussed shortly) is an acceptable measure of the impact of many commonly occurring noise environments. The various rating systems based on this A-weighted measurement differ only in how the time variation of the level is treated. Examples of rating procedures that use the statistical behavior of the A-weighted sound level are the *day-night averaged sound level* (L_{dn}), the *50-percentile exceeded sound level* (L_{50}), and the *community noise equivalent level (CNEL)*.

One exception to the use of A-weighted sound levels is the calculation of the impact of airport noise, where the *effective perceived noise level* (L_{EPN}), calculated from the instantaneous spectra, is used to make a *noise exposure forecast (NEF)*.

Excellent treatments of environmental noise and its control are contained in the books edited by Harris,[1] Beranek,[2] and Crocker.[3]

13.2 WEIGHTED SOUND LEVELS

The simplest and probably most widely used measure of environmental noise is the *A-weighted sound level* (L_A), expressed in dBA. (The reference pressure is 20 μPa and is usually left implicit.) A-weighting assigns to each frequency a "weight" that is related to the sensitivity of the ear at that frequency. For example, in a sound level meter the received signal is passed through a filter network with the dBA frequency characteristic, shown in Fig. 13.2.1, and the level of the filtered signal is then determined and displayed. The dBA frequency characteristic was originally designed to mirror the 40 phon equal-loudness-level contour. Accurate A-weighted levels can be obtained from octave-band levels by applying the corrections shown in Table 13.2.1 to each band level and then combining the corrected band levels. Table 13.2.2 displays A-weighted sound levels for some commonly encountered noises.

Other weightings have been proposed but few have gained widespread acceptance. Many sound level meters will allow selection of either A-weighting or C-weighting (designated as L_C and corresponding to the 90 phon equal-loudness-level contour). The frequency characteristic for C-weighting (also shown in Fig. 13.2.1) is nearly flat, rolling off somewhat at the frequency extremes. Although no single overall sound level can give information about the spectrum of a noise, measurement of both L_A and L_C will yield some information about the relative prominence of spectral components.

The L_A is in widespread use mainly because it is inexpensive to obtain and easier for most people to appreciate than are any of the other, more accurate but more complicated, noise rating procedures. For most environmental noises L_A correlates fairly well with the other rating procedures.[4]

[1]*Handbook of Acoustical Measurements and Noise Control, 3rd ed.*, ed. Harris, McGraw-Hill (1991); republished, Acoustical Society of America (1998).
[2]Beranek and Ver, *Noise and Vibration Control Engineering*, Wiley (1992).
[3]*Encyclopedia of Acoustics*, ed. Crocker, Wiley (1997).
[4]Botsford, *Sound Vib.*, **3**, 16 (1969).

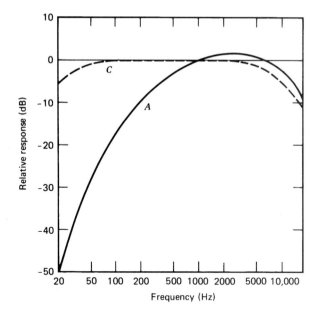

Figure 13.2.1 Filter characteristics for A- and C-weighted sound levels.

Table 13.2.1 Corrections to be added to octave-band levels to convert to *A*-weighted band levels

Center Frequency (Hz)	Correction (dB)
31.5	−39.4
63	−26.2
125	−16.1
250	−8.6
500	−3.2
1000	0
2000	+1.2
4000	+1.0
8000	−1.1

Table 13.2.2 *A*-weighted sound levels for some commonly encountered noises

A-Weighted Sound Level (dBA)	Source of Noise
110–120	Discotheque, rock-n-roll band
100–110	Jet flyby at 300 m (1000 ft)
90–100	Power mower,[a] cockpit of light aircraft
80–90	Heavy truck 64 km/h (40 mph) at 15 m (50 ft), food blender,[a] motorcycle at 15 m (50 ft)
70–80	Car 100 km/h (65 mph) at 7.6 m (25 ft), clothes washer,[a] TV audio
60–70	Vacuum cleaner,[a] air conditioner at 6 m (20 ft)
50–60	Light traffic at 30 m (100 ft)
40–50	Quiet residential—daytime
30–50	Quiet residential—nighttime
20–30	Wilderness area

[a] Measured at position of operator.

13.3 SPEECH INTERFERENCE

Noise decreases the intelligibility of speech by raising the listener's threshold of hearing while, at the same time, masking the information. This loss of information may be partially compensated for by moving closer, talking louder, or using electronic amplification.

Fortunately, speech is highly redundant. Usually, much of a sentence can be lost without seriously affecting intelligibility; meaning can still be extracted from context. To measure intelligibility, trained talkers recite, clearly and distinctly, specially selected words or sentences to trained listeners. Intelligibility is then rated according to the percent of correct responses. The intelligibility of isolated words is more strongly affected by noise but improves markedly with increasing numbers of syllables; disyllabic words are understood about twice as easily as monosyllabic words with the same background noise. Figure 13.3.1 shows the intelligibility of sentences and words as a function of the relative A-weighted sound levels of speech and noise. For a *sentence* intelligibility of better than 95%, the signal level must at least equal the noise level. Since it is difficult to reconstruct proper names from contexts, a paging system should have *word* intelligibility over 85%. This requires a signal-to-noise ratio of 6 dB—a specification seldom achieved in airport lobbies let alone bus depots.

The *voice level (VL)* is the *A-weighted SPL* recorded at a distance of 1 m in front of the speaker. For untrained voices, it appears that (within uncertainties of a few dBA) a quiet conversational voice corresponds to about $VL = 57$ dBA, a normal speaking voice to 64 dBA, a raised voice to 70 dBA, a very loud voice to 77 dBA, and a shout to about 83 dBA.

A measure of intelligibility suitable for field use is the *speech interference level (SIL)*. This is the arithmetic average of the noise levels (in dB, *not* dBA) in the four octave bands centered at 500, 1000, 2000, and 4000 Hz. In the absence of all other information, the A-weighted sound level can be used to get a *rough* estimate of the intelligibility of speech under most commonly encountered noise conditions. A *SIL* estimated by

$$SIL \approx L_A - 7 \tag{13.3.1}$$

will be in error by less than 4 dB for all but the most pathological noise spectra.

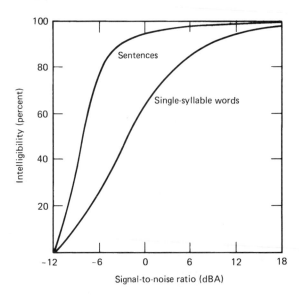

Figure 13.3.1   Percent of words and sentences correctly identified in the presence of background noise.

For people speaking at voice levels natural for their environment, and not trying to achieve more than *just reliable* communication (approximately 1 out of 20 sentences unintelligible), a set of criteria relating distance and the maximum *SIL* have been established.[5] For separation distances $r > 8$ m

$$SIL \leq -20 \log r + 58 \tag{13.3.2}$$

is required for the male voice. For smaller separations, disproportionately higher *SILs* can be accepted since people naturally progressively raise their voices as the *SIL* grows beyond 40 dB. For $r < 8$ m, the relationship for just-reliable communication at natural voice levels is

$$SIL \leq -29 \log r + 66 \tag{13.3.3}$$

with the voice level rising from normal for a $SIL = 40$ dB to very loud for a $SIL = 82$ dB. Approximate expressions for the minimum required voice levels to ensure just-reliable communication are

$$
\begin{aligned}
VL &\geq SIL + 20 \log r + 6 & r &> 8\,\text{m} \\
VL &\geq \tfrac{4}{3}(SIL + 20 \log r) - 13 & r &< 8\,\text{m}
\end{aligned}
\tag{13.3.4}
$$

For female voices, the *SILs* should be decreased by around 5 dB as noted in Section 11.10. If the required voice level is below 64 dBA, communication conditions are satisfactory or better. If the *VL* lies between 64 and 70 dBA then conditions are acceptable, between 70 and 77 dBA difficult, and above 77 dBA impractical to impossible.

13.4 PRIVACY

In addition to speech intelligibility, speech privacy in multifamily dwellings, private offices, open-plan offices, and classrooms is important not only to protect confidentiality but also to prevent intrusion on the privacy of others. The degree of privacy depends on the sound insulation of intervening walls and barriers and also on the background noise. The insulative properties of partitions will be discussed in Section 13.13; here we will consider only the effect of background noise. Figure 13.3.1 shows that if the A-weighted noise in the receiving room is 9 dBA above the A-weighted speech level transmitted through the wall, only about 10% of the words and 30% of the sentences will be understood. This should ensure adequate privacy with little distraction in an office, but an even lower speech level would be required in a dwelling unit.

Privacy can be enhanced by (1) decreasing the amount of sound that passes through the wall (build a better wall), (2) decreasing the speech level in the source room (add acoustic absorption), or (3) increasing the noise level in the receiving room (mask by unobtrusive noise like flow noise through a ventilator screen). A procedure for calculating the degree of privacy is discussed in Beranek.[6]

---

[5] ANSI S-3.14-1977(R-1986). See Tocci, *Encyclopedia of Acoustics*, Chap. 94, and Crocker, *Encyclopedia of Acoustics*, Chap. 80, for more detailed discussion.
[6] Beranek and Ver, *op. cit.*

13.5 NOISE RATING CURVES

A noise that is acceptable for speech interference may be annoying for other activities (reading, listening to music, or sleeping). While two continuous noises with the same frequency spectrum can be rated by comparing their A-weighted sound levels, time-varying noises with different spectra require a rating procedure that accounts not only for the intermittent nature of the noise but also for its subjective qualities.

The procedure generally accepted for rating steady broadband noise containing no significant tones, such as produced by ventilating systems and distant road traffic, consists of comparing measured octave-band levels with a family of criteria curves.

Several procedures for rating steady noise are in general use, differing only in the shapes of their criteria curves. One of the currently accepted methods uses the *balanced noise criterion (NCB)* curves suggested by Beranek.[7] These curves are shown in Fig. 13.5.1.

To determine the rating of a noise, its octave-band levels are plotted on the same graph as the criteria curves. The highest curve reached by any octave-band level of the noise gives the *NCB* rating of the noise. The dashed line in Fig. 13.5.1 is the spectrum of a noise that is rated at $NCB = 40$.

Determining the sonic quality of the noise spectrum requires two other procedures: (1) Calculate the *SIL* of the noise as described earlier. If the level of any octave band having center frequency at or below 1 kHz exceeds the *NCB* curve of value $SIL + 3$, the noise will appear "rumbly." (2) Calculate the average level of the octave bands centered at 125, 250, and 500 Hz and identify the *NCB* curve that intersects that level at 250 Hz. If any band level for the octaves with center frequencies equal to or greater than 1 kHz lies above that *NCB* curve, the noise will sound "hissy."

The maximum noise rating recommended for a room depends not only on the intended use of the room but also on such factors as the expectations of the users. For example, an urban dweller will tolerate significantly higher ambient noise levels than will a rural dweller. In fact, the urban dweller may be bothered by low noise levels because sounds otherwise masked may be clearly heard and prove

Table 13.5.1   Recommended acceptable noise levels in unoccupied rooms

Location	Noise Criteria (NCB)
Concert hall, recording studio	10–15
Music room, legitimate theater	25–30
Church, courtroom, conference room, hospital, bedroom	25–35
Library, private office, living room, classroom	30–40
Restaurant, movie theater, retail shop, bank	35–45
Gymnasium, clerical office	40–50
Shops, garage	50–60

distracting. Table 13.5.1 shows recommended acceptable noise ratings for a variety of different room uses.

[7]Beranek, *J. Acoust. Soc. Am.*, **86**, 650 (1989).

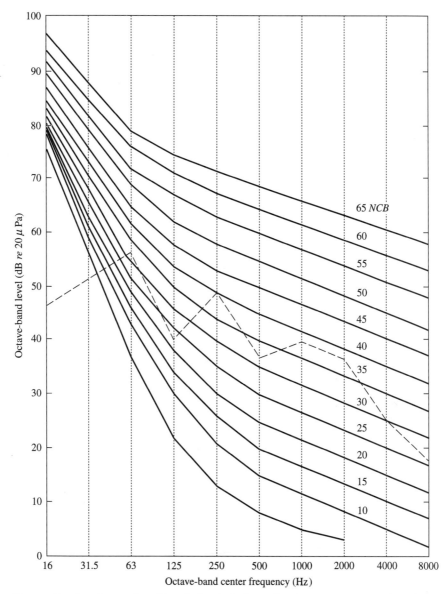

Figure 13.5.1   Octave-band levels for the balanced noise criteria curves *(NCB)*. (After Crocker, *op. cit.*)

13.6 THE STATISTICAL DESCRIPTION OF COMMUNITY NOISE

Figure 13.6.1 suggests the A-weighted sound levels for two time periods in a typical suburban environment. These curves illustrate the basic characteristics of most community noise: a fairly steady *residual noise level* associated with distant, unidentifiable sources on which are superimposed discrete local *noise events*. The residual noise level varies slowly with time, usually displaying diurnal, weekly, and seasonal cycles but with maximum excursions rarely exceeding about 10 dBA. The noise events vary in magnitude and duration, rising as much as 40 dBA above the residual level for seconds, minutes, or even longer.

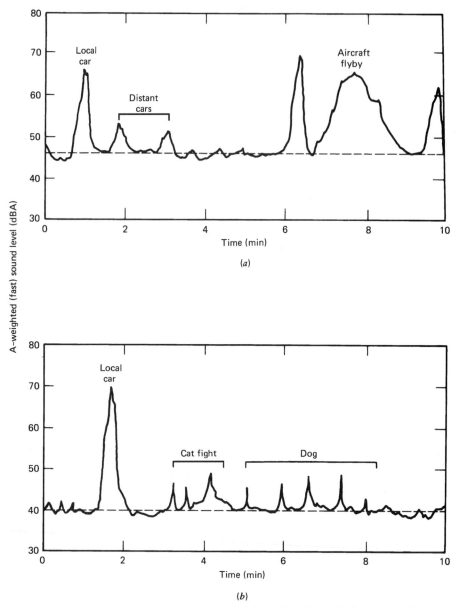

Figure 13.6.1   Typical community A-weighted sound levels in (*a*) daytime and (*b*) nighttime.

Continuous reading of the A-weighted sound level provides the basis for determining the statistics of community noise. From the recording of the A-weighted sound level over a period of time it is possible to construct a *histogram* (Fig. 13.6.2*a*), which displays the percent of the total sample time that the noise level spends within each increment of level, or a *cumulative distribution* (Fig. 13.6.2*b*), which displays the percent of the total sample time that the noise level spends above each value of the noise level. The statistical properties of the noise can then be determined from either of these graphs. (Continuous recording and sophisticated analysis equipment are not necessary; manual sampling of the

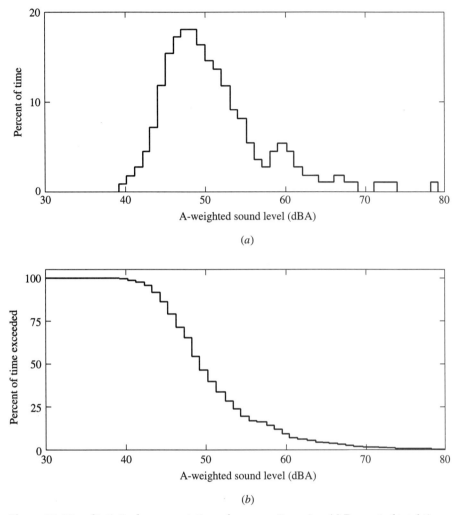

Figure 13.6.2   Statistical representation of community noise. (*a*) Percent of total time that the level is within each increment of level. (*b*) Percent of total sample time that the level is above each value of the level.

noise level at equal time intervals gives the same results but requires more effort.)[8]

Figure 13.6.2 shows that the distribution of community noise is not Gaussian but rises rather steeply and displays a long tail, representing very noisy single events of relatively rare occurrence. The length of this tail varies markedly with location, often being more extended for measurements made in the vicinity of an airport or factory.

The measure of the environmental impact of a noise should depend on the total energy received, the rate of occurrence of noise events, and the magnitudes of noisier single events. The following are some of the A-weighted quantities used in measuring the effects of environmental noise. (In this list and subsequent use of the quantities defined here, the subscript "*A*" is suppressed for notational economy.)

[8]Yerges and Bollinger, *Sound Vib.*, **7**, 23 (1973).

(a) *Equivalent continuous sound level* (L_{eq}). The steady-state sound that has the same A-weighted level as that of the time-varying sound averaged in energy over the specified time interval.

(b) *Daytime average sound level* (L_d). The L_{eq} calculated from 7 A.M. to 7 P.M.

(c) *Evening average sound level* (L_e). The L_{eq} calculated from 7 P.M. to 10 P.M.

(d) *Night average sound level* (L_n). The L_{eq} calculated from 10 P.M. to 7 A.M.

(e) *Hourly average sound level* (L_h). The L_{eq} calculated for any one-hour period.

(f) *Day–night averaged sound level* (L_{dn}). The 24 hour L_{eq} obtained after addition of 10 dBA to the sound levels from 10 P.M. to 7 A.M.

(g) *Noise exposure level* (L_{ex}). The level of the time integral of the squared, A-weighted sound pressure over a stated time referenced to $(1 \text{ s}) \times (20 \text{ } \mu\text{Pa})^2$.

(h) *x-percentile-exceeded sound level* (L_x). The fast, A-weighted sound level equaled or exceeded $x\%$ of the sample time. Most commonly used are L_{10}, L_{50}, and L_{90} (the levels exceeded 10%, 50%, and 90% of the time, respectively).

(i) *Single-event exposure level (SENEL)*. The L_{ex} determined for a single event.

(j) *Community noise equivalent level (CNEL)*. The 24 hour L_{eq} obtained after addition of 5 dBA to the sound levels from 7 P.M. to 10 P.M. and 10 dBA to the levels from 10 P.M. to 7 A.M.

Sets of measurements made at the same location on different days will show different results even when allowances are made for obvious changes such as

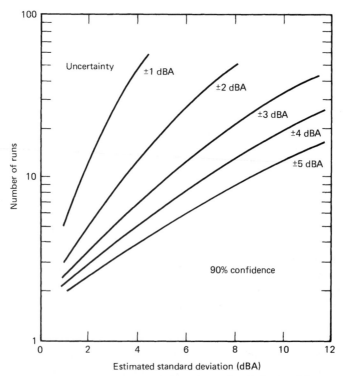

Figure 13.6.3 Uncertainty of the average level obtained from a number of runs for 90% confidence that the result lies within the desired uncertainty.

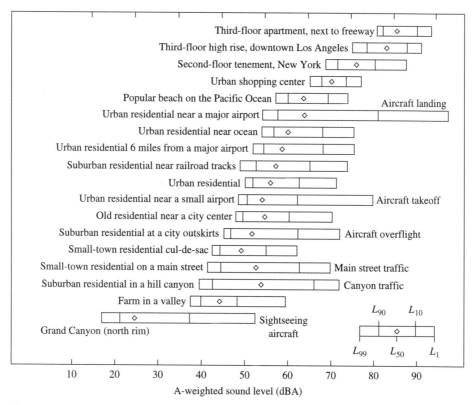

Figure 13.6.4 A-weighted sound levels measured in 1971 for the period 0700–1900 hours. (Reprinted with permission from Crocker, *Encyclopedia of Acoustics*, Chap. 80.)

weekday versus weekend traffic or an airport approach pattern that depends on the wind direction. For example, measurements of urban noise made on four different days may show a range for L_{50} of 9 dBA. Figure 13.6.3 displays, for an estimated standard deviation of any of the above levels L, the number of runs needed to determine (with 90% confidence) that the average of this L is known within the indicated uncertainty. For example, for a sample with an estimated standard deviation of 4 dBA, 13 runs are necessary to obtain the correct average of the desired L to within 2 dBA, nine times out of ten.

Figure 13.6.4 shows the A-weighted percentile-exceeded sound levels for a variety of noise environments.

13.7 CRITERIA FOR COMMUNITY NOISE

Originally, complaints about community noise were taken into court where a judge would decide, on the basis of common-law precedent, if the plaintiff's use or enjoyment of his or her property was harmed. As the pervasiveness of noise sources grew and the adverse effects of noise on the general public became more obvious, local governments either included noisemakers in their general ordinances defining "disorderly conduct" or enacted new ordinances forbidding "unnecessary" or "excessive" noise. Enforcement was left to the local police officers.

As our knowledge about community noise and its effects has become more sophisticated, so have our noise ordinances. A rating procedure, suitable when simplicity of understanding and enforcement are paramount, is exemplified by the noise ordinance of Gainesville, Florida.[9] This ordinance specifies the maximum allowable A-weighted noise level by district and time of day (Table 13.7.1), with the provision that these maximum levels are not to be exceeded for more than 3 cumulative minutes in any hour. Motor vehicles are treated separately: measurements made 15 m (50 ft) from the center of the lane in which the vehicle is traveling must not exceed 85 dBA for trucks and buses and 79 dBA for cars and motorcycles. Exempted from these restrictions are air conditioners, lawn mowers, and construction equipment "operating within the manufacturer's specifications."

Criteria from state and federal agencies are generally more complicated and they usually rely on one of the rating procedures discussed in the previous sections. For example, the U.S. Environmental Protection Agency (EPA) recommends[10] for the "protection of public health and welfare with adequate margin of safety" $L_{dn} \leq$ 55 dBA outdoors and $L_{dn} \leq$ 45 dBA indoors. The Federal Interagency Committee on Urban Noise (FICON) declares[11] an outdoor $L_{dn} \leq$ 65 dBA to be "considered generally compatible with residential development."

Table 13.7.1　Maximum allowable noise level limits for Gainesville, Florida

Location	Noise Level Limits (dBA)	
	Day	Night
Residential	61	55
Commercial	66	60
Manufacturing	71	65

Table 13.7.2 summarizes the recommendations of several federal and state agencies as to the suitability of various land uses subject to various levels of external noise. For structures in use during only part of the day, the appropriate level is L_{eq} measured during the period of use. For 24 hour use, L_{dn} is the appropriate measure.

The criteria for new and rehabilitated residential construction are spelled out in more detail by the U.S. Department of Housing and Urban Development (HUD). They recommend that interior levels with the windows open (unless there is sufficient mechanical ventilation) should not exceed (1) 55 dBA for more than 60 min in any 24 h period, (2) 45 dBA for more than 30 min from 11 P.M. to 7 A.M., and (3) 45 dBA for more than 8 h in any 24 h period.

The U.S. Department of Transportation (DOT) specifies the maximum noise level of a new highway for various land-use categories. These exterior levels, expressed as L_{10}, are (1) 60 dBA for land where quiet is of extraordinary significance, such as amphitheaters and open spaces dedicated to special qualities of serenity and quiet; (2) 70 dBA for residences, motels, hotels, public meeting rooms, schools,

[9]Schwartz, Yost, and Green, *Sound Vib.*, **8**, 24 (1974).
[10]*Information on Levels of Environmental Noise Requisite to Protect Public Health and Welfare with an Adequate Margin of Safety*, Report No. 550/9-74-004, EPA (1980).
[11]*Guidelines for Considering Noise in Land Use Planning and Control*, FICON Document 1981-338-006/8071 (1980).

Table 13.7.2 Compatibility of land use with noise environment[a]

| Facility | Outdoor L_{dn} or L_{eq} | | | | |
	65–70	70–75	75–80	80–85	85–90
Home	25[b]	30[b]	No	No	No
Hotel, motel, church, classroom, library	25[b]	30[b]	35[b]	No	No
Office, store, bank, restaurant	Yes	25	30	No	No
Outdoor music	No	No	No	No	No
Industry, manufacture	Yes	25[c]	30[c]	35[c]	No
Playground	Yes	Yes	No	No	No
Gymnasium	Yes	Yes	25	30	No

Source: Adapted from Air Force Environmental Planning Bulletin 125 USAF/PREVX Env. Planning Div., 1976.

Yes = Land use compatible with noise environment. Normal construction may be used.

No = Land use not compatible with noise environment even with special construction.

25/30/35 = Construction must provide the indicated outdoor-to-indoor noise level reduction in dBA.

[a] If facility is in use for only part of the day, use L_{eq} for the period of use.

[b] Use is discouraged and is acceptable only if alternative development options are not available.

[c] Noise reduction required only in areas sensitive to noise.

churches, libraries, hospitals, picnic areas, recreational areas, playgrounds, active-sports areas, and parks; and (3) 75 dBA for all other developed lands not included in the above categories.

Because of their importance, highway and aircraft noise will be discussed in separate sections.

*13.8 HIGHWAY NOISE

The noise radiated by moving land vehicles originates from the engine, drive train, and exhaust system, from the tire–road interaction, and from aerodynamically generated sound (only important above 80 km/h). For vehicles sold in the United States, the Federal Code of Regulations requires they pass a certification test consisting of a maximum-acceleration run with the noise measured at 15 m from the center of the lane and 1.2 m above the ground. For passenger and other light vehicles, these regulations require $L_A < 80$ dBA.

Measurements made on a large number of automobiles and light trucks under cruising conditions give typical A-weighted sound levels at 15 m (50 ft) of around

$$L_A = 71 + 32 \log(v/88) \tag{13.8.1}$$

where v is the vehicle's speed in kilometers per hour (88 km/h = 55 mph). Tire noise predominates at all but the lowest speeds.

Similar measurements for motorcycles give

$$L_A = 78 + 25 \log(v/88) \tag{13.8.2}$$

and for trucks

$$L_A = \begin{cases} 84 & v \leq 48 \text{ km/h} \\ 88 + 20 \log(v/88) & v > 48 \text{ km/h} \end{cases} \tag{13.8.3}$$

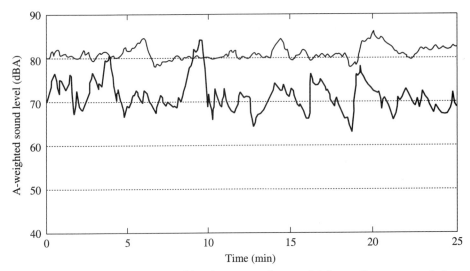

Figure 13.8.1   A-weighted sound levels measured near a highway. Lower curve is for a low traffic density and upper curve is for a high traffic density.

Figure 13.8.1 displays the A-weighted sound level with time for low and high traffic densities. At low traffic densities, the peaks in the sound level correspond to individual vehicles passing the receiver. At high traffic densities, these peaks coalesce, increasing the average noise level so that only the peaks of the noisiest vehicles can be identified. Thus, the fluctuation of traffic noise decreases as traffic density increases. Fluctuation will also decrease for a fixed traffic density as the receiver is moved away from the roadway, since at greater distances the peaks for the individual vehicles will be broader and smoother.

Highway noise is quantified by the usual descriptions of community noise, L_{10}, L_{50}, and L_{eq} being the most commonly encountered. Examples of proposed improved indices are (1) the *traffic noise index* $TNI = 4(L_{10} - L_{90}) + L_{90} - 30$ and (2) the *noise pollution level* $LNP = L_{eq} + 2.56\sigma$, where σ is the standard deviation of the sound level.

Elaborate procedures have been developed to predict highway noise based on the expected traffic density, composition, and road geometry. As a simplified example, consider a straight, two-lane road of infinite length, zero grade, and negligible truck traffic. The receiver is located at a distance d from the centerline of the nearest lane. Also needed for the calculation are the average speed v (in kilometers per hour) and the flow rate Q (in vehicles per hour). The A-weighted noise level L_{eq} is found from

$$L_{eq} = 39 + 10 \log Q + 22 \log(v/88) + \Delta L$$

$$\Delta L = \begin{cases} 0 & d \leq 15 \text{ m} \\ -a \log \left[\dfrac{d}{15} + \left(\dfrac{d-15}{75} \right)^2 \right] & d \geq 15 \text{ m} \end{cases} \qquad (13.8.4)$$

where $a = 13.3$ over ground and 10.0 if the line of sight from the road surface to the receiver is 10° or more above the slope of the terrain. From the values of L_{eq} determined for the times of interest, the desired sound levels (L_{10}, L_{50}, etc.) can be found.

As a numerical example, let us calculate L_{eq} for $Q = 6000/\text{h}$, $v = 88$ km/h (55 mph), $d = 61$ m (200 ft), and $a = 13.3$. From (13.8.4), $L_{eq} = 76$ dBA and $\Delta L = -8$ dBA. The equivalent continuous noise level is 68 dBA. For details on handling more realistic conditions consult Harris (*op. cit.*).

*13.9 AIRCRAFT NOISE RATING

Because of the large number of people severely affected by aircraft noise, and because of the great economic cost of reducing the impact of this kind of noise, aircraft noise has received more attention than any other environmental noise. The Federal Aviation Administration uses the annual average L_{dn} to evaluate airport projects funded by the federal government. The impact is measured by the number of people living in areas with $L_{dn} > 65$ dB.

A more complicated system for rating airport noise requires rating each class of aircraft for both takeoffs and landings. At each selected location in the vicinity of the airport, the 1/3-octave-band spectra for 24 bands with center frequencies from 50 to 10,000 Hz are obtained for each 0.5 s interval the aircraft noise is above the background noise. Even with modern equipment such as real-time analyzers with memory, this is a prodigious task when consideration is given to the number of locations that must be surveyed for each aircraft class and flight plan. Much reliance is therefore placed on computer simulations that model the source spectra of various aircraft classes and "fly" the aircraft along various flight paths.

Each spectrum is converted into a *tone-corrected perceived noise level* (L_{TPN}) by a procedure that takes into account details of the spectrum and adds a correction for dominant tones. The total effect of the flyby (still at one location) is then expressed as the *effective perceived noise level* (L_{EPN}), the maximum L_{TPN} plus a duration correction that accounts for the time the L_{TPN} is above a certain specified level. Readers desiring a more detailed description of the procedure should consult Harris (*op. cit.*).

Once the L_{EPN} is calculated at all locations around the airport for a given aircraft class and operation, contours of equal L_{EPN} can be constructed for superposition over a map of the airport and surrounding community. Such noise footprints are useful for evaluating the noise impact of new aircraft or changes in flight procedures.

The L_{EPN} also serves as the input for rating the total noise environment around an airport. The *noise exposure forecast (NEF)* for a given class of aircraft i on a flight path j is defined as

$$NEF_{ij} = L_{EPNij} + 10\log(N_d + 17N_n) - 88 \tag{13.9.1}$$

when N_d and N_n are the numbers of daytime and nighttime events of this type. Note that a nighttime event (10 P.M. to 7 A.M.) is considered to be 17 times more significant than the same event in the daytime. The total *NEF* at a given location is the combination of all aircraft classes and flight paths

$$NEF = 10\log\left[\sum_{i,j} antilog(NEF_{ij}/10)\right] \tag{13.9.2}$$

The *NEF* values at different locations are then connected to produce contours of equal *NEF*.

On the basis of *NEF* contours for existing conditions, decisions can be made on projected land use. For example, HUD funds cannot be used in areas with $NEF > 40$, but areas with $NEF < 30$ are acceptable. For areas with *NEF* between 30 and 40 special approval is needed. Predicted *NEF* contours can be used to estimate the impact on the community of changes such as the number and timing of flight operations or changes in the mix of aircraft.

The Noise Standard for California airports is somewhat simpler. It uses the measured hourly average A-weighted sound levels L_h and calculates a *community noise equivalent level*

$$CNEL = 10\log\left[\frac{1}{24}\left(\sum_{7\ A.M.}^{7\ P.M.} 10^{L_h/10} + 3\sum_{7\ P.M.}^{10\ P.M.} 10^{L_h/10} + 10\sum_{10\ P.M.}^{7\ A.M.} 10^{L_h/10}\right)\right] \tag{13.9.3}$$

According to the California Administrative Code, the *CNEL* in residential communities adjoining airports is not to exceed 65 dBA.

*13.10 COMMUNITY RESPONSE TO NOISE

The most difficult aspect of environmental acoustics (with the possible exception of getting people to agree on a rating system) is predicting community response to a given noise environment. The response to noise varies greatly from person to person and all attempts to quantify this subject have had to rely on the subjective judgments of the investigators. A discussion of the methodology of such investigations is beyond the intent of this book, and we will have to be content with presenting a few of the more interesting conclusions.

One approach to quantifying community response to noise begins with the A-weighted sound level, adds corrections for various noise characteristics, and then compares the corrected dBA to a scale of expected reaction. Table 13.10.1 displays the corrections of one of the simpler versions of this technique. If the corrected level is less than 45 dBA, no community reaction is to be expected; if it is between 45 and 55 dBA, sporadic complaints are to be expected; between 50 and 60 dBA, widespread complaints; between 55 and 65 dBA, threats of community action; and over 65 dBA, vigorous community action is certain. A similar approach is contained in an International Standardization Organization recommendation, ISO R1996 (1971).

For example, let's estimate the response to a dog kennel located in a suburban area with an ambient background level of 37 dBA. With the dogs barking, the noise level at the point of possible complaint is measured to be 72 dBA. The dogs bark at random times during the day and night for a total of 20 min per day. From Table 13.10.1, a correction of +5 is applied for intermittent character, −5 for duration less than 30 min, and 0 for

Table 13.10.1 Corrections to be added to the A-weighted sound level to produce a measure of community reaction

Noise Characteristics	Correction in dBA
Pure tone present	+5
Intermittent or impulsive	+5
Noise only during working hours	−5
Total duration of noise each day	
Continuous	0
Less than 30 min	−5
Less than 10 min	−10
Less than 5 min	−15
Less than 1 min	−20
Less than 15 s	−25
Neighborhood	
Quiet suburban	+5
Suburban	0
Residential urban	−5
Urban near some industry	−10
Heavy industry	−15

suburban neighborhood; this yields a total correction of 0 and a corrected level of 72 dBA. The expected reaction would be "vigorous community action."

Schultz[12] has developed a model relating the outdoor L_{dn} within a community to the percent of people "highly annoyed" (Fig. 13.10.1). Analyzing the results of surveys of community response to transportation noise (aircraft, railroad, and street traffic) conducted in several countries, he found the relationship between percent highly annoyed and L_{dn} to be

$$\text{Percent highly annoyed} = 0.036L_{dn}^2 - 3.27L_{dn} + 79 \qquad (13.10.1)$$

[12]Schultz, J. Acoust. Soc. Am., **64**, 377 (1978). Fidell, Barber, and Schultz, J. Acoust. Soc. Am., **89**, 221 (1991).

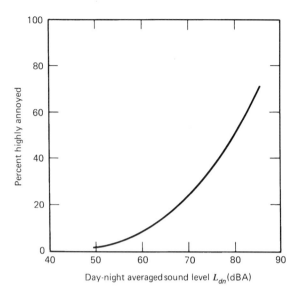

Figure 13.10.1 Estimate of the extent of public annoyance caused by transportation noise based on the day–night average sound pressure level L_{dn}.

with an uncertainty of about ± 5 dBA for $45 < L_{dn} < 85$ dBA. While this relationship is based on measurements of transportation noise, Schultz speculates that it may be applicable to community noise of other kinds.

13.11 NOISE-INDUCED HEARING LOSS

Hearing *loss* is quantified by specifying the *permanent threshold shift (PTS)* as a function of frequency. Hearing *impairment* is a broader term specifying the loss in the ability to understand speech. A commonly used classification of hearing impairment, shown in Table 13.11.1, is based on the average *PTS* for 500, 1000, and 2000 Hz measured in the lesser impaired of the ears.

Noise-induced hearing loss occurs in two ways: (1) *Trauma:* high-intensity sound, such as that from an explosion or jet engine, can rupture the eardrum, damage the ossicles, destroy the sensory hair cells, or cause collapse of a section of the organ of Corti. Such hearing loss is sudden and always associated with a specific noise event. (2) *Chronic:* noise levels below those necessary to cause trauma, if repeated often enough or continued for a long enough time, can cause dysfunction or destruction of the hair cells. This results from metabolic stress on the maximally stimulated cells. This type of hearing loss is more insidious than

Table 13.11.1 Classification of hearing impairment

Average Hearing Loss at 500, 1000, and 2000 Hz (dB)	Classification
Less than 25	Within normal limits
26–40	Mild or slight
41–55	Moderate
56–70	Moderately severe
71–90	Severe
More than 91	Profound

that brought on by trauma because the individual may not be aware of a slowly increasing loss.

Since controlled experiments on *PTS* in humans are unacceptable, most knowledge about hearing loss is obtained from field studies of workers who have been subjected to industrial noise for periods of time, and from inferences made from laboratory studies of induced *temporary threshold shift (TTS)*.

After cessation of a noise of intensity sufficient to cause *TTS* but not *PTS*, the ear recovers its initial threshold in a characteristic pattern. The *TTS* first decreases but then increases to a maximum at about 2 min (the bounce effect), after which the *TTS* decreases linearly with $\log t$ until it reaches 0 dB. Because of the complicated initial behavior, measurements of *TTS* are always taken 2 min after exposure ends.

For octave bands centered at 250 and 500 Hz, as long as the noise is below 75 dB there is no threshold shift no matter how long the exposure. For octave bands at 1, 2, and 4 kHz, there is no shift for levels below 70 dB. For band levels between 80 and 105 dB and for exposure times less than 8 h, the *TTS* increases linearly with $\log t$, the rate of increase being proportional to the level of the noise. For durations greater than 8 h, the *TTS* approaches an asymptotic value depending on the noise level. Pre- or postexposure to sound levels of 70 dB or lower has no effect on the amount of *TTS* or the rate of growth or decay.

For a given excitation frequency, the frequency of maximum *TTS* occurs from $\frac{1}{2}$ to 1 octave above the source frequency. For example, a pure tone of 700 Hz will produce a maximum *TTS* at 1 kHz.

Predicting the *TTS* for fluctuating or intermittent noises is difficult. For fluctuating noise, it appears that the average pressure level is important and not the total energy. For intermittent noise with an on-time between 250 ms and 2 min, the *TTS* is proportional to the fraction of on-time. For on-times less than 250 ms, the *TTS* is generally greater than that predicted by the above rule. For on-times greater than 2 min, predictions can be made considering the usual growth and recovery properties.

Studies of industrial workers exposed to the same noise for a number of years reveal great individual differences. However, the median *PTS* shows regular features depending on the intensity and length of exposure. In general, hearing loss first appears at frequencies near 4 kHz, the frequency region of the greatest acuity. With continued exposure, the severity of the hearing loss increases and extends to lower and higher frequencies. The *PTS* at 4 kHz increases with the A-weighted levels of the noise and with the exposure time up to 10 years, after which it appears to reach an asymptotic value. Figure 13.11.1 shows the median *PTS* for a 10 year exposure as a function of the A-weighted sound level.

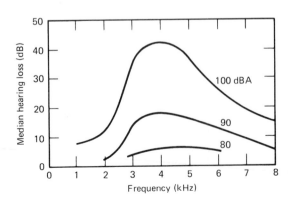

Figure 13.11.1 Median hearing loss after a 10 year exposure to industrial noise for several values of A-weighted sound pressure levels averaged over an 8 hour day, 5 days a week.

The effect of continuous industrial noise exposure for 8 h each working day for 10 years can be summarized as follows.

1. For the median individual, no hearing loss will occur if the sound level is below 80 dBA. The Environmental Protection Agency (EPA) accepts that a level of 70 to 75 dBA will produce a small but detectable *PTS* at 4 kHz in 1% to 10% of the population.
2. At 80 dBA, shifts in the median *PTS* between 3 and 6 kHz begin to appear.
3. At 85 dBA, median shifts of 10 dB occur between 3 and 6 kHz, with shifts of 15 to 20 dB in the susceptible 10% of the population.
4. At 90 dBA, median shifts of 20 dB are expected between 3 and 6 kHz, but the thresholds at 500, 1000, and 2000 Hz are still unaffected. Thus, by definition, there is no hearing impairment.
5. Above 90 dBA, shifts at 500, 1000, and 2000 Hz begin to appear, indicating the onset of hearing impairment.

In establishing criteria for noise exposure, it is necessary to determine the trade-off between level and duration. The experiments with *TTS* indicate that a 5 dB increase in level is equivalent to a doubling of exposure time.

On the bases of these and other findings, the Occupational Safety and Health Act (OSHA) of 1970 prescribes permissible exposures in industries doing business with the U.S. federal government. These levels are shown in Table 13.11.2. If the exposure consists of two or more periods at different noise levels, then the limit is that $\sum (t_i/T_i)$ not exceed unity, where t_i is the exposure time at a level whose total allowed exposure time is T_i.

The implicit philosophy behind these criteria should be made explicit. (1) These levels provide protection only for frequencies necessary for the understanding of speech; hearing loss at 4 kHz and above must be accepted as part of the job. (2) They assume an 8 hour day, 5 day week exposure time and assume that for the rest of the time no further ear damage occurs as might accompany recreational activities with high noise levels. (3) They are designed to protect only 85% of the exposed population, with financial compensation provided for the hearing impairment experienced by the more susceptible 15% of the population.

Table 13.11.2   Permissible daily noise exposure limits for industrial noise (OSHA)

Limiting Daily Exposure Time (h)	A-Weighted Sound Level Slow Response (dBA)
8	90
6	92
4	95
3	97
2	100
1.5	102
1	105
0.5	110
Less than 0.25	115

Table 13.11.3   Suggested daily noise exposure levels for nonoccupational noise

Limiting Daily Exposure Time	A-Weighted Sound Level Slow Response (dBA)
Less than 2 min	115
Less than 4 min	110
Less than 8 min	105
15 min	100
30 min	95
1 h	90
2 h	85
4 h	80
8 h	75
16 h	70

Although the OSHA criteria provide guidelines for judging the effects of nonoccupational noise, the above limitations must be taken under consideration when estimating the possible dangers to hearing from recreational or other nonoccupational noises. Table 13.11.3 shows recommended daily exposure times for nonoccupational noise.[13] For comparison with these levels, note that typical levels for a rock-n-roll band are 108 to 114 dBA, for a power mower 96 dBA, inside the cockpit of a light aircraft 90 dBA, and at 7.6 m (25 ft) from a motorcycle 86 dBA.

13.12 NOISE AND ARCHITECTURAL DESIGN

An important objective of architectural design is to provide sufficient acoustic isolation to prevent noise from interfering with the designated use of a space. In Section 13.5, noise criteria applicable to spaces intended for various uses were described. In this and the next three sections, procedures for meeting these criteria will be discussed.

The first line of defense against noise is city planning; zoning regulations should encourage the maximum possible separation between areas designated for noise-intensive use (heavy commercial, highways, airports, etc.) and noise-sensitive areas (residences, hospitals, parks, etc.).

Given an existing or potentially adverse acoustic environment, the architect can do much to alleviate economically the impact of the noise by considering acoustics from the start. Buildings can be located and oriented to provide noise barriers for each other; rooms requiring quiet should be placed on the side of a building away from major noise sources; within a building, noisy areas (such as kitchens, hallways, utility rooms, stairwells, and family rooms) should be separated by buffer zones from noise-sensitive areas (such as bedrooms, private offices, and study rooms).

To reduce the noise generated within the building, the architect should specify low-noise machinery (air conditioners, laundry facilities, water valves, etc.) mounted on properly designed and installed vibration isolators. Other considera-

[13]Cohen, Anticaglia, and Jones, *Sound Vib.*, **4**, 12 (1970).

tions, such as resilient floor coverings in halls and stairwells, will reduce the noise produced by nonmechanical sources.

The architect's final defense against intrusive noise is to acoustically isolate the room by specifying constructions that inhibit the transmission of structure-borne and airborne sounds.

The transmission of noise into a room is aided by the multitude of paths sound can find to penetrate the architect's defenses. The most prominent paths are (1) airborne noise outside the room that sets the common wall into vibration, which in turn radiates sound into the room, and (2) noise originating in the vibration of a solid structure (machinery, footfalls, etc.) that propagates along the structure and sets surfaces in the room into vibration. If the above paths are efficiently blocked by properly designed partitions and resilient mountings, then *flanking paths* can become important. Some flanking paths are obvious, such as the propagation through a false ceiling or crawl space and window-to-window transmission. Others are more insidious, such as porous cement block, poor seals between walls and ceiling or floor, gaps around wall penetrations, and back-to-back electrical outlets.

13.13 SPECIFICATION AND MEASUREMENT OF SOUND ISOLATION

The transmission loss (*TL*) for a partition is defined as

$$TL = 10 \log(\Pi_i/\Pi_t) \qquad (13.13.1)$$

where Π_i is the total power incident on the source side of the partition and Π_t is the total power transmitted through the partition. The transmission loss depends only on the frequency and the properties of the partition. A moment's thought will reveal that for a given partition (fixed *TL*), the noise reduction experienced between two rooms decreases as the area of the partition increases and increases as the absorptivity of the receiving room increases. (For the same intensity in the source room, a receiving room with little sound absorption will have a higher sound level than one with a large absorption.) A quantity of more direct architectural interest is the noise reduction

$$NR = 10 \log(I_1/I_2) = SPL_1 - SPL_2 \qquad (13.13.2)$$

where I_1 and I_2 are the intensities and SPL_1 and SPL_2 are the sound pressure levels in the source and receiver rooms, respectively.

The transmission loss and the noise reduction can be related: the power incident on a partition of area S is $\Pi_1 = I_1 S$. The power transmitted into the receiving room is equal to the rate at which energy is absorbed in the receiving room, $\Pi_2 = I_2 A$, where A is the sound absorption of the receiver room (Section 12.3). Combination shows that

$$TL = NR + 10 \log(S/A) \qquad (13.13.3)$$

Rather than measure *NR* for each frequency, SPL_1 and SPL_2 are usually specified to be spatially averaged 1/3-octave-band levels. Spatial averaging is used to reduce the effects of standing waves in both the source and receiving rooms.

Based on studies made with noise sources typical of multifamily dwellings, the *sound transmission class (STC)* provides a fairly successful single-number specification of the acoustic isolation characteristics of a partition.

To determine the *STC* of a partition, its *TL* is measured in the 16 contiguous 1/3-octave bands between 125 and 4000 Hz, inclusive. These measured values of *TL* are then compared to a reference contour consisting of three straight lines: a low-frequency segment that increases by 15 dB from 125 to 400 Hz, a middle segment that increases by 5 dB from 400 to 1250 Hz, and a horizontal segment at high frequencies (Fig. 13.13.1). The reference contour is chosen so that the maximum deficiency (deviation of the data below the contour) at any one frequency does not exceed 8 dB and the total deficiency at all frequencies does not exceed 32 dB. The *STC* of the partition is then the value of the *TL* corresponding to the intersection of the chosen reference contour with the 500 Hz ordinate.

Wall and ceiling/floor construction as well as door and window installations can be measured in the laboratory and their *TL* and *STC* values tabulated for use by the architect. Knowing the severity of the problem, the architect can chose the construction that will provide the required isolation. Table 13.13.1 gives some values of *STC* for representative constructions. More extensive compilations can be found in the literature.[14]

The *STC* of a composite structure, for example, walls with doors and windows, can be found from the *TL*s of the individual components. If S_i is the area of the individual component with transmission loss TL_i, its power transmission coefficient is

$$T_{\Pi i} = \text{antilog}(-TL_i/10) \qquad (13.13.4)$$

If S is the area of the entire wall, then the effective power transmission coefficient $T_\Pi(eff)$ is

$$T_\Pi(eff) = \frac{1}{S}\sum T_{\Pi i}S_i \qquad (13.13.5)$$

and the transmission loss for the composite structure is $10\log[1/T_\Pi(eff)]$. The *STC* for the composite structure can then be calculated by the usual procedure.

Sound transmission classes measured in the field are generally less than those obtained in the laboratory. This can usually be attributed to either flanking paths or poor quality work (improperly caulked joints, bridging between supposedly isolated elements). Even with properly constructed partitions, a 5 dB difference in *STC*s can be expected between field and laboratory measurements.

A procedure exists for quantifying the sound insulation between two rooms without band-level analysis. With a "pink" noise generator operating in the source room, a sound level meter is used to measure the C-weighted sound level (L_C) in the source room and the A-weighted sound level (L_A) in the receiving room. The *privacy rating (PR)* is then

$$PR = L_C - L_A + 10\log(2T_2) \qquad (13.13.6)$$

where T_2 is the reverberation time of the receiving room in seconds (see Section 12.3). It is claimed that the *PR* will be within a few decibels of the *STC*.

[14]Doelle, *Environmental Acoustics*, McGraw-Hill (1972).

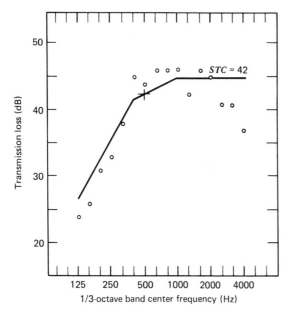

Figure 13.13.1 Determination of the sound transmission class *(STC)* from transmission loss *(TL)* measurements. The rating for this wall is *STC* = 42. (The *TL* at 4 kHz is 8 dB below the *STC* = 42 curve and the total deficiency is 30 dB.)

Table 13.13.1   Sound transmission class *(STC)* for representative wall constructions[a]

Construction	Mass per Unit Area (kg/m^2)	STC
1. 4 in. hollow block, $\frac{1}{2}$ in. plaster on both sides	115	40
2. 4 in. brick, $\frac{1}{2}$ in. plaster on both sides	210	40
3. 9 in. brick, $\frac{1}{2}$ in. plaster on both sides	490	52
4. 24 in. stone, $\frac{1}{2}$ in. plaster on both sides	1370	56
5. $\frac{3}{8}$ in. gypsum wallboard	8	26
6. $\frac{1}{2}$ in. gypsum wallboard	10	28
7. $\frac{5}{8}$ in. gypsum wallboard	13	29
8. Two $\frac{1}{2}$ in. gypsum wallboards bonded together	22	31
9. 2×4 studs on 16 in. centers, $\frac{1}{2}$ in. gypsum wallboard on both sides	21	33
10. Same as 9 but with $\frac{5}{8}$ in. gypsum wallboard on both sides	26	34
11. Same as 10 but with two sheets of $\frac{5}{8}$ in. gypsum wallboard on one side and one sheet on the other side	42	36
12. Same as 10 but with $\frac{1}{2}$ in. plaster over wallboard	68	46
13. Same as 9 but with a 2 in. isolation blanket	23	36
14. Same as 10 but with a 2 in. isolation blanket	29	38
15. Same as 11 but with a 2 in. isolation blanket	44	39
16. Same as 14 but with one side resiliently mounted	29	47
17. Same as 14 but with both sides resiliently mounted	29	49
18. Double row of 2×4 studs on 16 in. centers, $\frac{5}{8}$ in. gypsum wallboard on both sides, and 2 in. isolation blanket	37	57
19. Double row of 2×4 studs on 16 in. centers, two $\frac{5}{8}$ in. gypsum wallboard on both sides, no isolation blanket	60	58
20. Same as 19 but with 2 in. isolation blanket	60	62

[a] Numbers 1 to 15 are single-leaf construction. Numbers 16 to 20 are double-leaf construction.

Because of the different spectra associated with external noise, *STC* ratings cannot be used directly to predict isolation values of outside walls. However, a technique does exist that allows the *STC* to be used for this purpose. See Harris (*op. cit.*).

Besides their *STC*, floor/ceiling partitions should be considered for their *impact isolation class (IIC)*. This is a measure of their ability to isolate a room from noise produced by impact on the floor above. Again, see Harris (*op. cit.*).

13.14 RECOMMENDED ISOLATION

The amount of acoustic isolation required between two rooms depends on the noise level in the source room and the level of intrusive noise acceptable in the receiving room. Both these levels depend on the designated use of the rooms and the latter level depends on the ambient noise that will tend to mask the intrusive noise.

The recommendations of the Federal Housing Administration account for the difference in background noise by defining three grades of buildings.

Grade 1. Buildings with exterior nighttime noise levels lower than 40 dBA and recommended interior noise levels lower than 35 dBA.
Grade 2. Buildings with recommended interior noise levels of 40 dBA or lower.
Grade 3. Buildings with exterior nighttime noise levels of 55 dBA or higher and interior noise levels of 45 dBA or higher.

For walls separating different apartments, they recommend an *STC* of 55, 52, and 48 for building Grades 1, 2, and 3, respectively. For walls between rooms of the same dwelling unit, they recommend for a Grade 1 building the following: bedroom to bedroom *STC* = 48; living room to bedroom 50; bathroom to bedroom, kitchen to bedroom, and bathroom to living room 52. The recommendations for Grade 2 are 4 dB lower and those for Grade 3 are another 4 dB lower.

For walls separating apartment rooms from commonly shared service spaces (garages, laundries, party rooms), the following minimum requirements have been suggested: bedroom *STC* = 70, living room 65, and kitchen and bathroom 60.

Recommendations also exist for the impact isolation class for floors separating dwellings.

13.15 DESIGN OF PARTITIONS

Ignoring flanking and leakage, the basic mechanism of sound transmission through a wall is that sound in the source room forces the exposed surface to vibrate; this vibration is transmitted through the structure of the wall to the other surface, which in turn vibrates producing sound in the receiving room. If the two surfaces of the wall are rigidly connected so that they vibrate as a unit (a single-leaf partition), the transmission loss depends only on the frequency and the mass per unit area, stiffness, and intrinsic damping of the wall. If the partition consists of two unconnected walls separated by a cavity (a double-leaf partition), the transmission loss depends on the properties of the two walls and on the size of the cavity and its absorption.

(a) Single-Leaf Partitions

For a plane, nonporous, homogeneous, flexible wall the transmission loss is dependent on the density of the partition and the frequency of the noise. The energy ΔE_i that strikes area ΔS of the partition in the time interval Δt was found to be (12.2.2)

$$\Delta E_i = \frac{\mathscr{E}\Delta S \Delta r}{2} \int_0^{\pi/2} \sin \theta \cos \theta \, d\theta \tag{13.15.1}$$

The energy ΔE_t transmitted through the partition is the same integral, but with $T_{\Pi}(\theta)$ multiplying the integrand,

$$\Delta E_i = \frac{\mathscr{E}\Delta S \Delta r}{2} \int_0^{\pi/2} T_{\Pi}(\theta) \sin \theta \cos \theta d\theta \tag{13.15.2}$$

where the transmission coefficient is given by (6.7.4),

$$T_{\Pi}(\theta) = \frac{1}{1 + a^2 \cos^2 \theta} \tag{13.15.3}$$

$$a = \omega \rho_S / 2\rho_0 c = \pi f \rho_S / \rho_0 c$$

where ρ_S is the surface density of the partition and ρ_0 and c are for the air. Taking the ratio of (13.15.1) and (13.15.2) and integrating (13.15.1) gives an integral expression for the random incidence transmission coefficient,

$$\langle T_{\Pi} \rangle_\theta = \frac{\Delta E_t}{\Delta E_i} = 2 \int_0^{\pi/2} \frac{\sin \theta \cos \theta}{1 + a^2 \cos^2 \theta} \, d\theta \tag{13.15.4}$$

The change of variable $u = a \cos \theta$ allows immediate evaluation,

$$\langle T_{\Pi} \rangle_\theta = \frac{2}{a^2} \int_0^a \frac{u}{1 + u^2} \, du = \frac{1}{a^2} \ln(1 + a^2) \tag{13.15.5}$$

The transmission loss TL, defined in (13.13.1), is then

$$TL = -10 \log \langle T_{\Pi} \rangle_\theta = 20 \log a - 10 \log[\ln(1 + a^2)] \tag{13.15.6}$$

As seen from Fig. 13.15.1, values of ρ_S range from about 10 to 1000 kg/m^2 but cluster around 20 to 50 kg/m^2. Frequencies of interest run from about 60 Hz to 4 kHz, and $\rho_0 c$ for air is 415 Pa · s/m. Values of a thus span $4 < a < 3 \times 10^4$. Substitution shows that the last term on the right in (13.15.6) has values between about -5 and -13 dB, with the majority falling around -8 dB. As a conservative estimate, it is appropriate to assume the smallest deduction, so that the acoustic transmission through the partition is overestimated. Taking -5 dB and expressing $\rho_0 c$ and π numerically gives

$$TL = 20 \log (f \rho_S) - 47 \tag{13.15.7}$$

with frequency in Hz and surface density in kg/m^2. This *mass law* predicts that doubling the mass per unit area at a given frequency or doubling the frequency for a given mass per unit area increases the TL by 6 dB.

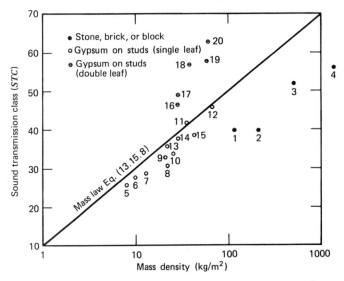

Figure 13.15.1   Sound transmission class for partitions and comparison with the mass law. Numbers denote constructions listed in Table 13.13.1. Single-leaf partitions are represented by numbers 1 to 15 while numbers 16 to 20 denote double-leaf construction.

A wall that obeys the mass law (13.15.7) throughout the frequency range from 125 to 4000 Hz has a sound transmission class, found by the procedure described in Section 13.13, of

$$STC = 20 \log \rho_S + 10 \qquad (13.15.8)$$

From Fig. 13.15.1, the STC for a single-leaf partition is invariably below that predicted by the mass law. Part of this is caused by the porosity of the material, as may be seen by the improvement when concrete block is sealed (note Table 12.5.1), but the remainder is related to the stiffness of the panel, which is neglected in developing the mass law.

For a stiff, homogeneous panel, flexural waves propagate along its surface with phase speed

$$c_p = \left(\frac{\pi^2}{3} \frac{Yt^2}{\rho} f^2 \right)^{1/4} \qquad (13.15.9)$$

where f is the frequency of excitation, Y the Young's modulus, t the thickness, and ρ the *volume* density of the panel. Since this propagation is dispersive, there exists a *coincidence frequency* f_c at which the wavelength of the flexural wave equals that of a wave of the same frequency in air:

$$f_c = \frac{c^2}{\pi t} \left(\frac{3\rho}{Y(1 - \sigma^2)} \right)^{1/2} \qquad (13.15.10)$$

where c is the speed of sound in air and σ is Poisson's ratio. For all frequencies above f_c there exists an angle of incidence θ (measured with respect to the normal) such that

$$\lambda_f = \lambda / \sin \theta \qquad (13.15.11)$$

where λ_f and λ are the wavelengths in the panel and air, respectively. For this particular angle, there is extremely good coupling of energy from the incident wave to the flexural wave, which in turn efficiently radiates energy into the receiving room. Therefore, for frequencies above f_c, the transmitted power averaged over all incident angles has a transmission loss that is less than predicted by the mass law. Since λ_f is proportional to $1/\sqrt{f}$ and λ to $1/f$, $\sin \theta$ is proportional to $1/\sqrt{f}$ so that at high frequencies only waves incident at near-normal angles can be coincident: the *TL* rises to the values given by the mass law. The depth of this coincidence dip depends on the intrinsic damping of the panel.

To obtain an *STC* close to that predicted by the mass law, a wall must be designed so that the coincidence frequency occurs at frequencies either below 125 Hz, requiring a thick wall with low density and high Young's modulus, or above 4000 Hz, requiring a thin, high-density wall with low Young's modulus. Examples of the effects of the *coincident dip* on walls of different constructions are displayed in Fig. 13.15.2.

A properly designed single-leaf partition still requires a large mass per unit area to obtain a respectable *STC*. For example, a 6 in. thick concrete wall with $\frac{1}{2}$ in. plaster on each side has a mass per unit area of 390 kg/m^2 and is rated at *STC* = 52; this is 3 dB shy of that recommended for party walls between luxury apartments. (Note the mass law predicts *STC* = 62 for this same wall.) To obtain high acoustic isolation without excessive weight, it is necessary to employ double-leaf construction.

(b) Double-Leaf Partitions

As seen in Fig. 13.15.1, the *STC* for a partition of double-leaf construction is considerably higher than that for a single-leaf partition of the same mass density. This effect is dramatically illustrated in Fig. 13.15.3, where the transmission loss curve for two sheets of $\frac{1}{2}$ in. gypsum board bonded together as a single leaf is

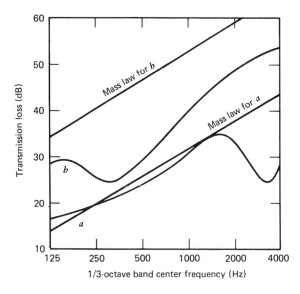

Figure 13.15.2 Effect of the properties of a wall on the coincidence frequency.
(*a*) Single $\frac{1}{2}$ inch gypsum wallboard (10 kg/m^2). Critical frequency = 2.6 kHz. *STC* = 28. (*b*) Lightweight foamed concrete (110 kg/m^2). Critical frequency = 200 Hz. *STC* = 35. (After Harris, *op. cit.*)

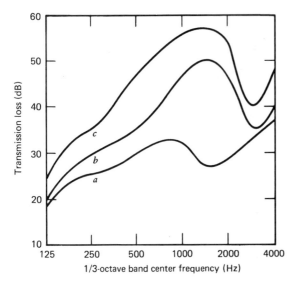

Figure 13.15.3   Transmission loss for single- and double-leaf partitions constructed from similar materials. (*a*) Two $\frac{1}{2}$ inch gypsum wallboards bonded together (22 kg/m^2). *STC* = 31. (*b*) Single $\frac{1}{2}$ inch gypsum wallboard on both sides of staggered 4 inch steel channel studs (21 kg/m^2). *STC* = 37. (*c*) Same as (*b*) but with 2 inch sound isolation blankets. *STC* = 45. (After Harris, *op. cit.*)

compared with that obtained for the same sheets used in a double-leaf construction. Also note the improvement when absorbing material is placed in the space between the boards.

To illustrate effects on the *STC* rating of various design changes, Table 13.15.1 outlines a sequence of partitions of similar construction. Without resilient mounting the addition of damping material adds only 4 dB to the *STC* rating; with resilient mounting a blanket adds 10 dB. With or without a blanket, the second resilient mounting adds only 1 or 2 dB.

To obtain even higher *STC* ratings, it is necessary to use a staggered-stud construction, which provides better vibration isolation, and more layers of gypsum to add mass. A wall that can provide an *STC* = 55 consists of two sets of 2 × 4 wood studs each with 24 in. spacing on separate 2 × 4 plates, spaced 1 in. apart, with the subfloor discontinuous between plates; one 2 in. isolation blanket; and one $\frac{5}{8}$ in. gypsum sheet nailed on one side and two nailed on the other side. If the single sheet of gypsum board is resiliently mounted instead of nailed, the *STC* increases to 60.

Table 13.15.1   Sound transmission class (*STC*) for partitions consisting of 2 × 4 studs on 16 in. centers, faced with $\frac{5}{8}$ in. gypsum wallboard that is either nailed directly to the studs or resiliently mounted on one or both sides, either with or without a 2 in. sound isolation blanket

Isolation Blanket	Resilient Mounting	STC
No	No	34
Yes	No	38
No	One	38
No	Two	39
Yes	One	47
Yes	Two	49

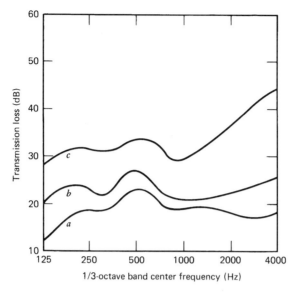

Figure 13.15.4 Transmission loss for $1\frac{3}{4}$ inch thick door constructions. (*a*) Hollow-core door, ungasketed (7 kg/m^2). $STC = 17$. (*b*) Hollow-core door, gasketed (7 kg/m^2). $STC = 24$. (*c*) Solid-core door, gasketed (20 kg/m^2). $STC = 26$. (After Doelle, *op. cit.*)

(c) Doors and Windows

Because of their low surface density and the gaps around their edges, doors and windows are acoustically weak elements of a partition. Figure 13.15.4 illustrates the advantages of solid-core doors over hollow-core doors and the value of an acoustical seal that includes an automatic threshold closer to seal the gap at the bottom. For windows, Fig. 13.15.5 illustrates the advantage of using double glazing with well-sealed edges and a minimum separation of 10 to 13 cm (4 to 5 in.).

(d) Barriers

An inexpensive means of preventing an outdoor noise source from being a disturbance at a particular location is to erect a barrier between the source and

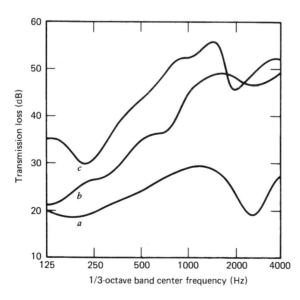

Figure 13.15.5 Transmission loss for different openable window constructions weatherstripped with sealed edges. (*a*) Single-pane window, 3 mm glazing (7.5 kg/m^2). $STC = 25$. (*b*) Double-pane window, 3 mm glazing with 10 cm air space (15 kg/m^2). $STC = 36$. (*c*) Double-pane window, 3 mm glazing with 20 cm air space (13 kg/m^2). $STC = 40$. (After Doelle, *op. cit.*)

receiver. Because of diffraction, this can be only partly successful, but often may be sufficient for the purpose. (The technique is more difficult to apply successfully indoors because sound can reach the receiver over a multitude of reflective paths. However, even here it can be helpful in isolating a source very close to an occupied space.)

If the *straight-line* path (through the barrier) between noise source and recipient is r, and the shortest possible *diffracted* path (the physical path of least-time-of-flight) is R, then the *Fresnel number* N_F is

$$N_F = 2(R - r)/\lambda \qquad (13.15.12)$$

where λ is the wavelength for each of the frequencies of concern. It has been shown[15] that the received power is reduced by a factor of about $20N_F$ when the barrier is introduced and the resultant increase in the transmission loss A_b is

$$A_b \sim 10\log(20N_F) \qquad (13.15.13)$$

Thus, if TL is the transmission loss between source and recipient in the absence of the barrier, then the transmission loss TL_b in the presence of the barrier is

$$TL_b = TL + A_b \qquad (13.15.14)$$

within accuracies of about ± 5 dB.

PROBLEMS

13.2.1. The band levels for a noise are 70 dB *re* 20 μPa at 31.5 Hz and decrease 3 dB for each octave. Find (*a*) the overall sound pressure level between 31.5 and 8000 Hz and (*b*) the A-weighted sound pressure level for this same frequency range.

13.2.2 The octave band levels measured by an instrument with perfectly flat frequency response are 78, 76, 78, 82, 81, 80, 80, 73, and 65 dB *re* 20 μPa for the octave bands centered at 31.5, 63, 125, 250, 500, 1000, 2000, 4000, and 8000 Hz respectively. Calculate the band levels that would be measured by a sound level meter set to (*a*) C-weighted and (*b*) A-weighted.

13.2.3 A sound level meter accidently set to A-weighted gave the following band levels for the octave bands centered at 31.5, 63, 125, 250, 500, 1000, 2000, 4000, and 8000 Hz: 75, 75, 62, 73, 78, 80, 81, 73, and 63 dB *re* 20 μPa. Calculate the band levels if the meter were set to C-weighted.

13.3.1. For the noise given in Problem 13.2.1, (*a*) find the speech interference level, and compare to the approximation given by (13.3.1). (*b*) Determine the voice level required for just-reliable face-to-face communication at 10 m, and classify the noise condition.

13.3.2. (*a*) For $SIL = 40$ dB, find the VL for just-reliable communication with a listener at 8 m. Estimate (*b*) the A-weighted level of the noise and (*c*) the A-weighted level of the signal at the listener's location. (*c*) From Fig. 13.3.1 determine the speech intelligibility for sentences. (*d*) Is this consistent with the discussion at the beginning of Section 13.3?

[15]Maekawa, *Appl. Acoust.*, **1**, 157 (1968)

13.5.1. For the noise of Problem 3.2.1, (a) determine the balanced noise criterion from Fig. 13.5.1. (b) A rough estimate of NCB can be obtained from $NCB \sim L_A - 8$. How good is that in this case? (c) Comment on the appropriateness of this noise for the rooms listed in Table 13.5.1.

13.6.1. For the community noise of Fig. 13.6.2, find L_{10}, L_{50}, and L_{90}.

13.6.2. Find the $SENEL$ for a single event where the A-weighted sound pressure level jumps to a constant 80 dB re 20 μPa for 25 s and then suddenly returns to the ambient level.

13.6.3. The hourly average sound level for a community noise is 60 dBA from 7 A.M. to 7 P.M., 55 dBA from 7 P.M. to 10 P.M., and 50 dBA from 10 P.M. to 7 A.M. Find (a) L_{eq} for the 24 hours, (b) L_{dn}, and (c) $CNEL$.

13.8.1. Calculate the A-weighted sound pressure levels for an automobile and a motorcycle under cruising conditions measured at a distance of 15 m for speeds of 44 and 88 km/h.

13.8.2. If, at a given distance from a straight, two-lane road of infinite extent, zero grade, and negligible truck traffic, the maximum equivalent sound level is to remain constant, find the ratio of the maximum vehicle flow rate (Q) if the maximum speed is reduced from 104 km/h (65 mph) to 88 km/h (55 mph).

13.11.1. A worker is exposed to noise levels of 92 dBA for 4 h and 90 dBA for 4 h each working day. Do these conditions exceed the OSHA recommendations?

13.13.1. In measuring the sound transmission class of a 30 m^2 partition, the band level in the source room was kept constant at 90 dB, and for each 1/3-octave band from 125 to 4000 Hz the band levels and total sound absorption A (in square meters) in the receiving room were: (band level, A) = (65, 15), (51, 15), (44, 16), (48, 17), (44, 19), (42, 20), (47, 25), (40, 28), (39, 30), (35, 32), (34, 36), (35, 45), (33, 53), (32, 60), (34, 60), and (37, 60). Find the STC of the partition.

13.15.1. (a) According to the mass law, what surface density is required to produce $TL = 20$ dB for a single-leaf partition at 100 Hz? (b) What type of construction will provide this TL?

13.15.2C. For a single-leaf partition with $\rho_S = 1000$ kg/m^2, plot, as a function of frequency from 60 Hz to 4 kHz, the TL given by (13.15.6) and compare to the approximation given by (13.15.7). Repeat for $\rho_S = 10$ kg/m^2. Comment on the range of applicability of the approximate expression.

13.15.3. Find the coincidence frequency for a $\frac{1}{4}$ in. thick lucite window.

13.15.4. (a) Show that the most effective place for a barrier of given height is halfway between the source and receiver. (b) Show that, for a fixed position and height, the increased transmission loss decreases with frequency as $10 \log \lambda$.

13.15.5C. It is desired to increase the transmission loss between a highway and a residence 100 m from the highway by erecting a barrier of height h midway between the highway and the house. For 100, 200, and 300 Hz, plot the increased transmission loss as a function of the height of the barrier. For each frequency, find the height required to increase the transmission loss by 5 dB.

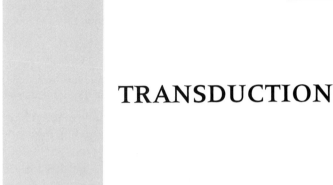

Chapter *14*

TRANSDUCTION

14.1 INTRODUCTION

The introduction to transduction theory presented in this chapter can be applied to any electroacoustic transducer, whether designed for use in air or water. Specific examples will be confined to transducers used in air. More detailed considerations of loudspeakers and electrostatic, piezoelectric, ferroelectric, and magnetostrictive transducers can be found in a number of textbooks.[1] Throughout this chapter, the symbols for the effective amplitudes of oscillatory electrical, mechanical, and acoustic quantities will have the subscript "e" suppressed.

14.2 THE TRANSDUCER
AS AN ELECTRICAL NETWORK

A transducer converting energy between electrical and mechanical forms can be treated as a *two-port network* that relates electrical quantities at one port to mechanical quantities at the other. These quantities are defined as

V = the voltage across the electrical inputs to the transducer
I = the current at the electrical inputs
F = the force on the radiating surface
u = the speed of the radiating surface

All of these are effective (rms) quantities
    If force is taken as the analog of voltage and speed the analog of current, the network is as shown in Fig. 14.2.1*a*. Under certain conditions it may be more useful to consider the *mechanical dual*, for which speed is the analog of voltage and force the analog of current. This network is shown in Fig. 14.2.1*b*. Which alternative is more convenient depends on the transducer.

[1]Hunt, *Electroacoustics*, Wiley (1954); republished, Acoustical Society of America (1982). *Physical Acoustics* IA, ed. Mason, Academic Press (1964). Camp, *Underwater Acoustics*, Wiley (1970). Wilson, *An Introduction to the Theory and Design of Sonar Transducers*, U. S. Gov't. Printing Office (1970).

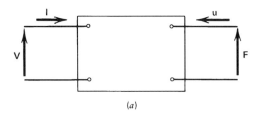

(a)

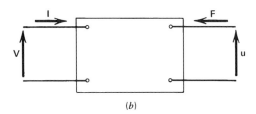

(b)

Figure 14.2.1   Two-port networks.
(a) Force the analog of voltage and speed
the analog of current. (b) The mechanical
dual with speed the analog of voltage
and force the analog of current.

Associated with the electrical and mechanical variables are a number of impedances that are measurable properties of the system:

$\mathbf{Z}_{EB} = \mathbf{V}/\mathbf{I}|_{u=0}$ = the blocked electrical impedance (Ω)

$\mathbf{Z}_{EF} = \mathbf{V}/\mathbf{I}|_{F=0}$ = the free electrical impedance (Ω)

$\mathbf{Z}_{mo} = \mathbf{F}/\mathbf{u}|_{I=0}$ = the open-circuit mechanical impedance (N $\cdot$ s/m)

$\mathbf{Z}_{ms} = \mathbf{F}/\mathbf{u}|_{V=0}$ = the short-circuit mechanical impedance (N $\cdot$ s/m)

(The unit of mechanical impedance is often defined as the *mechanical ohm*, where 1 mechanical ohm = 1 N $\cdot$ s/m.) Uppercase subscripts designate electrical impedances, lowercase subscripts mechanical impedances, and the second subscript refers to the constraint.

From the definition of $\mathbf{Z}_{EB}$ we have $\mathbf{V}(\mathbf{I},0) = \mathbf{Z}_{EB}\mathbf{I}$. If $\mathbf{u}$ is not zero, however, the value of $\mathbf{V}$ must be different for the same applied current. If $\mathbf{V}$ depends linearly on $\mathbf{u}$, as it does on $\mathbf{I}$, then

$$\mathbf{V} = \mathbf{Z}_{EB}\mathbf{I} + \mathbf{T}_{em}\mathbf{u} \tag{14.2.1}$$

where $\mathbf{T}_{em}$ is a *transduction coefficient*. Similarly, the force equation is generalized to

$$\mathbf{F} = \mathbf{T}_{me}\mathbf{I} + \mathbf{Z}_{mo}\mathbf{u} \tag{14.2.2}$$

Equations (14.2.1) and (14.2.2) are the *canonical equations* for the electromechanical behavior of a transducer.

If the electrical terminals are shorted so that $\mathbf{V} = 0$, then (14.2.1) can be solved to express $\mathbf{I}$ in terms of $\mathbf{u}$ and this result can be substituted into (14.2.2). The quantity $\mathbf{F}/\mathbf{u}$ is then $\mathbf{Z}_{ms}$ and further manipulation yields

$$\mathbf{Z}_{ms} = (1 - \mathbf{k}_c^2)\mathbf{Z}_{mo} \tag{14.2.3}$$

where a *coupling coefficient* $\mathbf{k}_c$ has been defined as

$$\mathbf{k}_c^2 = \mathbf{T}_{em}\mathbf{T}_{me}/\mathbf{Z}_{EB}\mathbf{Z}_{mo} \tag{14.2.4}$$

and often has a simple form and physical interpretation over a range of operating frequencies. Similarly, imposition of $\mathbf{F} = 0$ and manipulation reveals

$$\mathbf{Z}_{EF} = (1 - \mathbf{k}_c^2)\mathbf{Z}_{EB} \tag{14.2.5}$$

Under certain conditions of symmetry the canonical equations assume simpler forms and the networks of Fig. 14.2.1 can be represented by reversible three-terminal networks.

(a) Reciprocal Transducers

If $\mathbf{T}_{em} = \mathbf{T}_{me} = \mathbf{T}$, the transducer displays *electroacoustic reciprocity*. Crystal, ceramic, and electrostatic transducers are of this type. The canonical equations can be reduced to

$$\mathbf{V} = \mathbf{Z}_{EB}\mathbf{I} + \phi\mathbf{Z}_{EB}\mathbf{u}$$

$$\mathbf{F} = \phi\mathbf{Z}_{EB}\mathbf{I} + \mathbf{Z}_{mo}\mathbf{u} \tag{14.2.6}$$

$$\phi = \mathbf{T}/\mathbf{Z}_{EB}$$

where the *transformation factor* ϕ is real and constant for most frequencies of interest. This is a considerable simplification, since (14.2.6) allows use of the force–voltage and speed–current analogs of Fig. 14.2.1*a* and yields the equivalent circuits of Fig. 14.2.2. Linear circuit analysis can now be used. Figure 14.2.2*b* displays the physical significance of ϕ as the *turns ratio* of an ideal transformer linking the electrical and

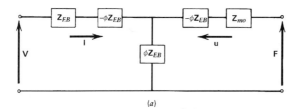

(a)

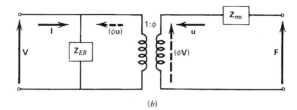

(b)

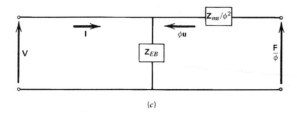

(c)

Figure 14.2.2   Equivalent circuits for a reciprocal transducer.

mechanical sides of the network. (Note that this transformer is conceptual only, since ϕ is not dimensionless.) The coupling factor defined in (14.2.4) becomes

$$\mathbf{k}_c^2 = \phi^2 \mathbf{Z}_{EB} / \mathbf{Z}_{mo} \tag{14.2.7}$$

which, when substituted into (14.2.3), gives

$$\mathbf{Z}_{ms} = \mathbf{Z}_{mo} - \phi^2 \mathbf{Z}_{EB} \tag{14.2.8}$$

As will be seen in Section 14.3, this coupling factor $\mathbf{k}_c$ has simple form and physical meaning for reciprocal transducers.

(b) Antireciprocal Transducers

Another kind of coupling displays *electroacoustic antireciprocity*, $\mathbf{T}_{em} = -\mathbf{T}_{me}$. Magnetostrictive, moving-coil, and moving-armature transducers are examples. The canonical equations and associated mechanical impedances now have a simple but different form

$$
\begin{aligned}
\mathbf{V} &= \mathbf{Z}_{EB}\mathbf{I} + \boldsymbol{\phi}_M \mathbf{u} \\
\mathbf{F} &= -\boldsymbol{\phi}_M \mathbf{I} + \mathbf{Z}_{mo}\mathbf{u} \\
\mathbf{Z}_{ms} &= \mathbf{Z}_{mo} + \boldsymbol{\phi}_M^2 / \mathbf{Z}_{EB} \\
\boldsymbol{\phi}_M &= \mathbf{T}_{em} = -\mathbf{T}_{me}
\end{aligned}
\tag{14.2.9}
$$

This transformation factor $\boldsymbol{\phi}_M$ is usually either real or complex with a small phase angle. The minus sign relating $\mathbf{T}_{em}$ and $\mathbf{T}_{me}$ precludes an equivalent circuit based on the force–voltage analog. However, there are two approaches that yield linear, three-terminal networks.

(1) THE MECHANICAL DUAL   Rearrangement of the canonical equations results in the symmetric pair

$$
\begin{aligned}
\mathbf{V} &= \mathbf{Z}_{EF}\mathbf{I} + \boldsymbol{\phi}_M \mathbf{Y}_{mo}\mathbf{F} \\
\mathbf{u} &= \boldsymbol{\phi}_M \mathbf{Y}_{mo}\mathbf{I} + \mathbf{Y}_{mo}\mathbf{F} \\
\mathbf{Y}_{mo} &= 1/\mathbf{Z}_{mo}
\end{aligned}
\tag{14.2.10}
$$

where $\mathbf{Y}_{mo}$ is the *open-circuit mechanical admittance*. These equations have the same form as (14.2.6), but with $\mathbf{F}$ and $\mathbf{u}$ exchanged and admittance appearing as impedance. A few useful equivalent circuits are shown in Fig. 14.2.3. Note the appearance of $\boldsymbol{\phi}_M$ as an *inverse* turns ratio of the transformer.

(2) SHIFTED FORCE AND SPEED   The change of variables $\mathbf{F}' = -j\mathbf{F}$ and $\mathbf{u}' = -j\mathbf{u}$ allows (14.2.9) to be rewritten as

$$
\begin{aligned}
\mathbf{V} &= \mathbf{Z}_{EB}\mathbf{I} + j\boldsymbol{\phi}_M \mathbf{u}' \\
\mathbf{F}' &= j\boldsymbol{\phi}_M \mathbf{I} + \mathbf{Z}_{mo}\mathbf{u}'
\end{aligned}
\tag{14.2.11}
$$

Comparison with (14.2.6) shows that (14.2.11) can be represented by the circuit in Fig. 14.2.2*a*, but with $j\boldsymbol{\phi}_M$ in place of $\phi\mathbf{Z}_{EB}$. This approach preserves the force–

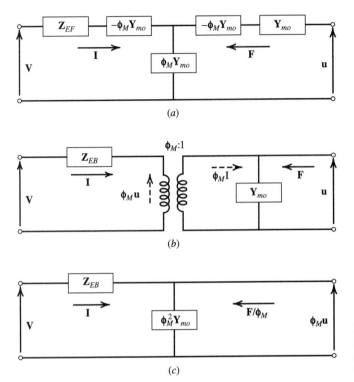

Figure 14.2.3
Equivalent circuits
for an antireciprocal
transducer.

voltage and speed–current parallels of the reciprocal transducer but is somewhat unphysical in shifting force and speed through phase angles of $\pi/2$.

In the case of an antireciprocal transducer, it is convenient to define a coupling coefficient $\mathbf{k}_m$ from (14.2.9)

$$\mathbf{Z}_{mo} = (1 - \mathbf{k}_m^2)\mathbf{Z}_{ms}$$
$$\mathbf{k}_m^2 = \boldsymbol{\phi}_M^2/\mathbf{Z}_{EB}\mathbf{Z}_{ms} \tag{14.2.12}$$

This coefficient differs from the $\mathbf{k}_c$ defined in (14.2.4) but has simple form and meaning for antireciprocal transducers at lower frequencies.

14.3 CANONICAL EQUATIONS FOR TWO SIMPLE TRANSDUCERS

To show application of the canonical equations, and the engineering approximations that simplify analysis, two transducers (one reciprocal and the other antireciprocal) will be analyzed in a little more detail.

(a) The Electrostatic Transducer (Reciprocal)

The electrostatic transducer can be modeled as a pair of capacitor plates, one held stationary while the other, the *diaphragm*, moves in response to mechanical or electrical excitation, as suggested in Fig.14.3.1a. The transducer is connected to an external circuit, Fig. 14.3.1b, where V_0 is a constant polarization voltage, C_B a

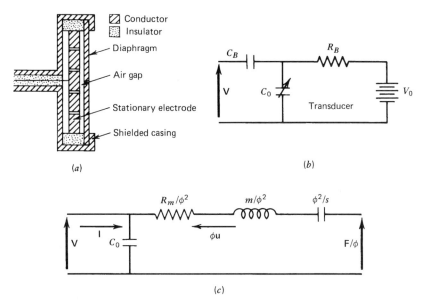

Figure 14.3.1 The electrostatic transducer as a representative reciprocal transducer. (*a*) Schematic of the construction of the transducer. (*b*) External circuit including the capacitance C_0 of the transducer. (*c*) Equivalent circuit.

capacitor blocking direct current from the electrical terminals, and R_B a resistor isolating the polarization source from the alternating current. When properly chosen so that $C_B \gg C_0$ and $R_B \gg 1/\omega C_B$, C_B and R_B can be ignored since they do not significantly alter the behavior of the transducer in the frequency range of interest. If a transient voltage is applied across the plates, the diaphragm moves in response to the changing charge. If the diaphragm experiences an incident pressure field, its resultant motion will generate an electrical signal.

The capacitance of the transducer with the plates at rest is $C_0 = \varepsilon S/x_0$, where ε is the dielectric constant of the material between the plates, S the surface area of the diaphragm, and x_0 the equilibrium spacing of the plates. For the moment, neglect any *leakage resistance* between the plates of the capacitor (this will be included later). The charge q_0 on the plates resulting from the equilibrium voltage V_0 is found from $q_0 = C_0 V_0$. When a sinusoidal voltage $\mathbf{V}$ is superimposed on V_0 the instantaneous charge $q_0 + \mathbf{q}$ and spacing $x_0 + \mathbf{x}$ are related by $\mathbf{V} + V_0 = (\mathbf{q} + q_0)(\mathbf{x} + x_0)/\varepsilon S$. If $|\mathbf{q}| \ll q_0, |\mathbf{x}| \ll x_0$, and $\mathbf{V}$ varies as $\exp(j\omega t)$ so that $\mathbf{u} = j\omega\mathbf{x}$ and $\mathbf{I} = j\omega\mathbf{q}$, linearization yields

$$\mathbf{V} = \frac{1}{j\omega C_0}\mathbf{I} + \frac{V_0}{j\omega x_0}\mathbf{u} \tag{14.3.1}$$

Comparison with (14.2.1) reveals

$$\mathbf{Z}_{EB} = 1/j\omega C_0$$
$$\mathbf{T}_{em} = V_0/j\omega x_0 \tag{14.3.2}$$

To obtain the second of the canonical equations, note that the change $\mathbf{f}$ in the force resulting from the change $\mathbf{q}$ of the charge on the fixed plates is $\mathbf{f} = -\mathbf{q}q_0/\varepsilon S =$

$-qV_0/x_0 = -IV_0/j\omega x_0$. Application of Newton's law then gives $\mathbf{Z}_{mo}\mathbf{u} = \mathbf{F} + \mathbf{f}$, or

$$\mathbf{F} = \frac{V_0}{j\omega x_0}\mathbf{I} + \mathbf{Z}_{mo}\mathbf{u} \tag{14.3.3}$$

Comparison with (14.2.2) shows that $\mathbf{T}_{me} = V_0/j\omega x_0$. Since the transduction coefficients are equal, $\mathbf{T} = \mathbf{T}_{em} = \mathbf{T}_{me}$, the electrostatic transducer displays electromechanical reciprocity with

$$\mathbf{T} = V_0/j\omega x_0$$
$$\phi = C_0 V_0/x_0 \tag{14.3.4}$$

The transformation factor is a real constant. The mechanical impedance under short-circuit conditions is

$$\mathbf{Z}_{ms} = \mathbf{R}_m + j(\omega m - s/\omega) \tag{14.3.5}$$

where R_m is the mechanical resistance, m the mass of the diaphragm, and s its (short-circuit) stiffness. This yields the equivalent circuit given in Fig. 14.3.1c, which has the form of Fig. 14.2.2c. Use of (14.2.8) and (14.3.2) gives

$$\mathbf{Z}_{mo} = \mathbf{R}_m + j(\omega m - s'/\omega)$$
$$s' = s + \phi^2/C_0 \tag{14.3.6}$$

Thus, the only effect on the mechanical impedance of opening the shorted electrical terminals is to alter the stiffness from s to s'. For frequencies low enough that $\omega \ll \sqrt{s/m}$ and $R_m \ll s/\omega$, (14.2.3) yields $\mathbf{k}_c^2 \approx 1 - s/s'$ so that $\mathbf{k}_c$ becomes pure real, $\mathbf{k}_c = k_c$. If we define the electrical capacitance equivalent to the short-circuit mechanical stiffness

$$C = \phi^2/s \tag{14.3.7}$$

then

$$k_c^2 = C/(C + C_0) \tag{14.3.8}$$

This is the ratio of stored mechanical energy to total stored energy for stiffness-controlled performance and negligible losses.

(b) The Moving-Coil Transducer (Antireciprocal)

This transducer consists of a diaphragm attached to a cylindrical coil of wire, the *voice coil*, that is suspended in a constant magnetic field B as suggested in Fig. 14.3.2. If an alternating current $\mathbf{I}$ is supplied to the coil, the interaction of the current and the magnetic field will induce a force on the coil moving the diaphragm. Conversely, forcing the coil to move in the magnetic field induces a voltage in the coil. If the diaphragm is *blocked*, the electrical impedance is

$$\mathbf{Z}_{EB} = R_0 + j\omega L_0 \tag{14.3.9}$$

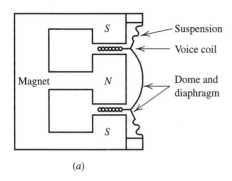

(a)

Figure 14.3.2   The moving-coil transducer as a representative antireciprocal transducer. (a) Schematic of the construction of the transducer. (b) Equivalent circuit.

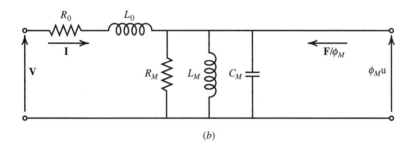

(b)

where R_0 and L_0 are the resistance and inductance of the voice coil. The current in the blocked voice coil generates a force on the diaphragm of complex amplitude $\mathbf{F}_e = Bl\mathbf{I}$, where l is the length of the wire in the voice coil and the direction of $\vec{B} \times \vec{l}$ is chosen as positive. The external force necessary to hold the diaphragm stationary is therefore $\mathbf{F} = -Bl\mathbf{I}$. If the electrical terminals of the transducer are open, then $\mathbf{I} = 0$ and there are no electrical effects. At frequencies low enough for the diaphragm to move as a unit, it is a simple mechanical oscillator with (open-circuit) mechanical impedance

$$\mathbf{Z}_{mo} = R_m + j(\omega m - s/\omega) \tag{14.3.10}$$

where m is the mass of the diaphragm, s the stiffness of the oscillator, and R_m the mechanical resistance of the system. The equation of motion is $\mathbf{F} = \mathbf{Z}_{mo}\mathbf{u}$. The linear equation expressing $\mathbf{F}$ in terms of $\mathbf{I}$ and $\mathbf{u}$ that satisfies these two special cases is

$$\mathbf{F} = -Bl\mathbf{I} + \mathbf{Z}_{mo}\mathbf{u} \tag{14.3.11}$$

This is one of the canonical equations. To obtain the other, note that by *Lenz's law* the motion of the coil in the magnetic field induces a voltage $Bl\mathbf{u}$ opposing the applied voltage. This leads to

$$\mathbf{V} = \mathbf{Z}_{EB}\mathbf{I} + Bl\mathbf{u} \tag{14.3.12}$$

These are the canonical equations for an *antireciprocal* transducer with transformation factor

$$\phi_M = Bl \tag{14.3.13}$$

that is real and constant. An equivalent circuit for this transducer is shown in Fig.14.3.2, where

$$R_M = \phi_M^2/R_m$$
$$L_M = \phi_M^2/s \tag{14.3.14}$$
$$C_M = m/\phi_M^2$$

For $R_0 \ll \omega L_0$ use of (14.2.9) yields

$$\mathbf{Z}_{ms} \to R_m + j[\omega m - (s + \phi_M^2/L_0)/\omega] \tag{14.3.15}$$

which shows the magnetically induced change in the low-frequency stiffness when the electrical terminals are shorted and a current can flow. If the frequencies for which (14.3.15) is true lie below those of mechanical resonance, and if losses are negligible, then the combination of (14.3.10) and (14.3.15) results in

$$\mathbf{Z}_{mo}/\mathbf{Z}_{ms} \approx 1 - (\phi_M^2/L_0)/(s + \phi_M^2/L_0) = 1 - L_M/(L_M + L_0) \tag{14.3.16}$$

and the coupling factor $\mathbf{k}_m$ is real and given by

$$k_m^2 = L_M/(L_M + L_0) \tag{14.3.17}$$

This is the ratio of stored mechanical energy to total stored energy for low frequencies and negligible losses, $R_0/L_0 \ll \omega \ll \sqrt{s/m}$.

14.4 TRANSMITTERS

As seen in Section 7.5, when a transducer acts as a source the force $\mathbf{F}$ on the diaphragm and the particle speed $\mathbf{u}$ of the adjacent medium are related by

$$\mathbf{F} = -\mathbf{Z}_r\mathbf{u} \tag{14.4.1}$$

where $\mathbf{Z}_r = R_r + jX_r$ is the *radiation impedance*.

   A couple of *sensitivities* and *sensitivity* levels were defined for sources and receivers in Section 5.12. These and others will now be related to the electroacoustic properties of the transducers. The sensitivity of a transmitter can be expressed as the ratio of (1) the acoustic pressure amplitude extrapolated back from the far field to 1 m from the source to (2) the amplitude of either the driving voltage V or driving current I,

$$\mathcal{S}_V = P_{ax}(1)/V$$
$$\mathcal{S}_I = P_{ax}(1)/I \tag{14.4.2}$$

(Remember that all such amplitudes are effective.) It is left as an exercise to show that

$$\mathcal{S}_V = \frac{P_{ax}(1)}{|\mathbf{u}|}\frac{T_{me}}{Z_{EB}|\mathbf{Z}_{ms} + \mathbf{Z}_r|}$$
$$\mathcal{S}_I = \frac{P_{ax}(1)}{|\mathbf{u}|}\frac{T_{me}}{|\mathbf{Z}_{mo} + \mathbf{Z}_r|} \tag{14.4.3}$$

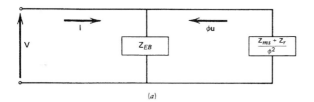

(a)

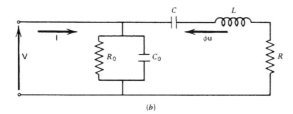

(b)

Figure 14.4.1 Equivalent circuits for a reciprocal transmitter.

The transmitting sensitivity level $\mathscr{S}\mathscr{L}$ is given in either case by

$$\mathscr{S}\mathscr{L} = 20\log(\mathscr{S}/\mathscr{S}_{ref}) \qquad (14.4.4)$$

where the reference sensitivity $\mathscr{S}_{ref}$ is usually 1 Pa/V for $\mathscr{S}_V$ and 1 Pa/A for $\mathscr{S}_I$. (Other conventions use 1 μbar or 1 μPa instead of 1 Pa.)

(a) Reciprocal Source

Use of $\mathbf{F} = -\mathbf{Z}_r\mathbf{u}$ in Fig. 14.2.2c yields Fig. 14.4.1a. The motion of the diaphragm arises from an apparent force of magnitude $-\phi\mathbf{V}$ acting on a mechanical impedance $\mathbf{Z}_{ms} + \mathbf{Z}_r$,

$$\mathbf{u} = -\phi\mathbf{V}/(\mathbf{Z}_{ms} + \mathbf{Z}_r) \qquad (14.4.5)$$

Over most of the frequency range of interest $\mathbf{Z}_{ms}$ is given by (14.3.5). The electrical leakage between the plates, neglected earlier, introduces a large resistance R_0 in parallel with C_0. The equivalent circuit is given by Fig. 14.4.1b with

$$C = \phi^2/s$$
$$L = (m + X_r/\omega)/\phi^2 \qquad (14.4.6)$$
$$R = (R_m + R_r)/\phi^2$$

Note that R_r and X_r cannot be further specified until the geometry of the source and the wavelength are known. However, at high frequencies $X_r \to 0$ and $R_r \to \rho_0 cS$, with S the area of the diaphragm and $\rho_0 c$ the characteristic impedance of the medium. At low frequencies $R_r \to 0$ and $X_r \to \omega m_r$, with m_r the radiation mass (Section 7.5).

The *input electrical admittance* of the circuit in Fig. 14.4.1 is

$$\mathbf{Y}_E = \mathbf{Y}_{EB} + \mathbf{Y}_{MOT}$$
$$\mathbf{Y}_{EB} = 1/R_0 + j\omega C_0 \qquad (14.4.7)$$
$$1/\mathbf{Y}_{MOT} = (R_m + R_r)/\phi^2 + j(\omega m - s/\omega + X_r)/\phi^2$$

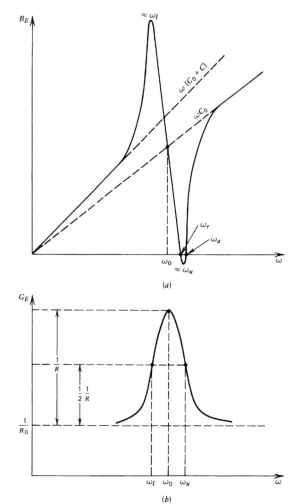

Figure 14.4.2   Reciprocal transmitter. (*a*) Input electrical susceptance. (*b*) Input electrical conductance.

where $\mathbf{Y}_{EB}$ is the *blocked electrical admittance* and $\mathbf{Y}_{MOT}$ the *motional admittance*. Solving for the real and imaginary parts of $\mathbf{Y}_E$ yields the *input electrical conductance* G_E and the *input electrical susceptance* B_E

$$\mathbf{Y}_E = G_E + jB_E$$
$$G_E = 1/R_0 + R/(R^2 + X^2)$$
$$B_E = \omega C_0 - X/(R^2 + X^2) \qquad (14.4.8)$$
$$R = (R_m + R_r)/\phi^2$$
$$X = \omega L - 1/\omega C = (\omega m + X_r - s/\omega)/\phi^2$$

Many properties of a reciprocal transmitter can be determined by measuring the input conductance and susceptance, shown as functions of frequency in Fig. 14.4.2, and then plotting them against each other as in Fig. 14.4.3*a*. At low frequencies this curve begins near $(1/R_0, 0)$ and as frequency increases develops a clockwise loop and then goes to $(1/R_0, \infty)$ as $\omega \to \infty$. If the mechanical resonance is reasonably sharp, this loop is very close to being a circle. If $G_{EB} = 1/R_0$ and $B_{EB} = j\omega C_0$ are

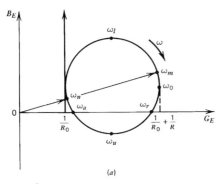

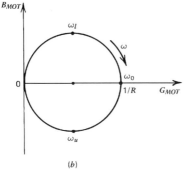

Figure 14.4.3 Reciprocal transmitter. (a) Input electrical admittance. (b) Motional admittance.

subtracted from G_E and B_E, respectively, the results are the *motional conductance* G_{MOT} and the *motional susceptance* B_{MOT}, where $\mathbf{Y}_{MOT} = G_{MOT} + jB_{MOT}$. For R_r and $m_r = X_r/\omega$ constant across the resonance, the plot of G_{MOT} against B_{MOT} results in a circle, as shown in Fig. 14.4.3b. The angular frequencies corresponding to the various performance points are

ω_0 = mechanical resonance, $X = 0$
ω_u = upper half-power for the motional branch, $B_{MOT} = -G_{MOT}$
ω_l = lower half-power for the motional branch, $B_{MOT} = G_{MOT}$
ω_m = point of maximum Y_E
ω_n = point of minimum Y_E
ω_r = electrical resonance, $B_E = 0$ and G_E large
ω_a = electrical antiresonance, $B_E = 0$ and G_E small

The last two, ω_r and ω_a, will not exist if the loop does not intercept the G_E axis. Application of the definitions for ω_l, ω_0, and ω_u to the motional branch results in formulas equivalent to those for the forced, damped harmonic oscillator,

$$\omega_0 = [s/(m + m_r)]^{1/2}$$

$$\omega_u\omega_l = \omega_0^2$$

$$\omega_u - \omega_l = (R_m + R_r)/(m + m_r)$$

$$Q_M = \omega_0/(\omega_u - \omega_l) = \omega_0 L/R = \omega_0(m + m_r)/(R_m + R_r)$$

(14.4.9)

with Q_M the *mechanical quality factor*. From Fig. 14.4.3a, the magnitude of the vector drawn from the origin to a point on the curve is Y_E. Its behavior as a function of frequency is shown in Fig. 14.4.4.

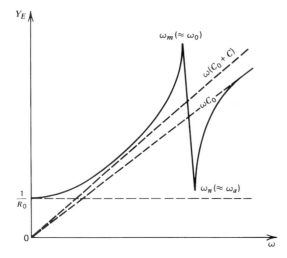

Figure 14.4.4   Input electrical admittance for a reciprocal transmitter.

The *electroacoustic efficiency* η is defined as the ratio of the acoustic power radiated to the total power consumed. It can be written as a product of ratios,

$$\eta = \frac{R_0}{R_0 + (R^2 + X^2)/R} \frac{R_r/\phi^2}{R} = \eta_{EM}\eta_{MA} \tag{14.4.10}$$

The first ratio measures the conversion of electrical energy into mechanical energy and is called the *electromechanical efficiency* η_{EM}. The second ratio $R_r/R\phi^2$ measures the conversion of mechanical energy into acoustic energy and is the *mechanoacoustic efficiency* η_{MA}. At mechanical resonance, we have $\omega = \omega_0$, the electromechanical efficiency simplifies to $R_0/(R_0 + R)$, and the efficiency η is maximized,

$$\eta_0 = \frac{R_0}{R_0 + R} \frac{R_r/\phi^2}{R} \tag{14.4.11}$$

It is possible to obtain estimates of the coupling coefficient from the properties of the transducer near its mechanical resonance rather than from its behavior for low frequencies. First, note from Fig. 14.4.2 (or Fig. 14.4.4) that the values of C_0 and $C_0 + C$ can be estimated from the slopes of the two dashed lines. Since in certain transducers, including those made with piezoelectric or ferroelectric materials, the value of C depends somewhat on frequency, this estimate of the electromechanical coupling coefficient may differ from that obtained at low frequencies. Consequently, this evaluation is called the *effective coupling coefficient*,

$$k_c^2(eff) = C(\omega_0)/[C_0 + C(\omega_0)] \tag{14.4.12}$$

In addition, if the transducer is of reasonably high quality and is operated in an unloaded condition so that R_r is sufficiently small, then it may be possible to treat R as very small. It can be shown analytically, and seen intuitively, that as R gets smaller, the diameter of the admittance circle gets larger, and the frequency interval for which the paired values of B_E and G_E lie on the circle gets smaller. This means that as $R \to 0$ we have $\omega_0 \to \omega_m$ and $\omega_a \to \omega_n$ and the antiresonance frequency can be estimated from

$$\omega_a C_0 \approx \frac{1}{\omega_a L - 1/\omega_a C_r} \tag{14.4.13}$$

Combination with $\omega_0 = \sqrt{1/LC}$ yields

$$k_c^2(eff) \approx 1 - (\omega_0/\omega_a)^2 \approx 1 - (\omega_m/\omega_n)^2 \tag{14.4.14}$$

Thus $k_c^2(eff)$ can be approximated in terms of the ratios of angular frequencies from Figs. 14.4.2 to 14.4.4.

(b) Antireciprocal Source

Substituting $\mathbf{F} = -\mathbf{Z}_r \mathbf{u}$ in Fig. 14.2.3b yields the circuit of Fig. 14.4.5a, and the motion of the diaphragm arises from an apparent force $\boldsymbol{\phi}_M \mathbf{I}$ acting on the mechanical impedance $\mathbf{Z}_m = \mathbf{Z}_{mo} + \mathbf{Z}_r$:

$$\mathbf{u} = \boldsymbol{\phi}_M \mathbf{I}/\mathbf{Z}_m = \boldsymbol{\phi}_M \mathbf{I}/(\mathbf{Z}_{mo} + \mathbf{Z}_r) \tag{14.4.15}$$

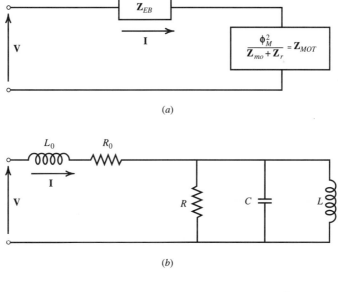

(a)

(b)

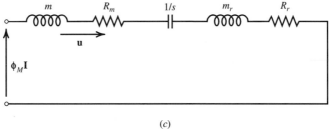

(c)

Figure 14.4.5 Three equivalent circuits for an antireciprocal transmitter.

with $\mathbf{Z}_{mo}$ given by (14.3.10). The circuit representing (14.4.15) is shown in Fig. 14.4.5c. The input electrical impedance is, from Fig. 14.4.5a,

$$\mathbf{Z}_E = \mathbf{Z}_{EB} + \mathbf{Z}_{MOT}$$
$$\mathbf{Z}_{MOT} = \boldsymbol{\phi}_M^2/(\mathbf{Z}_{mo} + \mathbf{Z}_r)$$

(14.4.16)

(While $\mathbf{Y}_E = 1/\mathbf{Z}_E$ and $\mathbf{Y}_{EB} = 1/\mathbf{Z}_{EB}$, note that $\mathbf{Y}_{MOT}$ is *not* the reciprocal of $\mathbf{Z}_{MOT}$!)

First, for simplicity assume that the transformation factor $\boldsymbol{\phi}_M$ is real and constant, $\boldsymbol{\phi}_M = \phi_M$. For most antireciprocal transmitters $\mathbf{Z}_{EB}$ is given by (14.3.9) and $\mathbf{Z}_{mo}$ by (14.3.10). If we use the definitions of (14.3.14) and further define

$$R_R = \phi_M^2/R_r$$
$$C_R = (X_r/\omega)/\phi_M^2$$

(14.4.17)

then Fig. 14.4.5a can be represented by Fig. 14.4.5b with

$$R = \frac{\phi_M^2}{R_m + R_r} = \frac{1}{1/R_M + 1/R_R}$$
$$L = L_M = \phi_M^2/s$$
$$C = (m + X_r/\omega)/\phi_M^2 = C_M + C_R$$

(14.4.18)

Thus, R is the parallel combination of R_M and R_R, and C the parallel combination of C_M and C_R. The motional impedance $\mathbf{Z}_{MOT} = R_{MOT} + jX_{MOT}$ has the form

$$R_{MOT} = \frac{1/R}{1/R^2 + (\omega C - 1/\omega L)^2}$$
$$X_{MOT} = -\frac{\omega C - 1/\omega L}{1/R^2 + (\omega C - 1/\omega L)^2}$$

(14.4.19)

A plot of X_{MOT} against R_{MOT} is a circle of diameter R (Fig. 14.4.6a) if X_r/ω and R_r are constant across the resonance. Adding R_{EB} to R_{MOT} displaces the circle to the right by R_0 and adding X_{EB} to X_{MOT} displaces each point of the circle upward by ωL_0. The resultant curve, Fig. 14.4.6b, has the same shape as that in Fig. 14.4.3a (with the exchange of admittance and impedance). From these figures, we can define angular frequencies in complete analogy with those for the reciprocal transducer:

ω_0 = mechanical resonance, $X_{MOT} = 0$
ω_u = upper half-power for the motional branch, $X_{MOT} = R_{MOT}$
ω_l = lower half-power for the motional branch, $X_{MOT} = -R_{MOT}$
ω_m = point of minimum Z_E
ω_n = point of maximum Z_E
ω_r = electrical resonance, $X_E = 0$ and R_E small
ω_a = electrical antiresonance, $X_E = 0$ and R_E large

Mechanical resonance is given by

$$\omega_0 = (1/LC)^{1/2} = [s/(m + m_r)]^{1/2}$$

(14.4.20)

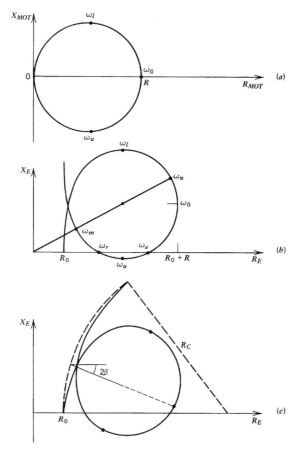

Figure 14.4.6 Impedance plots for an antireciprocal transmitter.

and the mechanical quality factor is

$$Q_M = \omega_0 RC = \omega_0(m + m_r)/(R_m + R_r) \qquad (14.4.21)$$

The acoustic power radiated by the transmitter is $R_r u^2$ and the total power consumed is $R_0 I^2 + (R_m + R_r)u^2$ so that the electroacoustic efficiency is

$$\eta = \frac{(R_m + R_r)u^2}{R_0 I^2 + (R_m + R_r)u^2}\frac{R_r}{R_m + R_r} = \eta_{EM}\eta_{MA}$$

$$= \frac{R_r}{R_0\phi_M^2/Z_{MOT}^2 + (R_m + R_r)} \qquad (14.4.22)$$

with the use of (14.4.15) and (14.4.16). At mechanical resonance this simplifies to

$$\eta_0 = \frac{R}{R_0 + R}\frac{R}{R_R} \qquad (14.4.23)$$

Reasoning similar to that resulting in (14.4.14) yields an effective coupling coefficient that is well estimated by

$$k_m^2(eff) \approx 1 - (\omega_0/\omega_r)^2 \approx 1 - (\omega_n/\omega_m)^2 \qquad (14.4.24)$$

for large R. Comparison of (14.4.24) with (14.4.14) (see also Figs. 14.4.6 and 14.4.3) shows that the roles of ω_m and ω_n and also ω_a and ω_r have been exchanged—a consequence of the differences between reciprocal and antireciprocal behavior.

Complications arise for some real antireciprocal devices. Resistance and capacitance in the windings and hysteresis losses in the magnetic material can result in a complex ϕ_M that yields a skewed impedance plot similar to Fig. 14.4.6c. The phase angle β of ϕ_M is called the *dip angle*. Moving armature and magnetostrictive transducers are two examples for which β is nonzero. The radius of curvature R_C is related to the losses within the magnetic material and determines the asymptotic behavior of the X_E versus R_E curve for frequencies removed from mechanical resonance.

The above developments show that the mechanical and acoustic properties of a transducer can be determined solely from electrical measurements made at the input terminals of the transducer. We illustrate this with consideration of a loudspeaker. Most larger loudspeakers have their resonances at low frequencies and can be blocked mechanically by placing the hand firmly on the cone near the voice coil. The values of R_0 and L_0 are easily determined from measurements made either with the speaker blocked or from the input electrical impedance at low and high frequencies for which Z_{MOT} becomes negligible. Determining the motional properties of the speaker can be accomplished in different ways. (1) The motional impedance circles can be determined for the speaker mounted on a large baffle and also unmounted (to remove the radiation loading). From these two circles, two sets of R, ω_0, and Q_M can be determined (see Fig. 14.4.6a). One set is given by (14.4.18) for R, (14.4.20) for ω_0, and (14.4.21) for Q_M and the other set by the same equations with $R_r = m_r = 0$. Then, if R_r or m_r can be *calculated* for the loaded transducer, the combination of these two sets of values yields R_m, m, s, and ϕ_M. (The same procedure can be used for an underwater transducer with the unloaded condition achieved by removing the transducer from the water.) (2) Another approach consists of measuring the impedance circles of the unloaded speaker, first without modification, and then with a known mass M affixed to the cone near the voice coil. (It is left as an exercise to show that these measurements allow determination of the parameters of the loudspeaker.) This latter technique has the advantage of not requiring the assumptions necessary to calculate the radiation impedance, but does assume that the additional mass M affects only the amplitude of the diaphragm vibration and not its mode of vibration.

The above procedures apply equally well to reciprocal transducers if admittance circles rather than impedance circles are measured.

14.5 MOVING-COIL LOUDSPEAKER

Figure 14.5.1 is a simplified drawing of a moving-coil loudspeaker. Note that the vibrating diaphragm (cone) is usually appreciably larger than the voice coil, to enhance the efficiency of radiation at the lower frequencies. Since a loudspeaker will become directive at high frequencies, a number of speakers with different diaphragm areas are often used in high-fidelity applications. A wide-range, high-fidelity system might consist of large, relatively massive speakers radiating the lower frequencies (subwoofers and woofers), smaller speakers for the mid-range frequencies (squawkers), and still smaller ones to radiate the highest frequencies (tweeters and supertweeters). These various speakers can be driven

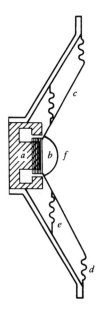

Figure 14.5.1 Schematic of the construction of a simple loudspeaker: *a*, magnet; *b*, voice coil; *c*, diaphragm; *d*, corrugated rim; *e*, spider supplying stiffness to system; and *f*, dome.

through electrical filtering networks that deliver to each its appropriate range of frequencies, or can be band-limited naturally through the characteristics of their own electrical, mechanical, and radiation impedances. The squawkers, tweeters, and supertweeters may be designed more along the lines of Fig. 14.3.2 with a small, or no, diaphragm. If the dome is sufficiently small there is good high-frequency dispersion.

As a simplified illustration of the analysis of a loudspeaker, assume a piston-like speaker is mounted in a sealed absorbent enclosure and radiates on one side as if it were on an infinite baffle but does not radiate from the enclosed side. (More realistic speaker enclosures will be discussed in a later section. For the moment, we will treat the mechanical properties of the air within the enclosure as part of the mechanical impedance of the speaker.) Let the speaker have the following physical and radiation properties:

Mass of the speaker diaphragm and voice coil, $m = 10$ g
Radius of the diaphragm, $a = 0.1$ m
Stiffness of the speaker, $s = 2000$ N/m
Mechanical resistance of the speaker, $R_m = 1$ N · s/m
Inductance of the voice coil, $L_0 = 0.2$ mH
Resistance of the voice coil, $R_0 = 5$ Ω
Length of wire in the voice coil, $l = 5$ m
Magnetic field, $B = 0.9$ T
Radiation resistance, $R_r = 13\, R_1(0.00366f)$ N · s/m
Radiation reactance, $X_r = 13\, X_1(0.00366f)$ N · s/m

The complications discussed after (14.4.24) can be neglected so that the transformation factor can be assumed real with value $\phi_M = 4.5$ T · m. For lower frequencies ($2ka < 1$), the radiation terms can be approximated by (7.5.13) and (7.5.14) within about 10% so that below 275 Hz, $R_r \approx 2.2 \times 10^{-5} f^2$ and $X_r \approx 0.02f$. For higher frequencies ($2ka > 4$), $R_1(2ka) \approx 1.0$ within 10% so that above 1100 Hz, $R_r \approx 13$ N · s/m. The frequency f_0 of mechanical resonance, determined from $(X_r + \omega_0 m - s/\omega_0) = 0$, is 62 Hz.

As sketched in Fig.14.5.2a, the resistance $R_r + R_m$ rises slowly from 1 N $\cdot$ s/m, increases rapidly between 100 and 1000 Hz as R_r begins to dominate, and then fluctuates around 14 N $\cdot$ s/m at higher frequencies. The magnitude of the reactance has large values at very low frequencies (stiffness controlled), reducing to zero at the resonance frequency, and then rising with higher frequencies (mass controlled). The magnitude Z_m of the mechanical impedance has a minimum value of 1.085 N $\cdot$ s/m at the resonance frequency and otherwise is nearly identical to the magnitude of the reactance.

The electroacoustic efficiency η is calculated from (14.4.22) and plotted in Fig. 14.5.2b. As frequency increases η rises rapidly to a maximum of about 6% at mechanical resonance. Below this frequency, R_r is proportional to f^2 and Z_m to $1/f$ so that η is proportional to f^4. In the interval between 200 and 700 Hz both R_r and Z_m increase and η falls gradually from about 2% to 1%. At higher frequencies the radiation resistance becomes nearly constant and Z_m increases roughly as f so that η falls roughly as $1/f^2$.

The resistive and reactive components of $\mathbf{Z}_{MOT}$ are calculated from (14.4.19) and plotted in Fig. 14.5.2c. The motional resistance R_{MOT} rises to a maximum value of $19\ \Omega$ at mechanical resonance. Away from resonance, the motional impedance is dominated by the reactance, which is positive below resonance and negative above.

Calculation of the input electrical resistance and reactance from (14.4.6) gives the curves plotted in Fig. 14.5.2d. The resistive component is identical to that of Fig. 14.5.2c except that all values have been increased by the $5\ \Omega$ resistance of the voice coil. At low frequencies the input resistance is primarily that of the voice coil. As frequency increases input resistance and reactance show the effect of the mechanical resonance. At a higher frequency the positive reactance of the voice coil cancels the negative motional reactance so that the input reactance vanishes. This defines the frequency f_r of *electrical resonance*, in this case 450 Hz. At frequencies below about 40 Hz and within about 200 to 2000 Hz the input impedance is primarily the resistance of the voice coil. (Because of this constancy, many loudspeaker manufacturers do not specify the frequency at which the speaker voice coil has its rated input impedance. In general, it is measured at a frequency somewhat above the resonance frequency.) Finally, at frequencies in excess of 4000 Hz the inductance L_0 of the voice coil dominates the input electrical impedance.

The input impedance circle is plotted in Fig. 14.5.2e. From this curve and the data supporting it the various frequencies identified in Fig. 14.4.6b can be measured.

Sketched in Fig. 14.5.2f are three curves showing computed values of the acoustic output of the speaker. Curve A was obtained from $\Pi = R_r u^2$ and (14.4.15) for an input current of 2 A. Curve B was obtained from the efficiency (14.4.22) for an input power of 20 W. Curve C was obtained from Π and (14.4.15) with $\mathbf{V} = \mathbf{IZ}_E$ for an input voltage of 10 V. These values of current, power, and voltage were chosen to yield nearly equal acoustic outputs at about 1 kHz. As the frequency of mechanical resonance at 62 Hz is approached, the three outputs differ widely. The output at resonance for an input of 2 A is seen to be some 20 times greater than that for an input of 10 V. The smoothest acoustic output is produced by the constant-voltage amplifier. Since most high-fidelity amplifiers employ large negative voltage feedback to reduce harmonic distortion, they have very low internal impedances and therefore maintain almost constant voltage output, the best of the three cases considered. For frequencies above about 700 Hz the acoustic output shown by curve C tends to fall off as $1/f^4$. This partially compensates for the

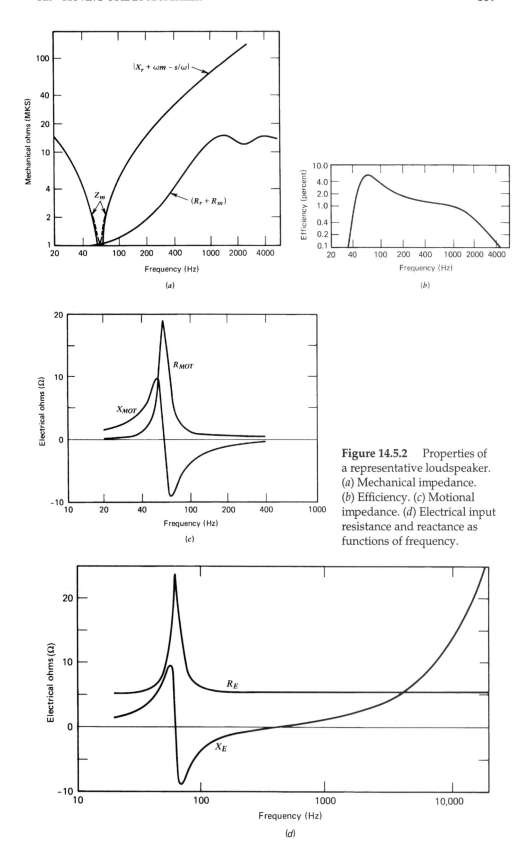

Figure 14.5.2 Properties of a representative loudspeaker. (*a*) Mechanical impedance. (*b*) Efficiency. (*c*) Motional impedance. (*d*) Electrical input resistance and reactance as functions of frequency.

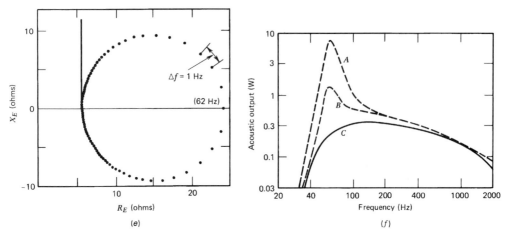

Figure 14.5.2 (*continued*)   (*e*) Electrical input impedance. (*f*) Computed output of the loudspeaker for constant current input (curve *A*), constant power input (curve *B*), and constant voltage input (curve *C*).

high-frequency directivity, which approaches $(ka)^2$, so that the source level on the acoustic axis of the speaker falls off only as $20 \log f$ and the high-frequency output on the axis is gradually attenuated. However, the actual behavior is somewhat more complicated:

1. At high frequencies the cone does not vibrate as a unit. The amplitude of vibration at the edge of the cone is relatively small, so that the radiation comes mostly from a central portion whose effective radius a_{eff} and effective mass m_{eff} gradually decrease with increasing frequency. This decrease in effective radius causes the radiation resistance R_r to decrease approximately as a_{eff}^2. Since the system is mass controlled at the higher frequencies, m_{eff} also decreases with a_{eff}, and η does not decrease as rapidly as for a rigid piston. The net result of these two effects is to produce a significant improvement in the output at frequencies above 700 Hz. Wide-range loudspeakers have their cones corrugated to take advantage of and enhance this effect.

2. At middle frequencies the time required for a displacement of the center of the cone to propagate to the rim is small compared to the period of vibration, so that the cone may be assumed to vibrate as a rigid surface. The speed of transverse waves on the cone is a function of its composition, thickness, stiffness, cone angle, and frequency. For cone materials commonly used in commercial loudspeakers, this speed is about 500 m/s. Consequently, it is reasonable to assume that the cone moves as a unit for frequencies below 500 Hz.

3. Maintaining a uniform acoustic output at low frequencies is more difficult. One method of improving the low-frequency response is to increase the radius of the speaker. The radiation resistance increases as the fourth power of the radius and increases the efficiency, but not so great as might be expected because the mass of the speaker also increases with the radius. The low-frequency response can also be enhanced by reducing the stiffness of the suspension system, thereby lowering the frequency of mechanical resonance. However, if the stiffness becomes too small the cone displacement becomes very large, and

the voice coil may move into nonuniform regions of the magnetic field. This can cause harmonic distortion of the acoustic output.

Except in the immediate vicinity of mechanical resonance, (14.4.16) and (14.4.22) yield

$$\eta \approx \frac{\phi_M^2}{|Z_{mo} + Z_r|^2} \frac{R_r}{R_0} \tag{14.5.1}$$

This simplified equation is useful in discussing the influence of different design parameters on performance. Since ϕ_M is directly proportional to the flux density in the air gap, increasing B will increase the efficiency of the speaker. The two most feasible methods of accomplishing this are to (1) use a more powerful field magnet and (2) decrease the width of the air gap to as small a value as practicable. An increase in the length of conductor forming the voice coil winding would be expected to improve the efficiency since $\eta \propto \phi_M^2/R_0 \propto l$. However, for any wire size there exists an optimum length beyond which the increase in mass and the decrease in B (due to the larger air gap required to accommodate the larger coil) are more significant than the gain resulting from the increase in l. If the mass of the conductor to be used in the voice coil is specified, no change in efficiency results from a change in wire size. The primary consideration in the choice of wire size is therefore its current carrying capacity per unit mass, and aluminum is superior to copper. If the volume occupied by the winding is the limiting factor, then copper is a better choice than aluminum.

*14.6 LOUDSPEAKER CABINETS

An important component of loudspeaker performance is the cabinet in which it is mounted. Three of the most commonly encountered enclosures are described in this section.

(a) The Enclosed Cabinet

A loudspeaker mounted in an enclosed cabinet can radiate acoustic energy only from the front of the speaker cone. The primary mechanical effect of such a cabinet is to contribute additional stiffness to that of the suspension system of the loudspeaker. The stiffness s_c of the cabinet at lower frequencies is that of a Helmholtz resonator

$$s_c = (\pi a^2)^2 \rho_0 c^2 / V \tag{14.6.1}$$

where a is the radius of the speaker and V the volume of the cabinet. This stiffness, added to the speaker stiffness, raises the mechanical resonance frequency of the system so that the low-frequency roll-off begins at a higher frequency than if the speaker were on an infinite baffle.

If the loudspeaker of Section 14.5 is mounted in a 0.05 m³ enclosed cabinet, the additional stiffness s_c is 2850 N/m and the frequency of mechanical resonance is raised from 62 to 96 Hz. Figure 14.6.1 shows that for frequencies below 80 Hz the output of the speaker mounted in this cabinet is much lower than for the same speaker mounted in an infinite baffle.

One way of improving the low-frequency response of an enclosed cabinet is to make the cabinet larger. Another (more practical) method is to reduce the stiffness and increase the mass of the cone. For example, the *acoustic suspension* speaker obtains its stiffness almost completely from the air within the sealed cabinet. With a total stiffness of 2850 N/m and

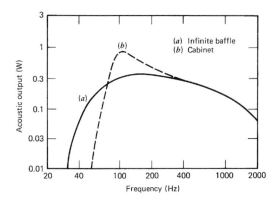

Figure 14.6.1 Acoustic output for constant input voltage of a loudspeaker (*a*) in an infinite baffle and (*b*) in a back-enclosed cabinet of volume $V = 5 \times 10^4 \text{ cm}^3$.

the mass of cone and voice coil increased to 45 g, the frequency of mechanical resonance of our example can be reduced to about 40 Hz. The greater mass, however, will strongly reduce the frequency response at the higher frequencies, so such modifications are most useful for woofers and subwoofers. If, in addition, the cabinet volume is filled with a sound-absorbing material, standing waves within the enclosure can be suppressed, but the apparent mechanical resistance of the speaker will be increased and its efficiency decreased by about 3 dB. Further, the high heat capacity of the material causes the compression and rarefaction of the air in the cabinet to become nearly isothermal. Under these circumstances, the stiffness from the cabinet is further reduced by $1/\gamma \sim 0.7$, providing about another 20% reduction in the resonance frequency.

(b) The Open Cabinet

Another common cabinet is the open-back enclosure typical of most radio and television sets. The equivalent circuit of Fig. 14.4.5c is modified as shown in Fig. 14.6.2. The mechanical impedance $\mathbf{Z}_{mc}$ acting on the back of the cone can be expressed as

$$\mathbf{Z}_{mc} = (\pi a^2)^2 \mathbf{Z}_{Ac} \tag{14.6.2}$$

where $\mathbf{Z}_{Ac}$ is the input acoustic impedance of the cabinet seen by the back of the speaker cone. It is similar to that of the open-ended pipe discussed in Section 10.2. At the fundamental resonance frequency of the cabinet, the motion of the speaker cone is enhanced. This can occur between 100 and 200 Hz, resulting in an intrinsic "boomy" quality. Below this cabinet resonance the system radiates more like a dipole and lower-frequency response will be further attenuated by around 6 dB per octave.

(c) Bass-Reflex Cabinet

One factor that limits the low-frequency output of all direct radiation loudspeakers is the inefficient coupling between the cone and the air. The use of a large cone increases the radiation resistance, thereby increasing this coupling, but at the expense of increased directivity. An alternative is to mount the speaker in a cabinet that allows radiation from the back of the cone to be added in phase with that from the front, thereby effectively increasing

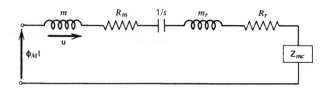

Figure 14.6.2 Equivalent circuit for a loudspeaker mounted in an open-backed cabinet.

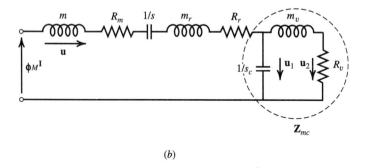

Figure 14.6.3   (*a*) Schematic of the construction of a bass-reflex cabinet. (*b*) Equivalent circuit.

(*a*)

(*b*)

the total radiation resistance. One cabinet of this type is the bass-reflex (or ducted-port) cabinet of Fig. 14.6.3. The equivalent circuit for this system is identical with that of Fig. 14.6.2 except for the mechanical impedance $\mathbf{Z}_{mc}$ presented to the speaker by the cabinet. For the lower frequencies this loading consists of the compliance $1/s_c$ of the cavity in parallel with the series combination of the inertance m_v and resistance R_v of the vent. (The inertance and resistance are in series since both experience the same particle speed u_v of the air in the vent. They are in parallel with the compliance since sealing the vent is equivalent to m_v becoming arbitrarily large; u_v must vanish and the system must reduce to the enclosed cabinet with stiffness s_c.) The acoustic impedances of these elements are found from the expressions for a flanged Helmholtz resonator whose neck is the vent, and the mechanical impedances experienced by the speaker are $(\pi a^2)^2$ times these quantities. (The assumption of a flange is plausible because of the surrounding cabinet face at higher frequencies and nearby walls at lower frequencies.) The mechanical impedance at the speaker cone is therefore given by

$$1/\mathbf{Z}_{mc} = j\omega/s_c + 1/(R_v + j\omega m_v)$$
$$s_c = (\pi a^2)^2 \rho_0 c^2 / V$$
$$R_v = (\pi a^2)^2 \rho_0 c k^2 / 2\pi$$
$$m_v = (\pi a^2)^2 \rho_0 (l_v + 1.7 a_v) / \pi a_v^2$$

(14.6.3)

where a_v is the radius of the circular vent. A rectangular vent of area S_v has an effective radius $a_v \sim \sqrt{S_v/\pi}$.

Above the fundamental resonance frequency of the cabinet, found from $X_{mc} = 0$, the impedance of the branch carrying $\mathbf{u}_2$ is larger than that of the branch carrying $\mathbf{u}_1$, and $\mathbf{u}$ and $\mathbf{u}_2$ are nearly 180° out of phase. Since $\mathbf{u}$ describes the motion of the *back* of the cone and $\mathbf{u}_2$ the motion of air out of the vent, radiation from the vent reinforces that from the front of the cone. At frequencies below this resonance the relative phase between $\mathbf{u}$ and $\mathbf{u}_2$ rapidly approaches zero and the system radiates like a dipole.

There are many considerations in designing a bass-reflex cabinet. However, if the area of the vent is approximately that of the speaker cone, and if the resonance frequency of the cabinet is somewhat lower than that of the speaker, the low-frequency response will be better than for the same speaker mounted on an infinite baffle. Compared to an open-back cabinet, a bass-reflex cabinet can extend and smooth the base response. The parameters of the cabinet can be adjusted further by filling the mouth of the vent with a *passive radiator* (a piston with properly chosen mass and stiffness). As mentioned for the enclosed cabinet, high-frequency response is smoothed if the cabinet is filled with absorptive material to reduce the effects of standing waves.

In the design of any speaker cabinet, the mechanical rigidity of its walls is important. An open-back wooden cabinet should have walls at least 1.3 cm thick. Requirements for enclosed and bass-reflex cabinets are, however, considerably more severe, since the acoustic pressures in these enclosures are relatively high. The resulting wall resonances usually occur at the lower audible frequencies and produce undesirable irregularities in the frequency response. The walls should therefore be thicker than those of an open-back cabinet and, if necessary, braced with stiffeners. Hollow walls filled with sand can also be used.

*14.7 HORN LOUDSPEAKERS

Attaching an appropriate horn to a piston-like source results in a marked increase in acoustic output at low frequencies. The horn acts as an acoustic transformer matching the impedance of the piston to that of the air. The low-frequency acoustic resistance at the throat of the horn is greater than that acting on a piston of equal area vibrating in an infinite baffle, and the acoustic output is consequently greater. At high frequencies the effect of the horn is almost negligible, for these frequencies are radiated in narrow beams and the confining effect of the walls of the horn is not significant.

The simplified analysis given here, while valid over a limited range of frequencies, leads to many conclusions of practical importance. Consider the volume element of length dx and area $S(x)$ within the horn of Fig. 14.7.1. Let ξ be the particle displacement parallel to the horn axis. As long as the fractional change in S is small over distances of a wavelength, the wave can be assumed uniform across the horn. The condensation is then

$$s = -\frac{1}{S}\frac{\partial(S\xi)}{\partial x} \tag{14.7.1}$$

With the help of the adiabat $p = \rho_0 c^2 s$, we have

$$p = -\frac{\rho_0 c^2}{S}\frac{\partial(S\xi)}{\partial x} \tag{14.7.2}$$

The force equation is $\partial p/\partial x = -\rho_0(\partial^2\xi/\partial t^2)$. Rearrangement (and use of the fact that S is independent of time) gives the approximate wave equation

$$\frac{\partial^2 p}{\partial t^2} = c^2\frac{1}{S}\frac{\partial}{\partial x}\left(S\frac{\partial p}{\partial x}\right) \tag{14.7.3}$$

(Note that c is the free field speed of sound and also that when S is constant this equation simplifies to the equation for a one-dimensional plane wave in free space.)

The most effective horns are those whose rate of flare dS/dx increases from throat to mouth. Hyperbolas, catenaries, and exponentials have been used. Because of its simplicity, we will investigate the exponential horn, characterized by a cross-sectional area $S(x) =$

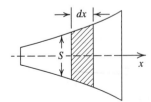

Figure 14.7.1   Volume element $S(x)\,dx$ of a horn loudspeaker.

$S_0 \exp(2\beta x)$, where x is the distance from the throat, S_0 the throat area, and 2β the *flare constant*. Conservation of energy suggests that the amplitude of the pressure wave should decrease with x at least approximately like $\exp(-\beta x)$. Substitution into (14.7.3) and solution of the resulting algebraic equations shows that

$$\mathbf{p} = e^{-\beta x}\left(\mathbf{A}e^{j(\omega t - \kappa x)} + \mathbf{B}e^{j(\omega t + \kappa x)}\right)$$
$$\kappa^2 = (\omega/c)^2 - \beta^2 \tag{14.7.4}$$

satisfies (14.7.3). The two terms in $\mathbf{p}$ represent waves traveling away from and toward the throat with amplitudes diminishing exponentially with distance away from the throat. Each has a phase speed

$$c_p = \omega/\kappa = c[1 - (\beta/k)^2]^{-1/2} \tag{14.7.5}$$

Since the phase speed c_p is a function of frequency, the air in an exponential horn is a dispersive medium. Waves will not propagate in the horn if the driving frequency is lower than the cutoff frequency found from $\kappa = \beta$,

$$f_c = \beta c/2\pi \tag{14.7.6}$$

At the cutoff frequency the phase speed becomes infinite, indicating that all parts of the medium within the horn move in phase.

Calculation of the acoustic impedance is straightforward from $\partial \mathbf{p}/\partial x = -\rho_0(\partial \mathbf{u}/\partial t)$ and gives

$$\mathbf{Z}(x) = \frac{\rho_0 c}{S(x)} \frac{1}{k} \frac{(\kappa + j\beta)\mathbf{A}e^{-j\kappa x} - (\kappa - j\beta)\mathbf{B}e^{j\kappa x}}{\mathbf{A}e^{-j\kappa x} + \mathbf{B}e^{j\kappa x}} \tag{14.7.7}$$

If the length L of the horn were infinite, there would be no reflected wave and $\mathbf{B}$ would be zero. It can be shown that the amplitude of the reflected wave is small compared to that of the incident wave and the horn may be treated as infinitely long whenever the radius a of the mouth is such that $ka > 3$. For instance, if the mouth of the horn opens into an infinite baffle, (14.7.7) predicts that, for $ka > 3$, $\mathbf{B}$ will be less than $\mathbf{A}/10$. If we set $\mathbf{B} = 0$ in (14.7.7), the acoustic impedance at the throat becomes

$$\mathbf{Z}(0) = (\rho_0 c/S_0 k)(\kappa + j\beta) = (\rho_0 c/S_0)\left\{[1 - (\beta/k)^2]^{1/2} + j\beta/k\right\} \tag{14.7.8}$$

To compare the throat impedance of an exponential horn with that acting on the piston when mounted in an infinite baffle, assume that the horn flares from a radius of 0.02 m at its throat to 0.4 m at its mouth and that its length is 1.6 m. The flare constant of this horn is $2\beta = 3.74$, which gives a cutoff frequency of about 100 Hz. The curves of Fig. 14.7.2 show the throat resistance R_0 and reactance X_0 as functions of frequency. Although calculated assuming the horn has infinite length, they are reasonably accurate for frequencies above 400 Hz. Below this frequency reflections from the mouth of the horn set up resonances that cause the horn resistance and reactance to fluctuate about the calculated values. The figure also shows the acoustic resistance and reactance for the 0.02 m piston mounted in an infinite baffle. For frequencies between 100 and 3000 Hz the resistance loading at the throat of the horn is considerably greater than that on the piston mounted in a baffle. Attaching the horn to the above piston speaker greatly enhances acoustic output at the lower frequencies.

Horns are found in high-fidelity applications. Large horns enhance the low-frequency response of woofers and smaller horns improve the efficiency of high-level tweeters. However, the principal applications of horns are in sound reinforcement systems for stadiums, large auditoriums, and public-address systems.

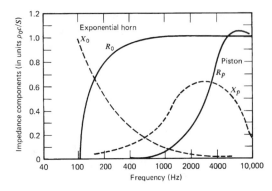

Figure 14.7.2 The acoustic resistance and reactance acting at the throat of an infinite exponential horn (R_0 and X_0) and on a piston mounted in an infinite baffle (R_p and X_p).

For more extensive treatment of the enhancement of the acoustic output of loudspeakers by properly designed cabinets and horns, the reader is referred to the texts by Beranek and Olsen.[2] For more sophisticated modern approaches utilizing filter theory see the *Journal of the Audio Engineering Society*.

14.8 RECEIVERS

The force experienced by the diaphragm of a receiver depends on the pressure field that existed in the absence of the receiver and the radiated pressure field produced by the motion of the diaphragm in response to the excitation. If we define

> $\mathbf{p}$ = the pressure at the field point in the *absence* of the receiver
>
> $\mathbf{p}_B$ = the pressure at the field point in the *presence* of the receiver
>   with *blocked* diaphragm

then the total force on the moving diaphragm is

$$\mathbf{F} = \langle \mathbf{p}_B \rangle S - \mathbf{Z}_r \mathbf{u} \tag{14.8.1}$$

where $\langle \mathbf{p}_B \rangle$ is the spatial average of $\mathbf{p}_B$ over the diaphragm of area S and $-\mathbf{Z}_r \mathbf{u}$ is the radiation force of the fluid on the diaphragm.

(a) Microphone Directivity

The relationship between $\langle \mathbf{p}_B \rangle$ and $\mathbf{p}$ depends on the diffraction properties of the blocked diaphragm and housing. The relation can be involved, but the results can be written using a *diffraction factor* $\mathscr{D}$,

$$\mathscr{D} = \langle \mathbf{p}_B \rangle / \mathbf{p} \tag{14.8.2}$$

The diffraction factor is a function of frequency and the orientation of the microphone with respect to the sound source. It approaches unity when the dimensions

---

[2]Beranek, *Acoustics*, McGraw-Hill (1954); republished, Acoustical Society of America (1986). Olsen, *Elements of Acoustical Engineering*, Van Nostrand (1947).

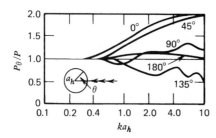

Figure 14.8.1 Sound pressure on the surface of a sphere of radius a_h ensonified by a plane wave incident from the right. The angle θ is measured from the direction of approach of the wave.

of the microphone become much less than a wavelength. Combination yields

$$\mathbf{F} = \mathscr{D}S\mathbf{p} - \mathbf{Z}_r\mathbf{u} \tag{14.8.3}$$

(1) DIFFRACTION   A brief consideration of the diffraction effects of a spherical housing will show roughly what to expect from more complicated housings. Figure 14.8.1 shows how the acoustic pressure at a point on the surface of a sphere of radius a_h varies with frequency for various angles of incidence of the acoustic waves. Here P_θ is the acoustic pressure amplitude at a point on the surface of the sphere whose polar angle to the direction of incidence of the plane wave is θ and P is the pressure amplitude in the undisturbed sound wave. For normal incidence, diffraction effects will increase the high-frequency response of a pressure-sensitive microphone by 6 dB. If a microphone is to be essentially nondiffractive throughout the audible range of frequencies, the microphone and its housing must be small, $ka_h \leq 1$. At 20 kHz, this requires $a_h \leq 0.3$ cm in air.

(2) PHASE INTERFERENCE ACROSS THE DIAPHRAGM   A second source of directivity in a pressure-sensitive microphone arises from the differences in phase among the forces exerted on its diaphragm by different segments of the incident wave when the direction of propagation is not perpendicular to its surface. Sensitivity to phase interference can be avoided by making the diaphragm small enough, $ka \leq \pi/4$, where a is the radius of the diaphragm. For phase interference effects to be negligible at 20 kHz in air, $a \leq 0.2$ cm.

(b) Microphone Sensitivities

The response of a microphone can be expressed in terms of the open-circuit output voltage amplitude V divided either by the pressure amplitude $P = |\mathbf{p}|$ at the point where the microphone is to be placed or by the amplitude of the average force per unit area F/S experienced by the diaphragm. Thus, the voltage sensitivities are

$$\mathcal{M}_o = V/P|_{I=0} = T_{em}\mathscr{D}S/|\mathbf{Z}_{mo} + \mathbf{Z}_r| \tag{14.8.4}$$
$$\mathcal{M}_o^D = V/(F/S)|_{I=0} = T_{em}S/Z_{mo}$$

Notice that at low frequencies for which $\mathbf{Z}_r \to 0$ and $\mathscr{D} \to 1$, we have $\mathcal{M}_o^D \to \mathcal{M}_o$. As mentioned in Section 5.12, it is convenient to express microphone sensitivities in dB,

$$\mathcal{ML} = 20\log(\mathcal{M}/\mathcal{M}_{ref}) \tag{14.8.5}$$

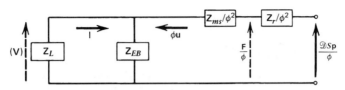

Figure 14.8.2   Equivalent circuit for a typical reciprocal receiver.

where $\mathcal{M}$ is either $\mathcal{M}_o$ or $\mathcal{M}_o^D$ for open-circuit voltage sensitivities and $\mathcal{M}_{ref}$ is the reference sensitivity level, 1 V/Pa. (Other references include 1 V/μbar in air and 1 V/μPa in water.)

(c) Reciprocal Receiver

For a reciprocal receiver (14.8.3) reveals that the force exerted on the moving face can be obtained from the equivalent circuit given in Fig.14.8.2, where the presence and orientation of the receiver are accounted for by adding an impedance $\mathbf{Z}_r$ to the mechanical side of the circuit of Fig. 14.2.2c and modifying the free field pressure at the diaphragm by the diffraction factor $\mathcal{D}$. Conventional microphones are usually small over most of the frequency range of interest, so that $\mathcal{D} \approx 1$. Then, with the help of (14.2.4) and (14.2.6),

$$\mathcal{M}_o = S\phi Z_{EB}/Z_{mo} = k_c^2 S/\phi \tag{14.8.6}$$

The open-circuit voltage sensitivity of a reciprocal microphone is uniform with frequency if diffraction is negligible.

(d) Antireciprocal Receiver

For open-circuit conditions, the first of (14.2.9) gives $\mathbf{V} = \boldsymbol{\phi}_M\mathbf{u}$. Combination of the equations in (14.2.9) with (14.8.3) and assuming that the wavelength is large compared to the dimensions of the receiver gives

$$\mathcal{M}_o \approx S\phi_M/Z_{mo} \tag{14.8.7}$$

The antireciprocal microphone sensitivity depends on frequency. If it is desired to obtain a sensitivity uniform with frequency, Z_{mo} must be manipulated through the introduction of additional mechanical elements to remain constant over the frequency range of interest.

14.9 CONDENSER MICROPHONE

The diaphragm of a typical condenser microphone consists of a stretched thin membrane, usually of steel, aluminum, or metallized glass, having a radius a, and separated by a small distance x_0 from a parallel rigid plate. The rigid plate is insulated from the remainder of the microphone, and the polarizing voltage V_0 is applied between it and the diaphragm. The *electret* microphone, encountered in many applications from inexpensive cassette recorders to high-fidelity recording, has a diaphragm of polarized plastic film aluminized on the outer surface. The intrinsic polarization of the plastic eliminates the necessity of having a polarizing voltage supply.

The canonical equations and equivalent circuit of a simple condenser microphone were developed in Section 14.3(a), and the open-circuit microphone sensitivity can be well approximated, except at high frequencies, by (14.8.6). The capacitance C_0 of the microphone in farads (F) is $\varepsilon S/x_0$. Substituting the *permittivity of free space* $\varepsilon_0 = 8.85\ \text{pF/m}$ gives

$$C_0 = 27.8\, a^2/x_0 \quad \text{pF} \tag{14.9.1}$$

Assume that the diaphragm acts like the forced membrane discussed in Section 4.8. For low driving frequencies such that $ka < 1$, we can neglect $\mathbf{Z}_r$ and in this limit the upper frequency for reasonably uniform response is given by (4.9.3). The average displacement amplitude of the diaphragm is found from (4.9.2),

$$\langle y \rangle = Pa^2/8\mathcal{T} \tag{14.9.2}$$

where P is the pressure amplitude in pascals and $\mathcal{T}$ is the tension in the diaphragm (in newtons per meter). If the mechanical stiffness s' of the diaphragm arises solely from the tension, then $PS = \langle y \rangle s'$ and $s' = 8\pi\mathcal{T}$. For most frequencies of interest the diaphragm is operating below its resonance frequency, so that $\mathbf{Z}_{mo} \rightarrow -j(s'/\omega)$. Furthermore, neglecting the leakage resistance, we have $\mathbf{Z}_{EB} = 1/j\omega C_0$. The transformation factor ϕ is given by (14.3.4). Collecting these results into (14.8.6), we obtain

$$\mathcal{M}_o \approx V_0 S/x_0 s' = V_0 a^2/8x_0\mathcal{T} \tag{14.9.3}$$

The design of a sensitive microphone requires large diaphragm area, high polarization voltage, small interelectrode spacing, and low stiffness. However, V_0/x_0 is limited by dielectric breakdown (arcing). Too large an area and too small a stiffness will degrade high-frequency response by lowering the mechanical resonance frequency.

As a numerical example, consider a microphone with an aluminum diaphragm 0.04 mm thick having a radius $a = 1$ cm and stretched to a tension $\mathcal{T} = 20,000\ \text{N/m}$. If the spacing x_0 between the diaphragm and the backing plate is 0.04 mm and the polarizing voltage V_0 is 300 V, then $\mathcal{M}_o = 4.7 \times 10^{-3}\ \text{V/Pa}$ and the sensitivity level is $\mathcal{ML} = -47$ dB *re* 1 V/Pa. The measured response of such a microphone agrees with the predicted response below 8 kHz. This is to be anticipated, since the limiting frequency predicted by (4.9.3) is 6.8 kHz. The fundamental frequency of the diaphragm, given by (4.4.12), is 2.4 times this frequency, or about 16 kHz. Near resonance the microphone may have a response about 5 to 10 dB above the low-frequency response. The exact amount of this increase depends on the magnitude of the damping forces. Above resonance the response falls off rapidly: the motion of the diaphragm becomes mass controlled so that the average displacement is no longer constant but decreases inversely with frequency.

Figure 14.9.1 shows the open-circuit voltage sensitivity of this condenser microphone. The rise in response near resonance could be reduced by constructing the microphone so that viscous damping, resulting from the motion of the air through special slots in the fixed backing plate, becomes significant near 16 kHz. The capacitance C_0 of the microphone in this example is only 69.5 pF so that $\mathbf{Z}_{EB}$

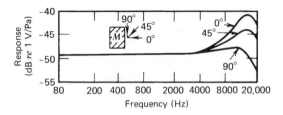

Figure 14.9.1 Free field sensitivity of a typical condenser microphone. Normal incidence is $\theta = 0°$.

is very high (e.g., 23 MΩ at 100 Hz). Consequently, the electrical terminals must be connected to a resistance in excess of 50 MΩ if the voltage across them is to be approximately equal to that generated in the microphone. Because of this high internal impedance, it is necessary to provide amplification in the immediate vicinity of the microphone. The preamplifier for a modern condenser microphone is usually mounted in the housing. The first input stage is an FET with an input impedance of about 500 MΩ. If the microphone is connected instead through a long cable to its preamplifier, electrical pickup (mostly 60 Hz line frequency and harmonics) can be a problem. Furthermore, the electrical capacitance of the cable is in parallel with C_0. Since the capacitance of ordinary shielded microphone cable is about 60 to 100 pF/m, the total capacitance of even a short run will be greater than C_0, thus decreasing the sensitivity of the microphone. Since $\mathbf{Z}_{EB}$ is capacitive, the primary effect of the cable capacitance is uniform attenuation across the entire audio bandwidth.

14.10 MOVING-COIL ELECTRODYNAMIC MICROPHONE

The *moving-coil* or *dynamic* microphone has a small metal coil attached to a light diaphragm. The incident pressure moves the diaphragm and coil in the radial field of a permanent magnet, thus generating a voltage. Basically, a moving-coil microphone is similar to a moving-coil loudspeaker. Indeed, the small loudspeakers of a typical intercom system are switched between acting as sources and as receivers.

The transformation factor for this antireciprocal device is given by (14.3.13). As before, we will discuss only the open-circuit voltage sensitivity (14.8.7),

$$\mathcal{M}_o \approx SBl/Z_{mo} \tag{14.10.1}$$

where it is assumed that the wavelength is long with respect to the dimensions of the receiver. The voltage sensitivity could be made independent of frequency by having R_m so large that the system is resistance controlled. However, high voltage sensitivity requires high speed amplitude with resultant small mechanical impedance. High sensitivity and flat response cannot both be obtained by using a mechanical system equivalent to a simple oscillator with $\mathbf{Z}_{mo}$ given by (14.3.10). Instead, additional mechanical elements must be introduced to modify the behavior of $\mathbf{Z}_{mo}$ at frequencies above and below mechanical resonance. A cross-sectional view of a moving-coil microphone having the desired mechanical characteristics is shown in Fig. 14.10.1. The diaphragm is a dome that vibrates as a rigid piston in the useful range of frequencies. The stiffness s and resistance R are contributed by the corrugated annulus supporting the diaphragm. The stiffness s_1 arises primarily

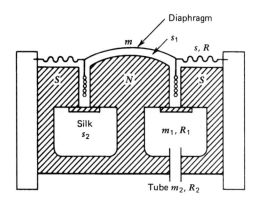

Figure 14.10.1 Schematic of a moving-coil microphone.

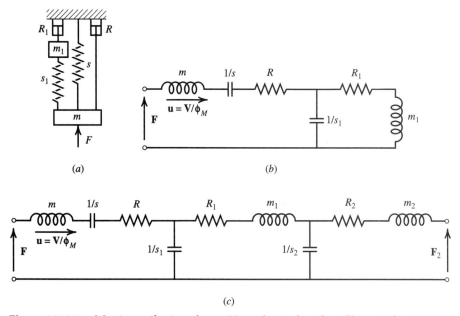

Figure 14.10.2   Moving-coil microphone: (a) mechanical analog, (b) equivalent circuit, and (c) more realistic equivalent circuit.

from compression of the air trapped in the chamber beneath the diaphragm. The mass m_1 and resistance R_1 come from viscous forces opposing the flow of air through the capillaries of the silk cloth. The stiffness s_2 of the air chamber below the silk cloth is relatively small, and for the moment we will neglect it and the influence of the tube. The equivalent mechanical system for the microphone is shown in Fig. 14.10.2a and the equivalent circuit in Fig. 14.10.2b. The open-circuit mechanical impedance is $\mathbf{Z}_{mo} = R_{mo} + jX_{mo}$ with

$$R_{mo} = R + \frac{s_1^2 R_1}{R_1^2 \omega^2 + m_1^2(\omega_1^2 - \omega^2)^2}$$

$$X_{mo} = \omega m - \frac{s}{\omega} - s_1\omega \frac{R_1^2 - m_1^2(\omega_1^2 - \omega^2)}{R_1^2 \omega^2 + m_1^2(\omega_1^2 - \omega^2)^2} \qquad (14.10.2)$$

$$\omega_1^2 = s_1/m_1$$

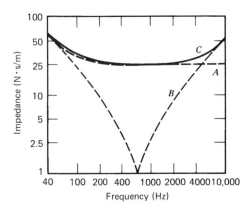

Figure 14.10.3   Input mechanical impedance for (curve A) a typical moving-coil microphone, (curve B) a simple oscillator with the same R, m, and s, and (curve C) the same simple oscillator but with R increased to match the impedance of curve A over the midfrequencies.

The mechanical elements R, m, s, R_1, m_1, and s_1 can be chosen so that Z_{mo} is fairly uniform over the audio range. Curve A of Fig. 14.10.3 shows Z_{mo} for the values $R = 1 \text{ N} \cdot \text{s/m}$, $R_1 = 24 \text{ N} \cdot \text{s/m}$, $s = 10^4 \text{ N/m}$, $s_1 = 10^6 \text{ N/m}$, $m = 0.6 \text{ g}$, and $m_1 = 0.3 \text{ g}$. For purposes of comparison, the mechanical impedance of the simple oscillator R, m, s alone has been plotted in curve B, and that of a similar oscillator for which $R = 25 \text{ N} \cdot \text{s/m}$ in curve C. It is evident that the mechanical system of curve A has enhanced response at the frequency extremes and has the most uniform response of the three curves. For a diaphragm of 5 cm² effective area attached to a coil of 10 m length and operating in a field of 1.5 T, the open-circuit voltage responses for these three mechanical systems are sketched in Fig. 14.10.4.

The low-frequency response of this microphone is further improved by the tube connecting the lower chamber to outer air. This modifies the equivalent electrical circuit to that of Fig. 14.10.2c. Because s_2 is small, s_2/ω becomes appreciable only at the lowest frequencies, where it allows the force $\mathbf{F}_2$ at the opening of the tube to affect the circuit. Analysis reveals that $\mathbf{F}_2$ then *increases* $\mathbf{u}$ and provides the sensitivity suggested by curve D of Fig. 14.10.4. The open-circuit voltage sensitivity level of this microphone is about $\mathcal{ML} = -72$ dB *re* 1 V/Pa.

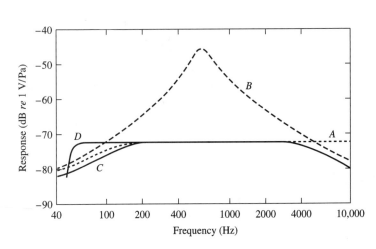

Figure 14.10.4 Sensitivity of the moving-coil microphone. Curves A to C are for the same mechanical systems identified in Fig. 14.10.3. Curve D shows the effect of adding a small tube connecting the chamber below the diaphragm to the outside air.

14.11 PRESSURE-GRADIENT MICROPHONES

All the above microphones are classified as pressure microphones. The acoustic pressure acts primarily on one side of the diaphragm and the resulting driving force is basically proportional to the pressure. It is also possible to construct *pressure-gradient* microphones for which the driving force is proportional to the difference between the pressures acting on the two sides of a diaphragm.

As an introduction to this kind of microphone, consider an acoustic wave incident on the front of a small plane diaphragm of mass m and area S mounted on a finite baffle. Let the angle of incidence θ be measured with respect to the normal to the diaphragm and baffle. If the diaphragm has dimensions much less than the wavelength, then the pressure on the front of the diaphragm can be written as $\mathbf{p}(t) = P\exp(j\omega t)$. The amount of time τ it takes for the pressure to be transmitted around the baffle to the back of the diaphragm is reasonably approximated by $\tau = (L/c)\cos\theta$, where L is some effective travel path at normal incidence. The net force $\mathbf{F}$ exerted on the diaphragm is given by $\mathbf{F}(t) = S[\mathbf{p}(t) - \mathbf{p}(t + \tau)]$, or

$$\mathbf{F} = PS(1 - e^{jkL\cos\theta})e^{j\omega t} \tag{14.11.1}$$

with $k = \omega/c$. If the piston is constrained to move normal to its face with a complex speed $\mathbf{u}$ and if it is mass controlled so that $\mathbf{F} = j\omega m\mathbf{u}$, then

$$\mathbf{u} = -(PS/j\omega m)(1 - e^{jkL\cos\theta})e^{j\omega t} \tag{14.11.2}$$

If a wire of length l is mounted on the diaphragm and there is a magnetic field of magnitude B whose field lines lie in the plane of the baffle and are also perpendicular to the wire, the voltage generated in the wire is $\mathbf{V} = Bl\mathbf{u}$. The open-circuit voltage sensitivity $|\mathbf{V}/\mathbf{p}|$ is then

$$\mathcal{M}_o = \frac{2BlS}{\omega m}\sin\left(\tfrac{1}{2}kL\cos\theta\right) \tag{14.11.3}$$

If the baffle is sufficiently small, $kL \ll 1$, this simplifies to

$$\mathcal{M}_o = \frac{BlSL}{cm}\cos\theta \tag{14.11.4}$$

In this limit the microphone has a directional factor $H(\theta) = |\cos\theta|$. The output voltage is proportional to $\cos\theta$ and, therefore, to the normal component of the particle velocity of the incident wave. Microphones with this property are termed *velocity* microphones. The microphone is *bidirectional*, favoring waves with angles of incidence near 0° and 180° and discriminating against those incident around ±90°. Substitution of $H(\theta)$ into (7.6.8) leads to directivity $D = 3$ and a corresponding directivity index $DI = 4.8$ dB. The baffle usually has rather irregular shape, so that effective pathlength L is quite difficult to calculate and may exhibit some dependence on the angle of incidence of the pressure field and also some frequency dependence. It is usually inferred from the measured sensitivity. Typical values are on the order of a couple of centimeters.

One example of a pressure-gradient microphone is the *ribbon* microphone of Fig. 14.11.1. It consists of a light, corrugated metal ribbon suspended in a magnetic field

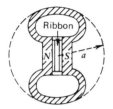

Figure 14.11.1   Simple velocity-ribbon microphone mounted in a baffle of radius a.

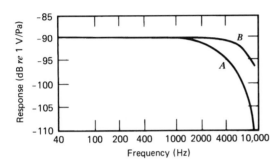

Figure 14.11.2   Computed (curve A) and measured (curve B) normal incidence sensitivities of a velocity-ribbon microphone.

and exposed to acoustic pressure on both sides. The stiffness of the suspension of the ribbon is so small that the frequency of mechanical resonance is below the audible range. As a result, the low-frequency response is quite flat. As the frequency increases into the kilohertz range, the response falls away from that predicted by (14.11.4) because L is no longer small compared to a wavelength. Instead, (14.11.3) must be used. At still higher frequencies diffraction will become significant.

Curve A of Fig. 14.11.2, computed from (14.11.3) for normal incidence, shows the open-circuit voltage response of a ribbon microphone having the parameters $m = 0.001$ g, $S = 5 \times 10^{-5}$ m^2, $l = 0.02$ m, $B = 0.5$ T, and $L = 3$ cm. Curve B shows the measured response of the microphone. The increased response in the region from 2 to 9 kHz is caused primarily by diffraction. The upper limit of usefulness is about 9 kHz.

The velocity microphone has three principal advantages. (1) Since it discriminates by a factor of 3 against background noise and reverberation, the distance between source and microphone can be $\sqrt{3}$ times that required with a nondirectional microphone for the same signal-to-noise ratio. (2) Its bidirectional characteristic permits sources to face each other across the microphone and have their sounds received with equal sensitivities. (3) The sharp null in the plane of the baffle makes it possible to orient the microphone so that a localized undesired source can be suppressed in the output.

A peculiarity of the velocity microphone is that if it is very close to a source, low-frequency response can be strongly enhanced. This arises because the ratio of particle velocity amplitude to acoustic pressure amplitude of a spherical wave increases as the source of sound is approached. Equation (5.11.9) predicts that the sensitivity level should be increased by $20 \log\{[1 + (kr)^2]^{1/2}/kr\}$. Some singers with light, bass-deficient voices may desire the enhanced richness this can bestow.

If the outputs of a pressure microphone placed together with a velocity microphone are combined in series, the resulting response favors the reception of sounds coming from one hemisphere. If the two axial sensitivities are equal, the

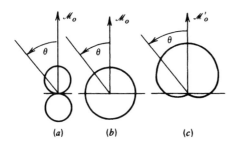

Figure 14.11.3   Directional response of various microphones: (*a*) velocity microphone, (*b*) pressure microphone, and (*c*) cardioid microphone.

combined sensitivity $\mathcal{M}'_o$ is

$$\mathcal{M}'_o = \mathcal{M}_o(1 + \cos\theta) \tag{14.11.5}$$

Figure 14.11.3 shows the directional responses of the individual and composite microphones. The combined directional characteristic is a cardioid of revolution, with the axis of rotation normal to the plane of the baffle. Microphones having a response of this type are known as *unidirectional* or *cardioid* microphones. The directivity of a cardioid microphone is also $D = 3$. Because of the large solid angle over which a cardioid microphone receives sound without appreciable discrimination, it is possible to cover widely separated sources with a single microphone. If such a microphone is placed near the front of a stage, its directional characteristic is exceptionally well suited to pick up sounds from the stage and exclude those from the audience.

*14.12 OTHER MICROPHONES

(a) The Carbon Microphone

The carbon microphone is widely used for telephone and radio communication where high electrical output, low cost, and durability are of greater importance than fidelity. Its operation results from the variation in resistance of a small enclosure filled with carbon grains, the *carbon button* (Fig. 14.12.1). As the diaphragm is displaced, the plunger varies the force applied to the button and hence the resistance from grain to grain, so that the total resistance across the button, nominally around 100 Ω, changes roughly linearly with pressure.

Below its fundamental mode of vibration, the diaphragm is stiffness controlled and for small displacements the resistance R of the button varies linearly with the displacement x of the center of the diaphragm. Then

$$R = R_0 + hx = R_0 + hSp/s \tag{14.12.1}$$

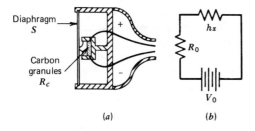

Figure 14.12.1   Carbon microphone. (*a*) Schematic of the construction of a carbon microphone. (*b*) Equivalent circuit.

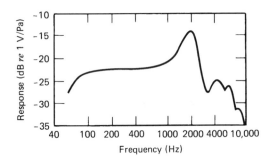

Figure 14.12.2   Sensitivity of a typical carbon microphone.

where R_0 is the quiescent resistance of the button, h its resistance constant (Ω/m), s the stiffness of the diaphragm, S its effective area, and p the acoustic pressure. The change in R causes the current to vary as $I = V_0/R$, where V_0 is the voltage of the battery. If $hSp/s \ll R_0$, then $1/R$ may be expanded and $I = (V_0/R_0)(1 - hSp/sR_0 + \cdots)$. The first term is a constant current V_0/R_0, the second term generates the desired output voltage $V = -V_0hSp/sR_0$ proportional to the acoustic pressure, and the unwritten higher-order terms generate harmonic distortion. The open-circuit voltage sensitivity is

$$\mathcal{M}_o = V_0 hS/sR_0 \tag{14.12.2}$$

The response increases as the battery voltage V_0 is increased or as the total resistance R_0 of the circuit is decreased. It also increases directly with the area S and inversely with the stiffness s of the diaphragm. However, high values of V_0 cause excessive heating and internal noise in the button, and reducing s or increasing S lowers the fundamental frequency of the diaphragm and restricts the useful frequency range.

A representative response curve is shown in Fig. 14.12.2. The peak near 2 kHz is from the fundamental frequency of the diaphragm. The uneven response at higher frequencies arises from higher modes of vibration. If the diaphragm is tightly stretched, its effective stiffness can be increased, raising the fundamental frequency. The region of relatively uniform response can be extended to about 8 kHz, but sensitivity is decreased.

While still popular in some applications, carbon microphones are being supplanted more and more by electret microphones.

(b) The Piezoelectric Microphone

Piezoelectric microphones employ crystals or ceramics that can become electrically polarized and produce voltages proportional to the strain. A nonmathematical discussion of their characteristics will suffice. (For further details of piezoelectric and ferroelectric transducers, consult the appropriate references.) Since the piezoelectric effect is reversible, a piezoelectric microphone will function as a source when an alternating voltage is applied to its terminals. These are reciprocal transducers and their equivalent circuits are the forms represented in Figs. 14.4.1 and 14.8.2.

Historically, quartz crystals have been of great significance as transducers. Today they are used principally as frequency standards in consumer goods (wrist watches) and in laboratories as ultrasonic sources and receivers for frequencies up through hundreds of megahertz. They are impervious to water and most corrosive materials, can be subjected to extreme temperatures, and are easily manufactured. Depending on the orientation at which the transducer element is cut from the crystal, it can generate longitudinal waves, shear waves, or combinations. Some cuts exhibit physical properties that are exceptionally temperature independent. Quartz transducers have relatively weak sensitivities.

Single crystals of Rochelle salt have been employed widely in microphones. Unfortunately, the crystals deteriorate in the presence of moisture and are permanently damaged if the internal temperature exceeds 46°C. Crystals cut from synthetic ammonium dihydrogen

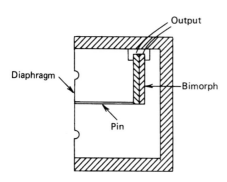

Figure 14.12.3 Schematic of the construction of a diaphragm-actuated crystal microphone.

phosphate (ADP) are somewhat less pressure-sensitive than Rochelle salt, but they can be raised to temperatures in excess of 93°C without deterioration and they exhibit much less temperature variation in their piezoelectric and dielectric properties. Other useful materials are sintered ceramics including barium titanate, lead zirconate, lead titanate, and mixtures of these and associated compounds. These *ferroelectric* ceramics are polarized (thereby behaving like piezoelectric materials) by application of a high electrostatic potential gradient of about 2000 kV/m while the temperature is above the Curie temperature of the ceramic (about 120°C, depending on composition) and maintenance of the external voltage during cooling. These ceramic microphones may be used interchangeably with crystal microphones. Their sensitivities are about 10 dB below those of Rochelle salt or ADP. On the other hand, ceramic microphones can withstand higher temperatures, are not as readily damaged by moisture or high humidity, and can be cast into a variety of shapes and sizes.

The magnitude of the strain-induced potential difference produced in a piezoelectric material depends on the type of deformation and its orientation relative to the axes of the crystal (or to the axis of polarization of the ceramic). Bending, shear, and longitudinal deformations have been utilized. The deformation may be brought about by the sound waves acting directly on the piezoelectric material, but a principal disadvantage is low sensitivity. As a consequence, piezoelectric microphones are often constructed in a manner similar to that shown in Fig. 14.12.3. Here, the sound waves act on a light diaphragm whose center is linked to an end or corner of the piezoelectric element by means of a driving pin. Although a single element could be used, two elements are usually sandwiched together to form a *bimorph*. In general, the bimorph has a smaller mechanical impedance than does a single element producing the same voltage output. Either a series or a parallel connection of the two elements may be used. The series connection gives a larger voltage output and the parallel connection gives a lower internal impedance.

The voltage output of a bimorph element is proportional to its strain, just as for the condenser microphone, so mechanically it must be stiffness controlled. The fundamental frequency of the entire system of diaphragm, pin, and bimorph must be above the desired range of relatively uniform response.

Piezoelectric microphones are widely used in public-address systems, sound level meters, and hearing aids. They have satisfactory frequency responses for such applications and are relatively high in sensitivity, low in cost, and small in size. Inexpensive diaphragm-actuated types are available and can cover from 20 Hz to 10 kHz with a variation in sensitivity level of less than ±5 dB. A typical average sensitivity level is −30 dB *re* 1 V/Pa. The electrical impedance of a piezoelectric microphone is essentially that of a dielectric capacitor. A typical value is 3000 pF. This is large compared to that of a condenser microphone, so these microphones may be connected to an audio amplifier by a cable of moderate length without an intervening preamplifier.

(c) Fiber Optic Receivers

Since the propagation of light in a single-mode optical fiber is a waveguide phenomenon, the phase speed of monofrequency light in the fiber depends on the index of refraction of the glass and the diameter of the fiber. The imposition of an acoustic pressure on the fiber changes the fiber's diameter, length, and index of refraction.[3] These changes result in a difference in phase of the light received at the end of a coil of fiber placed in an acoustic field and a similar coil isolated from the acoustic field. This phase change can be measured by interferometric techniques resulting in a receiver of high sensitivity and flexibility in geometric configuration.

*14.13 CALIBRATION OF RECEIVERS

Techniques for obtaining the sensitivity of a receiver fall into two general categories. *Absolute techniques* require measurement only of length, force, voltage, and so on. *Relative techniques* require *a priori* knowledge of the sensitivity of an available transducer used as a calibrator.

If a receiver of known sensitivity is on hand, any other receiver can be calibrated by *direct comparison*. In this method, (1) a transmitter produces a sound field in an anechoic environment. The anechoic environment is not necessary at low frequencies but becomes important at frequencies for which the wavelength of the sound is comparable to or less than the dimensions of the receiver, since for these wavelengths the orientation of the receiver with respect to the sound field must be known. (2) The receiver A of known sensitivity $\mathcal{M}_{oA}$ is placed in the sound field and its output voltage V_A is recorded as a function of frequency. (3) Receiver A is removed and the receiver X of unknown sensitivity $\mathcal{M}_{oX}$ is placed in exactly the same location (and orientation) and its output voltage V_X is recorded for the same frequencies as in step 2. (4) At each frequency, the sensitivity of the unknown is given by

$$\mathcal{M}_{oX} = \mathcal{M}_{oA} V_X / V_A \tag{14.13.1}$$

The most direct *absolute* technique consists of applying a known acoustic pressure to the receiver. This is usually accomplished by connecting the receiver to a chamber of known volume in which a piston of known area oscillates with known displacement amplitude. If the acoustic wavelength is much longer than the dimensions of the chamber, the pressure amplitude within the chamber can be calculated from the adiabat for the gas. While simple in concept, this technique is difficult in application because of the problem of accurately determining the volume of the chamber with the receiver in place. For certain popular makes of microphones, calibration devices, called *pistonphones*, are commercially available. However, a different coupler must be supplied for each type of microphone. The disadvantages of this technique are that the calibration can be performed at only one frequency (or at most a few frequencies) and the sensitivity derived is that based on the actual pressure at the microphone face and not that which exists before the microphone disturbs the sound field. However, the latter effect will be negligible for frequencies with wavelengths long compared to the dimensions of the receiver. The convenience and simplicity of this technique have made it quite popular.

A second technique for obtaining an absolute calibration is the *far-field reciprocity calibration*. To employ this technique, it is necessary to have a transmitter T, the microphone X to be calibrated, and a reversible transducer R. Neither the transmitting nor receiving response of the reversible transducer need be known. To avoid complications that obscure the discussion, assume that the dimensions of both X and R are much smaller than the shortest wavelength of interest. We will need to derive the relationship between $\mathcal{M}_{oR}$ and $\mathcal{S}_{IR}$ for a reversible transducer. Combination of (14.4.3) and (14.8.4) for frequencies such that $D \approx 1$, recognition that uS is the source strength of the reversible transducer, and use of (7.2.12) leads to

$$\mathcal{M}_{oR} / \mathcal{S}_{IR} = 2\lambda / \rho_0 c \tag{14.13.2}$$

---

[3] *Optical Fiber Sensors: Systems and Applications*, Vol. 2, ed. Culshaw and Dakin, Artech House (1989).

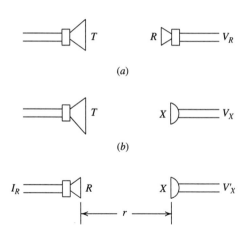

Figure 14.13.1   The three steps used in the reciprocity calibration of a microphone. (*a*) A reversible transducer *R* is placed a distance *r* from a transmitter *T* and the output voltage V_R is measured. (*b*) Next, *R* is replaced by the microphone *X* being calibrated and the output voltage V_X measured. (*c*) Finally, *T* is replaced by *R*, and the current I_R and the output voltage V'_X are measured.

The steps are as follows. (1) The reversible transducer *R* is placed at a specific position in the sound field of transmitter *T* (Fig. 14.13.1). The open-circuit voltage generated by *R* is $V_R = \mathcal{M}_{oR} P_T$. (2) The microphone *X* is substituted for *R* and its open-circuit output voltage is measured for the same incident sound field, so that $V_X = \mathcal{M}_{oX} P_T$. Elimination of P_T between these two equations gives

$$\mathcal{M}_{oX} = \mathcal{M}_{oR} V_X / V_R \tag{14.13.3}$$

(3) The reversible transducer *R* is substituted for the transmitter *T* and driven with a current I_R to generate a pressure $P_R = \mathcal{S}_{IR} I_R / r$ at the microphone *X*. The resultant open-circuit output voltage V'_X of the microphone is measured. Then $V'_X = \mathcal{M}_{oX} P_R$. Eliminating P_R between these two relations gives

$$V'_X = \mathcal{M}_{oX} \mathcal{S}_{IR} I_R / r \tag{14.13.4}$$

and then combining (14.13.2) to (14.13.4) gives our desired result,

$$\mathcal{M}_{oX} = \left(\frac{2\lambda r}{\rho_0 c} \frac{V_X V'_X}{V_R I_R} \right)^{1/2} \tag{14.13.5}$$

This method avoids measuring acoustic pressure amplitudes. It requires determining only voltages and currents and measuring a distance. [The results obtained from (14.13.5) can be converted to V/μbar by multiplying by 0.1.]

If two *identical reversible* microphones are available, the reciprocity principle makes it possible to calibrate both microphones by a single set of measurements involving only step (3). Effectively, since they are identical, $V_X = V_R$, so that (14.13.5) reduces to

$$\mathcal{M}_{oX} = \left(\frac{2\lambda r}{\rho_0 c} \frac{V'_X}{I_R} \right)^{1/2} \tag{14.13.6}$$

Finally, it is possible to self-calibrate a reversible transducer at high frequencies by using a short-pulse technique and electronic switching. The transducer is first driven with a short current pulse of amplitude *I* to generate a pressure pulse that is reflected from a perfectly reflecting plane surface back to the transducer. The transducer, now switched to operate as a receiver, detects the pulse and generates an output voltage *V*. The sensitivity is found from (14.13.6) by replacing V'_X with V and I_R with I,

$$\mathcal{M}_{oX} \left(\frac{2\lambda r}{\rho_0 c} \frac{V}{I} \right)^{1/2} \tag{14.13.7}$$

where r is the *round trip* distance (from transducer to reflector and back).

PROBLEMS

Unless explicitly stated otherwise, all amplitudes of oscillating quantities are assumed to be effective values.

14.2.1. Verify that the circuits of Figs. 14.2.2a and 14.2.2c yield the canonical equations (14.2.6) for a reciprocal transducer.

14.3.1. Assume an electrostatic transducer is used as a sound source. (a) Show that for sufficiently low frequencies the assumptions leading to (14.3.8) are equivalent to replacing the circuit of Fig. 14.3.1c with the parallel combination of C and C_0. (b) Recognizing that the energy E stored by charge q on a capacitance C is $E = q^2/C$, verify (14.3.8) as the specified ratio of energies.

14.3.2. From the canonical equations, find a formal relationship between $\mathbf{k}_c^2$ and $\mathbf{k}_m^2$. Compare the magnitudes of the coupling coefficients if the coupling is weak, $|\mathbf{k}_c| \ll 1$.

14.3.3. Assume a moving-coil transducer is used as a sound source. (a) Show that for sufficiently low frequencies the assumptions leading to (14.3.16) are equivalent to replacing the circuit of Fig. 14.3.2 with the series combination of L_M and L_0. (b) Recognizing that the energy E stored by current i in an inductance L is $E = Li^2$, verify (14.3.17) as the described ratio of energies.

14.4.1. A transmitter generates a sound pressure level at 1 m of 100 dB re 1 μbar for a driving voltage of 100 V. Find the sensitivity level in dB re 1 μbar/V.

14.4.2. A transmitter has a sensitivity level of 60 dB re 1 μbar/V. Find its sensitivity level re 1 μPa/V and re 20 μPa/V.

14.4.3. When the voice coil of a loudspeaker is blocked, the resistive component of its input impedance is 5 Ω. When the speaker is mounted on a large wall and driven at its frequency of mechanical resonance, the resistive component is found to be 10 Ω. When the speaker is removed from the wall to approximate eliminating the acoustic radiation loading, the input resistance is 12 Ω at mechanical resonance. What is the electroacoustic efficiency of the speaker at resonance when mounted on the wall?

14.4.4. A loudspeaker of 0.2 m radius is mounted on a large flat wall. When the speaker is being driven at 1000 Hz, the sound intensity at a point 5 m from the speaker on its axis is 0.1 W/m^2. (a) What is the intensity level at this point? (b) Assuming the speaker to have the same directional pattern as a flat piston of equal radius, what is the intensity level at this same distance on the wall? (c) In a direction 30° from the wall? (d) In a direction 60° from the wall? (e) What is the total acoustic output of the speaker in watts?

14.4.5. Derive (14.4.10) from its definition.

14.4.6. (a) Obtain the first identity in (14.4.22) from its definition in terms of powers. (b) Show that the two ratios are the electromechanical and mechanoacoustic efficiencies. (c) Verify the last identity with the help of the indicated equations.

14.4.7. Show in (14.4.23) that $\eta_{MA} = R/R_R$ and $\eta_{EM} = R/(R_0 + R)$ at mechanical resonance.

14.5.1. For the loudspeaker of Section 14.5, evaluate (a) k_m^2, (b) $k_m^2(eff)$, and (c) Q_M from the quantities given in the text.

14.5.2. The voice coil of a loudspeaker is 0.03 m in diameter and has 80 turns. Its blocked resistance is 3.2 Ω and its blocked inductance is 0.2 mH. It operates in a magnetic field of 1 T. The total mass of the cone and voice coil is 0.015 kg, the mechanical resistance R_m is 1 N $\cdot$ s/m, the radiation resistance R_r is 1 N $\cdot$ s/m, and the stiffness of the cone system is 1500 N/m. (a) Assuming the radiation reactance X_r to be negligible, what are the blocked electrical impedance Z_{EB}, the motional impedance Z_{MOT}, and the total electrical input impedance Z_E at 200 Hz? (b) What driving voltage is required to produce a displacement amplitude of the speaker cone of 0.1 cm at this frequency? (c) What acoustic output in watts will be produced by the driving voltage? (d) Calculate values for R, L, and C of the parallel electrical circuit having an impedance equivalent to the motional impedance of the above speaker.

14.5.3. (a) What is the frequency of mechanical resonance for the speaker of Problem 14.5.2? (b) What is the mechanical quality factor Q_M of the speaker cone system? (c) What is the rms displacement amplitude of the cone for an applied voltage of 5 V at resonance? (d) If the driving circuit is opened at an instant of maximum displacement, what will be the rms displacement amplitude at the end of 0.02 s?

14.5.4. A loudspeaker is mounted on an infinite baffle. It has a radius of 0.2 m, a moving mass of 0.04 kg, a voice coil having a resistance of 4 Ω, an inductance of 0.1 mH, and a transformation factor of 10 T $\cdot$ m. The suspension has a stiffness of 2000 N/m and a mechanical resistance of 2 N $\cdot$ s/m. (a) If an alternating voltage of 10 V and 200 Hz is applied to the voice coil, what is the acoustic power output of the speaker? Assume radiation loading on just one side of the speaker cone. (b) Assuming the radiated beam pattern to be that of a circular piston in an infinite baffle, what axial pressure level will be produced at a distance of 10 m?

14.5.5C. (a) Plot the motional impedance circle for the loudspeaker analyzed in this section. (b) Determine ω_u and ω_l from your plot, (c) calculate the mechanical quality factor from these data, and (d) compare with the value calculated from (14.4.21).

14.6.1. A loudspeaker has a radius of 0.15 m. When mounted on a wall its frequency of mechanical resonance is 25 Hz. When mounted in a sealed cabinet of 0.1 m^3 volume, its frequency of mechanical resonance is raised to 50 Hz. (a) What is the stiffness constant of the suspension? (b) What is the mass of the speaker cone? In each case consider the speaker cone to be loaded with radiation reactance on just one side.

14.6.2. A loudspeaker has mass = 0.01 kg, stiffness = 1000 N/m, mechanical resistance = 1.5 N $\cdot$ s/m, radius = 0.15 m, a voice coil of 1.5 cm radius having 150 turns of No. 34 copper wire, a flux density in the air gap of 0.8 T, and 0.4 mH inductance. The speaker is mounted in an enclosed cabinet 0.2 m $\times$ 0.5 m $\times$ 1.0 m. (a) Considering the speaker cone to be radiation loaded on just one side, what is the frequency of mechanical resonance? (b) If a voltage of 10 V is applied to the voice coil, what is the acoustic output in watts at the resonance frequency, at 200 Hz, and at 1 kHz?

14.6.3. Consider the loudspeaker of Problem 14.6.2 to be mounted in a bass-reflex cabinet. (a) If the vent is a circular hole of 0.15 m radius and negligible length, what must be the volume of the cabinet if its Helmholtz resonance frequency is to equal the mechanical resonance frequency of the speaker when mounted on an infinite baffle? (b) At 75 Hz what is the ratio of the acoustic output of the speaker when mounted in this cabinet to that when mounted on an infinite baffle? Assume the displacement of the speaker cone to be the same for both cases.

14.7.1. Verify that each term of (14.7.4) satisfies (14.7.3) when $S = S_0 \exp(2\beta x)$.

14.7.2. A small circular piston has a radius of 0.03 m and a mass of 0.002 kg. The stiffness of its suspension is such that its free oscillation frequency is 300 Hz. An exponential horn having a radius of 0.03 m at its throat, a length of 1.0 m, and a radius of 0.3 m at its mouth is fitted over the piston. (a) What is the new mechanical resonance frequency of the piston? Consider the mass loading of the piston by the horn to be that of an infinite horn having the same flare constant. (b) If the amplitude of the driving force acting on the piston is 5 N, what acoustic power will be radiated by this infinite horn at 300 Hz?

14.7.3. One watt of acoustic power is radiated at 250 Hz from an infinite exponential horn. The horn has a radius of 0.03 m at its throat and a flare constant $2\beta = 5$. (a) What is the cutoff frequency of the horn? (b) What is the peak volume velocity at the throat required to produce 1 W of acoustic output? (c) If the radius of the diaphragm in the driver is 0.05 m, what must be its peak displacement amplitude if it is to produce the above volume velocity?

14.7.4. The exponential horn of a tweeter has a radius of 0.01 m at its throat. The diaphragm has a radius of 0.03 m, a stiffness of 5000 N/m, and a mass of 0.001 kg. The voice coil has a resistance of 1.6 Ω, an inductance of 0.1 mH, and a transformation factor of 4 T $\cdot$ m. (a) What must be the flare constant of the horn if it is to have a cutoff frequency of 500 Hz? (b) What must be the peak volume-velocity amplitude at the throat of the horn, if the acoustic output is to be 0.2 W at 1 kHz? (c) What displacement amplitude of the diaphragm will produce this volume velocity at 1 kHz? (d) What is the efficiency of the speaker at this frequency? (e) What voltage must be applied to the voice coil to produce this acoustic output?

14.7.5. Assume that the traveling wave in the horn obeys the plane wave relationship between pressure and intensity. (a) How should the amplitude of the wave depend on the cross-sectional area of the horn? (b) On the basis of (a), determine the behavior of $P(x)$. Is this consistent with (14.7.4)? (c) For a conical horn, $S(x) = (x/a)S(a)$. If the throat is at $x = a$ and the pressure amplitude at the throat is $P(a)$, obtain $P(x)$. Is this consistent with spherical spreading?

14.7.6. Show for monofrequency excitation that (14.7.3) can be written as

$$\frac{\partial^2 p}{\partial x^2} + \frac{d(\ln S)}{dx}\frac{\partial p}{\partial x} + k^2 p = 0$$

with $\omega = kc$. (b) For an exponential horn, show that (1.6.3) is an analog of the above and find the analogs to p, x, 2β, k, and κ.

14.8.1. The receiving sensitivity level of a hydrophone is -80 dB re 1 V/μbar. (a) Express this level re 1 V/μPa. (b) What will be the output voltage if the pressure field is 80 dB re 1 μbar?

14.8.2. A microphone reads 1 mV for an incident effective pressure level of 120 dB re 1 μbar. Find the sensitivity level of the microphone re 1 V/μbar.

14.9.1. A condenser microphone diaphragm of 0.02 m radius and 0.00002 m spacing between diaphragm and backing plate is stretched to a tension of 10,000 N/m. (a) If the polarizing voltage is 200 V, what is the low-frequency open-circuit voltage response of the microphone in V/Pa? (b) What is the corresponding response level in dB re 1 V/Pa? (c) When acted on by a sound pressure of 1 Pa amplitude, what is the amplitude of the average displacement of the diaphragm? (d) What voltage will be generated in a load resistance of 5 MΩ by this microphone at 100 Hz when acted on by a wave of 10 μbar amplitude?

14.9.2. A condenser microphone having a diameter of 0.8 cm is to be used as a probe. The steel diaphragm is 0.001 cm thick and is stretched to the maximum allowable tension of 10,000 N/m. The spacing between the diaphragm and backing plate is 0.001 cm, and the polarizing voltage is 150 V. (*a*) What is the fundamental frequency of the diaphragm? (*b*) What is the open-circuit voltage sensitivity level *re* 1 V/μbar and *re* 1 V/Pa? Find the blocked input impedance at 10 kHz. (*c*) Considering diffraction to be equivalent to that for a sphere of equal diameter, what is the axial free-field response level *re* 1 V/Pa at this frequency?

14.10.1. A moving-coil loudspeaker is used both as a microphone and as a loudspeaker in an intercom system. The constants of the speaker are $m = 0.003$ kg, $a = 0.05$ m, $R_m = 10$ N·s/m, $s = 50,000$ N/m, $B = 0.75$ T, $l = 10$ m, $R_0 = 1$ Ω, and $L_0 = 0.01$ mH. Calculate its open-circuit voltage sensitivity level *re* 1 V/Pa at 1.1 kHz.

14.10.2. A moving-coil microphone has a moving element of 0.0002 m^2 cross section, 0.001 kg mass, 10,000 N/m stiffness, and 20 N·s/m mechanical resistance. The coil has a resistance of 5 Ω, a length of 5 m, and moves in a magnetic field of 1.0 T. (*a*) What is its open-circuit voltage sensitivity level *re* 1 μPa at 1 kHz? (*b*) Repeat at 100 Hz.

14.11.1. A velocity-ribbon microphone is constructed by mounting a thin aluminum strip in a circular baffle of 4 cm radius. The aluminum strip is 0.001 cm thick, 0.4 cm wide, and 2.5 cm long. The magnetic field has a flux density of 0.25 T. A plane wave of 250 Hz and 2 Pa acoustic pressure is incident normally on the face of the ribbon. (*a*) What voltage is developed in the ribbon? (*b*) What is the open-circuit voltage sensitivity level at this frequency? (*c*) What is the amplitude of the displacement of the ribbon under the above conditions?

14.11.2. For a standing plane wave $p = 2P \cos \omega t \sin kx$, derive an expression giving the net axial force acting on the diaphragm of an idealized pressure-gradient microphone. Let the diaphragm have cross-sectional area S and let the effective travel path from front to back be L. Consider only the case in which the surfaces of constant phase are parallel to the diaphragm. Show that this force is zero at the antinodes of pressure and maximized at the antinodes of velocity.

14.11.3. A microphone has a directivity such that the response in any direction making an angle θ with the principal axis is proportional to $\cos^2 \theta$. (*a*) Compute the numerical value of the directivity. (*b*) What is its directivity index?

14.11.4. (*a*) Obtain (14.11.3). (*b*) Obtain the directional factor. (*c*) Find the value of kL for which the open-circuit voltage sensitivity level of (14.11.3) has diverged by 3 dB from (14.11.4) at $\theta = 0°$. (*d*) For a ribbon microphone with a baffle radius of 3 cm, estimate the frequency associated with (*c*).

14.11.5. Show that the cardioid microphone has a directivity of 3.

14.12.1. Obtain expressions for the fractional second and third harmonic distortion in the output of a carbon microphone in terms of the bias voltage V_0 and the (peak) amplitude V_1 of the output voltage for the fundamental. Assume the distortions are small.

14.12.2. The open-circuit voltage response of a carbon microphone is -40 dB *re* 1 V/μbar when connected to a 12 V battery and its internal impedance is 100 Ω. Its diaphragm has an area of 0.001 m^2 and an effective stiffness of 10^6 N/m. (*a*) What is the value of the resistance constant h for this microphone? (*b*) For an incident wave of 100 μbar pressure amplitude, what will be the ratio of the second harmonic to fundamental voltage developed in this microphone?

14.12.3. A crystal microphone has an open-circuit voltage response level of -34 dB *re* 1 V/Pa and an internal capacitive impedance of 200,000 Ω at 400 Hz. (*a*) If a plane wave of this frequency with 70 dB pressure level *re* 20 μPa is incident on the microphone, what voltage will be generated in a 500,000 Ω resistor connected across the output terminals of the microphone? Assume that the crystal is stiffness controlled. (b) What power will be generated in this resistor? (c) If the area of the diaphragm is 0.0004 m², what is the ratio of this electrical power to the acoustic power incident on the microphone?

14.13.1. A condenser microphone has a diaphragm of 2 cm radius and the separation between the diaphragm and the backing plate is 0.002 cm. The diaphragm is stretched to a tension of 5000 N/m. The polarizing voltage is 200 V. (a) What is the low-frequency, open-circuit voltage response level *re* 1 V/Pa of the microphone? (b) Using the electroacoustical reciprocity theorem, compute the acoustic pressure level *re* 1 Pa produced by the microphone, acting as a loudspeaker, at a distance of 1 m when being driven by a current of 0.01 A at 1 kHz.

14.13.2. In a comparison calibration of a microphone, the standard microphone of sensitivity level −120 dB *re* 1 V/μbar gives an output voltage of 1 mV. The microphone to be calibrated gives an output of 0.2 mV when substituted for the standard. (a) What is the microphone sensitivity level of the unknown microphone? (b) What is the pressure level to which the microphones are being exposed?

14.13.3. A microphone is to be calibrated. From initial measurements, it is determined that its sensitivity is five times as great as that of a reversible transducer. When the reversible transducer is used as a source at a distance of 1.5 m from the microphone, the microphone is observed to have an open-circuit output of 0.001 V when a driving current of 1 A is supplied to the transducer at 500 Hz. (a) What is the open-circuit voltage response of the microphone? (b) What is the acoustic pressure acting on the microphone during the above experiment?

14.13.4. In a reciprocity calibration, the spacing between two identical reversible microphones is 2 m. At 2 kHz the measured open-circuit voltage output of one microphone is 0.0001 V for an input current of 0.01 A to the other. What is the open-circuit voltage response level of the microphones in dB *re* 1 V/Pa?

14.13.5. Derive (14.13.7) for the self-calibration of a reversible transducer directly from (14.13.2) et seq.

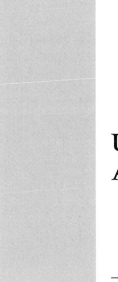

Chapter *15*

UNDERWATER ACOUSTICS

15.1 INTRODUCTION

The use of sound in water for transmitting and receiving information is of great interest both to humans and to the *Cetacea*. One of the earliest human applications of underwater sound was the installation of submerged bells on lightships. Underwater sounds from these bells could be detected at considerable distances by hydrophones mounted in the hulls of ships. If two such devices were located on opposite sides of a hull and the sounds received by each transmitted separately to the right and the left ears of a listener, the approximate bearing of the lightship could be determined. In 1912, Fessenden developed an electrodynamic transmitter that permitted underwater communication among vessels by Morse code. The safety of ocean navigation was enhanced by the introduction of the fathometer, which determined the depth of the water by measuring the time required for short pulses of sound to travel from the transmitter to the ocean bottom and return.

A very important application of underwater acoustics, known as sonar (SOund NAvigation and Ranging), has been detecting, tracking, and classifying submarines. This has required developing efficient conversion of electrical power into underwater sound, designing systems capable of detecting weak signals in the presence of noise, and studying the fundamental phenomena that affect the transmission of sound in the ocean.

For a discussion of underwater acoustics more comprehensive than this chapter, the inquisitive reader is referred to the book by Urick.[1]

15.2 SPEED OF SOUND IN SEAWATER

In Section 5.6 it was pointed out that the speed of sound in freshwater is a function of temperature and pressure. An additional factor in seawater is the salinity. An equation developed by Del Grosso[2] has widely been accepted as an

[1]Urick, *Principles of Underwater Sound*, 3rd ed., McGraw-Hill (1983).
[2]Del Grosso, *J. Acoust. Soc. Am.*, **56**, 1084 (1974).

accurate representation of the speed of sound in Neptunian[3] waters. The equation is appropriate for computer use. A reasonably accurate approximation to this equation is

$$c(T, S, P) = 1449.08 + 4.57Te^{-[T/86.9+(T/360)^2]} + 1.33(S - 35)e^{-T/120}$$
$$+ 0.1522Pe^{[T/1200+(S-35)/400]} + 1.46 \times 10^{-5}P^2e^{-[T/20+(S-35)/10]}$$

(15.2.1)

where c is the speed of sound in m/s, T the temperature in degrees Celsius, S the salinity in parts per thousand (ppt), and P the gauge pressure in atmospheres. This equation is valid for the open oceans but can also be applied to the low-salinity Black and Baltic Seas if the exponent $(S - 35)/400$ is replaced with its magnitude, although errors for the Black Sea will be up to about 60 cm/s. Otherwise, errors are typically within about 10 cm/s for depths down to 6 km, but can be up to 20 cm/s for some atypical ocean regions such as the Sulu Sea and the Halmahera, Caribbean, and East Indian Basins.

The relationship between pressure and depth is a function of latitude and the overlying densities in the water column. Formulas[4] exist to convert between pressure and depth, but for approximate calculations it can be assumed 1 atm $\approx$ 10 m of depth to within a percent at 45° latitude. With this approximation, P in (15.2.1) can be replaced with 100Z, where Z is the depth in kilometers. If a more accurate conversion is desired,

$$P = 99.5 (1 - 0.00263 \cos 2\varphi) Z + 0.239Z^2$$

(15.2.2)

where φ is the latitude in degrees, should suffice.

For routine estimations, a speed of sound of 1500 m/s and a nominal characteristic impedance $\rho_0 c = 1.54 \times 10^6$ Pa · s/m are adequate. These values will be used in calculating the acoustic pressure, particle velocity, and intensity in seawater. If *differences* in the speed of sound are required, (15.2.1) or a more accurate equation must be used.

15.3 TRANSMISSION LOSS

Transmission loss is defined as

$$TL = 10\log[I(1)/I(r)] = 20\log[P(1)/P(r)]$$

(15.3.1)

where $P(r)$ and $P(1)$ are the acoustic pressure amplitudes measured at distances r and 1 m from the sound source (as extrapolated back from the far field).

For example, the pressure amplitude of a damped spherical wave is

$$P(r) = (A/r)e^{-\alpha(r-1)}$$

(15.3.2)

and, for frequencies such that $\alpha \ll 0.1$ Np/m, (15.3.1) reduces to (8.3.7),

$$TL = 20\log r + ar$$
$$a = 8.7\alpha$$

(15.3.3)

---

[3]Leroy, *J. Acoust. Soc. Am.*, **46**, 216 (1969).
[4]Fofonoff and Millard, *UNESCO Tech. Pap. Mar. Sci.*, **44** (1983).

where a is the absorption coefficient in dB/m and the argument of the log is understood to be divided by 1 m. It is left as an exercise to show that if the sound is trapped between two parallel, perfectly reflecting surfaces, we have cylindrical spreading with absorption and the transmission loss is

$$TL = 10 \log r + ar \tag{15.3.4}$$

In general, it is convenient to separate the transmission loss into two parts,

$$TL = TL(geom) + TL(losses) \tag{15.3.5}$$

where $TL(geom)$ gives the attenuation from geometrical spreading and $TL(losses)$ represents the loss due to absorption and other nongeometrical effects such as scattering.

The absorption coefficient a for sound in seawater was discussed in Section 8.7. For instance, with pH = 8, S = 35 ppt, and T = 5°C, a = 0.063 dB/km at 1 kHz, 1.1 dB/km at 10 kHz, and 15 dB/km at 50 kHz. An approximation of (8.7.2) and (8.7.3) for seawater at these conditions adequate for the purposes of this chapter is

$$\frac{a}{F^2} = \frac{0.08}{0.9 + F^2} + \frac{30}{3000 + F^2} + 4 \times 10^{-4} \quad (dB/km)/(kHz)^2 \tag{15.3.6}$$

where F is the frequency in kHz and a is in dB/km. If it is desired to include a depth correction, multiplying the MgSO$_4$ loss term (the second on the right) by $\exp(-Z/6)$ with Z in km will be adequate for frequencies below about 100 kHz. Plotted in Fig. 15.3.1 are curves showing the transmission loss as a function of r for spherical spreading with absorption. At low frequencies and short ranges, TL is primarily from spherical divergence. As the frequency and range increase, curves B and C show that the absorption losses assume greater importance. Obviously, transmission to large distances requires low frequencies.

Transmission loss measurements usually deviate from the above simple formulas. (1) Divergence or convergence of the rays caused by refraction, interference associated with multipath propagation, and reflections from the surface and bottom of the sea affects the geometrical spreading. (2) Enhanced attenuation from

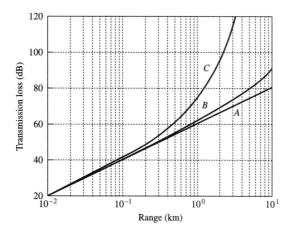

Figure 15.3.1 Transmission loss for spherical spreading in seawater with absorption. Curve A is for 1 kHz, B for 10 kHz, and C for 50 kHz.

diffraction and scattering caused by inhomogeneities in the water (see Section 8.10) affects the losses.

Although many factors limit the transmission of sound through seawater, acoustics is far superior to electromagnetism for transmitting energy through the oceans. In water, the lowest-frequency radio waves in commercial use, 30 kHz, have $a \sim 3$ dB/m, and higher frequencies are attenuated even more rapidly. Light passing through seawater is so rapidly attenuated by diffusion and scattering that the medium is essentially opaque over distances exceeding 200 m. Acoustics suffers only when compared with the transmission of radio and light waves through air.

15.4 REFRACTION

The most important phenomenon that alters the spherical spreading of sound in the ocean is the refraction that results from spatial variations in the sound speed induced by inhomogeneities in temperature, salinity, and pressure. Variations in salinity are important in regions where waters of differing salinities meet and near the surface where rain and evaporation have maximum effects. Variations in the speed of sound with depth are quite small. The change in pressure over a depth of 100 m (about 10 atm) increases the sound speed by 1.6 m/s, only about 0.1%. By contrast, variations in speed resulting from changes in temperature are quite large and are subject to large fluctuations, especially near the surface. Differences of more than 5°C are common in the first 100 m of the ocean. For temperatures near 15°C, a rise of 5°C increases the speed of sound by 16 m/s, about 1%.

Horizontal variations in the speed of sound are usually much less important than the variation with depth, but they can be significant near the mouths of large rivers, at the edges of large ocean currents such as the Gulf Stream, and in water close to melting ice packs. We will neglect these effects in the following discussions.

Given the dependences of temperature, salinity, and pressure on depth, the variation of c with depth can be calculated from (15.2.1); alternately, the speed of sound can be measured directly as a function of position. Figure 15.4.1 gives a representative speed of sound profile for the deep ocean. The most pervasive feature, found at all except the highest latitudes, is a distinct minimum. In the tropics, because of the heat provided by the sun, this minimum tends to lie deep. It rises toward the surface in the higher latitudes, sometimes reaching the surface in polar oceans. The depth at which this minimum occurs is called the *deep sound-channel axis*. Below this axis, the speed of sound increases until at great depths we find the *deep isothermal layer*, where the temperature remains a constant, between −1°C and 4°C, for most ocean basins. In this region the sound profile is nearly linear with a nominal positive gradient of about 0.016 (m/s)/m = 0.016 s^{-1}.

Above the deep sound-channel axis is the *main thermocline*. This region possesses negative gradients and responds slightly to seasonal changes but is a relatively stable feature of the profile, with characteristics determined primarily by latitude. Above this is the *seasonal thermocline*, also negative, which is responsive to seasonal variations. And finally, above this is the *surface layer*. This layer is quite dependent on the day-to-day, even hour-to-hour, variations in air and surface conditions. If there is sufficient surface-wave activity to mix the water near the surface, this becomes a *mixed layer*, which is isothermal and has a positive gradient of sound speed about 0.015 s^{-1}.

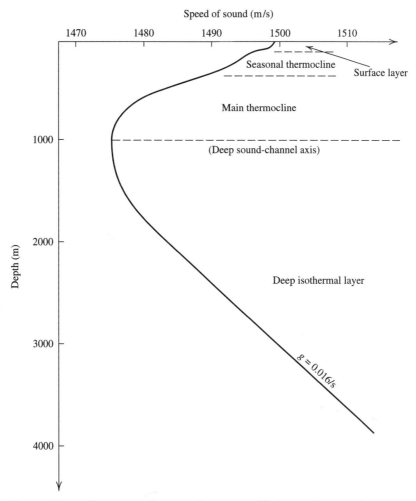

Figure 15.4.1 Representative sound-speed profile for midlatitude deep-ocean water.

The actual variations in c are very small compared with its magnitude. The profile of Fig. 15.4.1 has a maximum variation of about 30 m/s, about 2% of the nominal value. Nevertheless, this variation has an enormous influence on the propagation of sound in the ocean.

The path of a ray through a medium in which the speed of sound varies with depth can be calculated by application of Snell's law (5.14.16):

$$c/\cos\theta = c_0 \tag{15.4.1}$$

where θ is the *angle of depression* the ray makes with the horizontal at a depth where the speed of sound is c, and c_0 is the speed at a depth (real or extrapolated) where the ray would become horizontal ($\theta_0 = 0°$).

Complicated profiles such as that of Fig. 15.4.1 are often simplified by separation into thin layers, in each of which the gradient may be assumed constant. The advantage of this is that the path of a sound ray through a layer of water of constant sound-speed gradient g is an arc of a circle whose center lies at a *baseline depth*, where the sound speed in the layer extrapolates to zero. To show this,

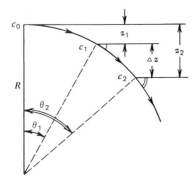

Figure 15.4.2   Diagram used in determining the relation between gradient g and the radius of curvature R of a sound ray.

consider a portion of a ray path with local radius of curvature R, as shown in Fig. 15.4.2. With the depth z chosen as positive downward, $\Delta z = R(\cos\theta_1 - \cos\theta_2)$. Combining the gradient $g = (c_2 - c_1)/\Delta z$ with Snell's law results in

$$R = -c_0/g = -c/g\cos\theta \tag{15.4.2}$$

and the ray path is a circle since c_0 and g are constants. The center of curvature of the circle lies at the depth where $\theta = 90°$, which corresponds to $c = 0$. For the situation illustrated in Fig. 15.4.2, the speed gradient is negative, so that R is positive. (Had g been positive, then in this geometry R would have been negative and the path would have curved upward.) In everything that follows, $R = |c_0/g|$ will be the *magnitude* of the radius of curvature and the signs of terms containing R explicitly will be assigned accordingly.

Once the baseline depth for each layer is determined by extrapolation of the isogradient profile in each layer, the ray path can be either traced graphically or computed. If the initial angle of depression of a ray is θ_1, reference to the geometry of Fig. 15.4.2 and use of (15.4.2) show that the changes in range Δr and depth Δz are

$$\Delta r = R(\sin\theta_1 - \sin\theta_2)$$
$$\Delta z = R(\cos\theta_2 - \cos\theta_1) \tag{15.4.3}$$

15.5 THE MIXED LAYER

Wave action can agitate the water near the surface, forming a *mixed layer* in which the pressure is the only agent affecting the sound speed. The positive sound-speed gradient in this layer traps sound near the surface. Once formed, the mixed layer remains until the sun begins to heat the upper portion, decreasing the gradient. This heating effect eventually culminates in a negative gradient that leads to a downward refraction and the loss of sound from the layer. Since this usually occurs in the afternoon, it is known as the *afternoon effect*. During the night, surface cooling and wave mixing allow the isothermal layer to be reestablished. It is rare for a positive gradient greater than about 0.015 s^{-1} to occur because this requires temperature increasing with depth, a dynamically unstable condition since, for constant salinity, the density would decrease with depth.

When the mixed layer is present, the sound profile near the surface can be modeled by two linear gradients, as shown in Fig. 15.5.1a, where D is the depth

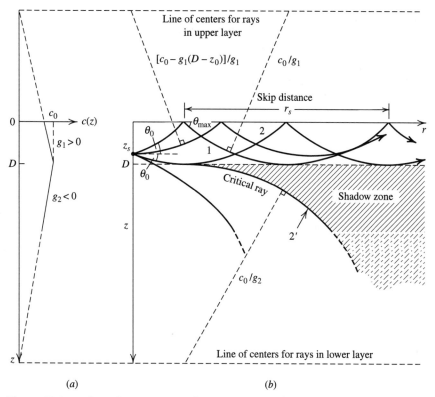

Figure 15.5.1 Sound transmission from a source within a mixed layer of depth D.

of the layer. Figure 15.5.1b shows representative rays from a source in the layer at depth z_s. A ray traveling upward reflects from the water–air interface with an angle of reflection equal to the angle of incidence, whereas a ray that intercepts the lower boundary of the layer continues on a path determined by the gradient below the layer. Note that the path continues smoothly across the change in gradient with no change in θ. All rays leaving the source with angles of elevation or depression between those of the rays labeled 1 and 2 will be confined to the mixed layer. Rays 1 and 2 have the same radius of curvature and both are tangent to the bottom of the layer. Ray 2′ is called the *critical ray* since it demarcates the inner boundary of the *shadow zone*, within which no rays are to be found. While this simple model suggests that there is no signal in the shadow zone, scattering from bubbles and the rough ocean surface, the presence of internal waves that cause D to fluctuate with horizontal distance, and the diffraction of sound into the shadow zone from its periphery contribute a weak, fluctuating ensonification of the shadow zone. For high kilohertz frequencies, signal levels are typically at least 40 dB less than those at the edges of the zone. For low kilohertz frequencies, the signal loss in the shadow zone is less severe; and for sufficiently low frequencies, the shadow zone may cease to exist because of strong diffraction and the breakdown of ray theory.

All rays whose angles of elevation or depression exceed those of rays 1 and 2 penetrate to greater depths and are lost from the layer. Rays between the limiting rays (1 and 2) initially spread spherically but ultimately are trapped in the layer and then spread cylindrically. The range at which the change from spherical to cylindrical spreading occurs is the *transition range* r_t. For small θ_0 this can be estimated by requiring that when $r = r_t$ the vertical extent of the beam subtending an angle $2\theta_0$ equals the layer depth D. This gives $r_t = D/2\theta_0$.

Another important parameter is the *skip distance* r_s. From Fig. 15.5.1*b*, $r_s = 2R\sin\theta_{max} \approx 2R\theta_{max}$, where $R = c_0/g_1$ is the radius of the ray that grazes the bottom of the layer. For small angles, Snell's law yields

$$1/c_0 = \left(1 - \tfrac{1}{2}\theta_{max}^2\right)\Big/ c(0) \qquad (15.5.1)$$

with $c(0)$ the sound speed at the surface ($z = 0$). For this isogradient layer,

$$c_0 = c(0)/(1 - D/R) \qquad (15.5.2)$$

Combination of the above equations yields

$$r_s = 2(2RD)^{1/2}$$
$$r_t = (r_s/8)[D/(D - z_s)]^{1/2} \qquad (15.5.3)$$

The nominal value of R for the mixed layer is $R = 1500/0.015 = 1.0 \times 10^5$ m.

A simple transmission loss model can now be constructed by assuming that for $r < r_t$ the geometric spreading is spherical and for $r > r_t$ the spreading is cylindrical. Since the transmission loss must be continuous across the transition range, a suitable form is

$$TL(geom) = \begin{cases} 20\log r & r \leq r_t \\ 10\log r + 10\log r_t & r > r_t \end{cases} \qquad (15.5.4)$$

While all rays trapped in the duct can rise to the surface to be reflected back downward, all rays cannot reach the bottom of the channel. (For example, a ray leaving the source horizontally can never attain a depth greater than the source depth.) If a receiver is at a depth z_r greater than that of the source, it can detect only those rays reaching depths equal to or deeper than itself. This means that the $TL(geom)$ between a source and a deeper receiver must be greater than that obtained from (15.5.4). Recall that acoustic reciprocity states that exchanging source and receiver cannot alter the observed transmission loss between them. If the source and receiver are exchanged, the source lies at the original receiver depth z_r and the transition range is now determined from (15.5.3) with z_r in place of z_s. This value exceeds that calculated earlier, so the transmission loss is greater. *The larger of source and receiver depths must be used in calculating r_t.*

Since the transmission loss depends on the depth of the source or receiver, whichever is greater, transmission loss in a given duct is minimized by having source and receiver shallow. However, neither can lie within a few wavelengths of the surface, or interference between it and its image can become important.

There are various contributions to *TL(losses)*. The absorption of sound by the seawater is accounted for by ar. There are also losses from the surface duct by scattering at the rough sea surface and by leakage out of the bottom of the duct from diffraction, internal waves, and irregularities in the speed of sound profile. These contributions can be parameterized as depending on the skip distance and a "loss per bounce" b. Combination yields $TL(losses) = ar + br/r_s$. Thus, the transmission loss is

$$TL = \begin{cases} 20 \log r + (a + b/r_s)r & r \leq r_t \\ 10 \log r + 10 \log r_t + (a + b/r_s)r & r > r_t \end{cases}$$ (15.5.5)

Although b is subject to considerable variation depending on the properties of the mixed layer, very rough but representative values can be obtained from an equation developed empirically by Schulkin,[5]

$$b = (SS)\sqrt{F}$$ (15.5.6)

where SS is the sea state, a rating of the roughness of the sea surface (see Table 15.9.1), and F is the frequency in kHz. The range of validity of this equation is for $3 < b < 14$ dB/bounce over around 2 to 25 kHz.

Because the trapping of sound in the mixed layer establishes a cylindrical waveguide, at lower frequencies normal-mode theory must be used instead of ray theory. The situation is not as simple as that studied in Sections 9.7 to 9.9, but the physics is the same. There exist normal modes each with its own cutoff frequency. This means that for sufficiently low frequencies many of the normal modes are evanescent and the ability of the mixed layer to carry energy will be reduced. As a consequence, the trapping of sound in the mixed layer does not occur for sound of sufficiently long wavelength. This cutoff frequency depends on the gradient *below* the mixed layer, but a rough approximation is

$$F \sim 200/D^{3/2}$$ (15.5.7)

where D is in meters and F is in kHz. For frequencies of about this value or lower, the mixed layer transmission loss model developed above is increasingly suspect, and normal-mode models provide more accurate predictions.

Rays whose angles of elevation or depression exceed θ_0 at the source are lost from the mixed layer and refract downward. If the water is deep enough, these rays will ultimately be refracted upward until they reach the surface. It often occurs that pairs of these refracted rays will intersect as they approach the surface, leading to areas of higher intensity levels at, and immediately below, the surface. This region of enhancement is the *convergence zone*. The range at which it occurs depends on the details of the sound-speed profile and varies from ocean to ocean, with *convergence zone ranges* varying between 15 and 70 km and widths (the distance over which there is significant enhancement of the signal) about 10% of the convergence zone range. A simple model for the transmission loss within the convergence zone assumes spherical spreading, absorption losses, and a *convergence gain G*,

$$TL = 20 \log r + ar - G$$ (15.5.8)

where r must be within the zone. The convergence gain is a complicated function of position in the convergence zone, but there is usually an abrupt increase in G as the zone is entered from the inside, with a gradual decrease as the outer edge of the zone is traversed. Determination of G requires numerical evaluation. Rays reflected from the bottom and returned to the surface also form a useful propagation path. If the depression angle of these rays is large, the ray paths are essentially straight lines and the transmission loss for *bottom bounce* propagation

---

[5]Schulkin, *J. Acoust. Soc. Am.,* **44,** 1152 (1968).

is well approximated by

$$TL = 20 \log r' + ar' + BL \qquad (15.5.9)$$

where r' is the actual distance traveled by the ray and BL is the *bottom loss* in dB. For simple bottoms, BL can be calculated from the angle the rays make with the bottom and the reflection coefficient at that angle for the bottom material. However, in most cases, BL must be measured *in situ*.

These bottom bounce rays aid in spanning the interval between the outer limits of propagation in the mixed layer and any convergence zone. However, the uncertainties in the reflective properties of the bottom limit the utility of bottom bounce propagation paths.

15.6 THE DEEP SOUND CHANNEL AND THE RELIABLE ACOUSTIC PATH

All rays originating near the axis of the deep sound channel and making small angles with the horizontal will return to the axis without reaching either the surface or the bottom, remaining trapped within this SOFAR channel (derived from "SOund Fixing And Ranging," an early application of this channel to locate by acoustic methods pilots downed at sea). Since the absorption of low frequencies in seawater is quite small, the low-frequency components of sound from explosive charges detonated in this channel can propagate to very large distances. Such signals have been detected at ranges in excess of 3000 km. The reception of these explosive signals by two or more widely separated arrays of hydrophones allows an accurate determination of the location of the explosion by triangulation.

A simple transmission loss model for sound propagation in the deep sound channel can be constructed by approximating the sound-speed profile in the vicinity of the axis by linear gradients, as indicated in Fig.15.6.1. The maximum angle θ_{max} with which the most steeply inclined trapped ray crosses the channel axis is found from Snell's law,

$$(c_{max} - \Delta c)/ \cos \theta_{max} = c_{max} \qquad (15.6.1)$$

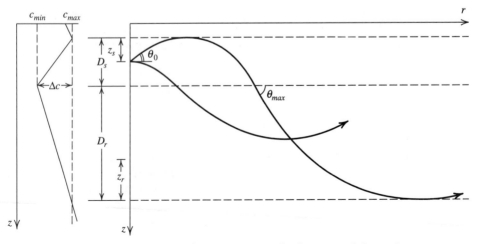

Figure 15.6.1   Sound transmission from a source in the deep sound channel.

where c_{max} is the greatest speed of sound found in the channel and Δc is the difference between c_{max} and the speed of sound c_{min} on the axis of the channel. For small angles,

$$\theta_{max} = (2\Delta c/c_{max})^{1/2} \tag{15.6.2}$$

For a source located above the axis and below the upper boundary, only those rays lying within some angle θ_0 above or below the horizontal will be trapped. Snell's law and the small-angle approximation yield

$$\theta_0 = \theta_{max}(z_s/D_s)^{1/2} \tag{15.6.3}$$

where z_s is the vertical distance between the top of the channel and the source. [Note that if the source were below the axis at a height z_r above the bottom of the channel, the trapping angle would be given by (15.6.3) with subscript s replaced with subscript r.] The skip distance r_s can be determined from Fig. 15.6.1 with a little geometry,

$$r_s = 2(D_s + D_r)(2c_{max}/\Delta c)^{1/2} \tag{15.6.4}$$

where $(D_s + D_r)$ is the vertical extent of the channel. For the profile of Fig. 15.4.1, $\Delta c \sim 30$ m/s and $(D_s + D_r) \sim 3000$ m so that $r_s \sim 60$ km. As in the mixed layer, the channel will be filled at a range $r_t = (D_s + D_r)/2\theta_0$. Manipulation reveals

$$r_t = (r_s/8)(D_s/z_s)^{1/2} \tag{15.6.5}$$

As with the mixed layer, the value of r_t must be the larger, calculated from the depth of the source z_s and D_s, or the depth of the receiver z_r and D_r. If the source and receiver are both above the channel axis, then r_t is calculated from the shallower of the two and D_s. If both are below the channel axis, the deeper is used with D_r.

Since it takes several skips for the acoustic energy to begin to be distributed fairly uniformly over the depth of the channel, the appropriate formula for the transmission loss,

$$TL = 10 \log r + 10 \log r_t + ar \tag{15.6.6}$$

will be applicable only after propagation distances of several skip distances. This may correspond to several hundred kilometers.

The expression (15.6.6) is valid for very long acoustic signals. Short tone bursts or explosive signals must be handled differently because of the dispersive behavior of the channel. In most oceans the speed of sound increases so rapidly with distance above and below the channel axis that those trapped rays having the greatest excursions provide the paths along which the energy travels fastest. The longest time t_{max} that can be taken for any ray to travel in this channel is thus less than or equal to r/c_{min}. Over long ranges, if the time of flight of the signal is t, the total signal of original duration τ will be *stretched* to $\tau + \Delta\tau$, where

$$\Delta\tau/t \sim [(D_s + D_r)/r_t]^2/24 \tag{15.6.7}$$

Again, the larger of the two r_t must be used.

The energy present in the original signal will be distributed over $\tau + \Delta\tau$ at a receiver at range r, but the distribution will not be uniform. Often the arrival is relatively weak at first, with the intensity increasing with time, followed by a sudden silence. The details of these effects depend strongly on the details of the speed of sound profile. If the sound speed does not increase sufficiently rapidly above and below the axis, the sound traveling straight down the axis may arrive first. As a rough estimate of how much time stretching can typically occur, for both source and receiver on the axis of the channel of Fig. 15.4.1, (15.6.7) gives $\Delta\tau/t = 0.007$ or about 9 s of time stretching over a distance of 2000 km.

Those rays leaving the source with angles of elevation greater than θ_0 can ensonify the surface and constitute a *reliable acoustic path* (RAP) from a deep source to the surface, and vice versa. The maximum available ensonified range at the surface increases with increasing source depth. If the rays are steep enough, they can be considered essentially straight lines and the transmission loss over this RAP is approximated by spherical spreading over the slant range r' with absorptive losses,

$$TL = 20\log r' + ar' \tag{15.6.8}$$

The geometry of Fig. 15.6.1 and the previous discussion concerning the convergence zone show that if the upper edge of the channel lies at the surface, then the range r_{cz} to the convergence zone is the skip distance in the sound channel. If the upper edge of the sound channel lies below the surface, then the convergence zone range must be increased by the appropriate additional propagation range. For example, if there is a mixed layer between the sound channel and the surface, then the convergence zone range may extend up to the sum of the skip distance for the channel plus that for the layer (depending on the source depth).

15.7 SURFACE INTERFERENCE

Whenever a nondirectional source of sound is present in the ocean, both surface- and bottom-reflected waves may combine with the direct wave. Depending on their relative phases, these waves may either reinforce or partially cancel each other. In deep water, when both source and receiver are near the surface, the bottom-reflected waves are relatively weak at short ranges. Under these circumstances the observed interference arises from the direct and surface-reflected waves.

If the surface is relatively smooth compared to a wavelength, then the reflected wave from a source at depth d acts as if it were emitted by an image located the same distance above the surface and of opposite phase. This is the acoustic doublet discussed in Section 6.8(b) and Problem 6.8.4. For application in underwater sound it is advantageous to recast those results, replacing the cylindrical coordinates (r, θ) with the horizontal range r from the doublet to the receiver at depth h. If both d and h are small compared to r,

$$P = (2A/r)\sin(khd/r) \tag{15.7.1}$$

where A is the pressure amplitude of the source at 1 m. The two waves reinforce each other when $khd/r = \pi/2, 3\pi/2, \ldots$ and cancel each other when $khd/r = \pi, 2\pi, \ldots$ (see curve A of Fig. 15.7.1). The most important practical situation occurs when $khd/r \ll 1$,

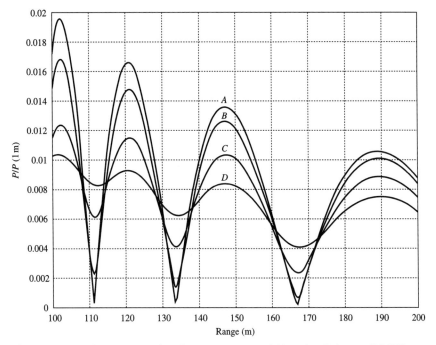

Figure 15.7.1   Pressure amplitudes at a receiver of 10 m depth from a 5.0 kHz source at 10 m depth for rms surface roughness $\sigma = 0$ m (A), 0.2 m (B), 0.4 m (C), and 0.6 m (D).

$$P \approx 2Akhd/r^2 \qquad (15.7.2)$$

For constant d and h, and r large, the pressure amplitude decreases as $1/r^2$. The transmission loss increases by 12 dB for each doubling of the range rather than by 6 dB as for spherical divergence.

An approximate criterion for the importance of surface roughness in modifying surface interference is

$$K = k\sigma \sin\theta_i \qquad (15.7.3)$$

where σ is the rms surface roughness (the standard deviation of the surface from being flat) and θ_i is the incident angle at the surface measured from the horizontal. If $K \ll 1$ the surface is smooth; if $K \gg 1$ the surface is rough. The surface roughness and angle of incidence determine the frequency below which there can be appreciable surface interference, and roughness becomes unimportant as $\theta_i \to 0$. However, for small grazing angles inhomogeneities near the surface and the bending of the rays because of the sound-speed profile can alter or destroy the interference effects at all but short ranges.

To characterize the effect of roughness, assume that rays scattered into angles other than θ_i are removed from the image. This means that the pressure amplitude of the signal appearing to emanate from the image can be reduced by a factor $\mu < 1$, which measures the importance of the surface roughness; μ is clearly a function of angle θ_i and can be approximated as $\mu \sim \exp(-2K^2)$. If the amplitude of the image is multiplied by μ, then (15.7.2) is generalized to

$$P = (A/r)[1 + \mu^2 - 2\mu \cos(2khd/r)]^{1/2} \qquad (15.7.4)$$

This model does not account for the presence of the scattered field. Scattering contributes an incoherent background that should be combined with the coherent interference field. The background leads to temporal fluctuations of the pattern and tends to "wash out" the interference nulls. For a more comprehensive treatment, see Clay and Medwin.[6]

15.8 THE SONAR EQUATIONS

In all applications of underwater acoustics the critical operation is detecting a desired acoustic signal in the presence of noise. If the level of the signal is the *echo level EL* and the level of the noise the *detected noise level DNL*, then the sonar equation is

$$\boxed{EL \geq DNL + DT}$$
(15.8.1)

which is no more than a restatement of the definition of detection threshold given in Chapter 11. The detection threshold DT is the value by which the echo level must exceed the detected noise level to give a 50% probability of detection for a specified probability of false alarm.

In signal processing applications, it is often conventional to use the detection threshold DT_1 appropriate for a bandwidth of $w = 1$ Hz. The relation between the two was obtained in Section 11.5,

$$DT = DT_1 - 10 \log w$$
[(11.5.1)]

This allows many equations to be expressed in spectrum levels rather than levels over the receiver bandwidth and is useful in narrowband processing. Since some authors use both DT and DT_1 with no notational differentiation in their texts, readers should be sure which is being used.

In underwater sound, the *detection index d* is used instead of the detectability index d',

$$d = (d')^2$$
(15.8.2)

(a) Passive Sonar

A passive sonar system listens for the noise produced by a *target*. The term "echo level" is not literally appropriate in this case, but the use is conventional. The sound radiated by the target at a source level SL [Section 7.6(c)] experiences a transmission loss TL on its way to the receiver. The echo level is then

$$EL = SL - TL$$
(15.8.3)

With a highly directive receiver, a passive system can determine the direction from which a signal arrives. If the target can be detected simultaneously on two or more such receivers separated by a known distance, the location of the target can be determined by triangulation.

Competing with the received signal is noise from a variety of sources. The oceans are filled with noise sources (breaking waves, snapping shrimp, surf, shipping,

---

[6]Clay and Medwin, *Acoustical Oceanography*, Wiley (1977).

etc.), which combine to produce broadband ambient noise. In addition, self-noise is produced by machinery on the receiving platform and by the motion of the water around it. The combined level of these sources is the noise level *NL*. If the receiver is directional, the detected noise level *DNL* is

$$DNL = NL - DI \tag{15.8.4}$$

where the directivity index *DI* [Section 7.6(*e*)] describes the ability of the receiver to discriminate against noise coming from undesired directions. Combining the above gives the passive sonar equation,

$$\boxed{SL - TL \geq NL - DI + DT_N} \tag{15.8.5}$$

where DT_N is the detection threshold for noise-limited performance. Since passive sonar detection is nearly always based on energy detection, the appropriate form is usually the second equation in (11.5.7)

(b) Active Sonar

For an active system, the signal is a pulse of acoustic energy that originates at the transmitter with a source level *SL*. This signal then travels to the target, accumulating a one-way transmission loss *TL*. At the target, a fraction of the incident signal, characterized by the target strength *TS*, is reflected toward the receiver and, after experiencing another transmission loss *TL'*, arrives at the receiver. For the *monostatic* case, source and receiver are at the same location so that $TL = TL'$ and the echo level is

$$EL = SL - 2TL + TS \tag{15.8.6}$$

Unless otherwise indicated, equations will be written for the monostatic case. To generalize to active *bistatic* echo ranging, replace *2TL* by (*TL* + *TL'*) and use the bistatic target strength appropriate for the geometry.

By measuring the time between the emission of a pulse and the reception of the echo, the distance from source to target to receiver can be found. If the locations of source and receiver are also known and either is highly directional, the bearing (and therefore location) of the target can be determined.

The detected noise level for an active system may be dominated by ambient or self-noise. Then, (15.8.1), (15.8.4), and (15.8.6) give the active sonar equation for *noise-limited* performance,

$$\boxed{SL - 2TL + TS \geq NL - DI + DT_N} \tag{15.8.7}$$

For active sonar there is an additional source of masking not present in a passive sonar—*reverberation*. Reverberation arises from the scattering of the emitted signal from unwanted targets such as fish, bubbles, and the sea surface and bottom. For this case, the detected noise level is the reverberation level *RL*,

$$DNL = RL \tag{15.8.8}$$

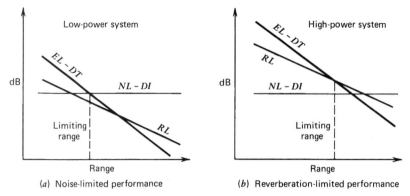

Figure 15.8.1   Noise- and reverberation-limited performances of a sonar system.

Combination of the above gives the active sonar equation for *reverberation-limited* performance,

$$SL - 2TL + TS \geq RL + DT_R \tag{15.8.9}$$

where DT_R is the detection threshold for reverberation-limited performance. An expression for DT_R will be developed later in Section 15.11.

Whether noise or reverberation dominates depends on the acoustic power, the range, and the speed of the target. The two possible situations are suggested in Fig. 15.8.1. Generally, low-power systems are noise-limited since the maximum detection range is achieved when the echo level falls below the level at which it can be extracted from the noise (Fig. 15.8.1a). Increasing the acoustic power of a system increases both the echo level and the reverberation level at a given range, but generally reverberation decreases with increasing range less rapidly than the echo level (Fig. 15.8.1b). If, as range increases, the echo level has diminished until it is buried in the reverberation, the system is reverberation limited.

To reduce the effect of reverberation, a notch filter can be used in the receiver to eliminate the energy in a narrow frequency band encompassing the reverberation. If the target is moving, the frequency of the echo will be different from that of the reverberation [see Section 15.9(c)] and the target can be detected more easily. However, the notch filter will also eliminate the echo if the target is stationary with respect to the water.

15.9 NOISE AND BANDWIDTH CONSIDERATIONS

From the sonar equation it is clear that sonar performance can be enhanced if the detected noise level is reduced. This can be accomplished by using knowledge of the frequency spectra of the ambient noise and of the target to select the bandwidth of the receiving system.

(a) Ambient Noise

The nominal shape of the ambient noise spectrum level *NSL* in the open ocean is sketched in Fig. 15.9.1. Between about 500 Hz and 20 kHz, the agitation of the local sea surface is the strongest source of ambient noise and can be characterized

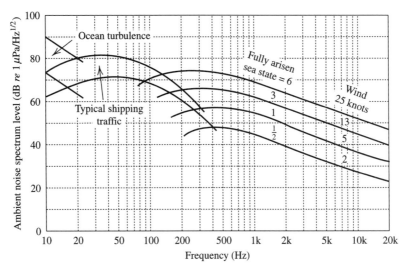

Figure 15.9.1 Deep-water ambient noise. [Adapted from Wenz, *J. Acoust. Soc. Am.*, **34**, 1936 (1962) and Perrone, *ibid.*, **46**, 762 (1969).]

by specifying the local wind speed. Relations among sea state, wave height, and representative wind speeds are given in Table 15.9.1. In this range of frequencies the noise spectrum level falls at about 17 dB/decade. At lower frequencies, the major contribution to the ambient noise is from distant shipping and biological noise. The indicated limits in the figure can be exceeded considerably if the ship traffic is heavy. Below about 20 Hz, ocean turbulence and seismic noise predominate. Above 50 kHz thermal agitation of the water molecules becomes an important noise source, and the noise spectrum level increases at 6 dB/octave. In shallow water, the noise levels can be much higher because of heavier shipping, nearby surf, higher biological noise, shore-based noises, off-shore drilling rigs, and so on.

The noise spectrum level in the figure was measured with omnidirectional receivers. The detected noise level sensed by a directional receiver depends on its orientation. With a directional receiver, the noise from the sea surface would be seen to arrive predominantly from the vertical direction, whereas the noise from shipping arrives more horizontally.

The detected noise level for ambient noise is (15.8.4). If the bandwidth w is small enough that NSL can be assumed constant over w, then

$$DNL = NSL + 10 \log w - DI \tag{15.9.1}$$

(b) Self-Noise

Self-noise is generated by the receiving platform and can interfere with the desired received signal. Self-noise can reach the receiver by transmission through the mechanical structure and by transmission through the water either directly from the source or by reflection from the sea surface. Self-noise usually tends to increase with increasing platform speed. At low frequencies and slow speeds, machinery noise dominates, whereas at high frequencies propeller and flow noise become important. As the speed is increased, these latter sources of noise assume more importance at all frequencies. At very low speeds, self-noise is usually less important than ambient noise or reverberation.

Table 15.9.1 Properties of the sea surface for deep waters far from land

Description	Wave Heights ($H_{1/3}$ in ft)	Sea State (SS)	12 Hour Wind (knots)	Average Wind Speed for Indicated Wave Heights — Fully Arisen Sea Wind (knots)	Required Time (h)	Required Fetch (nautical miles)
Sea is like a mirror.		0				
Ripples with the appearance of scales are formed, but without foam crests.		1/2	2	2	2	
Small wavelets, still short but more pronounced. Crests have a glassy appearance and do not break.	0–1	1	5	5	7	40
Large wavelets. Crests begin to break. Foam of glassy appearance. Perhaps scattered whitecaps.	1–2	2	9	9	11	100
Small waves, becoming longer; fairly frequent whitecaps.	2–4	3	14	13	14	150
Moderate waves, taking a more pronounced long form; many whitecaps are formed. (Chance of some spray.)	4–8	4	19	17	18	200
Large waves begin to form; white foam crests are more extensive everywhere. (Probably some spray.)	8–13	5	24	21	23	300

Wind speeds are measured 10 m above the sea surface. The relation between wind speed and wave height depends on the duration of the wind, the fetch, and the influence of near land masses and shallow waters.

A Rayleigh sea is defined in terms of the probability distribution $P(a)$ of the deviation a of points from the sea surface for Sea State 0:

$$P(a) = (a/\sigma^2)\exp[-\tfrac{1}{2}(a/\sigma)^2]$$

For a Rayleigh sea, the wave height is defined as $H_{1/3} = 4\sigma$ and can be estimated from $H_{1/3} \sim 0.4(SS)^2$ in ft or $H_{1/3} \sim 0.12(SS)^2$ in m, where $SS \approx 0.2 \times$ (12 h wind in kt).

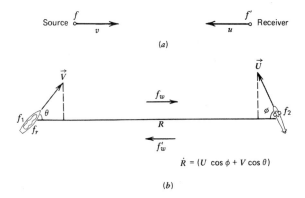

Figure 15.9.2 Diagrams used in deriving the relation between frequency shift and the speeds and directions of the source and receiver.

Self-noise is entered into the sonar equations as an equivalent isotropic noise spectrum level that expresses the masking level of the self-noise in the bandwidth of the receiver in terms of the level of an equivalent amount of ambient noise. With this convention, the detected noise level for self-noise-limited conditions is (15.9.1).

(c) Doppler Shift

Assume a source of frequency f is traveling with speed v in the water directly toward a receiver, and the receiver is traveling with a speed u in the water toward the source. (See Fig. 15.9.2a.) In a time interval τ the source will send $f\tau$ cycles into the water, and these will fill a distance $(c - v)\tau$ in the direction pointing toward the receiver. The wavelength λ_w of this sound in the water is $(c - v)/f$ and the frequency detected by an observer stationary in the water is $f_w = c/\lambda_w$ or

$$f_w = fc/(c - v) \tag{15.9.2}$$

In the same time interval τ, the receiver intercepts the number of wavelengths $(c + u)\tau/\lambda_w$. This number divided by τ is the number of cycles per second received. Thus, the frequency f' sensed by the receiver is $f' = (c + u)/\lambda_w$ and combination yields the *Doppler shift* equation

$$f' = f(c + u)/(c - v) \tag{15.9.3}$$

For the case $v \ll c$ and $u \ll c$, this simplifies to

$$f' = f[1 + (u + v)/c] \tag{15.9.4}$$

(1) PASSIVE SONAR   Let there be two vessels traveling in different directions with different velocities, and establish the geometry shown in Fig. 15.9.2b. If vessel 1 radiates a signal of frequency f_1, then by (15.9.2) the frequency f_w in the water of the signal traveling toward vessel 2 is

$$f_w = f_1[1 + (V/c)\cos\theta] \tag{15.9.5}$$

and vessel 2 will detect a frequency

$$f_2 = f_1(1 + \dot{R}/c)$$
$$\dot{R} = V\cos\theta + U\cos\phi \tag{15.9.6}$$

where $\dot{R} = -dR/dt$ is the *range rate*, the speed with which the two vessels are closing range. The Doppler shift for this passive case is

$$\Delta f = f_2 - f_1 = (\dot{R}/c)f_1 \tag{15.9.7}$$

If the vessels are approaching each other, $\dot{R}$ is positive and $f_2 > f_1$ (*up Doppler*). If they are receding from each other, $f_2 < f_1$ (*down Doppler*).

(2) ACTIVE SONAR   If the signal f_1 from vessel 1 is an active sonar pulse, the pulse travels in the water toward vessel 2 with f_w and vessel 2 will receive and reflect the sonar pulse at frequency f_2, just as before. Now, the echo traveling back toward vessel 1 will have a frequency in the water of

$$f'_w = f_2[1 + (U/c)\cos\phi] \tag{15.9.8}$$

and vessel 1 will receive an echo having a frequency

$$f'_1 = f_1(1 + 2\dot{R}/c) \tag{15.9.9}$$

The Doppler shift between the received echo and the generated sonar pulse is

$$\Delta f_1 = f'_1 - f_1 = (2\dot{R}/c)f_1 \tag{15.9.10}$$

On the other hand, reverberation comes from scatterers nearly stationary in the water, and the reverberation frequency f_r observed by vessel 1 is

$$f_r = f_1[1 + 2(V/c)\cos\theta] \tag{15.9.11}$$

The Doppler shift between the received echo and the reverberation in the direction of the target is

$$\Delta f_r = f'_1 - f_r = 2[(U/c)\cos\phi]f_1 \tag{15.9.12}$$

Vessel 1 can compare the received frequency f_1' of the echo either with its own sonar frequency f_1 or with the reverberation frequency f_r in the direction of the target.

(d) Bandwidth Considerations

In passive sonar, the Doppler shift of the frequency of the received signal from that of the source places a lower limit on the bandwidth of the receiver. Given the maximum expected range rate to be encountered, and if both approaching and receding targets are to be detected, the bandwidth must be twice the associated Doppler shift. Thus, a receiver for a passive sonar must have a total bandwidth w_p from (15.9.7) of

$$w_p = 1.33\,\dot{R}F \tag{15.9.13}$$

where w_p will be in Hz when $\dot{R}$ is in m/s and F is in kHz.

Comparing (15.9.7) with (15.9.10), we see that an active system designed to detect both up- and down-Doppler targets must have a total bandwidth w_a twice that of the passive case, so that

$$w_a = 2.67 \dot{R} F \qquad (15.9.14)$$

where w_a will be in Hz when $\dot{R}$ is in m/s and F is in kHz.

(In converting between metric and English units, note that 1 kt = 1.852 km/h = 0.5144 m/s.)

15.10 PASSIVE SONAR

The source level SL of the noise radiated by the target is obtained by extrapolating the radiated effective pressure amplitude from the far field to a distance 1 m from the acoustic center of the target. If the source spectrum level of a target is composed of a flat continuous spectrum out of which protrude tones, and if the intensity per unit bandwidth (at 1 m) for the continuous spectrum is $\mathcal{I}$ and the intensity of a tone (at 1 m) is I, then we must define a *source spectrum level SSL* for the continuous spectrum as the intensity level in a 1 Hz bandwidth,

$$SSL(cont) = 10\log(\mathcal{I} \cdot 1\,\mathrm{Hz}/I_{ref}) \qquad (15.10.1)$$

and a source level SL for the tone

$$SL(tone) = 10\log(I/I_{ref}) \qquad (15.10.2)$$

If the bandwidth w of the receiver includes the tone, the total intensity received is $\mathcal{I}w + I$ so that the overall SL is

$$SL = 10\log[(\mathcal{I}w + I)/I_{ref}] = 10\log(\mathcal{I}w/I_{ref}) + 10\log(1 + I/\mathcal{I}w) \quad (15.10.3)$$

If $\mathcal{I}w \gg I$, the contribution from the tone is negligible. For example, if the tone were 20 dB stronger than the continuous spectrum level, then for $w = 100$ Hz it would contribute 3 dB to the signal level and would be apparent to the ear or to a filter with $w \le 100$ Hz. If, however, the bandwidth of the filter were 400 Hz, the tone would add only 1 dB to the signal level received in that bandwidth. To use a nomogram in obtaining the overall SL within a bandwidth, refer back to the discussion in Section 11.3.

The noise radiated by a target in some direction depends on many parameters, including the orientation of the target, its mechanical state, speed, and depth. There are some general features present in almost all noise radiated from ships. The general broadband background tends to decrease at the higher frequencies at about 5 to 8 dB/octave. Thus, low-frequency signals predominate. Cavitation from the propellers adds a broadband contribution that is small at low frequencies, rises to a peak in some intermediate frequency, and falls off with increasing frequency. The region of maximum contribution shifts to lower frequencies with increasing speed or decreasing depth. Superimposed on this background are harmonic series contributed by machinery noise, engines, pumps, propellers, reduction gears, and other mechanical systems. Since the ship is a large acoustic source with individual noise generators at various locations, the various radiated noises can have individual directionalities that are functions of frequency and operating conditions.

The receiving system of a passive sonar can be either broadband to detect the total energy emitted by the target or narrowband so that detections are made on the tones. Thus, the passive sonar equation (15.8.5) can be written in two forms:

1. **Broadband Detection.** If the detector is broadband so that tones do not contribute significantly to the source level, but narrow enough that (15.10.1) applies,

$$SSL(cont) - TL \geq NSL - DI + DT_N \qquad (15.10.4)$$

2. **Narrowband Detection.** If the bandwidth of the receiver is small enough that $\mathcal{I}w \ll I$, (15.10.2) applies and

$$SL(tone) - TL \geq NSL + 10 \log w - DI + DT_N \qquad (15.10.5)$$

Detection of a tone would appear to be facilitated by having as narrow a bandwidth as possible. However, the bandwidth cannot be so narrow as to exclude a signal Doppler shifted by the motion of the target. Narrow bandwidth and broad frequency coverage are made possible by using a set of narrowband filters with contiguous bandwidths. This is called *parallel processing*. If each filter has bandwidth w, and there are n such filters, then the total bandwidth of the system is $w_T = nw$.

(a) An Example

A surfaced submarine is traveling at 4 kt. Its source spectrum level in the vicinity of 1 kHz is 120 dB re 1 μPa/Hz$^{1/2}$ and there are no tonals in this frequency region. The sea state is 3 and a mixed layer exists whose depth is 100 m. At what range will this submarine be detected by a 36 m deep hydrophone using broadband detection with a 100 Hz wide filter centered at 1 kHz? The directivity index is 20 dB and the detection threshold is 0 dB. *Solution:* The appropriate sonar equation is (15.10.4) where SSL, DI, and DT_N are given in the statement of the problem. The noise spectrum level at 1 kHz and sea state 3 is found from Fig. 15.9.1 to be $NSL = 62$ dB re 1 μPa/Hz$^{1/2}$. Substitution of these values into the sonar equation yields $TL < 78$ dB. Equation (15.5.7) shows that sound is trapped in this mixed layer if $f > 200$ Hz, so the transmission loss is given by (15.5.5). The skip distance and transition range (calculated with the receiver depth) are found from (15.5.3) to be 8660 m and 1350 m, respectively. Although the frequency is a little out of the range of (15.5.6), the loss per bounce can be estimated as about 3 dB/bounce and, from Fig. 8.7.1, the attenuation coefficient is 6×10^{-5} dB/m. Assuming the target is beyond the transition range when it is detected, insertion of these values into (15.5.5) yields

$$10 \log r + (4.1 \times 10^{-4})r \leq 47 \qquad (15.10.6)$$

Solution by trial and error results in $r = 13.4$ km, which is, as assumed, beyond the transition range. If r had turned out to be less than r_t, the problem would have had to be redone with the transmission loss given by the first equation of (15.5.5).

15.11 ACTIVE SONAR

Before discussing solutions of the active sonar equation, the concepts of target strength and reverberation must be developed further.

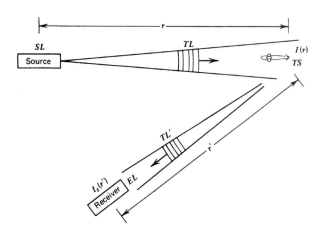

Figure 15.11.1 Diagram used in deriving the expression for target strength.

(a) Target Strength

When an acoustic wave ensonifies a target with intensity $I(r)$, the target scatters sound in all directions, some being sent in the direction of the receiver. As far as the receiver is concerned, the target has generated (albeit by reflection) an acoustic signal with source strength that is determined by extrapolating the scattered signal back to a distance $r' = 1$ m from the acoustic center of the target. (See Fig. 15.11.1.)

The apparent power Π radiated by the target as seen at the receiver is $I_s(r' = 1)$ multiplied by the area of the unit sphere centered on the acoustic center of the target. Thus, we have $\Pi = 4\pi I_s(1)$. Following Section 8.10, we define the *acoustic cross section* σ observed by the receiver as $\sigma = \Pi/I(r)$. This gives $\sigma I(r) = 4\pi I_s(1)$, or

$$I_s(r' = 1)/I(r) = \sigma/4\pi \tag{15.11.1}$$

The acoustic cross section is a strong function of the *orientation* of the target in the incident field and the *angle* between the directions from source to target and target to receiver. It is often assumed to be independent of the *distances* from source to target and from target to receiver, but it may be affected if the target is large or the range is small enough that the source, target, and/or receiver lie within each other's near fields or there is not complete ensonification of the full extent of the target.

The echo level at the receiver is

$$EL = 10\log[I_s(r' = 1)/I_{ref}] - TL'$$
$$= 10\log\left(\frac{I(r)}{I_{ref}}\frac{\sigma}{4\pi}\right) - TL' \tag{15.11.2}$$

Recognizing that $10\log[I(r)/I_{ref}]$ is $SL - TL$ and comparison with (15.8.6) reveals the relationship between σ and TS,

$$TS = 10\log(\sigma/4\pi) \tag{15.11.3}$$

If the first term on the right of (15.11.2) is interpreted as an *apparent source level* sL, then the echo level can be expressed in the form of (15.8.3) for passive sonar,

$$EL = sL - TL'$$
$$sL = 10\log[I(r)/I_{ref}] + TS \tag{15.11.4}$$

The target strength of a reflecting object is determined primarily by its size, shape, construction, orientation with respect to source and receiver, and the frequency of the incident sound. Let us examine two particularly simple cases.

(1) ISOTROPIC SCATTERING FROM A SMALL PERFECTLY REFLECTING SPHERE   Since the sphere is small ($ka \ll 1$), perfectly reflecting, and scatters uniformly in all directions, the pressure amplitude of the scattered spherical wave $P_s(r')$ on the surface of the sphere ($r' = a$) is equal to the pressure amplitude $P(r)$ of the incident wave so that $I_s(a) = I(r)$. The scattered intensity at distance $r' = 1$ m is related to that at $r' = a$ by $I_s(1)/a^2 = I_s(a)$. Substitution into (15.11.1) yields $\sigma = 4\pi a^2$. Now, substitution into (15.11.3) yields the target strength,

$$TS = 20 \log a \tag{15.11.5}$$

In this low-frequency limit a sphere of 1 m radius will have a target strength of 0 dB. A 0 dB target corresponds to one that reradiates sound with an apparent source level equal to the pressure level of the incident sound. For larger spheres in the low-frequency limit, the apparent source level is greater than the level of the incident sound. This is a consequence of the basic definition (15.11.1), which requires extrapolation back to $r' = 1$ m.

(2) BACKSCATTERING FROM A LARGE PERFECTLY REFLECTING SPHERE   For backscattering, we can assume the source and receiver are coincident (monostatic). For $ka \gg 1$ we can use ray acoustics to model the problem. The backscattered sound that reaches the receiver must arise from the portion of the spherical surface that is essentially perpendicular to the direction of the incident wave. This sound must be reflected back from this spherical cap into the infinitesimally small solid angle that encompasses the active element of the receiver as viewed from the target. In the ray limit, each ray of sound reflecting from the spherical cap must have equal angles of incidence and reflection. A sketch and a little geometry then shows that the reflected rays appear to originate at a point $a/2$ behind the spherical surface. Thus, these reflected (backscattered) rays appear to be coming from a sphere of radius $a/2$. Application of the same argument involving the incident and backscattered intensities as above leads immediately to the expression

$$TS = 20 \log(a/2) \tag{15.11.6}$$

In this high-frequency limit, a sphere of radius 2 m will exhibit a target strength of 0 dB. Note that the only portion of the target that contributes to the backscattering is the spherical cap transverse to the direction from target to receiver. The rest of the target makes no contribution. In this high-frequency limit, only this cap is important. Thus, an incoming torpedo whose nose has a radius of curvature a will have the above target strength regardless of the actual transverse size of the torpedo (as long as the width of the torpedo is many wavelengths of the detecting acoustic pulse).

(3) SCATTERING FROM IRREGULARLY SHAPED TARGETS   The target strength of an irregularly shaped object, such as a submarine, may be expected to depend on its orientations with respect to both the acoustic source and the receiver. For instance, the measured monostatic target strengths of World War II fleet submarines are at a minimum of about 10 dB for bow and stern aspects, increasing to about 25 dB on a beam aspect.

Little is known about the frequency dependence of the target strength for many targets. However, high frequencies lend themselves more favorably to classification of a target because the short wavelengths allow some of the structure of the target to be observed in the received echo, whereas for longer wavelengths much detail in the echo is lost. Very short high-frequency pulses will show reflections from various features of the target as discrete or overlapping returns. A pulse with length much less than or on the order of the apparent length of the target measured in the direction of propagation will return as a much longer echo with considerable amplitude modulation as it reflects from features first at the leading edge of the target and later at the trailing edge. A very long pulse will establish something closer to the target strength measured under continuous wave excitation.

For the bistatic geometry, the *bistatic angle* is defined as that formed by the intersection of the ray paths from source to target and target to receiver. (For the monostatic limit the bistatic angle is zero; if the target lies on a straight line between source and receiver, the bistatic angle is 180°.) A rough criterion borrowed from the radar community is that the bistatic target strength is approximated by the monostatic target strength measured in the direction of the bisector of the bistatic angle. How good this approximation is remains to be seen.

The wake generated by the propellers of the target can contribute to its detectability by providing a turbulent region filled with bubbles that can generate considerable scattered signal. The importance of the wake is a strong function of the speed of the target and its depth, tending to be very weak for a slow, deep submarine. A typical wake strength is between 0 and -30 dB for each meter of length illuminated.

(4) BACKSCATTERING FROM A LARGE FLAT SURFACE   The acoustic source often serves as a *fathometer* directed straight downward to reflect off the ocean bottom. Let the bottom be a distance z below the fathometer. The time $t = 2z/\langle c\rangle_z$ for the transmitted pulse to return from the bottom allows determination of the depth of the bottom. If the bottom is flat relative to a wavelength and consists of a thick homogeneous layer, then image theory predicts that the echo appears to be emitted by an image a distance z below the bottom with the source level SL of the fathometer reduced by $10 \log R_\Pi$, where R_Π is the power reflection coefficient of the bottom for normal incidence. The echo level EL received by the fathometer is then

$$EL = SL - 20 \log 2z - 2az + 10 \log R_\Pi \qquad (15.11.7)$$

(b) Reverberation

When a source ensonifies some portion of the ocean, there may be scattering from bubbles, particulate matter, fish, the sea surface and bottom, and any other inhomogeneities present. Some of this reverberation will compete with the echo from the target of interest. An essential step in obtaining the reverberation level RL is computing the volume V (or surface A) at the range of the target from which scattered sound can arrive at the receiver during the same time as the echo from the desired target. Clearly, this will depend on the pulse length, the directivities of source and receiver, and the geometry. Given this volume (or surface), the reverberation level can then be calculated directly. For the moment, we will assume that V (or A) is known and obtain the desired formulas for RL.

Figure 15.11.2 shows the target and the surrounding reverberation volume that can scatter sound to compete with the echo at the receiver. The intensity $I(r)$ of the

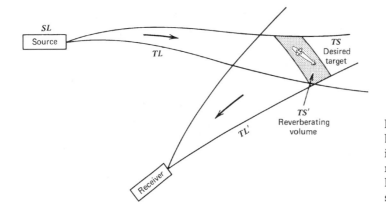

Figure 15.11.2
Diagram used
in deriving the
reverberation
level for volume
scatters.

signal illuminating this region is related to the source level SL of the transmitter by

$$10 \log[I(r)/I_{ref}] = SL - TL \tag{15.11.8}$$

where TL is the transmission loss from the source to the target.

From (15.11.3) and (15.11.4), each scatterer within the reverberating volume has an apparent source level given by

$$sL_i = 10 \log[I(r)/I_{ref}] + 10 \log(\sigma_i/4\pi) \tag{15.11.9}$$

where σ_i is the acoustic cross section of the ith scatterer. If the individual scatterers have random phases, the total scattered intensity is the sum of the individual intensities:

$$sL = 10 \log[I(r)/I_{ref}] + 10 \log\left(\sum_V \sigma_i/4\pi\right) \tag{15.11.10}$$

where the summation covers all the scatterers contained in V.

The *scattering strength* S_V for a *unit* volume is defined by

$$S_V = 10 \log\left(\sum_V \sigma_i/4\pi\right) - 10 \log V \tag{15.11.11}$$

so that

$$sL = 10 \log[I(r)/I_{ref}] + S_V + 10 \log V \tag{15.11.12}$$

Analogously, if the reverberation comes from a surface of area A, we can define a scattering strength S_A for unit surface area. The scattered sound level from the surface is then

$$sL = 10 \log[I(r)/I_{ref}] + S_A + 10 \log A \tag{15.11.13}$$

Reference to (15.11.4) shows that the last two terms in each of the above equations can be interpreted as target strengths for the reverberating region,

$$TS_R = \begin{cases} S_V + 10 \log V \\ S_A + 10 \log A \end{cases} \qquad (15.11.14)$$

The reverberation level at the receiver is $RL = sL - TL'$ and combination of this with the above equations results in

$$RL = SL - (TL + TL') + TS_R \qquad (15.11.15)$$

Another form of (15.8.9) is

$$TS \geq TS_R + DT_R \qquad (15.11.16)$$

This shows that reverberant interference with the desired signal is no more than the competition from undesired targets (scatterers). Equations (15.11.15) and (15.11.16) are independent of the source strength. Thus, once the SL is large enough that reverberation becomes more important than noise, there is no profit in increasing SL any further. Attention must be directed to increasing DI, thereby reducing V (or A) or reducing DT_R to obtain any improvement in detection range.

(1) VOLUME REVERBERATION   Table 15.11.1 gives approximate values for S_V for the ocean. Note that the deep scattering layer, a region of high biological activity,

Table 15.11.1   Approximate scattering strengths for deep ocean water

---

Volume Scattering
   Deep Scattering Layers (1 to 20 kHz)
   Scattering strength S_V ranges between -90 and -60 dB, with higher values tending to occur at higher frequencies; within the layer there is the possibility of strong peaks at specific frequencies and depths, corresponding to distinct biological species. Reverberation from the layer shows considerable structure, including sublayers at various depths.
   Water Volume (1 to 20 kHz)
   S_V varies between about -100 and -70 dB, with higher values tending to occur at higher frequencies and shallower depths.

Surface Reverberation Including Surface Layer (300 Hz to 4 kHz)
   At very low grazing angles and sea states between about 1 and 4, the scattering strength S_A ranges between about -55 to -45 dB. In this range of grazing angles the bubble layer lying just beneath the sea surface assumes greatest importance, especially at the higher frequencies. For grazing angles near about 40°, S_A increases to between -40 and -30 dB. Except near normal incidence, S_A tends to increase with frequency.

Bottom Reverberation (kHz range)
   Up to about 60°, grazing incidence scattering strength S_A is highly variable, depending on bottom and subbottom composition and roughness. S_A tends to increase with frequency, but not always. As grazing angles tend to 0°, S_A decreases very rapidly to become negligible. For grazing angles between about 20° and 60°, S_A lies between about -40 and -10 dB, the actual values being highly dependent on the bottom type.

Sources: Urick, *Principles of Underwater Sound,* 2nd. ed., McGraw-Hill (1975). Blatzler and Vent, *J. Acoust. Soc. Am.,* **41,** 154 (1967); Patterson, *ibid.,* **46,** 756 (1969); Scrimger and Turner, *ibid.,* **46,** 771 (1969); Urick, *ibid.,* **48,** 392 (1970); Hall, *ibid.,* **50,** 940 (1971); Brown and Saenger, *ibid.,* **52,** 944 (1972).

has a much higher S_V than ocean water in general. This layer possesses considerable structure, so its acoustic behavior is rather complicated.

Let us now return to the problem of finding V (or A). Because of the complexity of a general bistatic geometry, we will consider only the monostatic case. Assume that source and receiver have coaxial major lobes, and that the more highly directive of the two is described by an effective solid angle Ω_{eff}. The thickness of the reverberation volume V must be such that all the scatterers in it can contribute reverberation at the same time at the receiver. Let the source initiate a pulse of duration τ at time $t = 0$. Reverberation from scatterers at range r will reach the receiver during times $2r/c < t < 2r/c + \tau$. Reverberation from scatterers at range $r + L$ will be received during times $2(r + L)/c < t' < 2(r + L)/c + \tau$. If L is to be the thickness of V, then the received reverberation from the scatterers at range r must have just stopped when the reverberation received from those at $r + L$ just starts. This means that $2r/c + \tau = 2(r + L)/c$. Solving for L yields $L = c\tau/2$.

If the cross-sectional area of the reverberation volume that lies within Ω_{eff} is A_T, then the reverberation target strength is

$$TS_R = 10\log(A_T c\tau/2) + S_V \tag{15.11.17}$$

The geometrical transmission loss is simply the ratio of A_T to the area subtending Ω_{eff} at a distance of 1 m,

$$TL(geom) = 10\log(A_T/\Omega_{eff}) \tag{15.11.18}$$

so that the reverberation target strength TS_R is

$$TS_R = S_V + TL(geom) + 10\log(\Omega_{eff}c\tau/2) \tag{15.11.19}$$

and the reverberation level becomes

$$RL = SL - TL(geom) - 2TL(losses) + S_V + 10\log(\Omega_{eff}c\tau/2) \tag{15.11.20}$$

If it is more convenient to express (15.11.20) in terms of the directivity index, recall that Ω_{eff} can be replaced by $4\pi/D$.

The reverberation level diminishes as $TL(geom)$, whereas the echo level diminishes as $2TL(geom)$. Thus, RL diminishes with range slower than EL, as mentioned in the argument concerning Fig. 15.8.1. Clearly, decreasing Ω_{eff} and τ allows detection of a target at larger ranges. However, recall that TS can be reduced if the pulse length is less than the length of the target, and decreasing Ω_{eff} decreases the search rate.

(2) Surface Reverberation   If reverberation arises from a surface like the sea surface or bottom, then the reverberation target strength is

$$TS_R = 10\log A + S_A \tag{15.11.21}$$

Since there is very little horizontal bending of the rays, the surface area for an acoustic beam that grazes the surface at very small angles can be seen from Fig. 15.11.3 to be $A = r\theta(c\tau/2)$, where, in a monostatic coaxial geometry, θ is the smaller horizontal beam width of the source and receiver. This gives

$$TS_R = S_A + 10\log r + 10\log(\theta c\tau/2) \tag{15.11.22}$$

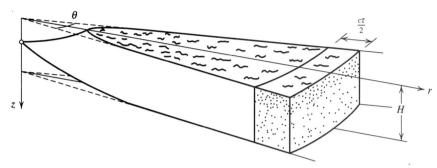

Figure 15.11.3 Diagram used in deriving the reverberation level for surface scatters.

The TS_R increases with range as $10 \log r$. The surface reverberation level becomes

$$RL = SL - 2TL + 10 \log r + S_A + 10 \log(\theta c\tau/2) \qquad (15.11.23)$$

As in the case of volume reverberation, detection ability decreases with range, but for surface reverberation the geometrical transmission loss is $10 \log r$.

Reverberation from the sea surface may involve a concentration of bubbles lying beneath the surface. It is convenient to lump the scattering within the layer of bubbles together with the scattering from the rough sea surface and treat the combination as occurring at the surface. The total effective surface scattering strength S_A must be the combination of the scattering S_S from the surface and the scattering S_B from the bubble layer,

$$S_A = 10 \log(10^{S_S/10} + 10^{S_B/10}) \qquad (15.11.24)$$

(or their levels are combined according to the nomogram of Section 11.3). Let $\sigma(a, z)$ be the acoustic cross section of a bubble of radius a and depth z, and $n(a, z)$ the number per unit volume of bubbles of radius a at depth z. Integrating $n\sigma/4\pi$ over all bubbles and the layer depth H gives the scattering strength of the bubbles,

$$S_B = 10 \log \left[\int_0^H \int_0^\infty n(a, z) \frac{\sigma(a, z)}{4\pi} da\, dz \right] \qquad (15.11.25)$$

Scattering from the rough sea surface is important for lower frequencies and larger grazing angles, whereas at higher frequencies and very low grazing angles the layer of bubbles becomes increasingly important.

For the ocean bottom, the scattering depends on frequency, angle of incidence, roughness of the bottom, and its composition. At lower frequencies, the composition and disposition of the deeper sublayers within the bottom become increasingly important. Table 15.11.1 gives representative values of surface scattering strengths S_A of ocean surface and bottom.

(c) Detection Threshold for Reverberation-Limited Performance

Basically, reverberation consists of a collective of incoherent, scattered replicas of the incident acoustic pulse. Correlation detection will be degraded since this undesired masking noise consists of myriads of pulses looking very much like the desired signal. Arguments have been made (e.g., see Urick[7]) that detection of an echo in reverberation should be treated like energy detection. Thus, the detection threshold for reverberation over the bandwidth w of the processor should be based on the equation for square-law detection given in (11.5.7). This expression must be modified, however, because of the frequency properties of reverberation. The preceding calculations for the reverberation level RL have been made under the assumption that all the scatterers are stationary with respect to each other and the water. In fact, this is usually not the case. For sea surface and volume reverberation, the scatterers may have significant relative motion. This means that the total reverberation for these cases will be distributed across a frequency interval determined by the complete range w_R of Doppler shifts resulting from the motions of the scatterers. This reverberation bandwidth w_R may exceed the bandwidth w of the receiving processors and may be distributed across several processors. A simple approach is to consider the reverberation spectrum level as fairly constant across its bandwidth. Then the reverberation level RL_w over the bandwidth w of each appropriate processor could be found by subtracting $10 \log w_R$ to obtain the reverberation spectrum level RSL and then adding $10 \log w$ to obtain the level within the processor bandwidth. Alternatively, a simpler (and equivalent) approach is not to modify the reverberation level within each processor, but rather to perform the equivalent conversion on the detection threshold of the processor. Thus, we leave the reverberation level RL as calculated, but convert the detection threshold given by (11.5.7) to DT_R according to

$$DT_R = 5 \log(d/w\tau) - 10 \log(w_R/w) \qquad (15.11.26)$$

If w_R is less than and entirely contained within w, then the bandwidth correction in the above equation should be ignored, and when the two bandwidths partially overlap, the equation should be modified appropriately.

If the target is stationary in the water and the scatterers have zero average velocity in the water, then the reverberation will have a center frequency equal to the center frequency of the target echo. Both will appear in the same processor, so that the performance will be reverberation limited if $(RL + DT_R)$ is significantly larger than $(NL - DI + DT_N)$, where DT_N is for square-law or correlation detection against noise, as appropriate for the processor. If, on the other hand, the target Doppler is large enough that the echo is removed from those processors occupied by the reverberation, then the performance will be noise limited.

(d) An Example

A stationary active sonar operates at 1 kHz with a source level of 220 dB *re* 1 μPa, a directivity index of 20 dB, a horizontal beam width of 10° (0.17 radian), and a pulse length of 0.1 s. Correlation detection is to be used and it is desired that the probabilities of detection and false alarm be 0.50 and 10^{-4}, respectively. Detection must be made with a single processor with a bandwidth wide enough not to reject a fast target. The target strength of the submarine is expected to be 30 dB and its speed may be up to 38 km/h (20.5 knots). Both the source and the submarine are in a mixed layer of 100 m depth with the source at 36 m and the submarine near the surface. The sea state is 3. The scattering strength for surface scatters is -30 dB, and volume scattering is negligible. Because of these maximum range rates,

[7]Urick, *op. cit.*

and because a single processor is being used, the reverberation bandwidth w_R can be assumed to be contained within the receiver bandwidth w. Both reverberation-limited and noise-limited cases must be investigated. The propagation conditions are the same as in the passive-sonar example [Section 15.10(a)]. If detection is made at $r > r_t$, we have

$$TL = 10 \log r + (4.1 \times 10^{-4})r + 31.2 \qquad (15.11.27)$$

For a range rate of 38 km/h the bandwidth at 1 kHz is found from (15.9.13) to be 28.1 Hz. Figure 11.4.2 reveals that the detection index must be 16 to provide the required probabilities of detection and false alarm, and (11.5.7) yields a detection threshold of $DT_N = 4.6$ dB. (1) First assume that the detection is noise limited. The detected noise level DNL is given by (15.9.1), and the noise spectrum level is 62 dB re 1 $\mu Pa/Hz^{1/2}$. Thus, $DNL(noise) = 56.5$ dB. Substitution of our known values into (15.8.7) gives $TL = 94.5$ dB and then solution of (15.11.27) for r by trial and error gives $r = 42$ km as the maximum detection range for noise-limited performance. (2) Now, assume that the detection is reverberation limited. Since w_R is assumed less than and contained by w, all the reverberation is within the processor bandwidth. The reverberation detection threshold DT_R is then 3.8 dB. Solving (15.8.9) and (15.11.23) for $10 \log r$ yields

$$10 \log r = 45.1 \qquad (15.11.28)$$

This gives $r = 32$ km for the maximum detection range under reverberation-limited performance. At this range, from (15.8.9) or (15.11.23) we have $RL = 66.2$ dB so that $(RL + DT_R) = 70$ dB exceeds $[DNL(noise) + DT_N] = 61$ dB by a substantial margin of 9 dB. Thus, the system is reverberation limited and detections under the specified probabilities can be made out to a range of about 32 km.

*15.12 ISOSPEED SHALLOW-WATER CHANNEL

In many isospeed environments it is possible to apply the powerful but cumbersome method of images. Solution by this method is usually expressed as a finite or infinite sum of the contributions of all the images. Profitable analysis is often accomplished only with the aid of digital computers. See, however, Tolstoy and Clay.[8]

As a consequence of the multiplicity of reflections taking place at both surface and bottom, predictions of sound transmission in shallow water are more complicated than for deep water. According to the method of images, the surface and bottom of the isospeed layer can be treated as interfaces across which sound propagates with pressure amplitude reduced by the reflection coefficient for that surface and angle of incidence. This gives a multilayered space (Fig. 15.12.1) wherein the various images of the source contribute to the field point along straight-line paths. The images have been labeled with the number of times the signal from the image to the field point reflects from the bottom. Simple transmission loss models can be obtained if some approximations are made.[9] The ocean surface is assumed to be a perfect reflector. If the inhomogeneities in the water and the roughness of the surface and bottom are assumed sufficient to randomize the phases of the various images, incoherent combination of the contributions can be assumed. Further,

[8]Tolstoy and Clay, *Ocean Acoustics*, McGraw-Hill (1966); republished, Acoustical Society of America (1987).
[9]This approach is adapted from a more complicated situation analyzed by Macpherson and Daintith, *J. Acoust. Soc. Am.*, **41**, 850 (1966).

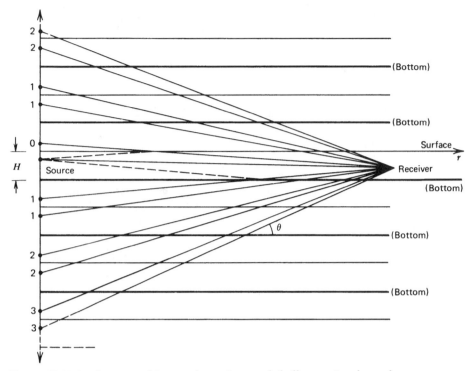

Figure 15.12.1　Source and images for an isospeed shallow-water channel.

at large ranges each set of four images whose contributions make i intersections with the bottom can be assumed to lie at an effective distance $r_i = [r^2 + (2iH)^2]^{1/2}$ from the field point. This allows the images to be summed in groups of four,

$$\frac{I(r)}{I(1)} = \frac{2}{r^2} + 4\sum_{i=1}^{\infty} \frac{R^{2i}(\theta_i)}{r^2 + (2iH)^2} = \frac{2}{r^2}(1 + 2S) \tag{15.12.1}$$

where S is

$$S = \sum_{i=1}^{\infty} \frac{R^{2i}(\theta_i)}{1 + i^2(2H/r)^2} \tag{15.12.2}$$

and each grazing angle of incidence θ_i is found from

$$\cos\theta_i = 1/[1 + i^2(2H/r)^2]^{1/2} \tag{15.12.3}$$

If the range from source to receiver is much greater than the channel depth, $H/r \ll 1$, the summation can be replaced by an integral,

$$S = \int_1^{\infty} R^{2u}(\theta)\cos^2\theta\, du \tag{15.12.4}$$

where the variable of integration u has replaced i and θ is a function of u. From (15.12.3) we see that $u = (r/2H)\tan\theta$, and changing the variable of integration from u to θ yields

$$S = \frac{r}{2H}\int_{\tan^{-1}(2H/r)}^{\pi/2} R^{(r/H)\tan\theta}\, d\theta \tag{15.12.5}$$

(a) Rigid Bottom

If the pressure reflection coefficient at the bottom is unity for all θ, the integrand of (15.12.5) is unity and for $r \gg H$,

$$S \approx \frac{r}{2H} \int_{2H/r}^{\pi/2} d\theta = \frac{\pi}{4} \frac{r}{H} - 1 \tag{15.12.6}$$

and substitution into (15.12.1) results in

$$TL(geom) \approx 10 \log r + 10 \log(H/\pi) \tag{15.12.7}$$

The divergence is cylindrical and there is a contribution to TL analogous to that of r_t for the surface duct if we identify r_t with H/π in (15.5.4). This will be compared later with (15.13.5), an estimate based on the incoherent combination of the terms resulting from a normal-mode model of this case.

(b) Slow Bottom

If the bottom has a slower speed of sound than the water, then the reflection coefficient is small for large grazing angles and grows to unity for small grazing angles. The major contribution to S will therefore come from the lower limit of integration. For small incidence angles the reflection coefficient can be written as $R(\theta) \sim \exp(-\gamma\theta)$, where γ is a parameter to be determined from the bottom characteristics. (See Problem 15.12.5.) Since most of the value of S comes from $\theta \sim 2H/r$, replace $\tan\theta$ with θ and let the upper limit of the integration become infinitely large. Then, (15.12.5) becomes

$$S \approx \frac{r}{2H} \int_{2H/r}^{\infty} e^{-(r/H)\gamma\theta^2} d\theta \tag{15.12.8}$$

A change of variable from θ to $x = \theta\sqrt{\gamma r/H}$ yields

$$S \sim \frac{1}{2}\left(\frac{r}{\gamma H}\right)^{1/2} \int_{2\sqrt{H\gamma/r}}^{\infty} e^{-x^2} dx \tag{15.12.9}$$

For large r the lower limit goes to zero and the integral goes to $\sqrt{\pi}/2$. Substitution into (15.12.1) gives

$$TL(geom) \sim 15 \log r + 5 \log(\gamma H/\pi) \tag{15.12.10}$$

The effect of a slow bottom is to provide a geometrical spreading between spherical and cylindrical.

(c) Fast Bottom

In this case the reflection coefficient is identically unity for all grazing angles less than the critical grazing angle θ_c. All paths that reflect from the bottom with larger grazing angles suffer reflection losses and attenuate more rapidly than cylindrically. Paths more grazing than critical will be trapped and spread cylindrically at large ranges. Because of these considerations, (15.12.6) can be evaluated by integrating from the lower limit up to θ_c and setting $R(\theta) = 1$ within that range. With the restriction $r \gg 2H/\theta_c$, the lower limit can be replaced with zero, and S becomes

$$S = \frac{r}{2H} \int_{0}^{\theta_c} d\theta = \frac{r}{2H}\theta_c \tag{15.12.11}$$

and thus

$$TL(geom) = 10 \log r + 10 \log(H/2\theta_c) \qquad (15.12.12)$$

Note that θ_c plays the same role that the limiting angle θ_0 plays in the mixed layer: the approximations applied to the acoustic fields obtained by the method of images have provided results similar to those based on ray theory. This will compare well with (15.13.6), which is based on a normal-mode approach.

*15.13 TRANSMISSION LOSS MODELS FOR NORMAL-MODE PROPAGATION

In many ocean propagation situations, particularly for low frequencies in a mixed layer or shallow water, ray theory fails and it is necessary to use the normal-mode theory as discussed in Sections 9.7 to 9.9. Figure 15.13.1 shows the differences between the transmission loss calculated by ray theory and normal-mode theory and compares these predictions to data gathered at sea.[10] The sound-speed profile (Fig. 15.13.1a) has a minimum at about 10 m so that the sound from a source at about 15 m will tend to be trapped in a shallow duct as shown by the ray traces in Fig. 15.13.1b. The transmission loss for a receiver at 71 m is shown in Fig. 15.13.1c. The scatter in the data is representative of what to expect from experiments conducted in the ocean. The transmission loss calculated from ray theory shows a pronounced surface interference effect at short ranges and a shadow zone beginning at around 7 km that is absent from the data. The normal-mode calculation shows good agreement with the data for all ranges. Figure 15.13.1d shows that good agreement is obtained at long ranges if only a few modes (10) are included in the calculation. However, reproducing the complicated behavior in the region of the surface interference requires many modes (40) to be calculated. As a rule of thumb, ray theory is useful at high frequencies and short ranges, while normal-mode theory is useful at low frequencies and long ranges.

Given the normal-mode solution Z_n for a particular sound-speed profile, the transmission loss can be calculated from (9.7.9), where the summation can be done either phase-coherently or randomly, depending on how greatly the irregularities of the medium affect the phase of each normal mode. If coherent phasing is plausible, the geometrical transmission loss is

$$TL(geom) = -20 \log \left| \sum_n (2\pi/\kappa_n r)^{1/2} Z_n(z_0) Z_n(z) e^{-j\kappa_n r} \right| \qquad (15.13.1)$$

If the assumption of random phasing is more appropriate, then

$$TL(geom) = -10 \log \left| \sum_n (2\pi/\kappa_n r) Z_n^2(z_0) Z_n^2(z) \right| \qquad (15.13.2)$$

The second of these equations is often the easier to calculate and amounts to a spatial smoothing. Note that TL is a function of both source and receiver depth. These forms reveal a distinct advantage of the normal-mode approach: once the set of eigenfunctions $Z_n(z)$ has been determined for a particular sound-speed profile, the range and depth dependence of the transmission loss can be calculated directly. Ray theory, on the other hand, requires a separate ray-tracing effort for each pair of positions of source and receiver.

Including the attenuation of sound from absorption and scattering in the normal-mode theory is not an easy matter. In general, each mode will have its own frequency-dependent attenuation coefficient α_n. If these can be determined, the total transmission loss is found

[10]Pedersen and Gordon, J. Acoust. Soc. Am., **37**, 105 (1965).

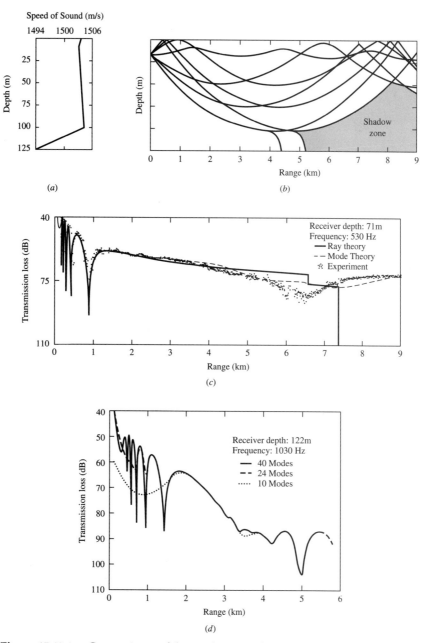

Figure 15.13.1   Comparisons of the predictions of ray and normal-mode theories compared to measured transmission loss. (*a*) Sound-speed profile. (*b*) Ray traces. (*c*) Transmission loss for a 530 Hz source at a depth of 17 m and a receiver at a depth of 71 m. (*d*) The results of mode theory as more modes are included in the calculation (122 m receiver at 1030 Hz). [Parts (*a*) and (*b*) adapted from Pedersen, *J. Acoust. Soc. Am.*, **34,** 1197 (1962) and parts (*c*) and (*d*) from Pedersen and Gordon, *J. Acoust. Soc. Am.*, **37,** 105 (1965).]

by multiplying each term in (15.13.1) by $\exp(-\alpha_n r)$ for coherent phasing or by multiplying each term in (15.13.2) by $\exp(-2\alpha_n r)$ for random phasing. Note that the transmission loss does not separate into a geometrical term plus a loss term unless all α_n are identical, a highly unlikely event. Since normal-mode theory is most useful at low frequencies, it is reasonable that we ignore losses in what follows.

(a) Rigid Bottom

It is instructive to obtain an estimate of the transmission loss for a fluid layer over a rigid bottom. (1) Let the channel be sufficiently nonideal that random phasing can be assumed. Then (15.13.2) yields

$$TL(geom) = -10 \log \left[\frac{4}{H^2} \sum_{n=1}^{N} \left(\frac{2\pi}{\kappa_n r} \right) \sin^2 k_{zn} z_0 \, \sin^2 k_{zn} z \right] \tag{15.13.3}$$

(2) Assume that the source and receiver are well away from either the top or bottom so that the values of the $\sin^2$ terms will fluctuate between 0 and 1 pseudorandomly as n varies from 1 to N. We can then take a statistical average and let the value of each $\sin^2$ term be $\frac{1}{2}$. Furthermore, if the frequency is sufficiently high, the number of modes will be large so that N can be approximated by $(H/\pi)(\omega/c_0)$ from (9.8.3); then use of this with (9.8.1) and (9.8.2) shows that κ_n can be replaced by $(\omega/c_0)[1 - (n/N)^2]^{1/2}$. The argument of the log can now be simplified to

$$\left(\frac{2\pi}{H^2 r} \right) \sum_{n=1}^{N} \frac{1}{\kappa_n} \approx \left(\frac{2\pi}{H^2 r} \right) \left(\frac{c_0}{\omega} \right) \sum_{n=1}^{N} \left[1 - \left(\frac{n}{N} \right)^2 \right]^{-1/2}$$

$$\approx \left(\frac{2\pi}{H^2 r} \right) \left(\frac{c_0}{\omega} \right) N \int_0^1 (1 - x^2)^{-1/2} \, dx \tag{15.13.4}$$

$$= \left(\frac{2\pi}{H^2 r} \right) \left(\frac{c_0}{\omega} \right) N \int_0^{\pi/2} d\theta = \frac{\pi}{Hr}$$

and thus

$$TL(geom) \approx 10 \log r + 10 \log(H/\pi) \tag{15.13.5}$$

This compares exactly with (15.12.7), which was obtained on the basis of incoherent summing of the individual images. Clearly, near the top boundary the expected values of $\sin^2$ terms will be less than $\frac{1}{2}$, so the transmission loss will increase above (15.13.5). Similarly, the transmission loss will decrease if source and receiver are near the bottom, where $\sin^2$ tends more toward 1.

(b) Fast Bottom

Application of the same procedure to the isospeed channel with fast bottom [described in Section 15.12(c)] gives the geometrical transmission loss under the assumption of phase incoherence. The result is

$$TL(geom) = 10 \log r + 10 \log \left(\frac{H}{2 \sin \theta_c} \right) \tag{15.13.6}$$

where $\sin \theta_c = [1 - (c_1/c_2)^2]^{1/2}$. This compares excellently with (15.12.12) for grazing critical angles $\theta_c < \pi/4$.

Problems

Unless otherwise indicated, assume *TL(geom)* is given by spherical spreading. Useful conversions can be found in Appendix A1.

15.2.1. From (15.2.1) and (15.2.2) compute the speeds of sound in seawater at a depth of 4000 m, a temperature of 4°C, and a salinity of 35 ppt at latitudes of 0°and 45°.

15.2.2C. Plot the speed of sound in water for $S = 0$ for various values of pressure from (15.2.1) and compare to that given for freshwater by (5.6.8).

15.2.3C. (*a*) Plot the speed of sound in surface seawater as a function of temperature for various values of salinity. Discuss the relative importance of temperature and salinity for realistic values of both parameters. (*b*) Repeat (*a*) for temperature and pressure for $S = 35$ ppt.

15.3.1. Compute the approximate sound absorption coefficient (dB/km) in seawater at frequencies of 100 Hz, 1 kHz, and 10 kHz and depths of 0 and 2000 m from (15.3.6) *et seq.* Compare with the values obtained from (8.7.3) for these frequencies and depths for seawater with $S = 35$ ppt, pH = 8, and temperatures of 5°C and 20°C.

15.3.2. A 30 kHz sonar transducer produces an axial sound pressure level of 140 dB *re* 1 μPa at a distance of 1000 m in seawater. Assume spherical spreading and losses. (*a*) What is the axial pressure level at 1 km? (*b*) At 2000 m? (*c*) At what distance will the axial pressure level be reduced to 100 dB? (*d*) At what distance will *TL(geom)* equal *TL(losses)*? (*e*) At what distance is the rate of transmission loss associated with spherical divergence equal to that associated with absorption?

15.3.3C. (*a*) Plot the attenuation in seawater as a function of frequency from (15.3.6) and compare to that obtained from (8.7.3) for the same conditions. (*b*) Investigate the limits of applicability of (15.3.6) by plotting (8.7.3) for various values of *T*, *S*, and pH and determining the values for which the deviation becomes unacceptable (see Leroy, *op. cit.*). How are these limits affected by frequency?

15.4.1. Show that $x = \sqrt{2c_0 d/g}$ gives the approximate horizontal distance *x* in which an initially horizontal ray will reach a depth *d* in a layer of water having a constant negative gradient of magnitude *g*.

15.4.2. (*a*) For a salinity of 35 ppt and at a latitude of 45°, verify that

$$c(T, 35, Z) \approx 1449.1 + 4.57T - 0.0526T^2 + 15.14Z + 0.181Z^2$$

is a reasonably accurate approximation for the speed of sound in the open ocean. *Hint:* For these conditions, $ZT < 20$ km $\cdot$ °C. (*b*) At a certain location in the midlatitudes, the speed of sound in seawater of 35 ppt salinity decreases uniformly from a value of 1500 m/s at the surface to 1480 m/s at a depth of 50 m. What is the gradient in depth? (*c*) What is the average temperature gradient? (*d*) What horizontal distance is required for a ray horizontal at the surface to reach a depth of 50 m? (*e*) What will be the downward angle of such a ray at this depth?

15.4.3C. (*a*) Develop a computer program to plot the ray paths in a horizontally stratified ocean for a source of arbitrary depth and an arbitrary number of layers each with its own constant sound speed. (*b*) Develop a computer program to plot the ray paths in a horizontally stratified ocean for a source of arbitrary depth and an arbitrary number of layers, each *i*th layer with its own constant gradient g_i and matching sound speeds with no discontinuities with the contiguous layers. (*c*) To compare the accuracy of these two models, find the number of constant-speed layers required to give reasonably accurate ray paths for a single layer of constant gradient.

15.4.4. Show that use of small-angle approximations (valid for $|\theta| < 20°$) and elimination of θ_2 from (15.4.3) provides a convenient parabolic relationship between range and depth increments along a ray:

$$\Delta z = \tan \theta_1 \, \Delta r - (g/2c_1)(\Delta r)^2$$

15.5.1. An isothermal layer of seawater has a temperature of 20°C, a salinity of 35 ppt, and extends to a depth of 40 m. A sonar transducer is located at a depth of 10 m in this isothermal layer. (*a*) In what horizontal distance will a ray, leaving the transducer in a horizontal direction, reach the surface of the water? (*b*) What is the downward angle from the transducer of a ray that will become horizontal at the bottom of the isothermal layer? (*c*) In what horizontal distance will the ray of part (*b*) reach the bottom of the isothermal layer? (*d*) The speed of sound decreases below the isothermal layer at a rate of 0.2 s^{-1}. What is the depth below the surface of a ray originally starting downward at an angle of 3° on reaching a horizontal distance of 2300 m?

15.5.2. A sonar transducer is at a depth of 5 m in shallow water having a flat bottom at a depth of 35 m. The speed of sound decreases linearly from a value of 1500 m/s at the surface to 1493 m/s at the bottom. (*a*) Calculate and plot the path for a ray leaving the transducer in a horizontal direction, until it strikes the bottom a second time. Assume the first reflection from the bottom to be specular. (*b*) Similarly calculate and plot the paths for rays initially 1° above and 1° below the horizontal.

15.5.3. A sonar transducer is at a depth of 10 m in water having a constant negative gradient of 0.2 s^{-1}. The speed of sound at the transducer depth is 1500 m/s. When the axis of the transducer is tilted down 6° from the horizontal, its beam appears to be centered on a target submarine at a horizontal distance of 1000 m. (*a*) What is the apparent depth of the submarine? (*b*) What is the true depth of the submarine?

15.5.4. A surface sound channel is formed by a layer of water in which the speed of sound decreases uniformly from 1500 m/s at the surface to 1498 m/s at a depth of 10 m and then increases uniformly to reach 1500 m/s at and below a depth of 100 m. (*a*) What is the maximum angle with which a ray may cross the axis of the channel and still remain in the channel? (*b*) What is the horizontal distance between crossings for such a ray in the upper part of the channel? (*c*) In the lower part of the channel? (*d*) Derive a general expression giving all angles of departure θ from a sound source on the axis of the above sound channel that will result in rays recrossing the channel axis at a distance of 3000 m.

15.5.5. The surface layer consists of isothermal water down to 100 m followed by a steep negative thermocline. If a submarine is just below the layer, calculate the range beyond which the submarine can try to hide from a surface sonar.

15.5.6. A source 7 m below the surface is generating a 3 kHz sound in a mixed layer of 100 m depth. The receiver is at the same depth. There is a loss of 6 dB/bounce. (*a*) Find the transition range. (*b*) Find the skip distance. (*c*) Find the approximate range for which the transmission loss is about 80 dB. (*d*) Find the minimum vertical beam width of the source that takes full advantage of the propagation characteristics of the layer.

15.5.7. Assume a mixed layer of depth 100 m and a 3.5 kHz source at a depth of 75 m. (*a*) Ignoring any "leakage" from the layer, calculate the transmission loss for a receiver at a depth of 50 m and a distance of 20 km. (*b*) Compare this answer with the observed value of about 90 dB and estimate the leakage coefficient in dB/km. From this, calculate the loss in dB/bounce.

15.6.1. By approximating the sound-speed profile of Fig. 15.4.1 with two linear segments, verify the estimated time stretching in the discussion after (15.6.7).

15.6.2. The sound-speed profile consists of straight lines connecting the following points: 1500 m/s at 0 m depth, 1514 m/s at 800 m, 1470 m/s at 1400 m, and 1562 m/s at 5400 m (bottom). (a) What is the depth of the axis of the deep sound channel? (b) What is the maximum angle at which a ray may leave a source on the axis of the channel and still be trapped? What is the cycle distance for this ray? (c) Assuming the receiver is also on the axis, find the transition range.

15.6.3C. A simple model for propagation in the deep sound channel consists of two layers each with the same absolute value of the sound-speed gradient. Plot the arrival time of rays leaving a source on the axis to a receiver on the axis at a distance of 2000 km as a function of elevation angle if $|g| = 0.025$ s^{-1}.

15.6.4C. A simple model for convergence zone propagation consists of two linear sound-speed gradients: with the sound speed 1500 m/s at the surface, 1475 m/s at 1000 m, and a gradient of 0.016 s^{-1} below 1000 m. The bottom is assumed deep enough not to interfere with any rays of interest. For a source at the surface, (a) plot, as a function of the angle of depression, the range at which a ray reaches the surface and determine the convergence zone range and the width of the convergence zone. (b) Determine the minimum water depth necessary for the convergence to exist.

15.6.5C. (a) For the sound-speed profile of Problem 15.6.4C with a flat, horizontal bottom at 3000 m and the source at the surface, plot, as a function of depression angle, the grazing angle of incidence of the ray on the bottom and the horizontal range at which a ray reaches the surface. (b) Compare the results of (a) to those obtained assuming the rays are straight lines and comment about the depression angles for which the straight-line approximation is applicable.

15.6.6C. If the bottom in Problem 15.6.5C is "red clay," use the straight-line approximation to plot the one-way transmission loss as a function of horizontal range for both the source and the target on the surface.

15.6.7C. For the sound-speed profile of Problem 15.6.4C and a source on the deep sound channel axis, (a) plot, as a function of elevation angle, the horizontal range at which a ray reaches the surface and compare this to the range calculated by assuming the rays are straight lines. (b) As a function of horizontal range, plot the transmission loss calculated from ray theory and compare to that calculated by assuming the rays are straight lines.

15.7.1. A source of spherically diverging waves of 1000 Hz is located 5 m below the surface in seawater. The source produces an rms pressure amplitude of 200 μbar at 1 m from the center of the source. Assuming 100% reflection ($\mu = 1$), what is the *SPL re* 1 μbar produced at a distance of 200 m and depths of (a) 1 m, (b) 5 m, and (c) 10 m? (d) What would be the sound pressure level produced by the direct wave alone at this distance? (e) Repeat the problem for $\mu = 0.5$.

15.7.2. (a) Assuming $\mu = 1$, derive an equation for the distance r expressed in terms of $k, h,$ and d beyond which the transmission loss caused by the surface interference effect exceeds that for spherical spreading by more than 10 dB. (b) What is this distance when $f = 500$ Hz, $d = 10$ m, and $h = 20$ m?

15.7.3. Consider a point source in water at a depth d below the surface. (a) What is the direction of the acoustic particle velocity on the surface? (b) If the pressure amplitude at 1 m from the point source alone is A, find the amplitude of the acoustic particle velocity on the surface in the limit $r \gg d$ and $kr \gg 1$.

15.7.4C. An omnidirectional 5 kHz source is at a depth of 10 m in isospeed water. For rms surface roughness from 0 to 0.6 m, plot the pressure amplitude (15.7.4) as a function of range.

15.8.1. A sonar at 1 kHz has a source level of 220 dB *re* 1 μPa and receives an echo of level 110 dB *re* 1 μPa from a target at 1 km distance. Calculate the target strength.

15.8.2. An active sonar system has a detection threshold of -3 dB and a source level of 220 dB *re* 1 μPa. If the detected noise level is 70 dB *re* 1 μPa and the one-way transmission loss is 80 dB, what is the weakest target it can detect 50% of the time?

15.8.3. A passive sonar system is designed to detect a tonal of level *SL*. If the noise has a spectrum level *NSL* constant over the bandwidth w, show that the appropriate sonar equation is

$$SL - TL = NSL - DI + DT_1$$

and express DT_1 in terms of d, w, and τ if the receiver uses energy detection.

15.8.4. A 2 kHz sonar first detects a bow aspect submarine at 5 km ($TS = 10$ dB). At what range will this same sonar first detect a bow aspect torpedo ($TS = -20$ dB)? Assume that the detection threshold is the same for both cases.

15.8.5. State what happens to $P(D)$, $P(FA)$, and detection index d, and explain qualitatively why it happens, in each of the following cases. (*a*) The target recedes and the received signal gets weaker (nothing else in the system changes). (*b*) With no other changes, the operator reduces the gain on the display (this is equivalent to raising the threshold).

15.8.6. How much must $w\tau$ change if DT is to be improved (decreased) 5 dB for (*a*) energy detection? (*b*) Correlation detection? (Assume d remains constant.)

15.9.1. Estimate the ambient noise level in deep water for sea state 3 and light shipping for a receiver with the following center frequencies and bandwidths: (*a*) 20 Hz and 0.1 Hz, (*b*) 200 Hz and 1 Hz, (*c*) 2000 Hz and 10 Hz, and (*d*) 20 kHz and 100 Hz.

15.9.2. A passive receiver operates at 500 Hz with a bandwidth of 100 Hz. Its directivity index is 10 dB. Find the detected noise level in a sea state 3.

15.9.3C. A submarine at 10 m depth traveling in a straight line at 4 kt is emitting an omnidirectional 5 kHz tone, which is being received by an omnidirectional hydrophone at 10 m depth. The distance of closest approach is 100 m. Plot the amplitude and frequency of the received signal as a function of time with $t = 0$ at the time of closest approach. Assume a constant speed of sound and a smooth sea surface.

15.10.1. A passive detection system uses energy detection. If the bandwidth is 200 Hz and the processing time is 10 s, find the $P(FA)$ for $P(D) = 0.5$ if the detection threshold must be -10 dB.

15.10.2. (*a*) Derive an expression for determining the optimum frequency F in kHz to be used in reaching a given detection range r with a line hydrophone passive detection system. Assume the continuous spectrum noise of the target to have the same dependence on frequency as the masking noise and that the sound absorption coefficient of the water is given by $a/F^2 = 0.01$ (dB/km)/(kHz)^2. (*b*) What frequency is optimum for reaching a range of 10 km? (*c*) What range will be reached most effectively at a frequency of 1 kHz?

15.10.3. A submarine radiates a monofrequency signal at 250 Hz with a source level of 150 dB *re* 1 μPa. The signal is received by an omnidirectional receiver in the presence of a sea state 3. (*a*) If the submarine is capable of 14.7 m/s (28.6 knots), estimate the requisite bandwidth of the receiver. (*b*) Evaluate the ambient *DNL* at the receiver. (*c*) It is desired to have detection at 10 km. What is the maximum allowed detection threshold? (*d*) If the false alarm rate of 0.2% at 50% probability of detection is acceptable, find the required observation time assuming energy processing.

(e) Repeat (b), (c), and (d) if the receiver consists of five parallel processors each of 1 Hz bandwidth. Assume the same overall false alarm rate.

15.10.4. A passive sonar system operates at 500 Hz. It must be able to detect a target generating a 500 Hz tone up to a range rate of 30 knots (15.4 m/s). (a) What must be the minimum bandwidth? (b) If the receiver contains 10 parallel filters of equal bandwidths, what is the bandwidth of each? (c) A detection is made in the highest frequency filter. What are the possible values for the range rate of the detected target?

15.10.5. (a) A conventional submarine at periscope depth is traveling at 4 knots (2.06 m/s). The sea state is 3 and the layer depth is 100 m. At what range will the submarine be detected by a 36 m deep hydrophone listening at 1 kHz if DI = 20 dB, DT = 0 dB, and the SSL of the submarine is 120 dB re 1 μPa/Hz$^{1/2}$ in this frequency range? (b) Repeat with DI = 0. (c) Repeat (a) with f = 100 Hz. Assume heavy shipping and SSL = 133 dB re 1 μPa/Hz$^{1/2}$.

15.10.6. A receiving platform has self-noise consisting of NSL = 120 dB re 1 μPa/Hz$^{1/2}$ between 200 Hz and 2 kHz, one 300 Hz tone of SPL = 140 dB re 1 μPa and another at 600 Hz of SPL = 160 dB re 1 μPa . Find the band levels between (a) 200 and 299 Hz, (b) 250 and 350 Hz, (c) 1 and 2 kHz, and (d) 200 Hz and 2 kHz. (e) What is the dominating contributor to the self-noise in each of the above bandwidths?

15.10.7. A passive sonar has 90 beams. Each beam is sent through the receiver, which has 20 parallel processors, each of 0.1 Hz bandwidth, and the integration time in each processor is 10 s (the processors for each beam operate simultaneously). (a) How long does it take to get one full bearing search if the beams are formed sequentially? (b) How many frequency-bearing cells are there? (c) There must be no more than one false alarm per hour. What is $P(FA)$? (d) Assume the processing is energy detection. Use Fig. 11.4.2 to find the detection threshold.

15.11.1. The bearing-range recorder of an active sonar divides a circle of 15 km radius into cells 3.6° wide and 100 m long. (a) What is the maximum repetition rate of the transmitted pulse (i.e., how many pulses per second) if an echo must be able to return from 15 km before the next ping is initiated? (b) If $P(D)$ is 0.5, what is $P(FA)$ if no more than one false alarm is experienced every ping? (c) Repeat (b) for the case of no more than one false alarm each minute. (d) What are the required values of d for (b) and (c)? Assume that Fig. 11.4.2 applies.

15.11.2. (a) Derive a general equation for the target strength of a flat ocean bottom at a depth of z meters below a fathometer with normal power reflection coefficient R_{Π}. (b) If the bottom has R_{Π} = 0.17, the SL of the fathometer is 100 dB re 1 μPa, and f = 30 kHz, what echo level will be returned from the bottom at z = 1000 m? (c) Explain in geometrical considerations why the target strength depends on the distance between fathometer and bottom.

15.11.3. One second after a transmitter with source level 220 dB re 1 μPa emits a pulse of 3 kHz, the reverberation level is 90 dB re 1 μPa. If the volume of water contributing to the reverberation at this instant is 10^7 m^3, calculate the scattering strength.

15.11.4. The DNL for ambient noise is 80 dB re 1 μPa. The DNL for reverberation is given by $113 - 10\log r - 10^{-4}r$ dB re 1 μPa. (a) Find the detected noise level for reverberation and the total detected noise level for the combination of ambient noise and reverberation for ranges of 500 m, 1 km, 2 km, 5 km, 10 km, and 20 km. (b) Within what range is reverberation more important than ambient noise? Assume w_R lies within w.

15.11.5. An active sonar operates at 1 kHz with a source level of 220 dB re 1 μPa, a directivity index of 20 dB, a horizontal beam width of 10°, and a pulse length of 0.1 s. Correlation detection is used and it is desired to have $P(D)$ = 0.50 with

$P(FA) = 10^{-4}$. The target strength of a submarine is expected to be 30 dB and its speed may be up to 10.6 m/s. Both the sonar and the submarine are in a mixed layer of depth 100 m with the submarine shallow and the source at 36 m. The sea state is 3, the scattering strength for surface scatterers is -20 dB, and volume scattering is negligible. For simplicity, assume that the detection threshold is the same for both noise-limited and reverberation-limited performance. Find the maximum range of detection.

15.11.6. A sonar is capable of echo ranging at 20 kHz out to a maximum detection range of 3 km on a given submarine target. (*a*) If the source level of the transducer is increased by 20 dB, what will be the new maximum detection range for the same target? (*b*) If the transmitted frequency is lowered to 10 kHz without changing the physical dimensions of the transducer or its acoustic power output from that leading to a 3 km detection range, what will be the new maximum detection range? Assume the transducer produces a searchlight type of beam and that the detection is masked by ambient noise.

15.12.1. For the isospeed shallow-water channel with perfectly reflecting surfaces, show that conservation of energy yields $TL(geom) = 10 \log r + 10 \log(H/2)$. Compare with (15.12.7) and comment.

15.12.2. Assume that the near-shore ocean can be considered to be modeled by two nonparallel planes: a horizontal pressure release surface and a sloping, rigid bottom. (*a*) Sketch the position and indicate the phase of the images representing the paths from a source that reflect once off the top; once off the bottom; first off the top then off the bottom; and first off the bottom and then off the top. (*b*) Show that all images will lie on a circle passing through the source and centered at the shore.

15.12.3. A 50 Hz source of spherical waves is located at a depth of 30 m in water having a quartz sand bottom at a depth of 60 m. The acoustic pressure produced at a distance of 1 m from the center of the source is 1000 Pa. (*a*) Compute the pressures produced at a horizontal distance of 150 m from the source at a depth of 30 m for the direct path and for the reflected paths coming from the first four images. (*b*) Considering the speed of sound to be a constant 1500 m/s, compute the phase differences between the above rays and the resultant pressure amplitude of their coherent combination. (*c*) Repeat (*b*) for the incoherent combination.

15.12.4. Find the range beyond which (15.12.12) becomes valid assuming a 90 m channel whose bottom is coarse silt.

15.12.5. Show for the case of a slow bottom that γ in (15.12.8) and (15.12.10) is well approximated by $2(\rho_2/\rho_1)[(c_1/c_2)^2 - 1]^{-1/2}$ for small grazing angles of incidence. *Hint:* Refer to Problem 6.4.8(*a*).

15.13.1. (*a*) Evaluate the transmission loss at a range of 20 km for an isospeed layer of seawater ($c_1 = 1.5 \times 10^3$ m/s) of depth 100 m carrying a 3.5 kHz signal. The bottom is a fluid with $\rho_2 = 1.25\rho_1$ and $c_2 = 1.6 \times 10^3$ m/s. Assume just the lowest mode is important at the range of interest. Let source and receiver both be at 50 m depth in the layer and comment on the results for other choices of depths. (*b*) Find the distance over which the lowest two normal modes will go through a relative phase shift of 180°.

15.13.2. Derive the random phase $TL(geom)$ of (15.13.6).

15.13.3C. A shallow-water channel of 100 m depth has a rigid bottom. A 200 Hz source is at a depth of 25 m. At a range of 100 m, plot the pressure amplitude as a function of depth for (*a*) each propagating mode, (*b*) the coherent sum of all propagating modes, and (*c*) the incoherent sum of all propagating modes.

15.13.4C. A shallow-water channel of 100 m depth has a rigid bottom. A 200 Hz source is at a depth of 25 m. For a receiver at a depth of 25 m, plot the transmission loss for

(a) each propagating mode, (b) the coherent sum of all propagating modes, and (c) the incoherent sum of all propagating modes.

15.13.5. Using the source and receiver depths and frequency in Fig. 15.13.1c, assume a constant sound speed of 1500 m/s and calculate the surface interference effect as a function of range from (6.8.7) and compare with that observed in the figure.

15.13.6. For the sound-speed profile of Fig. 15.13.1a, (a) calculate the gradient at a depth of 50 m and compare with that used for the mixed layer. Assuming the geometry of Fig. 15.13.1c and that the gradient of (a) extends all the way to the surface, calculate (b) the skip distance and (c) the transition range. (d) Estimate the cutoff frequency of the layer. (e) Calculate the transmission loss from (15.5.5) and compare with the curves. *Hint:* The strong interference effect suggests that the loss per bounce is no more than a couple of dB.

SELECTED NONLINEAR ACOUSTIC EFFECTS

16.1 INTRODUCTION

Nonlinear propagation of acoustic waver in the atmosphere, in water, and in pipes has been of theoretical interest for nearly 150 years. Recently, nonlinear effects have found practical applications in the design of precision fathometers, thermoacoustic heat engines, and acoustic compressors. While there have been many sophisticated solutions for nonlinear propagation, this chapter deals with a relatively straightforward *perturbation expansion* technique, valid when nonlinear effects are small. The examples demonstrate the approach and provide simple physical interpretations of the results. Those interested in investigating the field further should consult the references.[1]

16.2 A NONLINEAR ACOUSTIC WAVE EQUATION

Nonlinear acoustics problems have been analyzed with a number of different techniques including the method of characteristics, Riemann invariant analysis, finite-difference variations, generalizations of Burger's equation, and so forth. We will concentrate mainly on the perturbation expansion method, which is fairly successful when applied to cases that are not close to the formation of shock waves.

Consider an infinite homogeneous medium with no boundaries. The constitutive equations are those developed in Chapters 5 and 8 before linearization. It can be shown, and we will assume, that nonlinear contributions need be retained only through second order in acoustic variables to yield a nonlinear wave equation valid for small condensations, $|s| \ll 1$. Furthermore, for weakly nonlinear situations the nonlinear terms are small enough that linear, lossless acoustic relationships can be used to simplify them (any errors will be of third order or higher). Finally, previous work with the lossy linear wave equation revealed that terms describing

[1]Landau and Lifshitz, *Fluid Mechanics*, Addison-Wesley (1959). Beyer, *Nonlinear Acoustics*, Naval Ship Systems Command (1974); republished, Acoustical Society of America (1997). *Encyclopedia of Acoustics*, ed. Crocker, Vol. 2, Wiley (1997).

absorptive and dispersive effects are much smaller than the other linear terms, so
they can similarly be approximated using the linear acoustic relations.

For the adiabat, use (5.2.7),

$$p = \mathcal{P}_0\left[\gamma s + \tfrac{1}{2}\gamma(\gamma - 1)s^2 + \cdots\right] \tag{16.2.1}$$

For a perfect gas γ is the ratio of heat capacities, but for any general fluid it is an
empirical constant determined from the adiabatic compressibility. Also, $\mathcal{P}_0$ is the
hydrostatic pressure for a perfect gas, but otherwise is determined from (5.2.8).
Defining a dimensionless *scaled acoustic pressure q* by

$$q = p/\rho_0 c^2 \tag{16.2.2}$$

inverting the series in (16.2.1) and simplifying yields

$$s = q - \tfrac{1}{2}(\gamma - 1)q^2 \tag{16.2.3}$$

The nonlinear equation of continuity is (5.3.3),

$$\frac{\partial \rho}{\partial t} + \nabla \cdot (\rho \vec{u}) = 0 \tag{16.2.4}$$

When s is expressed in terms of q, this becomes

$$\frac{\partial q}{\partial t} + \nabla \cdot \vec{u} = \frac{\gamma - 1}{2}\frac{\partial q^2}{\partial t} - \nabla \cdot (q\vec{u}) \tag{16.2.5}$$

The appropriate force equation with acoustic losses is the nonlinear Navier-Stokes
equation (8.2.1),

$$\rho\left[\frac{\partial \vec{u}}{\partial t} + (\vec{u} \cdot \nabla)\vec{u}\right] = -\nabla p + \left(\tfrac{4}{3}\eta + \eta_B\right)\nabla(\nabla \cdot \vec{u}) - \eta\nabla \times \nabla \times \vec{u} \tag{16.2.6}$$

If this is rewritten to express ρ and p in terms of q, the rotational term discarded,
and the nonlinear terms grouped on the right side, we have

$$\frac{\partial \vec{u}}{\partial t} + c^2\nabla q - c^2\tau\nabla(\nabla \cdot \vec{u}) = -q\frac{\partial \vec{u}}{\partial t} - \tfrac{1}{2}\nabla u^2$$

$$\tau = \left(\tfrac{4}{3}\eta + \eta_B\right)/\rho_0 c^2 \tag{16.2.7}$$

after using the relation

$$(\vec{u} \cdot \nabla)\vec{u} = \tfrac{1}{2}\nabla u^2 - \vec{u} \times (\nabla \times \vec{u}) \tag{16.2.8}$$

Equations (16.2.5) and (16.2.7) must be combined to provide a nonlinear wave
equation. Proceed as in Chapter 5: differentiate (16.2.5) with respect to time,
take the divergence of (16.2.7), and subtract one from the other to eliminate
$\nabla \cdot (\partial \vec{u}/\partial t) = (\partial/\partial t)\nabla \cdot \vec{u}$. Next, simplify the loss and nonlinear terms using the linear,

lossless approximations of (16.2.5) and (16.2.7) and standard vector identities. With the introduction of a dimensionless *scaled particle velocity* $\vec{v}$,

$$\vec{v} = \vec{u}/c \tag{16.2.9}$$

the result is

$$c^2\left(1 + \tau\frac{\partial}{\partial t}\right)\nabla^2 q - \frac{\partial^2 q}{\partial t^2} = -\tfrac{1}{2}\frac{\partial^2}{\partial t^2}(\gamma q^2 + v^2) + \tfrac{1}{2}c^2\nabla^2(q^2 - v^2) \tag{16.2.10}$$

At this point, it is useful to examine the significance of q and v. From (16.2.3), the condensation and the scaled pressure are equal through first order terms. Furthermore, from linear acoustic propagation, we have seen that, except in the immediate vicinity of small sources, the amplitudes of p and u are related by $\rho_0 c$ and so q and v must be similar in magnitude. This means that, through first order, the amplitudes of s, q, and v are all equal to the peak *acoustic Mach number M*,

$$|q| = |p|/\rho_0 c^2 \approx |s| \approx |\vec{v}| = |\vec{u}|/c = M \tag{16.2.11}$$

The left side of (16.2.10) is the linearized, lossy wave equation (8.2.4). The right side contributes nonlinear inhomogeneous terms. Each nonlinear term can be viewed as an acoustic source distributed through space that is forcing a solution to the wave equation. The nonlinear terms of (16.2.10) cannot generate significant solutions unless they are themselves close to being solutions of the linear wave equation. Recognizing this fact, which can be verified on a case-by-case basis, we can simplify (16.2.10) to a much more convenient form,

$$c^2\square_L^2 q = -\beta\frac{\partial^2}{\partial t^2}(q^2)$$

$$c^2\square_L^2 \equiv c^2\left(1 + \tau\frac{\partial}{\partial t}\right)\nabla^2 - \frac{\partial^2}{\partial t^2} \tag{16.2.12}$$

$$\beta = \frac{\gamma + 1}{2}$$

where $\square_L^2$ is defined as the *lossy D'Alembertian*. Note that for $\beta = 0$ this equation becomes the linear, lossy wave equation. A parameter of nonlinearity B/A was defined in Section 5.2 and is related to β by $\beta = 1 + B/2A$. Values of B/A for a large number of fluids have been compiled[2] from the works of many investigators.

16.3 TWO DESCRIPTIVE PARAMETERS

Two important parameters can now be identified: (a) the *discontinuity distance* ℓ, which measures how rapidly nonlinear effects will develop when an initially monofrequency waveform propagates into a nonlinear medium, and (b) the *Goldberg number* Γ, which measures how strongly nonlinear effects will accumulate before absorption losses eventually cause the acoustic signal to decay away.

[2]Everbach, *Encyclopedia of Acoustics*, ed. Crocker, Chap. 20, Wiley (1997).

(a) The Discontinuity Distance

This parameter arose out of the studies of Earnshaw (ca. 1860). If the thermodynamic speed of sound (5.6.1) is calculated using the nonlinear adiabat (16.2.1), it is found to be a function of the instantaneous condensation,

$$c^2(s) = c^2[1 + (\gamma - 1)s] \tag{16.3.1}$$

where $c = c(0)$ is the speed of sound at zero condensation. The value $c(s)$ gives the speed with which a particular value of condensation (and thus also pressure and particle velocity) will propagate with respect to the surrounding fluid. The surrounding fluid is itself in motion with particle velocity $\tilde{u}$, so that the speed with which a stationary observer will see a particular value of an acoustic variable travel through space is the thermodynamic speed superimposed on the particle speed,

$$c_p = c(s) + u \approx c + \beta u \tag{16.3.2}$$

where we have used (16.2.12) and also assumed $\beta|u/c| \ll 1$. This means that for a *lossless* plane wave initially monofrequency at $x = 0$ and advancing in the $+x$ direction, the particle speed can be written as

$$u(x, t) = U \sin\left[\omega\left(t - \frac{x}{c + \beta u}\right)\right] \tag{16.3.3}$$

As the wave propagates to larger x, the *crests* of the wave (traveling with speed $c + \beta U$) overtake the troughs (traveling with speed $c - \beta U$) that lie ahead of them. Figure 16.3.1 shows the temporal behavior of the particle speed of an initially sinusoidal plane wave as a function of increasing distance from a source in air. As the distance from the source increases, the waveform steepens and the slope $(\partial u / \partial x)_{u=0}$ of the wave at the axis crossings between each trough and crest becomes increasingly greater. These axis crossings occur when the argument of the sine in (16.3.3) is an even integral multiple of π. Unlike a transverse wave (an ocean wave as it crests and breaks on the shore) the waveform cannot become triple valued, so discontinuities must eventually form at these axis crossings. Differentiation of (16.3.3) with respect to x and evaluation at the axis crossing, for which $u = 0$ and the argument of the cosine function is an even integral multiple of π, yields

$$\left(\frac{\partial u}{\partial x}\right)_{u=0} = U\left[-\frac{\omega}{c} + \frac{\omega x}{c^2}\beta\left(\frac{\partial u}{\partial x}\right)_{u=0}\right] \tag{16.3.4}$$

Solution for the derivative gives

$$\left(\frac{\partial u}{\partial x}\right)_{u=0} = -\frac{\omega/c}{1/U - \omega x\beta/c^2} \tag{16.3.5}$$

The derivative becomes infinite where the discontinuity first forms, and this gives the discontinuity distance

$$\ell = 1/\beta Mk \tag{16.3.6}$$

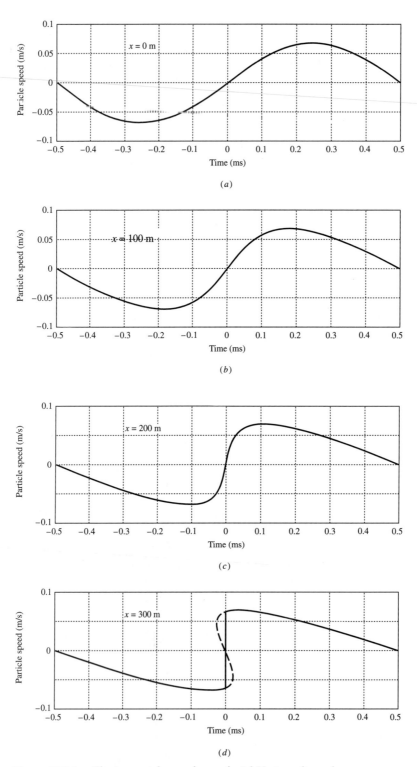

Figure 16.3.1   The temporal waveform of a 1 kHz traveling plane wave
in air without attenuation. (*a*) At $x = 0$ the particle-speed waveform
of the initially sinusoidal wave has an rms pressure amplitude of
$P = 20$ Pa ($SPL = 120$ dB *re* 20 μPa and a peak Mach number of
0.0002). The waveforms at (*b*) $x = 100$ m, (*c*) 200 m, and (*d*) 300 m show
increasing distortion. Since the particle speed cannot be triple valued, the
discontinuity distance lies between 200 and 300 m.

at which a *lossless* plane traveling wave of initial angular frequency $\omega = kc$ and initial amplitude specified by the peak acoustic Mach number will form a shock front. As an example, an acoustic wave propagating in air with sound pressure level $SPL = 120$ dB *re* 20 μPa has an effective pressure amplitude $P_e = 20$ Pa and a peak Mach number $M = \sqrt{2P_e/\rho_0 c^2} = 0.2 \times 10^{-3}$. For air, $\gamma = 1.4$ and $\beta = 1.2$. If the frequency is 1 kHz, then $k = 18.3$ m^{-1} and the discontinuity distance ℓ is about 230 m. For the same SPL but a frequency of 10 kHz, $\ell = 23$ m.

(b) The Goldberg Number

On physical grounds, it is clear that a shock front will not actually be initiated at the discontinuity distance. The presence of losses will cause the strength of the wave to weaken with distance, thus effectively increasing the value of ℓ. Since the absorption coefficient α increases with frequency, the higher harmonics are more rapidly attenuated than the lower, and this further retards the formation of a discontinuity. The *interaction* of nonlinear generation and loss accumulation must be an important descriptor of nonlinear propagation. The measure of the strength of the nonlinear effect is βM and the measure of the attenuation over distances of a wavelength is α/k (where both α and k are for the fundamental frequency of the waveform). The ratio of these quantities is the Goldberg number Γ,

$$\Gamma \equiv M\beta/(\alpha/k) = 1/\alpha\ell \tag{16.3.7}$$

For $\Gamma \ll 1$ the wave should decay away before nonlinear distortion becomes significant, and for $\Gamma \gg 1$ shock fronts should form before the wave has attenuated appreciably.

16.4 SOLUTION BY PERTURBATION EXPANSION

Assume that for some particular problem we know the solution q_1 in the infinitesimal-amplitude limit. It must satisfy the linear wave equation given by (16.2.12) with $\beta = 0$,

$$c^2 \square_L^2 q_1 = 0 \tag{16.4.1}$$

To determine the first correction to the linear approximation, try an additional contribution q_2. The trial solution is

$$q = q_1 + q_2 \tag{16.4.2}$$

If this is substituted into (16.2.12), and use made of the fact that q_1 satisfies (16.4.1), we are left with only q_2 on the left side. On the right side, we have the second time derivative of $(q_1^2 + 2q_1q_2 + q_2^2)$. It is plausible to assume $|q_2| \ll |q_1|$ so that $2q_1q_2$ and q_2^2 can be ignored. This leaves

$$c^2 \square_L^2 q_2 = -\beta \frac{\partial^2}{\partial t^2}(q_1^2) \tag{16.4.3}$$

We can solve for the first nonlinear correction $q_2 = q_2(hom) + q_2(part)$, where $q_2(hom)$ is the homogeneous solution to the linear wave equation for q_2 and $q_2(part)$

is the particular solution of (16.4.3). The coefficient of the homogeneous solution must be chosen so that the sum of homogeneous and particular solutions satisfies the appropriate boundary conditions for the pressure.

The next perturbation correction is found by substituting

$$q = q_1 + q_2 + q_3 \tag{16.4.4}$$

into (16.2.12). Now q_1 satisfies (16.4.1) and q_2 satisfies (16.4.3), which removes them from the left side. Assume that $|q_3| \ll |q_2|$ and retain only the lowest order terms on the right side. The result is

$$c^2 \square_L^2 q_3 = -\beta \frac{\partial^2}{\partial t^2}(2q_1 q_2) \tag{16.4.5}$$

Solution for q_3 proceeds as described above for q_2. If this process is continued, it is relatively straightforward to show that the nonlinear wave equation for the contribution q_n is

$$c^2 \square_L^2 q_n = -\beta \frac{\partial^2}{\partial t^2} \sum_{i=1}^{n-1} q_i q_{n-i} \tag{16.4.6}$$

16.5 NONLINEAR PLANE WAVES

Acoustic phenomena in the weakly nonlinear regime have been studied in almost every simple geometry (plane, cylindrical, spherical) with most imaginable boundary conditions. Because the mathematics very quickly gets out of hand for all but the simplest cases, we will illustrate the results only for three cases of one-dimensional plane waves.

(a) Traveling Waves in an Infinite Half-Space

Assume that a plane wave traveling in the $+x$ direction in free space is generated at $x = 0$ by a pressure $p(0, t) = P \sin(\omega t)$. The solution in the linear limit will be

$$q_1 = M e^{-\alpha x} \sin(\omega t - kx) \tag{16.5.1}$$

with $M = P/\rho_0 c^2$. For frequencies far below the relaxation frequency, $\omega/k \approx c$ and $\alpha \approx \omega^2 \tau/2c$ for the fundamental angular frequency ω. These are consistent with approximations made in the derivation of the nonlinear wave equation. Generation of q_1^2 and substitution into (16.4.3) gives

$$c^2 \square_L^2 q_2 = -\tfrac{1}{2}(2\omega)^2 \beta M^2 e^{-2\alpha x} \cos(2\omega t - 2kx) \tag{16.5.2}$$

The boundary condition on q_2 must be $q_2 = 0$ at $x = 0$, since we have specified that the source is oscillating at angular frequency ω. The particular solution is (see Problem 16.5.1)

$$q_2(part) = \tfrac{1}{4}M\frac{\beta M}{\alpha/k}e^{-2\alpha x}\sin(2\omega t - 2kx) \tag{16.5.3}$$

A homogeneous solution for angular frequency 2ω is

$$q_2(hom) = Ae^{-4\alpha x}\sin(2\omega t - 2kx) \tag{16.5.4}$$

The absorption coefficient at 2ω is 4α because it is proportional to the square of the frequency. The constant A is determined from the boundary condition on q_2. The result is

$$q_2 = \tfrac{1}{4}M\Gamma(e^{-2\alpha x} - e^{-4\alpha x})\sin(2\omega t - 2kx) \tag{16.5.5}$$

so the relative amplitude of the second harmonic with respect to the fundamental is

$$|p_2|/|p_1| = \tfrac{1}{4}\Gamma e^{-\alpha x}(1 - e^{-2\alpha x}) \tag{16.5.6}$$

For small distances ($\alpha x \ll 1$), expansion of the exponentials shows that $|p_2|/|p_1| \to \tfrac{1}{2}(x/\ell)$. The relative amplitude of the second harmonic grows linearly with x, the rate of growth depending inversely on ℓ. For large distances ($\alpha x \gg 1$), $|p_2|/|p_1| \to \tfrac{1}{4}\Gamma \exp(-\alpha x)$; the relative amplitude is proportional to the Goldberg number but also shows an exponential decay determined by the absorption. The competition between nonlinear generation and absorptive loss in limiting the growth of the second harmonic is apparent.

Generation of the next perturbation correction q_3 proceeds following (16.4.5). The product $q_1 q_2$ is expanded as a sum of the first and third harmonics and the particular solutions for each harmonic obtained. Combination with the homogeneous solutions to satisfy the boundary condition that $q_3 = 0$ at $x = 0$ gives

$$
\begin{aligned}
q_3 &= q_{31} + q_{33} \\
&= \tfrac{1}{8}M\Gamma^2\Big[-\Big(\tfrac{1}{4}e^{-\alpha x} - \tfrac{1}{2}e^{-3\alpha x} + \tfrac{1}{4}e^{-5\alpha x}\Big)\sin(\omega t - kx) \\
&\quad + \Big(\tfrac{1}{2}e^{-3\alpha x} - \tfrac{3}{4}e^{-5\alpha x} + \tfrac{1}{4}e^{-9\alpha x}\Big)\sin(3\omega t - 3kx)\Big]
\end{aligned} \tag{16.5.7}
$$

The solution q_3 contributes q_{33} to the third harmonic and q_{31} to the fundamental. The terms in q_{33} give the formation of the third harmonic by the interaction of the fundamental and second harmonic. For $\alpha x \ll 1$ the third harmonic has a relative amplitude $|q_{33}|/|q_1| \to \tfrac{3}{8}(x/\ell)^2$. This grows quadratically with the scaled distance (x/ℓ) and is independent of the losses. The correction q_{31} to the fundamental is subtractive, and for $\alpha x \ll 1$ has a relative amplitude $|q_{31}|/|q_1| \to \tfrac{1}{8}(x/\ell)^2$. This adjusts the fundamental amplitude for energy transferred to the second harmonic. For large distances $\alpha x \gg 1$, both q_{31} and q_{33} have relative amplitudes proportional to the square of the Goldberg number but also decay exponentially. Figure 16.5.1 shows the temporal behavior of an initially sinusoidal 1 kHz sound wave with $SPL = 120$ dB re 20 μPa in air at various distances from the source. Only the three lowest harmonics are included.

(b) Traveling Waves in a Pipe[3]

As developed in Section 8.9, walls introduce losses in addition to those present in the body of the fluid and also introduce phase-speed dispersion. Assuming the

[3]Coppens, *J. Acoust. Soc. Am.*, **49**, 306 (1971).

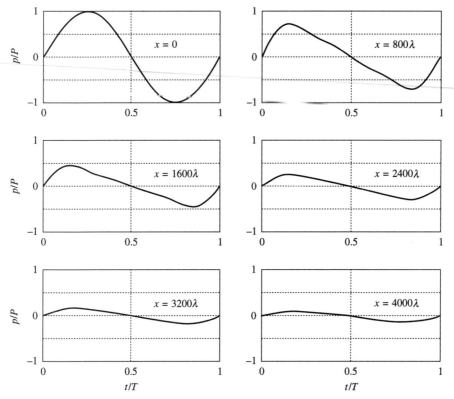

Figure 16.5.1 The temporal waveform of a 100 kHz traveling plane wave in dry air with $\alpha = 0.317$ s^{-1}. The initially sinusoidal wave has an rms pressure amplitude of 20 Pa (SPL = 120 dB re 1 μPa, a peak Mach number of 0.0002, and a Goldberg number of 3.19). The initial waveform is shown in the upper left with successive plots showing the waveform in incremental distances of 800λ. (Only the first three harmonics are included in the calculation.)

radius of the pipe is small enough and the frequency low enough that the bulk losses are much smaller than the losses at the walls, the total absorption coefficient is given by α_w in (8.9.19) and the phase speed by c_p in (8.9.20). This complicates the nonlinear solution because the absorption is proportional to $\sqrt{\omega}$ and the phase speed differs from c by a term proportional to $1/\sqrt{\omega}$. Each frequency component in the nonlinear term in (16.2.12) will have its own effective absorption coefficient and effective phase speed, while the lossy D'Alembertian must have the phase speed and absorption coefficient appropriate for each frequency in the linear limit. Thus, the linear solution for the wave generated in a pipe with $p = P\sin(\omega t)$ at $x = 0$ is

$$q_1 = Me^{-\alpha_1 x}\sin\left[\omega(t - x/c_1)\right] \tag{16.5.8}$$

where α_1 is given by (8.9.19) and c_1 is given by (8.9.20). In solving for the second harmonic, (16.2.12) is used with the D'Alembertian

$$c_2^2\Box_{L2}^2 = c_2^2\left(1 + \alpha_2\frac{2c}{(2\omega)^2}\frac{\partial}{\partial t}\right)\nabla^2 + (2\omega)^2 \tag{16.5.9}$$

where the angular frequency is 2ω, the absorption coefficient is $\alpha_2 = \sqrt{2}\alpha_1$ in the linear limit, and the phase speed is $c_2 = c(1 - \alpha_2/2k)$. The perturbation solution for the second harmonic is

$$q_2 = \frac{1}{\sqrt{2}-1} \frac{M\Gamma}{4} \left\{ e^{-\alpha_2 x} \sin\left[2\omega\left(t - \frac{x}{c_2}\right) - \frac{\pi}{4}\right] - e^{-\sqrt{2}\alpha_2 x} \sin\left[2\omega\left(t - \frac{x}{c_2'}\right) - \frac{\pi}{4}\right] \right\}$$

$$c_2/c = 1 - \alpha_2/2k$$

$$c_2'/c = 1 - \alpha_2/\sqrt{2}k$$

$$(16.5.10)$$

The first term on the right is the homogeneous solution and is the wave to be expected in the linear limit. The second on the right is the particular solution, having decay constant and phase speed different from the values in the linear limit. This is similar in form to the second harmonic for the plane traveling wave, in that there are two terms, each with its own exponential decay, but this expression is further complicated by the presence of separate phase speeds for each term. The actual observed phase speed of the sum is a function of distance, changing as the waves attenuate at their different rates.

(c) Standing Waves in a Pipe[4]

Any dimensional imperfections in a resonant cavity can perturb the resonance frequencies, shifting them away from their theoretical values calculated on the basis of perfect geometry and rigidity. This can be treated as a perturbation of the value of the phase speed appropriate for the particular standing wave. Theoretical prediction of the losses in sealed cavities, while in closer agreement with the observed losses than are the theoretical and observed phase speeds, also deviate somewhat from what is actually observed. As a consequence, for standing waves in cavities, it is usually better to measure the actual quality factors, the resonance frequencies, and the *nominal* dimensions. The wave equation appropriate for each standing wave can then be solved with the quality factor and effective phase speed appearing as empirical input.

Following the same arguments as above, it is clear that the only standing waves that can be appreciably excited, either linearly or nonlinearly, are those having frequencies very close to the resonance frequencies of the cavity, and this in turn means that significant nonlinear effects will be found only when the relevant resonance frequencies come close to being harmonics.

To sketch out the method of solution, begin with a narrow, rigid-walled pipe of length L rigidly capped at one end and driven with a massive piston, so it resembles a closed, closed pipe. For the plane wave modes, the lossy D'Alembertian for each frequency component of the nonlinear standing wave is

$$c_n^2 \Box_{Ln}^2 = \left[c_n^2 \left(1 + \frac{1}{n\omega Q_n} \frac{\partial}{\partial t}\right) \nabla^2 + (n\omega)^2 \right] \qquad (16.5.11)$$

[4]Coppens and Sanders, *J. Acoust. Soc. Am.*, **43**, 516 (1968), **58**, 1133 (1975); Coppens and Atchley, *Encyclopedia of Acoustics*, Chap. 22.

$n = 1, 2, 3, \ldots$. The linear, steady-state solution for the standing wave was obtained in Section 10.5. Within the assumption of weak absorption (high Q), we can approximate,

$$q_1 = M \cos(k_1 x) \sin(\omega t) \tag{16.5.12}$$

The angular frequency of the driven standing wave ω must be close to ω_1, the measured value at resonance. The Mach number M is based on the classical solution alone, and $k_1 = \omega_1/c_1$ with c_1 the value of the phase speed as determined from the nominal length of the pipe. (For example, if this standing wave is near the fundamental for the closed, closed pipe, then $k_1 = \pi/L$ with L the nominal length.)

Substituting (16.5.12) into (16.4.3) and again retaining just lowest order terms gives the equation for the first nonlinear correction,

$$\left[c_2^2 \left(1 + \frac{1}{2\omega Q_2} \frac{\partial}{\partial t} \right) \nabla^2 + (2\omega)^2 \right] q_2 = \tfrac{1}{4} M^2 \beta \frac{\partial^2}{\partial t^2} \cos(2k_1 x) \cos(2\omega t) \tag{16.5.13}$$

If this is now generalized to complex form,

$$\left[c_2^2 \left(1 + \frac{j}{Q_2} \right) \nabla^2 + (2\omega)^2 \right] \mathbf{q}_2 = \tfrac{1}{4} M^2 \beta \frac{\partial^2}{\partial t^2} \cos(2k_1 x) e^{j2\omega t} \tag{16.5.14}$$

solution is relatively straightforward. First, any homogeneous solution will exhibit exponential damping with time and will therefore disappear after a couple of decay times. The particular solution gives the steady-state response of the cavity. After extracting the real part of $\mathbf{q}_2$, we have the result

$$q_2 = \tfrac{1}{4} M^2 \beta Q_2 \cos \theta_2 \cos(2k_1 x) \sin(2\omega t + \theta_2)$$

$$\tan \theta_2 = Q_2 \left[\left(\frac{\omega_2}{2\omega} \right)^2 - 1 \right] \tag{16.5.15}$$

where Q_2 is the quality factor for the resonance at ω_2 and θ_2 describes how closely the second harmonic at 2ω is tuned to the resonance angular frequency ω_2. For fixed Mach number M the strength of the second harmonic is governed by the quality factor of the second harmonic and difference in driving angular frequency 2ω from its resonance angular frequency ω_2.

Solutions for higher order perturbation corrections are consistent with the above. The solution for q_3 contains a fundamental and third harmonic. The third harmonic amplitude depends on the quantity $Q_2 Q_3 \cos \theta_2 \cos \theta_3$ with θ_3 defined similarly to θ_2. The result is that higher harmonics are affected by the response curves of lower ones. The greatest nonlinear response is achieved when the harmonics of the fundamental match most closely the resonance frequencies of the higher driven standing waves.

16.6 A PARAMETRIC ARRAY

Intensive studies of the nonlinear interaction of high-intensity sound beams in the 1960s led to the development of practical parametric-array transducers. These acoustic sources have become useful as precision depth sounders because they provide highly directional sound beams of relatively low frequencies, but at modest strengths.

The following is a simple model of a parametric array evolved from the more elegant arguments of Westervelt.[5] Assume that a piston of radius a simultaneously transmits two *primary* beams of equal amplitudes and relatively high angular frequencies ω_1 and ω_2 into an unbounded medium. Define the *averaged angular frequency* $\omega = (\omega_2 + \omega_1)/2$, the *difference angular frequency* $\omega_d = |\omega_2 - \omega_1|$, and the downshift ratio ω/ω_d. Assume that $\omega_d \ll \omega$ and that the directivities of the two high-frequency beams are large enough that the primary fields are appreciably attenuated by absorption before their far fields develop. If the extent of the near field is designated by r_1 (as defined in Section 8.8), this latter condition can be stated as $\alpha r_1 > 1$, where α is the absorption coefficient associated with ω. Because of the strong attenuation experienced by each of the pressure waves in the near field and the spherical divergence that develops in the far field, the region of important nonlinear interaction lies in the near fields within the cross-sectional area $S \approx \pi a^2$ and extends along the axis of the source to a distance $L = 1/\alpha < r_1$. Assume that $L \gg a$.

Within this cylindrical volume $V = SL$, each primary beam can be approximated as a collimated plane wave of cross-sectional area S. (As seen in Section 7.4, there can be appreciable structure within the near field. Because later there will be an integration over this volume, it is plausible that the amplitude fluctuations, being nonperiodically distributed along the axis of the beam, will tend to wash out in the integration and can therefore be neglected.) The two primary beams are approximated by

$$q_{11}(z, t) = Me^{-\alpha_1 z} \cos(\omega_1 t - k_1 z)$$
$$q_{12}(z, t) = Me^{-\alpha_2 z} \cos(\omega_2 t - k_2 z)$$

(16.6.1)

within V and zero otherwise. (The coordinate system is the same as for the baffled piston in Section 7.4.) The total linear field in this approximation is $q_1 = q_{11} + q_{12}$.

For the first nonlinear correction, q_1^2 must be entered into the right side of (16.4.3) and the solution for q_2 sought. The right side contains terms of angular frequencies $2\omega_1$, $2\omega_2$, ω_d, and $\omega_1 + \omega_2$. All angular frequencies except the difference ω_d are nominally 2ω and therefore have absorption coefficients about four times those of the primaries. Consequently, these contributions to q_2 exist only in the immediate vicinity of the source. The field q_d with angular frequency ω_d, on the other hand, can propagate to large distances before it is significantly attenuated by absorption and will therefore be the dominant term. Thus, we have

$$c^2 \Box_L^2 q_d = +\omega_d^2 \beta M^2 e^{-2\alpha z} \cos(\omega_d t - k_d z)$$

(16.6.2)

The right side is zero for points outside V and $k_d = \omega_d/c$ is the propagation constant for the difference angular frequency. At locations exterior to the volume V the pressure wave q_d can be viewed as the field radiated by a distributed source occupying the volume V. This reduces the problem to an equivalent one in linear acoustics and can be couched in terms of complex quantities,

$$c^2 \Box_L^2 \mathbf{q}_d = +\omega_d^2 \beta M^2 e^{-2\alpha z} e^{j(\omega_d t - k_d z)}$$

(16.6.3)

[5]Westervelt, *J. Acoust. Soc. Am.*, **32**, 8 (1960).

The volume source is a distributed line of approximate length L and cross-sectional area S. It is amplitude-weighted by $\exp(-\alpha z)$ and has a complex phase given by $\exp(-jk_d z)$. Calculation for q_d proceeds according to the method of Section 7.10. If r' is the distance from a source point within V to the field point at (r, θ), where r is measured from the center of the piston face and θ is the angle between r and z, then

$$q_d(r,t) = \frac{\omega_d^2}{4\pi c^2} \int_V \beta M^2 e^{-2\alpha z} e^{j(\omega_d t - k_d z)} \frac{e^{-\alpha_d r'} e^{-jk_d r'}}{r'} \, dV \tag{16.6.4}$$

The quantity α_d is the absorption coefficient at the difference frequency. For $r \gg L$ and with the restriction $\lambda_d = 2\pi/k_d \gg a$ on the transverse extent of the volume,

$$r' \approx r - z\cos\theta \tag{16.6.5}$$

and the integral becomes

$$q_d(r,t) = \frac{\omega_d^2}{4\pi c^2} \beta M^2 \frac{S}{r} e^{-\alpha_d r} e^{j(\omega_d t - k_d r)} \int_0^L e^{-2\alpha z} e^{-jk_d z(1-\cos\theta)} \, dz \tag{16.6.6}$$

For $\alpha L > 1$ the upper limit can be extended to infinity and evaluation is immediate,

$$q_d(r,t) = \frac{\omega_d^2}{4\pi c^2} \beta M^2 \frac{S}{r} \frac{e^{-\alpha_d r} e^{j(\omega_d t - k_d r)}}{j[k_d(1-\cos\theta) - j(2\alpha)]} \tag{16.6.7}$$

Solving for the amplitude of the radiated pressure beam results in

$$|p_d(r,\theta,t)|/\rho_0 c^2 = M_{ax}(r)H(\theta)$$

$$M_{ax}(r) = \frac{\omega_d^2}{8\pi c^2} \beta \frac{M^2}{\alpha} \frac{S}{r} \tag{16.6.8}$$

$$H(\theta) = \frac{\alpha/k_d}{[\sin^4(\theta/2) + (\alpha/k_d)^2]^{1/2}}$$

The directional factor shows that there are no side lobes and that the -3 dB beam width is given by $2\theta_{1/2} \approx 4\sin^{-1}[(\alpha/k_d)^{1/2}]$. In seawater, for nominally 1 MHz carriers and a difference frequency of 100 kHz (a downshift ratio of 10), the beam width is about 2.4°. For the same downshift ratio, if the carriers are increased to 10 MHz, the beam width increases to about 7.6°. If the carriers are increased to 10 MHz and the difference frequency is kept at 100 kHz (for a downshift ratio of 100), the beam width becomes about 24°. Within the limiting assumptions, the lower the carrier frequencies and the larger the difference frequency, the smaller the beam width. For absorption proportional to the square of the frequency, the beam width varies as $\omega/\sqrt{\omega_d}$.

Figure 16.6.1 shows the beam pattern $[b = 20\log(H)]$ for a parametric array with $f = 1$ MHz for various values of the *downshift ratio* f/f_d. Note the narrow beam width and the complete absence of side lobes.

The axial response shows the $1/r$ dependence to be expected in the far field. The amplitude of the difference-frequency beam depends not only on the square

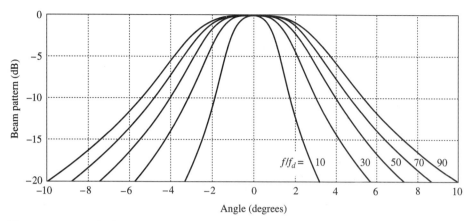

Figure 16.6.1   The beam pattern $[b = 20\log(H)]$ for a parametric array with $f = 1\,\text{MHz}$ for various values of the downshift ratio f/f_d. Note the narrow beam width and the complete absence of side lobes.

of the amplitude of the primary beams but also on the square of the difference frequency and is inversely proportional to the absorption coefficient of the primary frequencies. It can be shown (Problem 16.6.4) that the difference between the source level $SL(\omega_d)$ of the secondary beam and the source level $SL(\omega)$ of one of the primaries is

$$SL(\omega_d) - SL(\omega) = 20\log\left[\tfrac{1}{2}(\omega_d/\omega)^2\Gamma\right] \qquad (16.6.9)$$

The Goldberg number Γ is that for the primary beams.

The restrictions imposed in obtaining this model of a parametric array can be summarized as follows: $\alpha r_1 \geq 1$ so that the effective parametric array length L lies within the near fields of the primaries; fairly high directivity, $(ka)^2 > 10$, so that $L \gg a$; and fairly long wavelength for the difference frequency, $k_d a < 3$, so that phase errors in the volume integration arising from the approximation of r' are not significant.

The model also requires that there be relatively small losses of energy from the primary frequencies into nonlinearly generated harmonics. When this nonlinearly generated loss becomes significant, M can no longer be assumed constant along the axis within the primary near fields, but will rather be attenuated by nonlinearly generated losses. The effect is referred to as *saturation*. Since [from the discussion following (16.5.6)] the accumulation of second harmonics is initially about $\tfrac{1}{2}(r/\ell)$, we can prevent saturation from occurring by requiring $\tfrac{1}{2}(r/\ell) < \tfrac{1}{10}$. The geometrical arguments limit the downshift ratio to about $\omega/\omega_d > 10$, and saturation places an upper limit on the Goldberg number of about $\Gamma < \tfrac{1}{5}$. Together, these put a fairly conservative upper limit on the source level of the difference-frequency beam of about $-60\,\text{dB}$ compared to the source level of a primary.

More elaborate calculations show that somewhat narrower beam widths and much higher efficiencies are possible. Interested readers are referred to the literature.[6]

---

[6]Hamilton, *Encyclopedia of Acoustics*, Chap. 23.

PROBLEMS

16.2.1. Show that if all terms in $\omega\tau$ higher than first order are discarded in obtaining the solution to (16.2.12) in the *linear limit* for a plane traveling wave $p = P\exp(-\alpha x)\cos(\omega t - kx)$, the results are $k = \omega/c$ and $\alpha = \omega^2\tau/2c$. *Hint:* For ease, generalize p to complex form, substitute into the lossy wave (or Helmholtz) equation, and collect real and imaginary parts.

16.2.2. Assume that a term on the right side of (16.2.12) has the form $A\sin mkx \cos n\omega t$, where $\omega/k = c$ and m and n are not necessarily integers. (*a*) Show that the particular solution for this term has significant amplitude only when $m \approx n$. First outline the solution under the assumption $\tau = 0$. (*b*) Use this result to verify the assumption simplifying (16.2.10) to (16.2.12). *Hint:* Show for a linear lossless monofrequency standing or traveling wave q that $c^2\nabla^2 q^2 = c^2\nabla^2 v^2 = (\partial^2/\partial t^2)v^2 = (\partial^2/\partial t^2)q^2$ where q and v are related by the velocity potential.

16.2.3. Show that if

$$c^2\square_{LJ}^2 f = -\omega^2\beta e^{-\mu x}\cos(\omega t - kx)$$

then the particular solution through $O(\omega t)$ is

$$f = \tfrac{1}{2}\frac{\beta k}{\alpha - \mu}e^{-\mu x}\sin(\omega t - kx)$$

where $k = \omega/c$ and $\alpha = \omega^2\tau/2c$.

16.3.1. Calculate the discontinuity distances and the Goldberg numbers for the plane traveling waves in air of *SPL* 120 dB *re* 20 μPa and frequencies 0.1, 1, and 10 kHz in (*a*) moist and (*b*) dry air. (*c*) Repeat (*a*) and (*b*) for *SPL* 100 dB *re* 20 μPa.

16.3.2C. For the plane wave discussed at the end of Section 16.3(a), use (16.3.3) to plot the temporal waveform at appropriate distances to accurately determine the discontinuity distance and compare to the result calculated from (16.3.6).

16.3.3. For the plane wave discussed at the end of Section 16.3(a), calculate the Goldberg number assuming dry air. Will this wave develop into a shock wave?

16.5.1. Show that (16.5.3) is the particular solution to (16.5.2).

16.5.2C. For the plane wave traveling at 10 kHz in dry air discussed at the end of Section 16.3(a), plot the amplitudes of the first three harmonics out to the discontinuity distance. (*a*) Show that for $\alpha x \ll x/\ell \ll 1$ the third harmonic grows quadratically with scaled distance x/ℓ. (*b*) Show that for $\alpha x \ll x/\ell \ll 1$ the first harmonic decreases quadratically with scaled distance x/ℓ.

16.5.3C. For the wave in Problem 16.5.2C, plot the waveforms for q_1, q_2, q_3, and $q_1 + q_2 + q_3$ for several 2 m intervals from the origin out to the discontinuity distance. Comment on the growth of the higher order perturbation terms and the changes in the shape of the waveform with all three perturbation terms included.

16.5.4C. If the wave in Problem 16.5.2C is traveling in a rigid-walled pipe of 2 cm radius, plot the temporal behavior of the waveform at several distances from the source and determine if this wave forms a shock wave.

16.5.5C. A rigid-walled pipe of length L is terminated in a rigid cap and driven with a massive piston at the upper half-power point of the fundamental mode. Under the assumption of perfect geometry and ideal wall losses ($Q_n = Q_1\sqrt{n}$), an approximate solution valid for $\Gamma \sim 1.6$ can be obtained,

$$q \sim M\sum_n R_n \cos(nkx)\sin(n\omega t + \phi_n)$$

where $R_1 = 1, R_n \sim 0.6n^{-1.6}$ for $n \geq 2$, and $\phi_n \sim -(n-1)\pi/4$. (a) Plot the temporal behavior of the pressure waveform at $x/L = 0, \frac{1}{8}, \frac{1}{4}, \ldots, 1$. (b) Plot the spatial behavior at times $t/T = 0, \frac{1}{8}, \frac{1}{4}, \ldots, 1$, where T is the period of the fundamental. (c) Interpret the behaviors in terms of two near-shock fronts traveling in the pipe.

16.6.1. Obtain the right side of (16.6.3) from the classical solution given by (16.6.1).

16.6.2. Obtain (16.6.6) from (16.6.4) using the indicated geometrical approximations and the arguments of Chapter 7.

16.6.3. Evaluate the integral of (16.6.6) and obtain the far field difference beam (16.6.7).

16.6.4. Obtain (16.6.9) from (16.6.8) Use the expressions obtained in Section 7.4 to identify the assumed pressure amplitude P for each primary beam with the quantity $2\rho_0 c U_0$ for the near field, on-axis response. Relate P to the far field axial pressure $P_{ax}(1)$ for one of the primaries and then take 20 log of the ratio of the far field axial pressure amplitudes of the secondary and (one of the) primary beams.

16.6.5. For the simple model of the parametric array developed here, show that the half-power beam width can be written as $2\theta_{1/2} = 4(\beta M \Gamma k_d/k)^{-1/2}$.

16.6.6C. (a) Plot the beam patterns for a parametric array with an averaged carrier frequency of 1 MHz and for downshift ratios between 10 and 100. Measure the half-power beam widths and verify that, for a fixed frequency, the beam width varies as $f/\sqrt{f_d}$. (b) Plot the beam patterns for a parametric array for a fixed downshift ratio of 10 for carrier frequencies between 1 and 10 MHz and graphically determine the dependence of the beam width on the carrier frequency.

Chapter *17*

SHOCK WAVES
AND EXPLOSIONS

17.1 SHOCK WAVES

In Chapters 1 through 15 it has been assumed that the properties of the fluid are *continuous* functions of space and time so that the process of going to the limit of infinitesimal volume would be legitimate. However, as shown in Chapter 16, some acoustic processes have such high Mach numbers that shock waves can form. When a *shock wave* is generated, there are (nearly) discontinuous changes in the total pressure, density, and particle speed across the *shock front*. (Discontinuities previously have been treated as boundary conditions; for example, reflection from an interface and radiation from a source. The shock front is not amenable to such an analysis because mass flows across it.) In this chapter we focus our attention on normal plane and spherical shocks in air. Readers desiring information about more general cases, including underwater shocks, should begin with the references.[1]

There are two coordinate systems used to study shocks. One is useful in wind tunnels, in which air is driven by a fan to flow from region 1 (the *upstream*) toward the shock with the *shock speed* u_1 , passes through the shock front, and emerges into region 2 (the *downstream*) behind the shock with speed u_2. These coordinates, wherein the shock front is stationary, are called *shock coordinates*.

For the shock from an explosion, the observer is stationary with respect to the upstream fluid and the shock is approaching with speed u_1. Such coordinates are called *body coordinates*. After the shock passes, the stationary observer experiences the *back flow*, or *blast wind*, in which the fluid moves initially in the same direction as the shock with speed $u_b = u_1 - u_2$.

---

[1]Landau and Lifshitz, *Fluid Mechanics,* Addison-Wesley (1959). Kinney and Graham, *Explosive Shocks in Air,* 2nd ed., Springer-Verlag (1985) (accessible introductory level, but rife with typographical and calculational errors). Raspert, *Encyclopedia of Acoustics,* ed. Crocker, Chap. 31, Wiley (1997). Cole, *Underwater Explosions,* Princeton (1948).

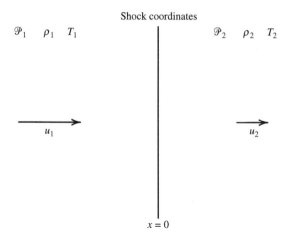

Shock coordinates

$\mathcal{P}_1 \quad \rho_1 \quad T_1$ $\qquad\qquad\qquad$ $\mathcal{P}_2 \quad \rho_2 \quad T_2$

u_1 $\qquad\qquad\qquad\qquad$ u_2

$x = 0$

Fig. 17.1.1 Normal shock quantities. Upstream: pressure $\mathcal{P}_1$, density ρ_1, temperature T_1, and particle speed u_1. Downstream: pressure $\mathcal{P}_2$, density ρ_2, temperature T_2, and particle speed u_2.

(a) The Rankine–Hugoniot Equations

Figure 17.1.1 shows a *normal, stationary* shock, as could be observed in a wind tunnel. The shock is *normal* because both fluid velocities are perpendicular to the plane of the shock and *stationary* because it is at rest with respect to the laboratory.

It is convenient to express the molar thermodynamic quantities in Appendix A9 in terms of properties per unit mass, designated as *specific*. Thus, the *specific internal energy e* is the internal energy per unit mass, E/m. The *specific heats* c_V and $c_\mathcal{P}$ are the heat capacities per unit mass, h the *specific enthalpy*, s the *specific entropy*, and $v = 1/\rho$ the *specific volume*. To remain consistent with the symbology used elsewhere in this book, we retain the symbol $\mathcal{P}$ to represent the *total* pressure. (The symbol usually used in this field is p, but we wish to reserve this to describe the fluctuation of pressure from its initial value.) The absolute temperature, however, will be represented by T (without the subscript K).

The upstream conditions (before the fluid passes through the shock) are $\mathcal{P}_1$, ρ_1, u_1, T_1. After passing through the shock, the downstream fluid conditions are $\mathcal{P}_2$, ρ_2, u_2, T_2. The conservation equations will be applied to a volume of fluid in region 1 with unit cross-sectional area and length $u_1 \Delta t$ whose leading edge is just about to enter the shock front. After a time Δt this element has passed entirely through the shock front. In this time a mass $\rho_1 u_1 \Delta t$ from region 1 passes through and emerges into region 2 with density ρ_2 and speed u_2. Conservation of mass for the fluid element requires $\rho_1 u_1 \Delta t = \rho_2 u_2 \Delta t$ or

$$\rho u = constant \tag{17.1.1}$$

The *total specific energy* e_{tot} of the element is the sum of its *specific internal energy* e and its *specific kinetic energy of translation* $u^2/2$, so that $e_{tot} = e + u^2/2$. For the *adiabatic* flow of an inviscid fluid there is no input of energy from heat conduction or chemical reactions and no thermoviscous losses so that $\delta Q = 0$. The increase of total specific energy $(e_{tot2} - e_{tot1})$ of the fluid element must equal the work $-(\mathcal{P}_2 u_2 - \mathcal{P}_1 u_1)$ done on it. Use of (17.1.1) gives the energy conservation equation, $e_{tot} + \mathcal{P}/\rho = constant$, which can be rewritten by using the definition of the specific enthalpy $h = e + \mathcal{P}/\rho$ to give

$$h + u^2/2 = constant \tag{17.1.2}$$

Since the forces acting on the sides of the element are perpendicular to the flow, the total impulse on the fluid element arises from the pressure difference on the two ends of the element. The mass per unit cross-sectional area of the element is $\rho u\,\Delta t$ so the momentum per unit area is $\rho u^2\,\Delta t$. The net impulse per unit area is $(\mathcal{P}_2 - \mathcal{P}_1)\,\Delta t$ so $\mathcal{P}_2 - \mathcal{P}_1 = \rho_1 u_1^2 - \rho_2 u_2^2$, and use of (17.1.1) gives

$$\mathcal{P}/\rho u + u = constant \qquad (17.1.3)$$

The above derivations assumed that the flow is uniform in regions 1 and 2. If the flow on either side of the shock front is nonuniform, Δt can always be taken small enough so that the equations still apply across the shock front.

Equations (17.1.1) through (17.1.3) constitute one form of the Rankine–Hugoniot equations. They contain a total of eight quantities, $\mathcal{P}$, ρ, u, and h for both regions. If the state of one region is completely known, then there are still four unknowns and three equations, so one more equation describing the state of the fluid is needed. For the rest of this chapter, we will assume the fluid is a perfect gas with a ratio of specific heats γ and a specific gas constant r. For notational convenience, we will use a number of additional equations and definitions, even though some are redundant:

$$
\begin{aligned}
h &= c_{\mathcal{P}} T \\
\mathcal{P} &= \rho r T \\
c^2 &= \gamma r T \\
M &= u/c
\end{aligned}
\qquad (17.1.4)
$$

In air, the ratio of specific heats will be taken as $\gamma = 7/5 = 1.4$. The local Mach number for the flow is M_1 where the stream speed is u_1 and M_2 where it is u_2.

Before proceeding, we can make some general observations with the help of a little thermodynamics. Each fluid element passing through the shock front undergoes an adiabatic but irreversible process. Consequently, the shocked air has larger entropy and higher temperature than the unshocked air. For a perfect gas, the internal energy is proportional to T and the enthalpy is $h = e + \mathcal{P}/\rho = e + rT$, so the enthalpy of the shocked air is increased. From (17.1.2) the speed u_2 of the shocked air is greater than u_1 for the unshocked air and (17.1.1) then shows that the density ρ_2 behind the shock must be less than ρ_1 in front. From (17.1.3) the pressure $\mathcal{P}_2$ is greater than $\mathcal{P}_1$ so that a shock front must *always* generate an overpressure; there can never be a shock front generating an underpressure.

(b) Stagnation and Critical Flow

If a stationary body is placed in the stream, the fluid is diverted to flow around it but there will be at least one spot on the leading edge, the *stagnation point*, where the fluid elements are brought adiabatically to rest with respect to the body. At this point $u = 0$ and the *stagnation temperature* T_0 is found from (17.1.2) and (17.1.4) to be related to temperature elsewhere by

$$\frac{T_0}{T} = 1 + \frac{\gamma - 1}{2}M^2 = 1 + \tfrac{1}{5}M^2 \qquad (17.1.5)$$

where M is the Mach number of the flow where the temperature is T. If, in addition, the flow outside any shocks is *isentropic* (adiabatic and reversible) so that the adiabat $\mathcal{P}/\mathcal{P}_0 = (\rho/\rho_0)^\gamma$ applies, then the *stagnation pressure* $\mathcal{P}_0$ is found to be

$$\frac{\mathcal{P}_0}{\mathcal{P}} = \left(1 + \frac{\gamma - 1}{2}M^2\right)^{\gamma/(\gamma-1)} = \left(1 + \tfrac{1}{5}M^2\right)^{7/2} \tag{17.1.6}$$

If the cross-sectional area of the flow reduces with distance, the speed of the flow will increase, and vice versa. The fluid achieves its *critical state* if and where the constriction is sufficient for the local flow speed u^* to equal the local speed of sound c^*. The associated temperature at this location is T^*. These *critical values* indicate the state of the fluid when it is at the shock threshold. Since the critical state has $M = 1$, (17.1.5) gives $T_0/T^* = (\gamma + 1)/2 = 6/5$ and substitution into (17.1.6) relates T^* with T anywhere else in an adiabatic flow,

$$\frac{T^*}{T} = \frac{2}{\gamma + 1}\left(1 + \frac{\gamma - 1}{2}M^2\right) = \frac{M^2 + 5}{6} \tag{17.1.7}$$

In addition to their physical significance, the critical sound speed c^* and the critical temperature T^* serve as convenient links between upstream and downstream quantities.

(c) Normal Shock Relations

Relating fluid speeds, Mach numbers, and thermodynamic properties on both sides of the shock front is a matter of fairly tedious algebraic manipulation of (17.1.1) to (17.1.7), and use of the critical sound speed $c^* = \sqrt{\gamma r T^*}$. The conservation of energy equation applies throughout the adiabatic flow and relates the critical sound speed to sound speed anywhere else in the flow,

$$c^2 + (\gamma - 1)u^2/2 = (\gamma + 1)(c^*)^2/2 \tag{17.1.8}$$

Note this is a stream constant. The momentum equation can be applied directly across the shock front and gives

$$c_1^2/u_1 + \gamma u_1 = c_2^2/u_2 + \gamma u_2 \tag{17.1.9}$$

Application of (17.1.8) across the shock and combination with (17.1.9) results in the *Prandtl relation*,

$$u_1 u_2 = (c^*)^2 \tag{17.1.10}$$

If (17.1.7) is rewritten in terms of c^* and c, applied across the shock, the two equations multiplied together, and then the Prandtl relation used, the upstream and downstream Mach numbers can be related:

$$M_2^2 = \frac{(\gamma - 1)M_1^2 + 2}{2\gamma M_1^2 - (\gamma - 1)} = \frac{M_1^2 + 5}{7M_1^2 - 1} \tag{17.1.11}$$

Inverting to obtain M_1 in terms of M_2 gives exactly the same equation with the subscripts exchanged. The ratio of temperatures follows from dividing (17.1.8) by c^2, applying the result to both regions 1 and 2, eliminating $(c^*)^2$ from the pair, and then eliminating M_2 with the help of (17.1.11):

$$\frac{T_2}{T_1} = \frac{\left[(\gamma - 1)M_1^2 + 2\right]\left[2\gamma M_1^2 - (\gamma - 1)\right]}{(\gamma + 1)^2 M_1^2} = \frac{(M_1^2 + 5)(7M_1^2 - 1)}{36M_1^2} \quad (17.1.12)$$

The remaining ratios of interest follow fairly directly from the above with the help of the perfect gas law and conservation of mass:

$$\frac{u_2}{u_1} = \frac{(\gamma - 1)M_1^2 + 2}{(\gamma + 1)M_1^2} = \frac{M_1^2 + 5}{6M_1^2} \quad (17.1.13)$$

$$\frac{\mathcal{P}_2}{\mathcal{P}_1} = \frac{2\gamma M_1^2 - (\gamma - 1)}{(\gamma + 1)} = \frac{7M_1^2 - 1}{6} \quad (17.1.14)$$

$$\frac{\rho_2}{\rho_1} = \frac{(\gamma + 1)M_1^2}{(\gamma - 1)M_1^2 + 2} = \frac{6M_1^2}{M_1^2 + 5} \quad (17.1.15)$$

In body coordinates, M_1 is the Mach number of the approaching shock front and the speed of the front is $u_1 = M_1 c_1$. The fluid immediately behind the shock is being pulled along with speed $u_b = u_1 - u_2$. Expression of u_b in terms of the Mach number of the shock front gives

$$\frac{u_b}{c_1} = \frac{2(M_1^2 - 1)}{(\gamma + 1)M_1^2} = \frac{5(M_1^2 - 1)}{6M_1} \quad (17.1.16)$$

This flow u_b of the shocked fluid is the initial value of the blast wind.

Figure 17.1.2 shows some of these properties for air.

(d) The Shock Adiabat

Analogous to the perfect gas adiabat $\mathcal{P}_2/\mathcal{P}_1 = (\rho_2/\rho_1)^\gamma$, there is a *shock adiabat* that describes the pressure–density relationship for a shock process. Combination of

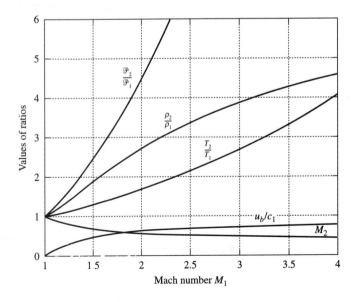

Fig. 17.1.2 Ratios of parameters across a normal shock for air ($\gamma = 1.4$) as functions of the upstream Mach number M_1.

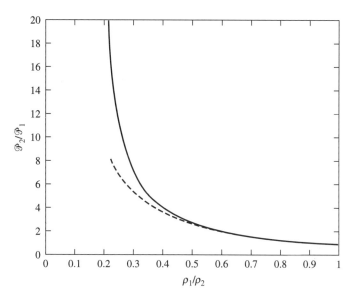

Fig. 17.1.3 The adiabats for air ($\gamma = 1.4$). The solid line is the shock adiabat. The dashed line is the reversible adiabat. For small values of the pressure ratio, the shock and reversible adiabats are almost identical.

the previous relations shows that

$$\frac{\rho_2}{\rho_1} = \frac{1 + \dfrac{\gamma + 1}{\gamma - 1}\dfrac{\mathscr{P}_2}{\mathscr{P}_1}}{\dfrac{\gamma + 1}{\gamma - 1} + \dfrac{\mathscr{P}_2}{\mathscr{P}_1}} = \frac{1 + 6\dfrac{\mathscr{P}_2}{\mathscr{P}_1}}{6 + \dfrac{\mathscr{P}_2}{\mathscr{P}_1}} \tag{17.1.17}$$

If plots are made of the pressure ratio $\mathscr{P}_2/\mathscr{P}_1$ as a function of the density ratio ρ_1/ρ_2 for both the adiabat and the shock adiabat (see Fig. 17.1.3), it is seen that they are virtually identical for $\mathscr{P}_2$ only slightly in excess of $\mathscr{P}_1$. For weak shocks ($M \approx 1$) the shock adiabat is essentially the adiabat, so the process is nearly (but not exactly) isentropic and can be well described by linear acoustics. As the pressure ratio exceeds unity the density ratio for the shock adiabat exceeds that for the adiabat, indicating an entropy increase (see Problems 17.1.5 and 17.1.6) and as the pressure ratio gets arbitrarily large the density ratio for the adiabat goes to zero while the density ratio of the shock adiabat approaches an asymptotic limit of $(\gamma - 1)/(\gamma + 1)$. This implies that shock waves are good ways to study the properties of matter at high temperatures and pressures, but not high densities.

We have assumed that a shock is a true mathematical discontinuity. The properties of the fluid can be spatially varying on each side of this discontinuity, and the Rankine–Hugoniot equations apply to quantities measured infinitesimally close to each side of the discontinuity. This is appropriate as long as the fluid is inviscid. When viscosity is included,[2] a shock has a finite thickness δ but this thickness is so small that the shock can be treated mathematically as a discontinuity. For example, in dry air at 20°C and $\mathscr{P}_1 = 1$ atm, the width of the shock front between the points where the overpressure has risen from 12% to 88% of its final value is approximately $\delta \sim 2 \times 10^{-7}\,\mathscr{P}_1/(\mathscr{P}_2 - \mathscr{P}_1)$ m. Since the mean free path in air at these conditions is about 6×10^{-8} m, for a shock with $\mathscr{P}_2 = 2\mathscr{P}_1$ the shock thickness is around three mean free paths in the unshocked air.

[2] Landau and Lifshitz, *op. cit.*

17.2 THE BLAST WAVE

When an explosive charge is detonated in free space, there is an extremely rapid chemical reaction that converts the solid or liquid mixture of the charge into a highly compressed mixture of reacting gases at very high temperature. The outward expansion of this *firebull* generates a spherically expanding shock front that propagates out into the quiescent atmosphere originally at a hydrostatic pressure $\mathcal{P}_1$. This shock front is the *blast wave*. At a range r from the origin of the explosion, the passage of the blast wave results in a sudden overpressure that evolves as a function of time. If time t is measured from the instant of arrival of the shock front, then the initial overpressure $p(0) = \mathcal{P}_2 - \mathcal{P}_1$, where $\mathcal{P}_2$ is the total instantaneous pressure just behind the shock front. As time advances, the pressure shows a fairly sharp decrease, falling to the initial value $\mathcal{P}_1$, smoothly undershooting this value (because of the inertia of the outward flow of air), and then slowly recovering back to $\mathcal{P}_1$. A striking feature of what is in actuality a fairly complicated process is that it can be approximated fairly well by a relatively simple formula,

$$p(t)/p(0) = [1 - (t/\tau)]e^{-b(t/\tau)} \tag{17.2.1}$$

where τ is a duration that depends on the distance from the explosion to the receiver and on the strength or *yield* of the explosion. See Fig. 17.2.1 for a representative profile. The decay parameter b will be seen to depend only on the Mach number M_1.

Accompanying this blast wave is the associated particle velocity. The initial value $u(0)$ of this particle speed has previously been identified by the quantity u_b in (17.1.16). It can be seen (Problem 17.2.1) that $p(0)/u(0) = \rho_1 u_1$. But by (17.1.1) this is a constant of the flow so that at each point in space the overpressure and blast wind must be proportional. Consequently, the instantaneous value of the blast wind $u(t)$ follows the same time dependence,

$$u(t)/u(0) = [1 - (t/\tau)]e^{-b(t/\tau)} \tag{17.2.2}$$

Knowing the initial value of the overpressure $p(0)$, we can find the Mach number M_1 and then use (17.1.16) to obtain the associated initial value of the blast wind, $u(0)$. An interesting feature of (17.2.1) and (17.2.2) is that debris is first forced

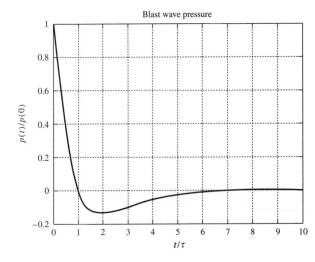

Fig. 17.2.1   The overpressure behind a blast wave for $b = 1.0$.

outward but then can be sucked back inward; there are situations for which debris may end up closer to the explosion than from where it was generated.

The major effects of the explosion, either in terms of moving earth or damaging structures, arises from the strongest portion of the blast wave, that which occurs over the duration τ. Consequently, it is convenient to define the *impulse per unit area* $\mathcal{J}/S$ for the explosion as

$$\frac{\mathcal{J}}{S} = \int_0^\tau p(t)\, dt = p(0)\frac{\tau}{b}\left(1 - \frac{1 - e^{-b}}{b}\right) \tag{17.2.3}$$

17.3 THE REFERENCE EXPLOSION

For reasons that will become apparent when the *scaling laws* of the next section are considered, we will present the physical parameters governing the evolution of the blast wave as it propagates outward for the two reference explosions (chemical and nuclear). Mainly for historical reasons, both references are based on standard masses of TNT. The energy liberated by the explosion of a mass of TNT is *defined* to be 4610 kJ/kg. This is slightly in error, low by about 1%, but was adopted as an engineering convenience.

(a) The Reference Chemical Explosion

The reference chemical explosive is 1 kg TNT detonated in the standard atmosphere, $\mathcal{P}_1 = 1$ atm at 15°C, away from any boundaries. Examination of the experimental data yields reasonably good Bode equation fits for the physical parameters of the explosion (see Fig. 17.3.1). The shock overpressure is related to the undisturbed hydrostatic pressure by

$$\frac{p(0)}{\mathcal{P}_1} = \frac{9.7}{R^{9/4}}\left[1 + \left(\frac{R}{7.2}\right)^3\right]^{5/12} \tag{17.3.1}$$

where the range R is confined to the interval $1 < R < 500$ m. The Mach number M_1 for the shock front is found from (17.1.7),

$$M_1 = \left(\frac{6}{7}\frac{p(0)}{\mathcal{P}_1} + 1\right)^{1/2} \tag{17.3.2}$$

The duration τ in milliseconds at each range R in meters is

$$\tau(\text{ms}) = \frac{0.85R^4}{\left[1 + \left(\frac{R}{0.915}\right)^{25/4}\right]^{1/2}\left[1 + \left(\frac{R}{8.54}\right)^{21/8}\right]^{1/3}} \tag{17.3.3}$$

and b has the value

$$b = \frac{3.7}{R^{3/2}}\left[1 + \left(\frac{R}{3.4}\right)^6\right]^{1/6}\left[1 + \left(\frac{R}{40}\right)^{3/2}\right]^{1/4} \tag{17.3.4}$$

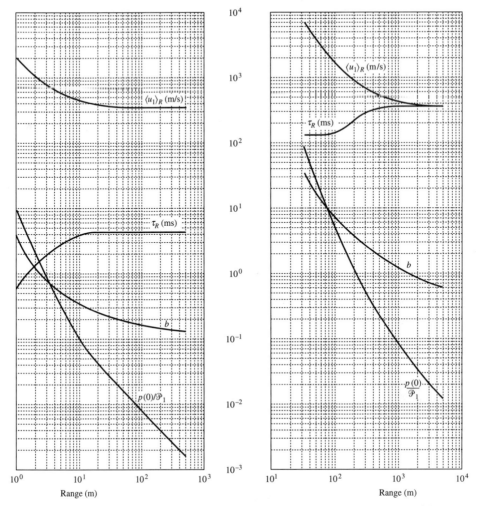

Fig. 17.3.1 The parameters of the reference chemical explosion consisting of 1 kg TNT detonated in the standard atmosphere.

Fig. 17.3.2 The parameters of the reference nuclear explosion with a yield of 1 ktonne TNT detonated in the standard atmosphere.

The average value of the propagation speed of the shock front in traveling from the center of the explosion to the field point at R is

$$\langle u_1 \rangle_R = \frac{1960}{R^{9/10}} \left[1 + \left(\frac{R}{5.2} \right)^{14/5} \right]^{2/7} \left[1 + \left(\frac{R}{70} \right)^2 \right]^{1/20} \tag{17.3.5}$$

in meters per second.

(b) The Reference Nuclear Explosion

The equivalent quantities for the nuclear explosion in an unbounded standard atmosphere are presented for a yield of 1 ktonne (1000 kg) TNT and are valid for ranges $35 < R < 5000$ m (see Fig. 17.3.2).

The overpressure is given by

$$\frac{p(0)}{\mathcal{P}_1} = \frac{4 \times 10^6}{R^{16/5}} \left[1 + \left(\frac{R}{165} \right)^{22/25} \right]^{5/2} \tag{17.3.6}$$

the duration by

$$\tau(\text{ms}) = 125 \left[1 + \left(\frac{R}{110} \right)^{9/2} \right]^{1/5} \Big/ \left[1 + \left(\frac{R}{360} \right)^{3} \right]^{3/10} \tag{17.3.7}$$

the decay parameter by

$$b = \frac{6800}{R^{3/2}} \left[1 + \left(\frac{R}{105} \right)^{9/2} \right]^{1/6} \left[1 + \left(\frac{R}{1200} \right)^{5} \right]^{1/8} \tag{17.3.8}$$

and the averaged propagation speed by

$$\langle u_1 \rangle_R = \frac{1 \times 10^6}{R^{7/5}} \left[1 + \left(\frac{R}{170} \right)^{5/2} \right]^{2/5} \left[1 + \left(\frac{R}{1200} \right)^{9/5} \right]^{2/9} \tag{17.3.9}$$

in meters per second.

17.4 THE SCALING LAWS

In the absence of specific laws governing a physical process, it is possible to predict behavior based on some basic observations of selected experimental data and knowledge of the physical parameters involved in the process.

For description of the blast wave, we need to isolate the parameters that describe (1) the strength of the explosion, (2) the atmosphere into which the blast wave is advancing, (3) the strength of the shock front, and (4) the propagation of the shock front. The explosive strength is provided by the equivalent mass of TNT. This is expressed in terms of the *yield* W,

$$W = m/m_{ref} \tag{17.4.1}$$

where m is the equivalent mass of TNT liberating the same energy as the explosive and m_{ref} is the mass of TNT in the reference explosion. The atmosphere can be described by the equilibrium density ρ_1 and the speed of sound c_1. The shock front is described by the Rankine–Hugoniot relations, so all we need is the Mach number M. The propagation of the blast wave requires specifying the time t at which a certain property of it arrives at range r. From these quantities we can construct three dimensionless combinations of parameters:

$$M \qquad m/\rho_1 r^3 \qquad c_1 t/r \tag{17.4.2}$$

(It can be seen that any other dimensionless combinations of these six parameters are simply combinations of the three listed above.)

For our purposes, we will restrict attention to the standard atmosphere. This simplifies the problem and removes the complications of scaling for the atmospheric properties. The appropriate *reference* explosion data are used to predict at what range R and time T some portion of the blast wave with Mach number M

will arrive. For an actual explosive with different mass m, the equivalent portion of the shock wave with Mach number M will arrive at range r at time t. The values of the dimensionless parameters in (17.4.2) must remain the same for both explosions. The second ratio gives us the relation between R for the reference explosion and r for that with mass m,

$$r = W^{1/3}R \qquad\qquad (17.4.3)$$

and the third ratio gives us the relevant time,

$$t = W^{1/3}T \qquad\qquad (17.4.4)$$

Thus, the same event will be observed at range r and time t for the actual explosion with yield W as observed for the reference explosion at range R and time T when the respective ranges and times are related by (17.4.3) and (17.4.4). The range R is the *scaled distance* and T the *scaled time* for the explosion with yield W.

The normalized blast wind profile (17.2.2) is a function of the dimensionless ratio t/τ. Point by point on the blast wind profile, the ratios $u(t)/u(0)$ at distance r and $u(T)/u(0)$ at scaled distance R must be the same, where t and T are the elapsed times since the arrival of the shock front at the respective distances r and R. Since the ratio of elapsed time to duration is the same for the actual explosion and the reference explosion at the respective locations r and R, the scaling does not affect the shape of the blast wave profile. This means that b is the same for both explosions. It does not scale and is determined uniquely by the value of the Mach number and the kind of explosion—chemical or nuclear. It follows at once that the same behavior holds for the normalized blast wind overpressure (17.2.1).

The average shock speed is the average value of u over the travel time and must therefore scale the same way. The product of the first and third ratios in (17.4.2) shows that ut/r is an invariant with the consequence

$$\langle u_1 \rangle_r = (r/R)(T/t)\langle u_1 \rangle_R = \langle u_1 \rangle_R \qquad\qquad (17.4.5)$$

17.5 YIELD AND THE SURFACE EFFECT

Determining the strength of a particular explosive can be done in many ways, and each way will usually give a slightly different answer. Some, like the *sand crush test* (how finely a standard sand is crushed) and the *plate dent test* (how deep a dent a 20 g sample impresses into a $\frac{5}{8}$ in. prepared steel plate), measure the explosive's *brisance*, or ability to generate a very high initial impulse and shatter or crush contiguous material. Others, like the *ballistic pendulum test* (a bob containing the test charge and fitted with a projectile recoils after detonation) and the *Trauzl block test* (a lead container is filled with sand and the charge, and the cavity volume change measured after detonation), determine something closer to the total power of the explosive. One of the more popular tests is simply to compare the peak overpressure, duration, average shock speed, and so on of the test charge with a standard mass of TNT. This allows the explosive to be rated in terms of its TNT equivalent mass. For example, the atomic bombs used in WWII were rated as equivalent to 18 ktonne (18×10^6 kg) of TNT. Since we have seen that the properties of the explosion scale as $W^{1/3}$, the mass equivalent need not be known to particularly high accuracy; 10% error in TNT equivalent mass causes about a 3% error in scaling.

When there is a boundary close to the explosive charge, the rated yield may have to be adjusted for any *surface effect* and/or *cratering*. If the charge is close to the ground but does not gouge out any crater, then the shock front can reflect from the ground as if it were a perfectly rigid surface. If the scaled altitude Z is not too large compared to the scaled horizontal range X (measured from ground zero to the field point at the slant range R), the reflected wave will quickly merge with the unreflected wave and give the appearance of a doublet with the image source in phase with the actual source. In this case, there is effectively twice as much explosive, so the actual yield used to calculate the explosion should be doubled. If, on the other hand, there is cratering and the altitude is still sufficiently small, a reasonable approximation is to multiply the rated yield by about 1.5. As the altitude gets larger, there is less merging of the two blast waves until very large distances are reached: at short ranges X the direct and reflected shock fronts maintain their identities, joined together just at the ground. As time passes and the range increases, the two shocks begin to merge into one, forming a Mach stem whose height gradually increases with increasing X, and it is only at very large distances that the two shock fronts can merge into a single one. This process is considerably more complicated to investigate and is beyond our purposes.

Conservative criteria for the relationship between altitude and horizontal range, both in meters, allowing use of the surface effect approximation are

$$0 < Z < 0.5X^{1/3} \qquad 1 < X < 500 \text{ m} \tag{17.5.1}$$

for chemical explosions, and

$$0 < Z < 16X^{1/3} \qquad 100 < X < 3000 \text{ m} \tag{17.5.2}$$

for nuclear explosions. The scaled distances are based on the yield assuming no surface effect. If the results show that there is a surface effect, then the distances should be rescaled using whatever yield is appropriate between $1.5W$ and $2W$.

Problems

17.1.1C. For the isentropic flow of a perfect gas ($\gamma = 1.4$), tabulate and plot $\mathcal{P}/\mathcal{P}_0$, ρ/ρ_0, and T/T_0 for M between 0 and 4.0.

17.1.2C. For normal stationary shock in a perfect gas ($\gamma = 1.4$), tabulate and plot M_2, $\mathcal{P}_2/\mathcal{P}_1$, ρ_2/ρ_1, and T_2/T_1 for M_1 between 1.0 and 4.0.

17.1.3. A normal shock with overpressure 800 mbar travels into air at rest at 22°C and 985 mbar pressure. (*a*) Find the Mach number of the shock front. (*b*) Find the temperature behind the shock. (*c*) Find the stagnation pressure and stagnation overpressure (with respect to $\mathcal{P}_1$) of the blast wind. (*d*) Evaluate the ratio of densities across the shock. (*e*) With the help of (A9.24) find the changes in molar and specific entropies across the shock.

17.1.4. A normal shock traveling into air at 15°C and 1 atm ambient pressure generates a blast wind with $u_b = 436$ m/s. (*a*) Find the overpressure behind the shock. (*b*) Find the Mach number of the shock.

17.1.5. (*a*) From (A9.24) show that an acoustic process is isentropic. *Hint:* Express the volume in terms of density and recall that acoustic processes obey the adiabat. (*b*) Show that a shock generates an increase in entropy. (*c*) Show that if it were possible to have a shock with $\mathcal{P}_2 < \mathcal{P}_1$ the entropy would decrease.

17.1.6. From the fact that $\mathcal{P}_2 > \mathcal{P}_1$, show from (17.1.11) to (17.1.15) that (a) the density and the temperature increase across the shock, (b) the stream Mach number in front of the shock is supersonic and the stream Mach number behind the shock is subsonic, (c) the stagnation pressure decreases across the shock, and (d) the stagnation temperature is unchanged across the shock.

17.1.7. (a) Write (17.1.1) and (17.1.3) for a normal plane shock that moves with speed U into a fluid moving toward the shock with a speed u and brings the fluid to rest. (b) Find the total pressure and temperature at a rigid wall after the normal reflection, in atmospheric air, of a Mach 2.0 shock wave. (c) Show that for very weak shocks, this reflection gives the same results as expected from acoustics.

17.1.8. (a) Write (17.1.1) and (17.1.3) for a standing oblique plane shock where the upstream velocity makes an angle α measured from the shock front. Let β be the angle between the downstream velocity and the shock front and δ the angle the downstream velocity makes with the initial direction of flow. *Hint:* Choose a coordinate system moving parallel to the shock front that turns the oblique shock into a normal shock. (b) Find the total pressure, temperature, and direction of the flow downstream of an oblique shock in atmospheric air with Mach number 2.0 and $\alpha = 40°C$. (c) Show how this solution can be used to describe the supersonic flow at the leading edge of an infinite wedge.

17.2.1. Show that the initial overpressure and blast wind are related by $p(0)/u(0) = \rho_1 u_1$.

17.2.2. The accumulating impulse $\mathcal{I}(t)$ from the blast wave acting on an airborne object of cross-sectional area S is

$$\mathcal{I}(t) = \int_0^t p(t) \, S \, dt$$

(a) Evaluate the integral to find $\mathcal{I}(t)$. (b) Verify the impulse per unit area $\mathcal{I}/S$ defined in (17.2.3). (c) Determine the total impulse $\mathcal{I}(\infty)$. (d) From (c) find the conditions under which light debris could end up closer to the explosion than it started.

17.2.3. (a) Show that for $0 < t/\tau \ll 1$ a plot of $\ln p$ against t is a straight line. (b) What is the intercept at $t = 0$ and the slope? (c) If the time record (17.2.1) is also known, indicate how the decay parameter can quickly be found from (b).

17.2.4 (a) Show that $p(0)/u(0)$ is proportional to the shock Mach number and evaluate for standard conditions. (b) For weak shocks, does $p(0)/u(0)$ approach the value observed in linear acoustics?

17.3.1. A 160 lb sofa is pushed from the top of a 60 story building (each story is 15 ft high). The terminal speed is 180 ft/s. (a) What energy is liberated when the sofa hits the ground? (b) What is the TNT equivalent?

17.3.2. Two automobiles, each weighing about 2000 lb and traveling at 60 mph, collide head on. (a) Calculate the energy released in the collision. (b) Estimate the TNT equivalent.

17.3.3. Show that for an explosive charge of radius a that the time it takes the shock front to reach range r is given by

$$t_a = \frac{1}{c_1} \int_a^r \frac{1}{M_1} \, dr$$

17.3.4. Show that the Mach number M_1 at range r can be found from the average propagation speed $\langle u_1 \rangle_r$ from the explosion to that range by the formula

$$M_1 = \frac{\langle u_1 \rangle_r}{c_1} \left(1 - \frac{d(\ln\langle u_1 \rangle_r)}{d(\ln r)} \right)^{-1}$$

17.3.5. A chemical explosion with an apparent mass of 1 kg TNT is detected by a number of sensors at different sites. Given what each sensor detected, determine the others of range r, Mach number M_1, overpressure ratio $p(0)/\mathcal{P}_1$, averaged speed of arrival $\langle u_1 \rangle_R$, duration τ, impulse per unit area $\mathcal{J}/S$, arrival time t_a, and decay parameter b at each site. Sensor A records an arrival time of 4.0 ms; sensor B a shock speed u_1 of 500 m/s; sensor C an overpressure of 60 mbar; sensor D is 1.3 m from the explosion; sensor E a duration of 2.87 ms; sensor F an impulse per unit area of 0.838 bar · ms; sensor G an averaged shock front speed of 369 m/s; reconstruction of data from sensor H indicates a decay parameter of 0.13.

17.3.6. An explosive is detonated at ground level in air at 25° C and pressure 950 mbar. At a sensor site some distance from the blast a peak overpressure of 0.80 bar and a (positive pressure) duration $\tau = 2.0$ ms are recorded. The impulse per unit area (for the positive pressure) is 0.45 bar · ms. Determine (a) the decay parameter b, (b) the Mach number M_1 of the shock front, (c) the blast wind $u(0)$, and (d) the temperature just behind the shock front.

17.3.7C. For the reference chemical explosion, tabulate and plot $p(0)/\mathcal{P}_1$, M_1, τ, b, and $\langle u_1 \rangle_R$ for R between 1 and 500 m.

17.3.8C. Plot the blast wave profile for a reference chemical explosion for several values of R between 1 and 500 m.

17.3.9C. For the reference nuclear explosion, tabulate and plot $p(0)/\mathcal{P}_1$, M_1, τ, b, and $\langle u_1 \rangle_R$ for R between 35 and 5000 m.

17.3.10C. Plot the blast wave profile for a reference nuclear explosion for several values of R between 35 and 5000 m.

17.3.11. For the reference chemical explosion, obtain approximations valid at large R for (a) the spatial dependence of the initial overpressure $p(0)$ and (b) the average shock speed. (c) Are these consistent with the results from linear acoustics?

17.3.12. Do Problem (17.3.11) for the reference nuclear explosion.

17.5.1. A supply of chemical explosives detonates at ground level in the standard atmosphere, forming a crater. A monitoring station at $r = 50$ m detects a blast wave 0.053 s after detonation that arrives with $u_1 = 512$ m/s. The duration is about 28 ms. Find (a) the Mach number and initial blast wind overpressure, (b) the initial blast wind, (c) the average speed of the shock front over the path, (d) the scaled distance and decay parameter, (e) the impulse per unit area at the station, (f) the effective mass of the supply in kg TNT, and (g) the actual equivalent mass of TNT (detonating in free space).

17.5.2. A 40 ktonne nuclear weapon is detonated at an altitude of 600 m in the standard atmosphere. The monitoring site is on the ground, 6000 m from ground zero. (a) Determine the initial overpressure at the site. (b) How long after detonation does the shock front reach the site? (c) What is the duration τ of the blast wave? (d) What is the initial value of the blast wind? (e) Find the impulse per unit area.

17.5.3. A TV film crew comes across a disaster in the making at a rocket fuel manufacturing plant. Barred from entry, they record from the public highway. Later analysis of the videotape reveals that after several rather small explosions, each of which took 11.01 s to reach the camera (measured from the times of reception of the flash and the accompanying sound), a major explosion occurred whose travel time was determined to be 10.55 s. (a) How distant was the crew from the plant? (b) What is the apparent yield of the major explosion? Assuming some cratering, estimate the actual equivalent mass of TNT that exploded. (c) Evaluate the peak overpressure. (d) Find the initial value of the blast wind.

A1 CONVERSION FACTORS AND PHYSICAL CONSTANTS

(a) Conversions between SI and CGS

Quantity	Multiply SI	by	to obtain CGS
Length	meter (m)	10^2	centimeter (cm)
Mass	kilogram (kg)	10^3	gram (g)
Time	second (s)	1	second (s)
Force	newton (N)	10^5	dyne
Energy	joule (J)	10^7	erg
Power	watt (W)	10^7	erg/s
Volume density	kg/m^3	10^{-3}	g/cm^3
Pressure	pascal (Pa)	10	dyne/cm^2
Speed	m/s	10^2	cm/s
Energy density	J/m^3	10	erg/cm^3
Elastic modulus	Pa	10	dyne/cm^2
Coefficient of viscosity	Pa · s	10	dyne · s/cm^2
Volume velocity	m^3/s	10^6	cm^3/s
Acoustic intensity	W/m^2	10^3	erg/(s · cm^2)
Mechanical impedance	N · s/m	10^3	dyne · s/cm
Specific acoustic impedance	Pa · s/m	10^{-1}	dyne · s/cm^3
Acoustic impedance	Pa · s/m^3	10^{-5}	dyne · s/cm^5
Mechanical stiffness	N/m	10^3	dyne/cm
Magnetic flux density	tesla (T)	10^4	gauss

(b) Other Conversions ($\equiv$ designates exact conversions)

1 lb (mass) $\equiv$ 0.45359237 kg
1 in. $\equiv$ 2.54 cm
1 ft $\equiv$ 0.3048 m
1 yd $\equiv$ 0.9144 m
1 fathom $\equiv$ 1.8288 m
1 mi (U.S. statute) $\equiv$ 1.609344 km
1 mi (international and U.S. nautical) $\equiv$ 1 nm $\equiv$ 1.852 km = 6076 ft
1 mph $\equiv$ 0.44704 m/s $\equiv$ 1.609344 km/h
1 knot $\equiv$ 1 nm/h $\equiv$ 1.852 km/h = 0.5144 m/s = 1.1508 mph
1 bar $\equiv$ 1×10^5 Pa $\equiv$ 1×10^6 dyne/cm^2 = 14.5037 psi
1 kgf/m^2 $\equiv$ 9.80665 Pa
1 ft H$_2$O (39.2°F) = 2.98898×10^3 Pa
1 in. Hg (32°F) = 3.38639×10^3 Pa
1 lbf/in.^{2} (psi) = 6.89476×10^3 Pa
1 atm = 1.01325 bar = 14.6959 psi (lbf/in.^{2}) = 1.03323×10^4 kgf/m^2
    = 33.8995 ft H$_2$O (39.2°F) = 29.9213 in. Hg (32°F)
°C = K $-$ 273.15 = $\frac{5}{9}$(°F $-$ 32)

(c) Physical Constants

Acceleration of gravity	g	9.80665 (standard)	m/s^2
Avogadro constant	A	6.022×10^{26}	$kmol^{-1}$
Boltzmann constant	k_B	1.3807×10^{-23}	J/K
Gas constant	$\mathscr{R}$	8.3145	$J/(mol \cdot K)$
		8.3145×10^3	$J/(kmol \cdot K)$
Molecular weight	M		
Dry air		28.964	kg/kmol
H_2O		18.016	kg/kmol
Specific gas constant	r		
Dry air		287.06	$J/(kg \cdot K)$
H_2O (gas)		461.50	$J/(kg \cdot K)$

A2 COMPLEX NUMBERS

Let x and y be real functions and define $j = \sqrt{-1}$. Then from

$$\sin x = x - \frac{x^3}{3!} + \frac{x^5}{5!} - \cdots$$

$$\cos x = 1 - \frac{x^2}{2!} + \frac{x^4}{4!} - \cdots$$

$$e^x = 1 + x + \frac{x^2}{2!} + \frac{x^3}{3!} + \cdots$$

we obtain *Euler's identity*

$$\boxed{e^{j\theta} = \cos\theta + j\sin\theta}$$

and thus

$$\cos\theta = \frac{e^{j\theta} + e^{-j\theta}}{2} \qquad \sin\theta = \frac{e^{j\theta} - e^{-j\theta}}{2j}$$

If

$$\mathbf{f} = x + jy = Ae^{j\theta}$$

then

$$\text{Re}\{\mathbf{f}\} = x = A\cos\theta$$

$$\text{Im}\{\mathbf{f}\} = y = A\sin\theta$$

$$|\mathbf{f}| = \sqrt{x^2 + y^2}$$

$$\theta = \tan^{-1}(y/x)$$

$$\mathbf{f}^* = x - jy = Ae^{-j\theta}$$

If

$$\mathbf{f} = Fe^{j(n\omega t + \theta)} \qquad \mathbf{g} = Ge^{j(n\omega t + \phi)}$$

then

$$|\mathbf{fg}| = |\mathbf{f}||\mathbf{g}| = FG$$

$$\langle \text{Re}\{\mathbf{f}\}\text{Re}\{\mathbf{g}\}\rangle_T = \tfrac{1}{2}\text{Re}\{\mathbf{fg}^*\} = \tfrac{1}{2}\text{Re}\{\mathbf{f}^*\mathbf{g}\} = \tfrac{1}{2}FG\cos(\theta - \phi)$$

In the last expression, $T = 2\pi/\omega$. If $n = 0$ then the factors of $\tfrac{1}{2}$ must be deleted.

A3 CIRCULAR AND HYPERBOLIC FUNCTIONS

Let $z = x + jy$.

$$\sinh z = (e^z - e^{-z})/2 \qquad \cosh z = (e^z + e^{-z})/2$$

$$\tanh z = \sinh z / \cosh z \qquad \coth z = 1/\tanh z$$

$$\frac{d}{dz}\sinh z = \cosh z \qquad \frac{d}{dz}\cosh z = \sinh z$$

$$\sin(jy) = j\sinh y \qquad \sinh(jy) = j\sin y$$

$$\cos(jy) = \cosh y \qquad \cosh(jy) = \cos y$$

$$\sin z = \sin x \cosh y + j\cos x \sinh y \qquad \sinh z = \sinh x \cos y + j\cosh x \sin y$$

$$\cos z = \cos x \cosh y - j\sin x \sinh y \qquad \cosh z = \cosh x \cos y + j\sinh x \sin y$$

$$\sin^2 z + \cos^2 z = 1$$

$$\cosh^2 z - \sinh^2 z = 1$$

In the following equations, the companion relationships can be obtained by differentiation with respect to z_1 and/or z_2.

$$\sin(z_1 + z_2) = \sin z_1 \cos z_2 + \cos z_1 \sin z_2$$

$$\sinh(z_1 + z_2) = \sinh z_1 \cosh z_2 + \cosh z_1 \sinh z_2$$

$$2\sin z_1 \sin z_2 = \cos(z_1 - z_2) - \cos(z_1 + z_2)$$

$$2\sinh z_1 \sinh z_2 = \cosh(z_1 + z_2) - \cosh(z_1 - z_2)$$

Useful summations are

$$\sum_{n=0}^{N-1} \cos(n\theta) = \frac{\sin(N\theta/2)\cos[(N-1)\theta/2]}{\sin(\theta/2)}$$

$$\sum_{n=0}^{N-1} \sin(n\theta) = \frac{\sin(N\theta/2)\sin[(N-1)\theta/2]}{\sin(\theta/2)}$$

A4 SOME MATHEMATICAL FUNCTIONS

In this appendix, $z = x + jy$ with x and y real, the index ν is a real number, and $l, m,$ and n are real integers. Any other restrictions on these quantities will be explicitly noted. We will quote relationships pertinent to the text. For complete properties, see Abramowitz and Stegun, *Handbook of Mathematical Functions*, Dover (1965).

(a) Gamma Function

The gamma function, while not necessary, is convenient. For arguments with positive real part, it is given by

$$\Gamma(z) = \int_0^\infty t^{z-1}e^{-t}\, dt \qquad z = 0$$

Useful equalities are

$$\Gamma\left(\tfrac{1}{2}\right) = \sqrt{\pi} \qquad \Gamma\left(n + \tfrac{1}{2}\right) = \frac{1 \cdot 3 \cdot 5 \cdots (2n - 1)}{2^n} \sqrt{\pi}$$

$$\Gamma(1) = 1 \qquad \Gamma(n + 1) = n!$$

$$\Gamma(z + 1) = z\Gamma(z)$$

(b) Bessel Functions, Modified Bessel Functions, and Struve Functions

The differential equation

$$z^2 \frac{d^2w}{dz^2} + z\frac{dw}{dz} + (z^2 - \nu^2)w = \frac{4(z/2)^{\nu+1}}{\sqrt{\pi}\,\Gamma\left(\nu + \tfrac{1}{2}\right)}$$

has *homogeneous* solutions that are any linear combinations $AJ_\nu(z) + BY_\nu(z)$ of the Bessel function of the first kind $J_\nu(z)$ and the Bessel function of the second kind $Y_\nu(z)$ [also known as Weber's function or Neumann's function and sometimes notated as $N_\nu(z)$]. The specific combinations

$$H_\nu^{(1)}(z) = J_\nu(z) + jY_\nu(z) \qquad H_\nu^{(2)}(z) = J_\nu(z) - jY_\nu(z)$$

are the Bessel functions of the third kind, the Hankel functions. The *particular* solution of the differential equation is the Struve function $H_\nu(z)$. The index ν designates the *order* of the functions.

In the rest of this appendix, all functions are understood to be of argument z unless otherwise written.

For integral orders,

$$J_{-n} = (-1)^n J_n \qquad Y_{-n} = (-1)^n Y_n$$

The Wronskian for J_ν and Y_ν is

$$W\{J_\nu, Y_\nu\} = J_{\nu+1}Y_\nu - J_\nu Y_{\nu+1} = 2/\pi z$$

Series expansions for orders 0 and 1, useful for small z, are

$$J_0 = 1 - \frac{z^2}{2^2} + \frac{z^4}{2^2 \cdot 4^2} - \frac{z^6}{2^2 \cdot 4^2 \cdot 6^2} + \cdots$$

$$J_1 = \frac{z}{2} + \frac{2z^3}{2 \cdot 4^2} - \frac{3z^5}{2 \cdot 4^2 \cdot 6^2} + \cdots$$

$$Y_0 = \frac{2}{\pi}\left\{\left[\ln\left(\frac{z}{2}\right) + \gamma\right]J_0 + \frac{z^2}{2^2} - \frac{z^4}{2^2 \cdot 4^2}\left(1 + \frac{1}{2}\right) + \cdots\right\}$$

$$Y_1 = -\frac{2}{\pi}\frac{1}{z} + \cdots$$

$$H_0 = \frac{2}{\pi}\left(z - \frac{z^3}{1^2 \cdot 3^2} + \frac{z^5}{1^2 \cdot 3^2 \cdot 5^2} - \cdots\right)$$

$$\mathbf{H}_1 = \frac{2}{\pi}\left(\frac{z^2}{1^2 \cdot 3} - \frac{z^4}{1^2 \cdot 3^2 \cdot 5} + \frac{z^6}{1^2 \cdot 3^2 \cdot 5^2 \cdot 7} - \cdots\right)$$

where $\gamma = 0.57721\ldots$ is Euler's constant.

Approximations useful for large z and $|\arg z| < \pi$ are

$$J_\nu \to \sqrt{2/\pi z}\,\cos(z - \nu\pi/2 - \pi/4)$$

$$Y_\nu \to \sqrt{2/\pi z}\,\sin(z - \nu\pi/2 - \pi/4)$$

$$H_\nu^{(1)} \to \sqrt{2/\pi z}\,\exp[\,j(z - \nu\pi/2 - \pi/4)]$$

$$H_\nu^{(2)} \to \sqrt{2/\pi z}\,\exp[-j(z - \nu\pi/2 - \pi/4)]$$

$$\mathbf{H}_\nu - Y_\nu \to \frac{1}{\pi}\frac{\Gamma\left(\frac{1}{2}\right)}{\Gamma\left(\nu + \frac{1}{2}\right)}\left(\frac{z}{2}\right)^{\nu-1}$$

Let C_ν represent any linear combination of Bessel functions of order ν. Then some recurrence and differential relations for the Bessel and Struve functions are

$$C_{\nu-1} + C_{\nu+1} = \frac{2\nu}{z}C_\nu \qquad C_{\nu-1} - C_{\nu+1} = 2\frac{d}{dz}C_\nu$$

$$\frac{d}{dz}C_0 = -C_1 \qquad \frac{d}{dz}(z^\nu C_\nu) = z^\nu C_{\nu-1} \qquad \frac{d}{dz}\left(\frac{1}{z^\nu}C_\nu\right) = -\frac{1}{z^\nu}C_{\nu+1}$$

$$\frac{d}{dz}\mathbf{H}_0 = \frac{2}{\pi} - \mathbf{H}_1 \qquad \frac{d}{dz}(z^\nu \mathbf{H}_\nu) = z^\nu \mathbf{H}_{\nu-1}$$

Useful integral representations are

$$J_0(z) = \frac{2}{\pi}\int_0^{\pi/2} \cos(z\cos\theta)\,d\theta \qquad \mathbf{H}_0(z) = \frac{2}{\pi}\int_0^{\pi/2} \sin(z\cos\theta)\,d\theta$$

$$J_n(z) = \frac{(z/2)^n}{\sqrt{\pi}\,\Gamma\left(n + \frac{1}{2}\right)}\int_0^\pi \cos(z\cos\theta)\sin^{2n}\theta\,d\theta = \frac{(-j)^n}{2\pi}\int_0^{2\pi} e^{jz\cos\theta}\cos n\theta\,d\theta$$

The arguments of J_ν for which the function has real zeros and extrema are real and defined as $j_{\nu n}$ and $j'_{\nu n}$. Relevant evaluations include

$$J_\nu(j_{\nu n}) = 0 \qquad J'_\nu(j_{\nu n}) = J_{\nu-1}(j_{\nu n}) = -J_{\nu+1}(j_{\nu n})$$

$$J'_\nu(j'_{\nu n}) = 0 \qquad J_\nu(j'_{\nu n}) = \frac{j'_{\nu n}}{\nu}J_{\nu-1}(j'_{\nu n}) = \frac{j'_{\nu n}}{\nu}J_{\nu+1}(j'_{\nu n})$$

With these defined arguments, normalizations for orthogonal Bessel functions of the first kind are facilitated with

$$\int_0^1 J_\nu(j_{\nu m}t)J_\nu(j_{\nu n}t)t\,dt = \tfrac{1}{2}[J'_\nu(j_{\nu n})]^2\delta_{nm}$$

$$\int_0^1 J_\nu(j'_{\nu m}t)J_\nu(j'_{\nu n}t)t\,dt = \frac{1}{2}\frac{(j'_{\nu n})^2 - \nu^2}{(j'_{\nu n})^2}[J_\nu(j'_{\nu n})]^2\delta_{nm}$$

The *modified* Bessel function I_ν satisfies the differential equation

$$z^2 \frac{d^2 w}{dz^2} + z \frac{dw}{dz} - (z^2 + \nu^2)w = 0$$

and is related to J_ν by

$$I_n(z) = j^{-n} J_n(jz)$$

All other needed relationships follow from the substitution jz for the argument z in the preceding equations for $J_\nu(z)$. For example,

$$I_{\nu-1} - I_{\nu+1} = \frac{2\nu}{z} I_\nu \qquad I_{\nu-1} + I_{\nu+1} = 2\frac{d}{dz} I_\nu$$

$$\frac{d}{dz} I_0 = I_1 \qquad \frac{d}{dz}(z^\nu I_\nu) = z^\nu I_{\nu-1} \qquad \frac{d}{dz}\left(\frac{1}{z^\nu} I_\nu\right) = \frac{1}{z^\nu} I_{\nu+1}$$

(c) Spherical Bessel Functions

The *spherical* Bessel functions $j_n(z)$ of the first kind, $y_n(z)$ of the second kind, and $h_n(z)$ of the third kind, and any linear combination of them satisfy the differential equation

$$z^2 \frac{d^2 w}{dz^2} + 2z \frac{dw}{dz} + [z^2 - n(n+1)]w = 0$$

They are related to the Bessel functions by

$$j_n = \sqrt{\pi/2z}\, J_{n+1/2}$$

$$y_n = \sqrt{\pi/2z}\, Y_{n+1/2}$$

$$h_n^{(1,2)} = \sqrt{\pi/2z}\, H_{n+1/2}^{(1,2)}$$

Explicit forms for j_n are

$$j_0 = \frac{\sin z}{z} \qquad\qquad\qquad j_1 = \frac{\sin z}{z^2} - \frac{\cos z}{z}$$

$$j_2 = \left(\frac{3}{z^3} - \frac{1}{z}\right)\sin z - \frac{3}{z^2}\cos z \qquad j_{n+1} = \frac{2n+1}{z} j_n - j_{n-1}$$

(d) Legendre Functions

The Legendre function $P_l^m(z)$ of *degree l* and *order m* is a solution of the differential equation

$$(1 - z^2)\frac{d^2 w}{dz^2} - 2z \frac{dw}{dz} + \left(l(l+1) - \frac{m^2}{1 - z^2}\right)w = 0$$

The greatest order for any degree is limited by $m \leq l$. While the Legendre functions are in general rather complicated functions of z, our interest is restricted to their behavior for real arguments x lying in the interval $|x| \leq 1$.

For spherical standing waves, $x = \cos\theta$. For the zeroth order Legendre functions, $m = 0$ and the order superscript is suppressed. These are the *Legendre polynomials*. They can be obtained from

$$P_n(x) = \frac{1}{2^n n!} \frac{d^n}{dx^n} (x^2 - 1)^n$$

and the lowest four integral degrees are

$$P_0 = 1 \qquad\qquad P_1 = \cos\theta$$
$$P_2 = \tfrac{1}{2}(3\cos^2\theta - 1) \qquad P_3 = \tfrac{1}{2}(5\cos^3\theta - 3\cos\theta)$$

Higher orders, the *associated* Legendre functions, can be obtained from

$$P_l^m(x) = (-1)^m (1 - x^2)^{m/2} \frac{d^m}{dx^m} P_l(x)$$

so that

$$P_1^1 = -\sin\theta \qquad P_2^1 = -3\sin\theta\cos\theta \qquad P_3^1 = -\tfrac{3}{2}\sin\theta(5\cos^2\theta - 1)$$
$$P_2^2 = 3\sin^2\theta \qquad\qquad P_3^2 = 15\sin^2\theta\cos\theta$$
$$P_3^3 = -15\sin^3\theta$$

Two recurrence relations are

$$(l - m + 1)P_{l+1}^m = (2l + 1)xP_l^m - (l + m)P_{l-1}^m$$
$$P_l^{m+1} = (1 - x^2)^{-1/2}[(l - m)xP_l^m - (l + m)P_{l-1}^m]$$

A5 BESSEL FUNCTIONS: TABLES, GRAPHS, ZEROS, AND EXTREMA

(a) Table: Bessel and Modified Bessel Functions of the First Kind of Orders 0, 1, and 2

x	$J_0(x)$	$J_1(x)$	$J_2(x)$	$Y_0(x)$	$Y_1(x)$	$Y_2(x)$	$I_0(x)$	$I_1(x)$	$I_2(x)$
0	1.0000	0	0	$-\infty$	$-\infty$	$-\infty$	1.0000	0	0
0.1000	0.9975	0.0499	0.0012	-1.5342	-6.4590	-127.6448	1.0025	0.0501	0.0013
0.2000	0.9900	0.0995	0.0050	-1.0811	-3.3238	-32.1571	1.0100	0.1005	0.0050
0.3000	0.9776	0.1483	0.0112	-0.8073	-2.2931	-14.4801	1.0226	0.1517	0.0113
0.4000	0.9604	0.1960	0.0197	-0.6060	-1.7809	-8.2983	1.0404	0.2040	0.0203
0.5000	0.9385	0.2423	0.0306	-0.4445	-1.4715	-5.4414	1.0635	0.2579	0.0319
0.6000	0.9120	0.2867	0.0437	-0.3085	-1.2604	-3.8928	1.0920	0.3137	0.0464
0.7000	0.8812	0.3290	0.0588	-0.1907	-1.1032	-2.9615	1.1263	0.3719	0.0638
0.8000	0.8463	0.3688	0.0758	-0.0868	-0.9781	-2.3586	1.1665	0.4329	0.0844
0.9000	0.8075	0.4059	0.0946	0.0056	-0.8731	-1.9459	1.2130	0.4971	0.1083
1.0000	0.7652	0.4401	0.1149	0.0883	-0.7812	-1.6507	1.2661	0.5652	0.1357
1.1000	0.7196	0.4709	0.1366	0.1622	-0.6981	-1.4315	1.3262	0.6375	0.1671
1.2000	0.6711	0.4983	0.1593	0.2281	-0.6211	-1.2633	1.3937	0.7147	0.2026
1.3000	0.6201	0.5220	0.1830	0.2865	-0.5485	-1.1304	1.4693	0.7973	0.2426
1.4000	0.5669	0.5419	0.2074	0.3379	-0.4791	-1.0224	1.5534	0.8861	0.2875
1.5000	0.5118	0.5579	0.2321	0.3824	-0.4123	-0.9322	1.6467	0.9817	0.3378
1.6000	0.4554	0.5699	0.2570	0.4204	-0.3476	-0.8549	1.7500	1.0848	0.3940
1.7000	0.3980	0.5778	0.2817	0.4520	-0.2847	-0.7870	1.8640	1.1963	0.4565

(continued)

x	$J_0(x)$	$J_1(x)$	$J_2(x)$	$Y_0(x)$	$Y_1(x)$	$Y_2(x)$	$I_0(x)$	$I_1(x)$	$I_2(x)$
1.8000	0.3400	0.5815	0.3061	0.4774	−0.2237	−0.7259	1.9896	1.3172	0.5260
1.9000	0.2818	0.5812	0.3299	0.4968	−0.1644	−0.6699	2.1277	1.4482	0.6033
2.0000	0.2239	0.5767	0.3528	0.5104	−0.1070	−0.6174	2.2796	1.5906	0.6889
2.1000	0.1666	0.5683	0.3746	0.5183	−0.0517	−0.5675	2.4463	1.7455	0.7839
2.2000	0.1104	0.5560	0.3951	0.5208	0.0015	−0.5194	2.6291	1.9141	0.8891
2.3000	0.0555	0.5399	0.4139	0.5181	0.0523	−0.4726	2.8296	2.0978	1.0054
2.4000	0.0025	0.5202	0.4310	0.5104	0.1005	−0.4267	3.0493	2.2981	1.1342
2.5000	−0.0484	0.4971	0.4461	0.4981	0.1459	−0.3813	3.2898	2.5167	1.2765
2.6000	−0.0968	0.4708	0.4590	0.4813	0.1884	−0.3364	3.5533	2.7554	1.4337
2.7000	−0.1424	0.4416	0.4696	0.4605	0.2276	−0.2919	3.8417	3.0161	1.6075
2.8000	−0.1850	0.4097	0.4777	0.4359	0.2635	−0.2477	4.1573	3.3011	1.7994
2.9000	−0.2243	0.3754	0.4832	0.4079	0.2959	−0.2038	4.5027	3.6126	2.0113
3.0000	−0.2601	0.3391	0.4861	0.3769	0.3247	−0.1604	4.8808	3.9534	2.2452
3.1000	−0.2921	0.3009	0.4862	0.3431	0.3496	−0.1175	5.2945	4.3262	2.5034
3.2000	−0.3202	0.2613	0.4835	0.3071	0.3707	−0.0754	5.7472	4.7343	2.7883
3.3000	−0.3443	0.2207	0.4780	0.2691	0.3879	−0.0340	6.2426	5.1810	3.1027
3.4000	−0.3643	0.1792	0.4697	0.2296	0.4010	0.0063	6.7848	5.6701	3.4495
3.5000	−0.3801	0.1374	0.4586	0.1890	0.4102	0.0454	7.3782	6.2058	3.8320
3.6000	−0.3918	0.0955	0.4448	0.1477	0.4154	0.0831	8.0277	6.7927	4.2540
3.7000	−0.3992	0.0538	0.4283	0.1061	0.4167	0.1192	8.7386	7.4357	4.7193
3.8000	−0.4026	0.0128	0.4093	0.0645	0.4141	0.1535	9.5169	8.1404	5.2325
3.9000	−0.4018	−0.0272	0.3879	0.0234	0.4078	0.1858	10.3690	8.9128	5.7983
4.0000	−0.3971	−0.0660	0.3641	−0.0169	0.3979	0.2159	11.3019	9.7595	6.4222
4.1000	−0.3887	−0.1033	0.3383	−0.0561	0.3846	0.2437	12.3236	10.6877	7.1100
4.2000	−0.3766	−0.1386	0.3105	−0.0938	0.3680	0.2690	13.4425	11.7056	7.8684
4.3000	−0.3610	−0.1719	0.2811	−0.1296	0.3484	0.2916	14.6680	12.8219	8.7043
4.4000	−0.3423	−0.2028	0.2501	−0.1633	0.3260	0.3115	16.0104	14.0462	9.6258
4.5000	−0.3205	−0.2311	0.2178	−0.1947	0.3010	0.3285	17.4812	15.3892	10.6415
4.6000	−0.2961	−0.2566	0.1846	−0.2235	0.2737	0.3425	19.0926	16.8626	11.7611
4.7000	−0.2693	−0.2791	0.1506	−0.2494	0.2445	0.3534	20.8585	18.4791	12.9950
4.8000	−0.2404	−0.2985	0.1161	−0.2723	0.2136	0.3613	22.7937	20.2528	14.3550
4.9000	−0.2097	−0.3147	0.0813	−0.2921	0.1812	0.3660	24.9148	22.1993	15.8538
5.0000	−0.1776	−0.3276	0.0466	−0.3085	0.1479	0.3677	27.2399	24.3356	17.5056
5.1000	−0.1443	−0.3371	0.0121	−0.3216	0.1137	0.3662	29.7889	26.6804	19.3259
5.2000	−0.1103	−0.3432	−0.0217	−0.3313	0.0792	0.3617	32.5836	29.2543	21.3319
5.3000	−0.0758	−0.3460	−0.0547	−0.3374	0.0445	0.3542	35.6481	32.0799	23.5425
5.4000	−0.0412	−0.3453	−0.0867	−0.3402	0.0101	0.3439	39.0088	35.1821	25.9784
5.5000	−0.0068	−0.3414	−0.1173	−0.3395	−0.0238	0.3308	42.6946	38.5882	28.6626
5.6000	0.0270	−0.3343	−0.1464	−0.3354	−0.0568	0.3152	46.7376	42.3283	31.6203
5.7000	0.0599	−0.3241	−0.1737	−0.3282	−0.0887	0.2970	51.1725	46.4355	34.8794
5.8000	0.0917	−0.3110	−0.1990	−0.3177	−0.1192	0.2766	56.0381	50.9462	38.4704
5.9000	0.1220	−0.2951	−0.2221	−0.3044	−0.1481	0.2542	61.3766	55.9003	42.4273
6.0000	0.1506	−0.2767	−0.2429	−0.2882	−0.1750	0.2299	67.2344	61.3419	46.7871
6.1000	0.1773	−0.2559	−0.2612	−0.2694	−0.1998	0.2039	73.6628	67.3194	51.5909
6.2000	0.2017	−0.2329	−0.2769	−0.2483	−0.2223	0.1766	80.7179	73.8859	56.8838
6.3000	0.2238	−0.2081	−0.2899	−0.2251	−0.2422	0.1482	88.4616	81.1000	62.7155
6.4000	0.2433	−0.1816	−0.3001	−0.1999	−0.2596	0.1188	96.9616	89.0261	69.1410
6.5000	0.2601	−0.1538	−0.3074	−0.1732	−0.2741	0.0889	106.2929	97.7350	76.2205
6.6000	0.2740	−0.1250	−0.3119	−0.1452	−0.2857	0.0586	116.5373	107.3047	84.0208
6.7000	0.2851	−0.0953	−0.3135	−0.1162	−0.2945	0.0283	127.7853	117.8208	92.6150
6.8000	0.2931	−0.0652	−0.3123	−0.0864	−0.3002	−0.0019	140.1362	129.3776	102.0839
6.9000	0.2981	−0.0349	−0.3082	−0.0563	−0.3029	−0.0315	153.6990	142.0790	112.5167
7.0000	0.3001	−0.0047	−0.3014	−0.0259	−0.3027	−0.0605	168.5939	156.0391	124.0113
7.1000	0.2991	0.0252	−0.2920	0.0042	−0.2995	−0.0885	184.9529	171.3834	136.6759
7.2000	0.2951	0.0543	−0.2800	0.0339	−0.2934	−0.1154	202.9213	188.2503	150.6296
7.3000	0.2882	0.0826	−0.2656	0.0628	−0.2846	−0.1407	222.6588	206.7917	166.0035
7.4000	0.2786	0.1096	−0.2490	0.0907	−0.2731	−0.1645	244.3410	227.1750	182.9424
7.5000	0.2663	0.1352	−0.2303	0.1173	−0.2591	−0.1864	268.1613	249.5844	201.6055
7.6000	0.2516	0.1592	−0.2097	0.1424	−0.2428	−0.2063	294.3322	274.2225	222.1684
7.7000	0.2346	0.1813	−0.1875	0.1658	−0.2243	−0.2241	323.0875	301.3124	244.8246
7.8000	0.2154	0.2014	−0.1638	0.1872	−0.2039	−0.2395	354.6845	331.0995	269.7872
7.9000	0.1944	0.2192	−0.1389	0.2065	−0.1817	−0.2525	389.4063	363.8539	297.2914
8.0000	0.1717	0.2346	−0.1130	0.2235	−0.1581	−0.2630	427.5641	399.8731	327.5958

(b) Graphs: Bessel Functions of the First Kind of Orders 0, 1, 2, and 3

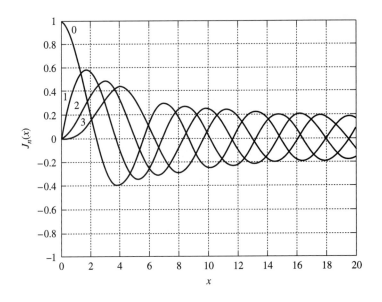

(c) Zeros: Bessel Functions of the First Kind, $J_m(j_{mn}) = 0$

	j_{mn}					
$\begin{matrix} & n \\ m & \end{matrix}$	0	1	2	3	4	5
0	—	2.40	5.52	8.65	11.79	14.93
1	0	3.83	7.02	10.17	13.32	16.47
2	0	5.14	8.42	11.62	14.80	17.96
3	0	6.38	9.76	13.02	16.22	19.41
4	0	7.59	11.06	14.37	17.62	20.83
5	0	8.77	12.34	15.70	18.98	22.22

(d) Extrema: Bessel Functions of the First Kind, $J'_m(j'_{mn}) = 0$

	j'_{mn}				
$\begin{matrix} & n \\ m & \end{matrix}$	1	2	3	4	5
0	0	3.83	7.02	10.17	13.32
1	1.84	5.33	8.54	11.71	14.86
2	3.05	6.71	9.97	13.17	16.35
3	4.20	8.02	11.35	14.59	17.79
4	5.32	9.28	12.68	15.96	19.20
5	6.41	10.52	13.99	17.31	20.58

(e) Table: Spherical Bessel Functions
of the First Kind of Orders 0, 1, and 2

x	$j_0(x)$	$j_1(x)$	$j_2(x)$	x	$j_0(x)$	$j_1(x)$	$j_2(x)$
0	1.000	0	0	4.0000	−0.1892	0.1161	0.2763
0.1000	0.9983	0.0333	0.0007	4.1000	−0.1996	0.0915	0.2665
0.2000	0.9933	0.0664	0.0027	4.2000	−0.2075	0.0673	0.2556
0.3000	0.9851	0.0991	0.0060	4.3000	−0.2131	0.0437	0.2435
0.4000	0.9735	0.1312	0.0105	4.4000	−0.2163	0.0207	0.2304
0.5000	0.9589	0.1625	0.0164	4.5000	−0.2172	−0.0014	0.2163
0.6000	0.9411	0.1929	0.0234	4.6000	−0.2160	−0.0226	0.2013
0.7000	0.9203	0.2221	0.0315	4.7000	−0.2127	−0.0426	0.1855
0.8000	0.8967	0.2500	0.0408	4.8000	−0.2075	−0.0615	0.1691
0.9000	0.8704	0.2764	0.0509	4.9000	−0.2005	−0.0790	0.1521
1.0000	0.8415	0.3012	0.0620	5.0000	−0.1918	−0.0951	0.1347
1.1000	0.8102	0.3242	0.0739	5.1000	−0.1815	−0.1097	0.1170
1.2000	0.7767	0.3453	0.0865	5.2000	−0.1699	−0.1228	0.0991
1.3000	0.7412	0.3644	0.0997	5.3000	−0.1570	−0.1342	0.0811
1.4000	0.7039	0.3814	0.1133	5.4000	−0.1431	−0.1440	0.0631
1.5000	0.6650	0.3962	0.1273	5.5000	−0.1283	−0.1522	0.0453
1.6000	0.6247	0.4087	0.1416	5.6000	−0.1127	−0.1586	0.0277
1.7000	0.5833	0.4189	0.1560	5.7000	−0.0966	−0.1634	0.0106
1.8000	0.5410	0.4268	0.1703	5.8000	−0.0801	−0.1665	−0.0060
1.9000	0.4981	0.4323	0.1845	5.9000	−0.0634	−0.1679	−0.0220
2.0000	0.4546	0.4354	0.1984	6.0000	−0.0466	−0.1678	−0.0373
2.1000	0.4111	0.4361	0.2120	6.1000	−0.0299	−0.1661	−0.0518
2.2000	0.3675	0.4345	0.2251	6.2000	−0.0134	−0.1629	−0.0654
2.3000	0.3242	0.4307	0.2375	6.3000	0.0027	−0.1583	−0.0780
2.4000	0.2814	0.4245	0.2492	6.4000	0.0182	−0.1523	−0.0896
2.5000	0.2394	0.4162	0.2601	6.5000	0.0331	−0.1452	−0.1001
2.6000	0.1983	0.4058	0.2700	6.6000	0.0472	−0.1368	−0.1094
2.7000	0.1583	0.3935	0.2789	6.7000	0.0604	−0.1275	−0.1175
2.8000	0.1196	0.3792	0.2867	6.8000	0.0727	−0.1172	−0.1244
2.9000	0.0825	0.3633	0.2933	6.9000	0.0838	−0.1061	−0.1299
3.0000	0.0470	0.3457	0.2986	7.0000	0.0939	−0.0943	−0.1343
3.1000	0.0134	0.3266	0.3027	7.1000	0.1027	−0.0820	−0.1373
3.2000	−0.0182	0.3063	0.3054	7.2000	0.1102	−0.0692	−0.1391
3.3000	−0.0478	0.2848	0.3067	7.3000	0.1165	−0.0561	−0.1396
3.4000	−0.0752	0.2622	0.3066	7.4000	0.1214	−0.0429	−0.1388
3.5000	−0.1002	0.2389	0.3050	7.5000	0.1251	−0.0295	−0.1369
3.6000	−0.1229	0.2150	0.3021	7.6000	0.1274	−0.0163	−0.1338
3.7000	−0.1432	0.1905	0.2977	7.7000	0.1283	−0.0033	−0.1296
3.8000	−0.1610	0.1658	0.2919	7.8000	0.1280	0.0095	−0.1244
3.9000	−0.1764	0.1409	0.2847	7.9000	0.1264	0.0218	−0.1182
				8.0000	0.1237	0.0336	−0.1111

(f) Graphs: Spherical Bessel Functions of the First Kind of Orders 0, 1, and 2

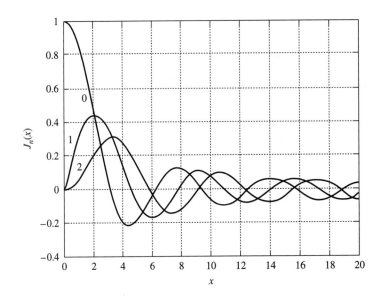

(g) Zeros: Spherical Bessel Functions of the First Kind, $j_m(\zeta_{mn}) = 0$

	ζ_{mn}					
n $\diagdown$ m	0	1	2	3	4	5
0	—	3.14	6.28	9.42	12.57	15.71
1	0	4.49	7.73	10.90	14.07	17.22
2	0	5.76	9.10	12.32	15.51	18.69
3	0	6.99	10.42	13.70	16.92	20.12
4	0	8.18	11.70	15.04	18.30	21.53
5	0	9.36	12.97	16.35	19.65	22.90

(h) Extrema: Spherical Bessel Functions of the First Kind, $j_m'(\zeta_{mn}') = 0$

	ζ_{mn}'				
n $\diagdown$ m	1	2	3	4	5
0	0	4.49	7.73	10.90	14.07
1	2.08	5.94	9.21	12.40	15.58
2	3.34	7.29	10.61	13.85	17.04
3	4.51	8.58	11.97	15.24	18.47
4	5.65	9.84	13.30	16.61	19.86
5	6.76	11.07	14.59	17.95	21.23

A6 TABLE OF DIRECTIVITIES AND IMPEDANCE FUNCTIONS FOR A PISTON

x	Directivity Functions ($x = ka \sin\theta$)		Impedance Functions ($x = 2ka$)	
	Pressure	Intensity	Resistance	Resistance
	$\dfrac{2J_1(x)}{x}$	$\left(\dfrac{2J_1(x)}{x}\right)^2$	$R_1(x)$	$X_1(x)$
0.0	1.0000	1.0000	0.0000	0.0000
0.2	0.9950	0.9900	0.0050	0.0847
0.4	0.9802	0.9608	0.0198	0.1680
0.6	0.9557	0.9134	0.0443	0.2486
0.8	0.9221	0.8503	0.0779	0.3253
1.0	0.8801	0.7746	0.1199	0.3969
1.2	0.8305	0.6897	0.1695	0.4624
1.4	0.7743	0.5995	0.2257	0.5207
1.6	0.7124	0.5075	0.2876	0.5713
1.8	0.6461	0.4174	0.3539	0.6134
2.0	0.5767	0.3326	0.4233	0.6468
2.2	0.5054	0.2554	0.4946	0.6711
2.4	0.4335	0.1879	0.5665	0.6862
2.6	0.3622	0.1326	0.6378	0.6925
2.8	0.2927	0.0857	0.7073	0.6903
3.0	0.2260	0.0511	0.7740	0.6800
3.2	0.1633	0.0267	0.8367	0.6623
3.4	0.1054	0.0111	0.8946	0.6381
3.6	0.0530	0.0028	0.9470	0.6081
3.8	+0.0068	0.00005	0.9932	0.5733
4.0	−0.0330	0.0011	1.0330	0.5349
4.5	−0.1027	0.0104	1.1027	0.4293
5.0	−0.1310	0.0172	1.1310	0.3232
5.5	−0.1242	0.0154	1.1242	0.2299
6.0	−0.0922	0.0085	1.0922	0.1594
6.5	−0.0473	0.0022	1.0473	0.1159
7.0	−0.0013	0.00000	1.0013	0.0989
7.5	+0.0361	0.0013	0.9639	0.1036
8.0	0.0587	0.0034	0.9413	0.1219
8.5	0.0643	0.0041	0.9357	0.1457
9.0	0.0545	0.0030	0.9455	0.1663
9.5	0.0339	0.0011	0.9661	0.1782
10.0	+0.0087	0.00008	0.9913	0.1784
10.5	−0.0150	0.0002	1.0150	0.1668
11.0	−0.0321	0.0010	1.0321	0.1464
11.5	−0.0397	0.0016	1.0397	0.1216
12.0	−0.0372	0.0014	1.0372	0.0973
12.5	−0.0265	0.0007	1.0265	0.0779
13.0	−0.0108	0.0001	1.0108	0.0662
13.5	+0.0056	0.00003	0.9944	0.0631
14.0	0.0191	0.0004	0.9809	0.0676
14.5	0.0267	0.0007	0.9733	0.0770
15.0	0.0273	0.0007	0.9727	0.0880
15.5	0.0216	0.0005	0.9784	0.0973
16.0	0.0113	0.0001	0.9887	0.1021

A7 VECTOR OPERATORS

In these relations the scalars f and g and the vectors $\vec{A}$ and $\vec{B}$ can be functions of time as well as space. The magnitude of $\vec{A}$ is A.

$$\nabla^2 f = \nabla \cdot (\nabla f)$$

$$\nabla^2 \vec{A} = \nabla(\nabla \cdot \vec{A}) - \nabla \times \nabla \times \vec{A}$$

$$\nabla \times (\nabla f) = 0$$

$$\nabla \cdot (\nabla \times \vec{A}) = 0$$

$$\nabla(fg) = f\nabla g + g\nabla f$$

$$\nabla \cdot (f\vec{A}) = f\nabla \cdot \vec{A} + \vec{A} \cdot \nabla f$$

$$\nabla \times (f\vec{A}) = (\nabla f) \times \vec{A} + f(\nabla \times \vec{A})$$

$$\nabla(\vec{A} \cdot \vec{B}) = (\vec{A} \cdot \nabla)\vec{B} + (\vec{B} \cdot \nabla)\vec{A} + \vec{A} \times (\nabla \times \vec{B}) + \vec{B} \times (\nabla \times \vec{A})$$

$$\nabla \times (\vec{A} \times \vec{B}) = \vec{A}(\nabla \cdot \vec{B}) - \vec{B}(\nabla \cdot \vec{A}) + (\vec{B} \cdot \nabla)\vec{A} - (\vec{A} \cdot \nabla)\vec{B}$$

$$(\vec{A} \cdot \nabla)\vec{A} = \tfrac{1}{2}\nabla(\vec{A} \cdot \vec{A}) - \vec{A} \times \nabla \times \vec{A}$$

$$\vec{A} \cdot \frac{d\vec{A}}{dt} = A\frac{dA}{dt}$$

(a) Cartesian Coordinates

$$dV = dx\, dy\, dz$$

$$\nabla f = \hat{x}\frac{\partial f}{\partial x} + \hat{y}\frac{\partial f}{\partial y} + \hat{z}\frac{\partial f}{\partial z}$$

$$\nabla \cdot \vec{A} = \frac{\partial A_x}{\partial x} + \frac{\partial A_y}{\partial y} + \frac{\partial A_z}{\partial z}$$

$$\nabla^2 f = \frac{\partial^2 f}{\partial x^2} + \frac{\partial^2 f}{\partial y^2} + \frac{\partial^2 f}{\partial z^2}$$

$$\nabla \times \vec{A} = \hat{x}\left(\frac{\partial A_y}{\partial z} - \frac{\partial A_z}{\partial y}\right) + \hat{y}\left(\frac{\partial A_z}{\partial x} - \frac{\partial A_x}{\partial z}\right) + \hat{z}\left(\frac{\partial A_x}{\partial y} - \frac{\partial A_y}{\partial x}\right)$$

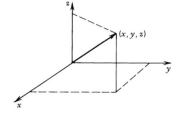

(b) Cylindrical Coordinates

$$dV = r\, dr\, d\theta\, dz$$

$$\nabla f = \hat{r}\frac{\partial f}{\partial r} + \hat{\theta}\frac{1}{r}\frac{\partial f}{\partial \theta} + \hat{z}\frac{\partial f}{\partial z}$$

$$\nabla \cdot \vec{A} = \frac{1}{r}\frac{\partial}{\partial r}(rA_r) + \frac{1}{r}\frac{\partial}{\partial \theta}A_\theta + \frac{\partial}{\partial z}A_z$$

$$\nabla^2 f = \frac{1}{r}\frac{\partial}{\partial r}\left(r\frac{\partial f}{\partial r}\right) + \frac{1}{r^2}\frac{\partial^2 f}{\partial \theta^2} + \frac{\partial^2 f}{\partial z^2}$$

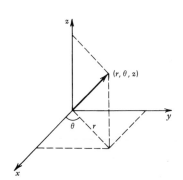

(c) Spherical Coordinates

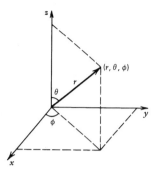

$$dV = r^2 \sin\theta \, dr \, d\theta \, d\phi$$

$$\nabla f = \hat{r}\frac{\partial f}{\partial r} + \hat{\theta}\frac{1}{r}\frac{\partial f}{\partial \theta} + \hat{\phi}\frac{1}{r\sin\theta}\frac{\partial f}{\partial \phi}$$

$$\nabla \cdot \vec{A} = \frac{1}{r^2}\frac{\partial}{\partial r}(r^2 A_r) + \frac{1}{r\sin\theta}\frac{\partial}{\partial \theta}(A_\theta \sin\theta) + \frac{1}{r\sin\theta}\frac{\partial A_\phi}{\partial \phi}$$

$$\nabla^2 f = \frac{1}{r^2}\frac{\partial}{\partial r}\left(r^2\frac{\partial f}{\partial r}\right) + \frac{1}{r^2\sin\theta}\frac{\partial}{\partial \theta}\left(\sin\theta\frac{\partial f}{\partial \theta}\right) + \frac{1}{r^2\sin^2\theta}\frac{\partial^2 f}{\partial \phi^2}$$

A8 GAUSS'S THEOREM AND GREEN'S THEOREM

(a) Gauss's Theorem in Two- and Three-Dimensional Coordinate Systems

Gauss's theorem is a special case of the transport theorem. In three dimensions, it is

$$\int_V \nabla \cdot \vec{F}\, dV = \int_S \vec{F} \cdot \hat{n}\, dS$$

where $\hat{n}$ is the outward unit vector normal to the surface S of the volume V. In words, it states that the total outward flux of $\vec{F}$ through a surface S is equal to the totality of the divergence of $\vec{F}$ in the enclosed volume V. In electromagnetism it relates the integral over a closed surface of the normal component of the electrostatic field to the total enclosed charge.

A special case is the two-dimensional form,

$$\int_S \nabla \cdot \vec{F}\, dS = \int_C \vec{F} \cdot \hat{n}\, dl$$

where $\hat{n}$ is the unit vector normal to the perimeter C of the surface S in a two-dimensional coordinate system.

(b) Green's Theorem

Green's theorem

$$\int_V (U\nabla^2 V - V\nabla^2 U)\, dV = \int_S (U\nabla V - V\nabla U) \cdot \hat{n}\, dS$$

is a consequence of the vector identities

$$\nabla \cdot (U\nabla V) = \nabla U \cdot \nabla V + U\nabla^2 V$$
$$\nabla \cdot (V\nabla U) = \nabla V \cdot \nabla U + V\nabla^2 U$$

and Gauss's theorem. To prove this, take the difference of the above identities and integrate it over the volume V within the surface S,

$$\int_V (U\nabla^2 V - V\nabla^2 U)\, dV = \int_V \nabla \cdot (U\nabla V - V\nabla U)\, dV$$

and then apply Gauss's theorem to the right side with $\vec{F} = U\nabla V - V\nabla U$ to convert the volume integral of $\nabla \cdot \vec{F}$ into the surface integral over S of $\vec{F} \cdot \hat{n}$.

A9 A LITTLE THERMODYNAMICS AND THE PERFECT GAS

(a) Energy, Work, and the First Law

For notational simplicity, the absolute temperature T_K will be written without the subscript. The First Law of Thermodynamics states that heat, like work, is a form of energy, *thermal energy*. Consequently, the change in internal energy dE of a thermodynamic system can be expressed as the sum of the thermal energy δQ added to the system and the work $\delta W = -\mathscr{P}\, dV$ done on the system,

$$dE = \delta Q + \delta W = \delta Q - \mathscr{P}\, dV \qquad\qquad (A9.1)$$

This is a statement of the conservation of energy, which was originally postulated from empirical observations. The minus sign appears because for negative dV the system has been compressed and must have gained energy from the work done on it. The work δW done on a system and the thermal energy δQ given to the system as it is taken from an initial to a final state both depend on the specifics of the process. They are *path dependent*. For example, if a system is taken from state 1 with pressure $\mathscr{P}_1$, volume V_1, and temperature T_1 to state 2 with $\mathscr{P}_2$, V_2, T_2, the amounts of work and thermal energy required will depend on whether the system is first compressed and then heated, or first heated and then compressed, even though the final internal energy must be the same in both cases. The internal energy E is a *state function*; its value depends only on the state ($\mathscr{P}$, V, T) of the system and not how it got there.

Let the thermodynamic system have a mass M, where M is the *molecular weight in grams*. This amount of material is defined as 1 *mole*. (If the mass is the molecular weight in kilograms, then the amount is 1 *kmol*.) If thermal energy ΔQ is added to a 1 mole system whose volume is held constant and the temperature is changed by ΔT, then the molar *heat capacity at constant volume* is defined as

$$C_V = \lim_{\Delta T \to 0} \left(\frac{\Delta Q}{\Delta T} \right)_V \qquad\qquad (A9.2)$$

with units $J/(\text{mol} \cdot \text{K})$. Since $\Delta V = 0$ and no work is done on the system, $dE = \delta Q$ over each step of the process and (A9.1) gives

$$\Delta E = C_V \, \Delta T \qquad (\Delta V = 0)$$

$$C_V = \lim_{\Delta T \to 0} \left(\frac{\Delta E}{\Delta T} \right)_V = \left(\frac{\partial E}{\partial T} \right)_V \qquad\qquad (A9.3)$$

Analogously, introduce ΔQ into a 1 mole system under the constraint that the pressure remain fixed. The associated temperature change ΔT is then related to ΔQ by the (molar) *heat capacity at constant pressure*,

$$C_{\mathscr{P}} = \lim_{\Delta T \to 0} \left(\frac{\Delta Q}{\Delta T} \right)_{\mathscr{P}} \qquad\qquad (A9.4)$$

Use of (A9.1) now gives

$$\Delta E = C_{\mathscr{P}}\Delta T - \mathscr{P}\Delta V \qquad (\Delta\mathscr{P} = 0)$$

$$C_{\mathscr{P}} = \left(\frac{\partial E}{\partial T}\right)_{\mathscr{P}} + \mathscr{P}\left(\frac{\partial V}{\partial T}\right)_{\mathscr{P}} \qquad\qquad (A9.5)$$

For a system containing 1 mole of a single substance, the *equation of state* relates the pressure, volume, and temperature so that the internal energy is a function of only two of the variables. Thus, the energy can be considered as a function of T and V so that

$$\Delta E = \left(\frac{\partial E}{\partial T}\right)_{V}\Delta T + \left(\frac{\partial E}{\partial V}\right)_{T}\Delta V \qquad\qquad (A9.6)$$

If some process now changes the temperature of a system, but holds the pressure constant, we can write

$$\left(\frac{\partial E}{\partial T}\right)_{\mathscr{P}} = \left(\frac{\partial E}{\partial T}\right)_{V} + \left(\frac{\partial E}{\partial V}\right)_{T}\left(\frac{\partial V}{\partial T}\right)_{\mathscr{P}} \qquad\qquad (A9.7)$$

We can relate C_V and $C_{\mathscr{P}}$ by combining (A9.3), (A9.5), and (A9.7),

$$C_{\mathscr{P}} - C_V = \mathscr{P}\left(\frac{\partial V}{\partial T}\right)_{\mathscr{P}} + \left(\frac{\partial E}{\partial V}\right)_{T}\left(\frac{\partial V}{\partial T}\right)_{\mathscr{P}} \qquad\qquad (A9.8)$$

This will yield a simple result when applied to a perfect gas.

(b) Enthalpy, Entropy, and the Second Law

Two other thermodynamic quantities important in acoustics and fluid flow are the *enthalpy H* and the *entropy S*. (There are two more, the *Gibbs function* $G = H - TS$ and the *Helmholtz function* $A = E - TS$, that need not concern us here.) The enthalpy is defined as

$$H = E + \mathscr{P}V \qquad\qquad (A9.9)$$

Since E is a state function and the product $\mathscr{P}V$ is a function only of the state of the system, the enthalpy is also a state function. Taking the differential of H and using the First Law gives $dH = \delta Q + V d\mathscr{P}$. For an *isobaric* process, $d\mathscr{P} = 0$ at each step of the process so that $dH = \delta Q$ at each step of the process and $\Delta H = \Delta Q$. Then (A9.4) gives

$$C_{\mathscr{P}} = \lim_{\Delta T \to 0}\left(\frac{\Delta H}{\Delta T}\right)_{\mathscr{P}} = \left(\frac{\partial H}{\partial T}\right)_{\mathscr{P}} \qquad\qquad (A9.10)$$

In making a transition between an initial and a final state, a system can proceed in a way that cannot be undone, an *irreversible* process. Examples are the free expansion of a gas, a nuclear detonation, and the diffusion of one gas into another. If, however, a system in equilibrium is acted upon so slowly that it remains essentially in equilibrium as it transits from initial to final states, the process is *reversible*. The slow compression of a gas in a perfectly insulated container is an example. Reversible processes can be described in terms of the entropy S. If δQ_{rev} is the thermal energy added to a system during an infinitesimal reversible process at temperature T, then the change in the entropy is defined by

$$dS = \frac{\delta Q_{rev}}{T} \tag{A9.11}$$

The Second Law of Thermodynamics, also postulated from empirical observations, states that any heat engine must operate between two thermal reservoirs of different temperatures. It is equivalent to asserting that the entropy of a system is a state function. Thus, for example, if a system starts in state 1 and by whatever process ends up in state 2, the entropy change can be calculated by ignoring the actual process and instead finding a reversible way of accomplishing the same change in state (e.g., first moving reversibly at constant T_1 to V_2 and then reversibly to T_2 at constant V_2). The entropy and internal energy are related by

$$dE = T\,dS - \mathcal{P}\,dV \tag{A9.12}$$

(c) The Perfect Gas

A perfect gas can be considered as a collection of infinitesimally small, rigid particles that exert forces on each other only when they collide (e.g., a collection of perfectly elastic, rapidly moving billiard balls). The absence of interparticle forces means that there can be no potential energy. The internal energy of the system is then just the sum of the kinetic energies of all the particles, and application of the *kinetic theory of gases* reveals two important results.

1. The energy of the gas is a function *only* of temperature, which has the immediate consequence

$$\left(\frac{\partial E}{\partial V}\right)_T = 0 \tag{A9.13}$$

so that (A9.8) becomes

$$C_{\mathcal{P}} - C_V = \mathcal{P}\left(\frac{\partial V}{\partial T}\right)_{\mathcal{P}} \tag{A9.14}$$

2. The pressure, volume, and temperature of a mole of a perfect gas are related by the equation of state,

$$\boxed{\mathcal{P}V = \mathcal{R}T} \tag{A9.15}$$

where $\mathcal{R}$ is the *universal gas constant*

$$\mathcal{R} = 8.3145 \ \text{J}/(\text{mol} \cdot \text{K}) \tag{A9.16}$$

If there are n moles, then $\mathscr{P}V = n\mathscr{R}T$. In terms of the density ρ, we have $\rho V = M$ so that

$$\mathscr{P} = \rho r T \qquad r = \mathscr{R}/M \tag{A9.17}$$

where r is the gas constant for the particular gas in question.

Two important conclusions about the thermodynamic behavior of perfect gases can now be drawn.

1. Since E is a function only of T, the heat capacities are related by (A9.14) and use of the equation of state (A9.15) reveals

$$\left(\frac{\partial V}{\partial T}\right)_{\mathscr{P}} = \left(\frac{\partial}{\partial T}\frac{\mathscr{R}T}{\mathscr{P}}\right)_{\mathscr{P}} = \frac{\mathscr{R}}{\mathscr{P}} \tag{A9.18}$$

so that

$$\boxed{C_{\mathscr{P}} - C_V = \mathscr{R}} \tag{A9.19}$$

2. For an *adiabatic process* there is no gain or loss of thermal energy so that $\Delta Q = 0$. We then have

$$\Delta E = -\mathscr{P}\Delta V \tag{A9.20}$$

Now, (A9.6), (A9.13), and (A9.3) yield $\Delta E = (\partial E/\partial T)_V\,\Delta T = C_V\,\Delta T$ so that $-\mathscr{P}\Delta V = C_V\Delta T$ for a perfect gas. Use of (A9.15) gives

$$-\mathscr{R}\Delta V/V = C_V\Delta T/T \tag{A9.21}$$

Integration of both sides gives $-\mathscr{R}\ln(V/V_0) = C_V\ln(T/T_0)$ or

$$(V_0/V)^{\mathscr{R}} = (T/T_0)^{C_V} \tag{A9.22}$$

Use of (A9.15) yields the *adiabat*

$$\boxed{\begin{array}{c} \mathscr{P}/\mathscr{P}_0 = (\rho/\rho_0)^{\gamma} \\ \gamma = C_{\mathscr{P}}/C_V \end{array}} \tag{A9.23}$$

where γ is defined as the *ratio of heat capacities*.

For a perfect gas with constant heat capacities between states 1 and 2, use of $\Delta E = C_V\Delta T$ and (A9.1) gives $\delta Q_{rev} = C_{\mathscr{P}}\,dT - V\,d\mathscr{P}$. Division by temperature to obtain the entropy, use of (A9.15) to eliminate V, and then direct integration of temperature and pressure between the two states gives

$$S_2 - S_1 = C_V\left[\ln\left(\frac{\mathscr{P}_2}{\mathscr{P}_1}\right) + \gamma\ln\left(\frac{V_2}{V_1}\right)\right] \tag{A9.24}$$

(a) Solids

Solid	Density (kg/m^3) ρ_0	Young's Modulus (Pa) Y $\times 10^{10}$	Shear Modulus (Pa) $\mathcal{G}$ $\times 10^{10}$	Adiabatic Bulk Modulus (Pa) $\mathcal{B}$ $\times 10^{10}$	Poisson's Ratio σ	Speed (m/s) c		Characteristic Impedance $(Pa \cdot s/m)$ $\rho_0 c$	
						Bar	Bulk	Bar $\times 10^6$	Bulk $\times 10^6$
Aluminum	2700	7.1	2.4	7.5	0.33	5150	6300	13.9	17.0
Brass	8500	10.4	3.8	13.6	0.37	3500	4700	29.8	40.0
Copper	8900	12.2	4.4	16.0	0.35	3700	5000	33.0	44.5
Iron (cast)	7700	10.5	4.4	8.6	0.28	3700	4350	28.5	33.5
Lead	11300	1.65	0.55	4.2	0.44	1200	2050	13.6	23.2
Nickel	8800	21.0	8.0	19.0	0.31	4900	5850	43.0	51.5
Silver	10500	7.8	2.8	10.5	0.37	2700	3700	28.4	39.0
Steel	7700	19.5	8.3	17.0	0.28	5050	6100	39.0	47.0
Glass (Pyrex)	2300	6.2	2.5	3.9	0.24	5200	5600	12.0	12.9
Quartz (X-cut)	2650	7.9	3.9	3.3	0.33	5450	5750	14.5	15.3
Lucite	1200	0.4	0.14	0.65	0.4	1800	2650	2.15	3.2
Concrete	2600	—	—	—	—	—	3100	—	8.0
Ice	920	—	—	—	—	—	3200	—	2.95
Cork	240	—	—	—	—	—	500	—	0.12
Oak	720	—	—	—	—	—	4000	—	2.9
Pine	450	—	—	—	—	—	3500	—	1.57
Rubber (hard)	1100	0.23	0.1	0.5	0.4	1450	2400	1.6	2.64
Rubber (soft)	950	0.0005	—	0.1	0.5	70	1050	0.065	1.0
Rubber (rho-c)	1000	—	—	0.24	—	—	1550	—	1.55

(b) Liquids

Liquid	Temperature (°C) T	Density (kg/m³) ρ_0	Isothermal Bulk Modulus (Pa) $\mathcal{B}_T$ ×10⁹	Ratio of Specific Heats γ	Speed (m/s) c	Characteristic Impedance (Pa·s/m) $\rho_0 c$ ×10⁶	Coefficient of Shear Viscosity (Pa·s) η ×10⁻³	Specific Heat [J/(kg·K)] $c_{\mathcal{P}}$ ×10³	Thermal Conductivity [W/(m·K)] κ	Prandtl Number Pr
Water (fresh)	20	998	2.18	1.004	1481	1.48	1.00	4.19	0.603	6.95
Water (sea)	13	1026	2.28	1.01	1500	1.54	1.07			
Alcohol (ethyl)	20	790	—	—	1150	0.91	1.20			
Castor (oil)	20	950	—	—	1540	1.45	960			
Mercury	20	13600	25.3	1.13	1450	19.7	1.56	0.14	8.21	0.0266
Turpentine	20	870	1.07	1.27	1250	1.11	1.50			
Glycerin	20	1260	—	—	1980	2.5	1490			
Fluid-like sea bottoms										
Red clay		1340		—	1460	1.96				
Calcareous ooze		1570		—	1470	2.31				
Coarse silt		1790		—	1540	2.76				
Quartz sand		2070		—	1730	3.58				

(c) Gases

Gas (at 1 atm)	Temperature (°C) T	Density (kg/m^3) ρ_0	Ratio of Specific Heats γ	Speed (m/s) c	Characteristic Impedance (Pa·s/m) $\rho_0 c$	Coefficient of Shear Viscosity (Pa·s) η $\times 10^{-5}$	Specific Heat [J/(kg·K)] $c_\mathcal{P}$ $\times 10^3$	Thermal Conductivity [W/(m·K)] κ	Prandtl Number Pr
Air	0	1.293	1.402	331.5	429	1.72			
Air	20	1.21	1.402	343	415	1.85	1.01	0.0263	0.710
O$_2$	0	1.43	1.40	317.2	453	2.00	0.912	0.0245	0.744
CO$_2$ ($f \ll f_M$)	0	1.98	1.304	258	512	1.45	0.836	0.0145	0.836
CO$_2$ ($f \gg f_M$)	0	1.98	1.40	268.6	532				
H$_2$	0	0.090	1.41	1269.5	114	0.88	14.18	0.168	0.743
Steam	100	0.6	1.324	404.8	242	1.3			

A11 ELASTICITY AND VISCOSITY

(a) Solids

Application of an external force to a body distorts the body until internal forces arise to counterbalance the applied force and the body assumes a new shape. The applied force per unit area is the *stress* and the fractional changes in dimensions are the *strains*. For small strains of an isotropic solid, we can make two pivotal assumptions: (1) the stress and strains are linearly related (*Hooke's law*), and (2) individual stresses cause individual strains and the results combine linearly. The various stress–strain relationships can all be expressed in terms of *Young's modulus* Y and *Poisson's ratio* σ. [For a complete development, see Feynman, Leighton, and Sands, *The Feynman Lectures on Physics*, Vol. 2, Chap. 38, Addison-Wesley (1965).]

(1) LONGITUDINAL COMPRESSION OF A THIN ROD   The stress–strain relationship for a thin rod under longitudinal compression or extension is developed in Section 3.3. When a compressive force f is applied to the ends of a bar of cross-sectional area S and length l, the bar will shorten by a small amount Δl. The strain $\Delta l / l$ is proportional to the applied stress f/S,

$$f/S = -Y\,\Delta l/l \qquad\qquad (A11.1)$$

where the constant of proportionality Y is *Young's modulus.* (The minus sign is consistent with a positive pressure leading to a reduction in volume.)

(2) POISSON'S RATIO   The change in length of the thin bar is accompanied by a change in the transverse dimensions. If all transverse dimensions are labeled r, the reaction to the change in length Δl is a proportionate change in the transverse dimensions

$$\Delta r/r = -\sigma\,\Delta l/l \qquad\qquad (A11.2)$$

where the proportionality constant σ is *Poisson's ratio.*

(3) UNIFORM VOLUME COMPRESSION   Uniform compression is developed in Chapter 5 and gives the stress–strain relationship (5.2.6)

$$p = -\mathscr{B}\,\Delta V/V \qquad\qquad (A11.3)$$

where $s = \Delta\rho/\rho = -\Delta V/V$ and $\mathscr{B}$ is the *bulk modulus.* We can easily relate $\mathscr{B}$, Y, and σ. Impose a uniform compression sequentially on a block of material of length l, width w, and thickness t. If $p = f/S$ is applied to the sides delimiting l, the fractional change in length will be given by (A11.1). If the same compressive stress is applied to the sides delimiting w, then l will be pushed back out a little and the magnitude of Δl will be diminished by a factor $(1 - \sigma)$. When the final stress is applied to the sides delimiting t, the length will again be pushed back a little more and Δl will be diminished by a total factor of $(1 - 2\sigma)$, so that the actual change Δl is

$$\Delta l/l = -(1 - 2\sigma)f/YS \qquad\qquad (A11.4)$$

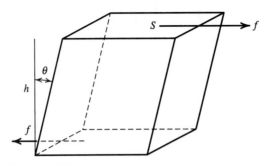

Figure A11.1

Thus, the direct compression is reduced linearly by each of the compressions in the two transverse directions. By symmetry, exactly the same must occur for Δw and Δt. The total change in fractional volume is just the sum of the fractional changes in the dimensions,

$$\Delta V/V = \Delta l/l + \Delta w/w + \Delta t/t = -3(1 - 2\sigma)p/Y \qquad (A11.5)$$

where p has replaced f/S. Direct comparison of (A11.3) and (A11.5) shows that

$$\mathscr{B} = Y/3(1 - 2\sigma) \qquad (A11.6)$$

   (4) SHEAR   If an antiparallel couple of forces f are applied to the top and bottom surfaces, each of area S, of a cube of height h, the cube will deform as shown in Fig. A11.1. Dividing the displacement of the upper surface with respect to the lower by the height gives the strain. For small strains this is well approximated by the angle of deformation θ. The stress is f/S. The applied stress is proportional to the strain,

$$f/S = \mathscr{G}\theta \qquad (A11.7)$$

where $\mathscr{G}$ is the *modulus of rigidity* or *shear modulus*. This is identical with (3.13.2), where $\theta = r(d\phi/dx)$ and $S = dw\,dr$. The element does not rotate, so the net torque is zero. This requires induced vertical forces of the same magnitude f. Vector combination of these direct and induced four forces shows that they are equivalent to a pair of compressional forces acting along one diagonal L and a pair of extension forces acting along the other. Geometry shows that these compression and extension stresses have the same magnitude as the shearing stresses. Just as before, the change in length of each diagonal is the sum of the strains from the direct compression along the diagonal and the lateral extension perpendicular to it. The effects enhance each other, so

$$\Delta L/L = -(1 + \sigma)f/YS \qquad (A11.8)$$

for the compressed diagonal. A little geometry reveals that $\Delta L/L = \theta/2$ and thus

$$\theta = -2(1 + \sigma)f/YS \qquad (A11.9)$$

Comparison of (A11.7) and (A11.9) gives us

$$\mathscr{G} = Y/2(1 + \sigma) \qquad (A11.10)$$

(5) Longitudinal Bulk Compression   If a compressive stress is applied to the ends of a bar that is restricted so that the transverse dimensions cannot change, then the bar cannot "relax" against the stress by partially compensating with an increase in cross-sectional area. Under this constraint, a greater stress will be required to achieve the same longitudinal strain. The result is a larger constant of proportionality between stress and strain. We can generalize the previous case of bulk compression to consider *different* values of compressive pairs of forces in the three directions. The discussion is straightforward if we assume that the material element is a cube of dimension h. Let us compress the cube in the x direction with a pair of forces f_x and require no expansion or contraction in the y and z directions. By symmetry, the transverse forces in the y and z directions necessary to prevent any transverse expansion or contraction must equal each other, and we label them f_T. In the x direction we must have

$$\Delta h/h = -(f_x - 2\sigma f_T)/YS \tag{A11.11}$$

and in each of the transverse dimensions, for no change of length,

$$0 = -[f_T - \sigma(f_T + f_x)]/YS \tag{A11.12}$$

Solving this equation for f_T/S and substituting into (A11.11) gives us the relationship between the applied stress and the resultant strain for longitudinal bulk compression,

$$\Delta h/h = -[(1 + \sigma)(1 - 2\sigma)/(1 - \sigma)]f_x/YS \tag{A11.13}$$

This can be expressed in simpler form with the help of the bulk modulus $\mathscr{B}$ and modulus of rigidity $\mathscr{G}$. Using (A11.6) and (A11.10) results in

$$f_x/S = \left(\mathscr{B} + \tfrac{4}{3}\mathscr{G}\right)\Delta h/h \tag{A11.14}$$

(b) Fluids

The molecules of a fluid have sufficient kinetic energies to migrate from current nearest neighboring molecules to others. This mobility manifests itself in the appearance of additional forces that generate macroscopic effects.

(1) Shear Viscosity   If a fluid is in a state of nonuniform macroscopic motion (acoustic propagation, laminar or turbulent flow, etc.) there can be a diffusion of momentum among neighboring fluid elements caused by the migration of molecules from one element to another. This diffusion of momentum gives rise to internal forces that reduce the relative motions of adjacent elements and bring the fluid back to a state of uniform motion or to rest. The diffusion of momentum occurs whether the fluid is in shear or in longitudinal relative motion, or in both. Shear viscosity is an important mechanism of energy transformation from collective (acoustic) to random (thermal). If a fluid is subjected to a *shear stress* τ, it will respond by developing an attendant shearing motion. The diffusion of momentum between neighboring lamina of the fluid will lead to a local steady-state velocity, or a *rate of deformation* when the frictional forces counterbalance the shearing forces. In the case of *simple shear*, such as the flow between two infinite parallel plates, the velocity of the fluid is parallel to the stress. If the parallel plates extend in the y and z directions and one is moving uniformly in the y direction,

then both the shear stress τ and the fluid velocity u are in the y direction and the magnitude of the stress is proportional to $\partial u / \partial x$, the spatial change in u with respect to the (transverse) x direction,

$$\tau = \eta \frac{\partial u}{\partial x} \tag{A11.15}$$

where the proportionality constant η is the *coefficient of shear viscosity* (Pa · s).

For more complicated motion, obtaining the relationship between the force arising from shear viscosity and the motion is nontrivial. Developing the stress and rate of strain tensors within a body possessing shear viscosity and then calculating the internal body force per unit mass results in

$$\vec{F}_S(\vec{r}, t) = \tfrac{4}{3}\eta \nabla(\nabla \cdot \vec{u}) - \eta \nabla \times \nabla \times \vec{u} \tag{A11.16}$$

as the contribution to the inhomogeneous wave equation of Section 5.14. The interested reader is referred to Temkin, *Elements of Acoustics,* Wiley (1981).

(2) VOLUME VISCOSITY   For a fluid at rest, the bulk modulus $\mathcal{B}$ is defined exactly the same as for a solid. This modulus expresses the purely compressive (spring-like) properties of the fluid. The frictional effects arising from the transport of momentum between adjacent fluid elements are accounted for by the shear viscosity. In addition to this, however, there are other mechanisms separate from the diffusion of momentum that can lead to losses in some fluids. (1) Under new thermodynamic conditions (higher pressure and temperature for a compression), some groups of molecules may accommodate a different nearest neighbor configuration. This takes time, so that the equilibrium volume (for a constant pressure change) lags behind the instantaneous volume. This, although arising from a totally separate mechanism, introduces the same kind of an adjustment toward equilibrium, requiring a finite duration of time as does shear viscosity. (2) The changing conditions may lead to changes in the equilibrium between ionized and un-ionized concentrations of a compound (e.g., magnesium sulfate in sea water). This can result in a relaxation effect similar to the above, particularly because of the association and dissociation of ionized and un-ionized compounds with the adjacent ionized and un-ionized water complexes. These provide examples of processes that depend on the instantaneous thermodynamic state of the fluid element, and that require some time to adjust to new conditions. In effect, all act like *structural* changes that require a finite time during which the fluid attempts to find a new equilibrium volume in response to the external stimulus. These adjustments, like the momentum diffusion, exhibit themselves as internal friction-like forces but depend only on the externally generated temporal changes in the local density. Consequently, by the linearized equation of continuity, they are functions only of $\nabla \cdot \vec{u}$, the rate of strain of the fluid element, but not on the flow. Since the forces on fluid elements arise from the gradient of the pressure, and thus the gradient of the density, it is reasonable to introduce them into Euler's equation as a body force per unit volume (see Section 5.14),

$$\vec{F}_B(\vec{r}, t) = \eta_B \nabla(\nabla \cdot \vec{u}) \tag{A11.17}$$

where η_B is the *coefficient of bulk viscosity* (also known as the coefficient of volume viscosity, expansion coefficient of viscosity, and so forth).

A12 THE GREEK ALPHABET

A	**A**	α	**α**	alpha		N	**N**	ν	**ν**	nu
B	**B**	β	**β**	beta		Ξ	**Ξ**	ξ	**ξ**	xi
Γ	**Γ**	γ	**γ**	gamma		O	**O**	o	**o**	omicron
Δ	**Δ**	δ	**δ**	delta		Π	**Π**	π	**π**	pi
E	**E**	ε	**ε**	epsilon		P	**P**	ρ	**ρ**	rho
Z	**Z**	ζ	**ζ**	zeta		Σ	**Σ**	σ	**σ**	sigma
H	**H**	η	**η**	eta		T	**T**	τ	**τ**	tau
Θ	**Θ**	θ	**θ**	theta		Υ	**Υ**	υ	**υ**	upsilon
I	**I**	ι	**ι**	iota		Φ	**Φ**	ϕ	**φ**	phi
K	**K**	κ	**κ**	kappa		X	**X**	χ	**χ**	chi
Λ	**Λ**	λ	**λ**	lambda		Ψ	**Ψ**	ψ	**ψ**	psi
M	**M**	μ	**μ**	mu		Ω	**Ω**	ω	**ω**	omega

ANSWERS TO
ODD-NUMBERED PROBLEMS

1.2.1. (a) $(1/2\pi)\sqrt{2s/m}$. (b) $(1/2\pi)\sqrt{s/2m}$. (c) $(1/2\pi)\sqrt{s/2m}$. (d) $(1/2\pi)\sqrt{2s/m}$.

1.2.3. (b) $(1/2\pi)\sqrt{g/L}$.

1.3.1. $-(U/\omega_0)\sin(\omega_0 t)$.

1.5.3. (a) $AB\cos(2\omega t+\theta+\phi)$. (b) $(A/B)\cos(\theta-\phi)$. (c) $AB\cos(\omega t+\theta)\cos(\omega t+\phi)$. (d) $2\omega t+\theta+\phi$. (e) $\theta-\phi$.

1.6.1. $R_m = 1.0$ kg/s, $\omega_d = 9.85$ rad/s, $A = 0.0402$ m, $\phi = -5.8°$.

1.6.5. $1.005, -5.74°$; $1.021, -11.5°$; $1.048, -17.5°$.

1.7.1. (a) $-(\omega F/Z_m)\sin(\omega t-\phi)$. (b) $\omega_0/(1-R_m^2/2\omega_0^2 m^2)^{1/2}$.

1.7.3. $\omega^2 mL/Z_m$, where $Z_m^2 = R_m^2 + [\omega(M+m)-s/\omega]^2$.

1.10.5. $\frac{1}{2}f_0(\partial\Theta/\partial f)_{f0}$.

1.12.1. (a) $x = F(1/s - 1/m\omega^2)\sin\omega t$.

1.12.3. (b) $[s(1/m+1/M)]^{1/2}$. (c) Unchanged.

1.15.1. $U(t) = (F/Z)\exp(j\omega t)$.

1.15.7. $(gm/s)\cos\left(\sqrt{s/m}\,t\right)$.

1.15.9. (c) $\Delta w\,\Delta t = 4\pi$. (d) Yes.

1.15.11. (b) $\frac{1}{2}[\delta(w-\omega)+\delta(w+\omega)]$, $\frac{1}{2}j[\delta(w-\omega)-\delta(w+\omega)]$.

2.3.1. (a) $\partial^2 y/\partial t^2 = [T/\rho_L(x)]\,\partial^2 y/\partial x^2$. (b) $\partial^2 y/\partial t^2 = g(\partial/\partial x)(x\,\partial y/\partial x)$.

2.8.1. (a) 4.0 cm, 1.5 cm/s, 0.48 Hz, 3.1 cm, 2 cm^{-1}. (b) 0.

2.8.3. $0.835A$.

2.9.1. $\rho_L c(1-j\cot kL)$.

2.9.3. (a) $-j2\rho_L c\cot\left(\frac{1}{2}kL\right)$. (c) $(F/2kT)\sin\left(\frac{1}{4}kL\right)/\cos\left(\frac{1}{2}kL\right)$.

2.9.5. (a) $\frac{1}{2}nc/L$. (b) $\frac{1}{2}nc/L$. (c) $\frac{1}{2}(n-\frac{1}{2})c/L$. (d) No.

2.9.7. $[(n-\frac{1}{2})/2L]\sqrt{T/\rho_L}$.

2.9.9C. (a) $0.65\pi, 1.57\pi, 2.54\pi$. (b) $\left(n-\frac{1}{2}\right)\pi$.

2.9.11. (a) $\tan kL = -\pi(m/m_l)$, $m_l = \rho_L\lambda/2$. (b) $m_l > \pi m$. (c) $m \ll m_l/\pi$, $m \gg m_l/\pi$.

2.10.1. $A_n = [9h/(n\pi)^2]\sin(n\pi/3) = 0.79h, 0.198h, 0, -0.049h.$

2.11.1. $0.65\pi, 1.57\pi, 2.54\pi, \ldots.$

2.11.3. Yes.

2.11.5C. (a) $0.27\pi, 1.09\pi, 2.05\pi.$ (b) $n\pi.$ (c) Disagree.

2.11.7C. (c) 10 s.

2.11.9. (d) $\beta/\omega = \alpha/k$ through $O(\alpha/k)^2.$ (e) Equal through $O(\alpha/k)^3.$

3.4.1. (b) 2525 Hz. (d) $A_1 = 2.1 \times 10^{-4}, A_3 = -2.3 \times 10^{-5}, A_5 = 8.3 \times 10^{-6}$ m.

3.5.1. (a) 6.8 kHz. (b) 0.185 m. (c) 1.91. (d) 15.9 kHz.

3.5.3. (a), (b) No.

3.6.1. $0.35M.$

3.6.3. $Z_{m0} = -\mathbf{Z}_{m0}.$

3.6.5. (a) $\tan kL = -(mkL/\rho_L L - sL/SYkL)/(1 + msL/\rho_L SY).$ (b) 456 Hz.
(c) No nodes on the bar.

3.7.3. (a) $A = [F/(YSk\sin kL)]\cos k(L - x).$ (b) $\mathbf{Z}_{m0} = j\rho_0 cS \tan kL.$ (c) $Z_{m0} = \rho_0 cS.$

3.7.5. (a) $(n/2)(c/L).$ (b) Equal.

3.8.1. $a/2.$

3.10.1. (b) $f_n = n^2\pi\kappa c/L^2 = 20.2, 80.9, 182, \ldots$ Hz; the 1,4,9, ... harmonics.

3.11.1. $\frac{1}{2}.$

3.11.5. (a) 14.4 Hz. (b) $A = 0.0250$ m, $B = -0.0183$ m, $g = 1.876$ m$^{-1}.$

3.12.1. (a) 179 Hz. (b) 0.033 m.

3.13.1. (a) 148 N·m. (b) 2980 m/s. (c) 1490 Hz.

4.3.1. (a) $0.406A.$ (b) $0.5 = \sin(\pi x/a)\sin(\pi z/a).$ (c) No.

4.3.3. (a) $(\mathcal{T}/\rho_S)^{1/2}\sqrt{5}/4L.$ (b) $f_{nm} = (\mathcal{T}/\rho_S)^{1/2}[(2n - 1)^2 + 4m^2]^{1/2}/4L,$
$n = 1, 2, \ldots, m = 1, 2, 3, \ldots, \mathbf{y}_{nm} = A_{nm}\sin\left[\left(n - \tfrac{1}{2}\right)\pi x/L\right]\sin(m\pi z/L)\exp(j\omega t).$

4.4.3. (a), (b) 10.4, 13.8 kHz.

4.5.1. (a) 5.42 kHz. (b) 1.36×10^{-2} cm$^3.$

4.5.3. (a) 11.1 kHz. (b) $5.55 \times 10^{-3}, 1.28 \times 10^{-2}$ m. (c) 6.24×10^{-6} m.

4.7.1. (a), (b) 153, 194 Hz.

4.7.3. (a) 3.83, 5.14, 6.38, 7.59, 8.77. (b) $s = -0.238, -0.148, -0.083, -0.037, 0;$
not uniform, but regular $[s \approx 0.21\ln(n/6)].$

4.8.1. $f_n = j_{1n}/2\pi a.$

4.9.3. 68.6%.

4.9.5C. (b) 0.57.

4.10.1. $A_{nm} = 2/\sqrt{L_x L_y}.$

4.10.3. (a) $\sqrt{3/8}.$ (b) (3,1). (c) Both, neither, (1,2). (d) $f_{62} = f_{24} = 2f_{31}, f_{71} = f_{34} = 2.10f_{31},$
$f_{93} = f_{36} = 3f_{31}.$

4.11.1. (a) 1.23 kHz. (b) Double the frequency. (c) Quarter the frequency.

4.11.3. (a) $-0.0025.$ (b) $y_2 = A_2\cos(\omega_2 t + \phi_2)J_0(6.3r/a) - 0.0025I_0(6.3r/a).$ (d) 0.38.

4.11.5. (*b*) Doubled.

4.11.7. No cylindrically symmetric normal modes.

5.2.1. (*a*) $\mathcal{B} = \mathcal{P}_0\gamma$. (*b*) $\mathcal{B} \propto T_K$.

5.2.3. $\approx \gamma$

5.6.1. (*a*) 1260 m/s. (*b*) Yes. (*c*) 4.1 C°.

5.6.3. (*a*) No, yes. (*b*) $c = \sqrt{rT_K}$. (*c*) 344, 291 m/s.

5.7.3. (*a*) P/ρ_0c^2. (*b*) $|s|$.

5.7.5. (*b*) 0.0065, 0.029.

5.9.1. (*a*) $(P/c^2)\exp[j(\omega t - kx)]$. (*b*) $(P/\rho_0c)\exp[j(\omega t - kx)]$. (*c*) $j(P/\rho_0\omega)\exp[j(\omega t - kx)]$. (*d*) $(P^2/\rho_0c^2)\cos^2(\omega t - kx)$. (*e*) $P^2/2\rho_0c$.

5.9.3. (*a*) $(P/c^2)\cos(kx)\sin(\omega t)$. (*b*) $-(P/\rho_0c)\sin(kx)\sin(\omega t)$. (*c*) $-(P/\rho_0\omega)\cos(kx)\sin(\omega t)$. (*d*) $P^2/4\rho_0c^2$. (*e*) 0.

5.10.3. $j\rho_0c\tan(kx)$.

5.11.1C. $0.2 < ka < 7$, ρ_0c, $\frac{1}{2}\rho_0c$.

5.11.3. (*a*) $-j(A/\rho_0cr)[\sin(kr)+(1/kr)\cos(kr)]\exp(j\omega t)$. (*b*) $j\rho_0c\cos(kr)/[\sin(kr)+(1/kr)\cos(kr)]$. (*c*) $(A/r)^2(1/\rho_0c)\cos(kr)[\sin(kr) + (1/kr)\cos(kr)]\cos(\omega t)\sin(\omega t)$. (*d*) 0.

5.11.5. 0.30 m, phase and direction of the particle velocity.

5.12.1C. (*b*) No.

5.12.3. (*a*) 4.8×10^{-3} W/m², 96.8 dB *re* 10^{-12} W/m². (*b*) 7.7×10^{-6} m. (*c*) 4.82×10^{-3} m/s. (*d*) 1.41 Pa. (*e*) 97 dB *re* 20 μPa.

5.12.5. (*b*) 6.75 kW/m². (*c*) 59.7.

5.12.7. (*b*) 430, 378 Pa·s/m. (*c*) +13.8%. (*d*) +0.56, 0 dB.

5.12.9. 60 dB *re* 1 μbar/V.

5.12.11. (*a*) -180 dB *re* 1 V/μPa. (*b*) 1.0 V.

5.13.1C. 1.1.

5.13.5. (*c*) $\theta = \tan^{-1}(k_z/k_r)$.

5.14.1. (*b*) $c_0/(g\cos\theta_0)$, yes. (*c*) 115, 1.51 km.

5.14.3. (*a*) $Z = 2X - X^2$ with $Z = \varepsilon z/(\sin^2\theta_0)$ and $X = \varepsilon x/(2\sin\theta_0\cos\theta_0)$.
(*b*) $(4/\varepsilon)\cos\theta_0\sin\theta_0$, $(1/\varepsilon)\sin^2\theta_0$. (*c*) $c_0\cos\theta_0/(1 - \frac{2}{3}\sin^2\theta_0)$, $1 + (\theta_0^2/6)$.
(*d*) 1%. (*e*) 1467, 1467.1, 1467.3, 1468.9, 1474.4 m/s.
(*f*) Paths above the axis, 0, 7.6, 30.4, 189.9, 753.8 m and 0, 1.74, 3.49, 8.68, 17.1 km; paths below the axis, twice the previous values.

5.14.5. 4200 m.

5.14.7. 4.00, 3.41, 2.00, 0.59, 0.00, 0.59, 2.00, 3.41, 4.00.

5.14.11C. (*b*) 4.6°, 49.3 km. (*c*)–(*d*) 4.6, 3.6 km.

5.14.13C. 0.048 s.

5.15.1. (*a*) $\vec{u} \cdot \vec{v}$. (*b*) $\nabla \cdot \vec{u}$. (*c*) $\vec{u} \cdot \nabla f$. (*d*) $f\nabla \cdot \vec{u} + \vec{u} \cdot \nabla f$.

6.2.1. (*a*) 0.0281 Pa. (*b*) 1.69, 1.9 μW/m². (*c*) 29.5 dB. (*d*) 66.5 Pa, 1.5 mW/m², 0.5 dB.
(*e*) 0.109.

6.2.3. (*a*) 5.53×10^{-4}, 1.10×10^{-3}. (*b*) 2, 1.10×10^{-3}. (*c*) -65, -30, $+6$, -30 dB.

6.2.5C. $(1 + R)/(1 - R)$.

6.3.3. (a) 1 dB. (b) 0.2. (c) 0.2 dB, 0.05.

6.4.3. (a) 11.9°. (b) 0.83.

6.4.5. (a), (b) 46, 146 Pa. (c) 0.212. (d) 47.7°.

6.6.1. (a) 73°. (b) 0.30. (c) 0.53.

6.7.1. (b) $|k_2 L - n\pi| \ll 1$. (c) $k_2 L \ll 1$ consistent with (b) for $n = 0$.

6.8.3. (c) $\theta = \sin^{-1}(n\pi/kd)$.

6.8.5. (a) $2A/r$. (b) $2(A/\rho_0 c)(d/r^2)[1 + (kr)^2]^{1/2}/kr$. (c), (d) Plots with $r = \sqrt{R^2 + d^2}$.

7.1.1. (a) 0.628 W. (b) 5 W/m², 64.4 Pa, 0.863 m/s; 1.37×10^{-3} m, 1.37×10^{-2}, 4.52×10^{-4}, 2.52×10^{-3}. (c) 0.216 W/m², 13.4 Pa, 4.79×10^{-2} m/s, 7.62×10^{-5} m, 1.52×10^{-4}, 9.41×10^{-5}, 1.40×10^{-4}.

7.1.3. (a) $\rho_0 c(1 + j)/2$, $\frac{1}{2}$. (b) $\propto \omega^2$, constant.

7.1.5C. (c) 0.316.

7.2.1. (a) $\frac{1}{4}$. (b) 4.

7.2.3. (a) $(2A/r)\exp[j(\omega t - kr)]$. (b) Yes.

7.3.1. (b) 5, 5, 6. (c) Partial, full, partial.

7.4.1. (a) $\sin\theta_1 = 3.83/ka$. (b) $r/a = (ka/4\pi)[1 - (2\pi/ka)^2]$. (c) Impossible.

7.4.3. 14.8°.

7.5.1. (a) $(1/2\pi)\sqrt{s/m}$. (b) Mass-controlled when $\omega m \gg s/\omega$ and $\gg R_m + S\rho_0 c$, compliance-controlled when $s/\omega \gg \omega m$ and $\gg R_m + S\rho_0 c$.

7.5.3. (a) $2\pi a^2 \rho_0 c \cos\theta_a \exp(j\theta_a)$ with $\cot\theta_a = ka$. (b) Identical. (c) 3.

7.6.3. (a) 0.0328 m/s. (b) 0.0556 kg. (c) 18.6°. (d) 25 dB.

7.6.5. $(Pr/\omega)(2\pi c/\Pi\rho_0)^{1/2}$, $\Pi k/P\pi r$.

7.6.7. 13.2.

7.6.9C. (b) 6.83 kHz.

7.8.5. (a) 2.42 m. (b) Straight line at $\theta = 90°$. (c) 42°. (d) 19 dB.

7.8.7. (b) $\cos\theta_0 \sim |\theta_0 - \theta_1|$.

7.8.9C. Smooth transition between (7.8.20) and (7.8.21).

7.9.1. $\sin\left(\frac{1}{2}kL_y \sin\theta\right)\sin\left(\frac{1}{2}kL_x \sin\phi\right) / \left(\frac{1}{4}k^2 L_y L_x \sin\theta \sin\phi\right)$.

7.10.3. (a) $(Q/4\pi r)\exp[j(\omega t - kr)]\{3 + 2[-(kd\sin\theta)^2/2! + (kd\sin\theta)^4/4! - \cdots]\}$.

7.10.7. $-(Aa^4/12r)\exp[j(\omega t - kr)]$, 0, $-(Aa^4/180r)(ka)^2 \exp[j(\omega t - kr)]$.

8.2.1. (a) 3.2×10^{-10} s. (b) 490 MHz. (c) 3.2×10^{-12} Np·s²/m, within 7%.

8.2.3. (a) **p**, **s** in phase; **p**, **u** phase angle $\tan^{-1}(\alpha/k)$. (b) $(P^2/2\rho_0 c)\exp(-2\alpha x)$ through $O(\alpha/k)$.

8.2.5. (b) $\mathbf{u} = U_0 \exp(-\alpha z)\exp[j(\omega t - kz)]$. (c) $\sqrt{2\eta\omega/\rho_0}$, $\sqrt{\rho_0\omega/2\eta}$. (d) $\sqrt{2\eta/\rho_0\omega}$, 6.9×10^{-4}, 6.9×10^{-5} m.

8.3.1. (a) $(P_0/\rho_0 c^2)[1 - \exp(-t/\tau)]$. (b) $c\sqrt{1 + j\omega t}$.

8.3.3. (b) $20\log r + a(r - 1)$. (c) 0.0115 Np/m.

8.3.5C. 0.308, 0.53°.

8.4.3. 1.83×10^{-10}, 1.73×10^{-10} s; 1.06.

8.6.1. $f = 1/2\pi\tau_M$.

8.6.3. (a) 3.18×10^{-5} s. (b) and (c)

Frequency(kHz)	α_M(Np/m)	α(Np/M)
1	0.00123	0.00124
2	0.00441	0.00446
5	0.0160	0.0163
7	0.0212	0.0219
10	0.0256	0.0270

(d) Unimportant.

8.6.5. $C_\mathscr{P} = 35.2$ J/(mol·K), $C_V = 26.8$ J/(mol·K), $C_e = 20.3$ J/(mol·K), $C_i = 6.5$ J/(mol·K).

8.7.1. (b), (c) 2.6, 47 dB.

8.9.1C. (b) 8.3×10^{-5}, 8.3×10^{-6}; 9.9×10^{-5}, 9.9×10^{-6}.

8.9.3. (a)–(c) 1.4×10^{-4}, 4.0×10^{-2}, 4.0×10^{-2} Np/m. (d) -0.71 dB.

8.9.5. Free: 1.2×10^{-4}, 1.2×10^{-2}, 1.2 dB/m. Tube: 0.82, 2.60, 9.4 dB/m.

8.10.1. (a)–(c) 0.0104, 0.0016, 0.0042 dB/m.

8.10.3. 0.054.

8.10.7. (a) 46.1 kHz, (b) 6.23×10^{-5}, 1.14×10^{-5} m². (c) 36.9 m⁻³. (d) Same.

9.2.1. 60.8, 66.2, 70.9, 89.9, 93.4, 97.0, 114.5, 121.6, 132.4, 138.5 Hz.

9.2.5. (1,0,0) (0,1,0) (0,0,1) out, (1,1,0) (1,0,1) (0,1,1) in, (1,1,1) out, (2,0,0) (0,2,0) (0,0,2) in phase.

9.3.3. (a) (0,1,1) 20.1, (0,2,1) 33.3, (0,0,2) 41.8, (0,3,1) 45.9, (1,0,1) 57.2 Hz. (c) None.

9.3.5C. 462 Hz.

9.4.1. (a), (b) 568 Hz, node at 0 cm, (antinode at 20 cm); 912, 0, (20); 1230, 14, (0, 20). (c) Only the last one (0,0,1).

9.5.1. 37.5, 45.1, 62.5, 83.9, 106.8 Hz.

9.5.3. 5.73, 9.14, 12.3, 13.2, 15.2 kHz.

9.5.5. (a) 10.6, 16.8, 21.2, 23.7 kHz. (b) $f_{12} = f_{21}, f_{13} = f_{31}$. (c) (1,1), (3,1), (1,3). (d) No.

9.6.1. 1.

9.8.1. (a) $Z_n(z) = A_n \cos k_{zn}z$, $k_{zn} = n\pi/H$, $n = 0, 1, 2, \ldots$, $A_0 = \sqrt{1/H}$, $A_n = \sqrt{2/H}$ for $n \geq 1$. (b) (9.8.5) with $\omega_n = n\pi c/H$. (c) $N = [(H/\pi)(\omega/c)]$. (d) (9.8.4) with $(2/H)$ replaced with A_n^2, sum from $n = 0$ to N, κ_n as in (b), and A_n and k_{zn} as in (a).

9.8.3. (a) 2. (b) $0.254/\sqrt{r}$ Pa.

9.9.15. (a) 120 dB. (b) 31 km.

10.2.1. 0.218 m.

10.3.1. (a) $0.029 + j13.2$ N·s/m. (b) 62 N. (c) 0.32 W.

10.3.3C. (b) 0.44.

10.4.1. (*a*) 0.25. (*b*) 4.44 × 10^6 Pa·s/m.

10.4.3. (*a*) 12.3. (*b*) 0.38, 1.24 m.

10.4.5. 546 − *j*1349 Pa·s/m.

10.5.1. (*a*) $\alpha = [(SWR)_1 - (SWR)_2]/[(SWR)_1(SWR)_2(x_1 - x_2)]$.
(*b*), (*c*) 2.77 × 10^{-2}, 2.14 × 10^{-2} Np/m. (*d*) Very roughly around 5% or 50%.

10.5.3C. (*b*) 28.6 from plot, 29.1 from equation.

10.6.3. (*a*) 438 Hz, 2.55π. (*b*) 617 Hz, 3.60π.

10.8.1. (*a*) 1.94 cm. (*b*) 0.34 μbar. (*c*), (*d*) 380, 452 Hz.

10.9.1. (*c*) Yes.

10.10.1. (*a*) $4S_1^2/(S_1 + S_2)^2$. (*b*) $S_1 > S_2$. (*c*) S_1/S_2 for $S_1 > S_2$, S_2/S_1 for $S_1 < S_2$.

10.10.3. (*a*) 0.33. (*b*) $2S_1P/S_2$. (*c*) 6.

10.10.5. (*a*) 0.49 m^3. (*b*) 0.5.

10.10.9. (*a*) Plot of $x^2/(1 + x^2)$ with $x = 2\omega m/\rho_0 cS$. (*b*) 0.62 kg.

10.10.11. (*a*) Band stop. (*b*) 6.90 × 10^{-4} m^3. (*c*) 0.64.

10.10.13. $2(\sqrt{2} - 1)$.

10.11.5. (*a*) $S_1 = 0.1(1 + 0.11/L)$ m^2 with $L < 0.05$ m to ensure $kL < 1$ at 1 kHz. (*b*) 333 Hz.

11.2.1. (*a*) 3, 1.260. (*b*) 2, 1.414. (*c*) 12, 1.059.

11.3.1. (*a*)–(*c*) 6 × 10^{-6}, 6 × 10^{-6}, 3.6 × 10^{-5} W/m^2.

11.3.3. (*a*) 150.4 dB *re* 1 μPa, 150.4 dB *re* 1 μPa/$Hz^{1/2}$

160	150
170	150
(*b*) 153	153
160.4	150.4
170	150
(*c*) 160.4	160.4
163	153
170.4	150.4

11.3.5. (*a*) $128 - 20 \log f$ dB *re* 20 μPa/$Hz^{1/2}$. (*b*) −6 dB /octave. (*c*) 77 dB *re* 20 μPa.

11.4.1. $A_{S,N}$ increased by 0.7σ.

11.4.3. 0.03.

11.5.1. If $\tau > T_S$, DT' is constant; if $\tau < T_S$, DT' increases with decreasing τ.

11.5.3. 2, 3 dB.

11.6.1. $z = 3\tan^{-1}[\frac{2}{3}(4 - \log f)]$.

11.7.1. (*a*), (*b*) 22, 53 dB *re* 20 μPa.

11.7.3. 90 dB *re* 20 μPa.

11.8.1. (*a*) 2.1 sone. (*b*), (*c*) 40, 85 dB *re* 10^{-12} W/m^2.

11.8.3. (*a*) 85 dB *re* 10^{-12} W/m^2. (*b*) 78 sone. (*c*) 97 dB *re* 10^{-12} W/m^2.

11.8.7. (*a*) $I_T \sim 5 \times 10^{-12}$ W/m^2. (*b*) $N = 460(I - I_T)^{1/3}$. (*d*) $P_T \sim 4.6 \times 10^{-5}$ Pa, $N = 460(\sqrt{I} - \sqrt{I_T})^{2/3}$. (*e*) Pressure choice slightly better.

12.3.1. (*a*) 3.01×10^{-7} W/m^2. (*b*) 1.51×10^{-5} W.

12.3.3. (*a*) 1.79 s. (*b*) 8.7×10^{-6} W. (*c*) 2.4×10^{-7} W/m^2.

12.3.5. (*a*) $(Sc/8V)\bar{a}$. (*b*) $(S/8V)\bar{a}$.

12.3.7. (*a*) 0.020. (*b*) 0.52. (*c*) 0.15 s.

12.4.1. (*a*) 11.6 ft. (*b*) 0.23.

12.4.3. (*a*) 0.99, 0.97, 0.95, 0.90, 0.72. (*b*) $0.161V/S$, 0. (*c*) 13.8, 0, Eyring.

12.4.5. (*a*) $0.67 \le L_M \le 1.3$.

12.5.1. (*a*)–(*c*) 1.0, 0.18, 0.13 s.

12.6.1. (*a*) 1.4×10^{-4} W. (*b*) 2205 ft^2. (*c*) 0.2 s.

12.6.3. (*a*) 77.6 dB *re* 20 μPa. (*b*) 0.142 s. (*c*) 5.83 ft^2.

12.7.1. Choice (*a*) is better.

12.8.1. 13.8.

12.8.3. (*a*) 1.6 s, good agreement. (*b*) 30 ms (side), 20 (ceiling), sit further back in the hall. (*c*) 12.5 m, 47. (*d*) $\bar{a} = 0.30$, $\bar{a}_E = 0.26$

12.9.1. (*a*) 2.2 kHz. (*b*) 0.113 s. (*c*) 116 dB *re* 20 μPa. (*d*) 141, 187, 234 Hz.

12.9.3. (*a*) 54.3 Hz. (*b*) 1.56 s. (*c*) 259 Hz.

12.9.5. (*a*) 0.35. (*b*), (*c*) 0.28, 0.46 s.

12.9.7. (*a*) 6.4. (*b*) 0.64. (*c*) 0.018 s.

13.2.1. (*a*) 73.0 dB *re* 20 μPa. (*b*) 60.8 dBA.

13.2.3. 39, 50, 62, 73, 78, 80, 81, 74, 64 dB *re* 20 μPa.

13.3.1. (*a*) 53.5 vs. 53.8 dB *re* 20 μPa. (*b*) 80 dBA, impractical.

13.5.1. (*a*) 53. (*b*) 52.8, very good. (*c*) Shops, garages.

13.6.1. 58, 48, 43 dBA.

13.6.3. (*a*)–(*c*) 57.5, 59.6, 60 dBA.

13.8.1. 61, 71, 67, 78 dBA.

13.11.1. Yes.

13.13.1. 51.

13.15.1. (*a*) 22.4 kg/m^2. (*b*) Table 13.13.1, items 8, 10, 13.

13.15.3. 6.1 kHz.

13.15.5C. 3.68, 2.60, 2.12 m.

14.4.1. 60 dB *re* 1 μbar/V.

14.4.3. 14.3%.

14.5.1. (*a*) 0.98. (*b*) 0.98. (*c*) 4.7.

14.5.3. (*a*) 50.3 Hz. (*b*) 2.37. (*c*), (*d*) 1.9×10^{-3}, 5×10^{-4} m.

14.5.5C. 4.77 vs. 4.67.

14.6.1. (*a*) 2.37×10^3 N/m. (*b*) 0.085 kg.

14.6.3. (*a*) 0.69 m^3. (*b*) 1.08.

14.7.3. (*a*) 136 Hz. (*b*) 4.0×10^{-3} m^3/s. (*c*) 3.2×10^{-4} m.

14.7.5. (*a*) $1/\sqrt{S}$. (*b*) Yes. (*c*) $P(x) = (a/x)P(a)$, yes.

14.8.1. (*a*) -180 dB re 1 V/μPa. (*b*) 1.0 V.

14.9.1. (*a*) 0.05 V/Pa. (*b*) -26 dB *re* 1 V/Pa. (*c*) 5×10^{-9} m. (*d*) 0.028 V.

14.10.1. -50 dB *re* 1 V/Pa.

14.11.1. (*a*) 5.4×10^{-5} V. (*b*) -91 dB *re* 1 V/Pa. (*c*) 5.5×10^{-6} m.

14.11.3. (*a*) 5. (*b*) 7 dB.

14.12.1. $V_1/2V_0$, $(V_1/2V_0)^2$.

14.12.3. (*a*) 1.17×10^{-3} V. (*b*) 2.7×10^{-12} W. (*c*) 6.7×10^{-4}.

14.13.1. (*a*) -20 dB *re* 1 V/Pa. (*b*) $+4$ dB *re* 1 Pa.

14.13.3. (*a*) 5.0×10^{-3} V/Pa. (*b*) 0.2 Pa.

15.2.1. 1529.66, 1529.83 m/s.

15.3.1.

	100 Hz	1 kHz	10 kHz
(15.3.6) 0 km	0.98×10^{-3} dB/km	5.25×10^{-2} dB/km	1.09 dB/km
2 km	0.95×10^{-3}	4.97×10^{-2}	0.81
0 km 5°C	1.23×10^{-3} dB/km	6.31×10^{-2} dB/km	1.13 dB/km
20°C	0.71×10^{-3}	5.22×10^{-2}	0.74
2 km 5°C	1.18×10^{-3}	5.90×10^{-2}	0.84
20°C	0.68×10^{-3}	4.96×10^{-2}	0.57

15.5.1. (*a*) 1.36 km. (*b*) 1.5°. (*c*) 2.39 km. (*d*) 310 m.

15.5.3. (*a*), (*b*) 115, 184 m.

15.5.5. 4.34 km.

15.5.7. (*a*) 80 dB. (*b*) 4.5 dB/bounce.

15.7.1. (*a*)–(*d*) -13.6, 0, 4.8, 0 dB *re* 1 μbar. (*e*) -5.6, -1.2, 2.4, 0 dB *re* 1 μbar.

15.7.3. (*a*) Perpendicular to the surface. (*b*) $2Ad/\rho_0cr^2$.

15.8.1. 10 dB.

15.8.3. $5\log(wd/\tau)$

15.8.5. (*a*) $P(D)$ decreases, $P(FA)$ remains the same, d decreases.
(*b*) $P(D)$ decreases, $P(FA)$ decreases, d remains the same.

15.9.1. (*a*)–(*d*) 58, 65, 67, 60 dB *re* 1 μPa.

15.10.1. 1.3×10^{-5}.

15.10.3. (*a*) 4.9 Hz. (*b*) 75 dB *re* 1 μPa. (*c*) -5 dB. (*d*) 20 s. (*e*) 68 dB *re* 1 μPa, 2 dB, 5.2 s.

15.10.5. (a)–(c) 13.5, 0.76, 8.0 km.

15.10.7. (a) 15 min. (b) 1800. (c) 1.39×10^{-4}. (d) 5.8 dB.

15.11.1. (a) 20 s. (b) 6.67×10^{-5}. (c) 2.22×10^{-5}. (d) 16, 18.

15.11.3. −88 dB.

15.11.5. 2.7 km.

15.12.1. Discrepancy of 2 dB is consistent with assumptions.

15.12.3. (a) 6.67, 6.19, 6.19, 2.81, 2.81 Pa. (b) 0, 42, −186, −325, −325°, 12.6 Pa. (c) 11.7 Pa.

15.13.1. (a) 80 dB. (b) 31 km.

15.13.5. Good agreement.

16.3.1. (a), (b)

f (kHz)	ℓ (m)	Γ (moist air)	Γ (dry air)
0.1	2.27×10^3	3.1	12
1	2.27×10^2	2.4	9.0
10	22.7×10^1	22	2.2

(c) ℓ increased and Γ decreased by a factor of 10.

16.3.3. 15.8, yes.

17.1.3. (a) 1.302. (b) 351.8 K. (c) 2002 mbar, 1017 mbar. (d) 1.52. (e) 0.17 J/(mol·K).

17.1.7. (a) $\rho_1 U = \rho_2(U + u)$, $\mathcal{P}_1 + \rho_1 U^2 = \mathcal{P}_2 + \rho_2(U + u)^2$. (b) 15.1 atm, 735 K.

17.2.3. (a) $\ln p \approx \ln p(0) - [(b + 1)/\tau]t$. (b) $\ln p(0)$, $-(b + 1)/\tau$. (c) $p(t) = 0$ when $t = \tau$, solve for b from (b).

17.3.1. (a) 1.09×10^5 J. (b) 23.7 g TNT equivalent.

17.3.5.

Station	R (m)	M	$p(0)/\mathcal{P}_1$	t_a (ms)	$\langle u_1 \rangle_R$ (m/s)	τ (ms)	$\mathcal{J}/S$ (bar·ms)	b
A	3.05	1.303	0.813	4.0	761	1.67	0.52	0.75
B	2.42	1.47	1.35	2.65	913	1.377	0.66	1.01
C	14.7	1.025	0.0592	36.1	408	3.915	0.105	0.30
D	1.3	2.37	5.39	0.84	1557	0.767	1.03	2.50
E	6.38	1.077	0.187	12.9	495	2.87	0.24	0.44
F	1.87	1.746	2.39	1.65	1134	1.10	0.838	1.46
G	38	1.009	0.022	103	369	4.18	0.041	0.21
H	400	1.001	0.002	1166	343	4.206	0.003	0.13

17.3.11. (a) $\propto 1/R$. (b) 343 m/s. (c) Both consistent.

17.5.1. (a) 1.506, 1.50 bar. (b), (c) 238, 943 m/s. (d) 1.06. (e) 15.2 bar·ms. (f),(g) 10, 15 tonne.

17.5.3. (a) 3.776 km. (b) 250, 374 tonne TNT. (c) 13.7 mbar. (d) 3.3 m/s.

INDEX

GLOSSARY OF SYMBOLS

(continued from front endpapers)

r	specific gas constant; characteristic acoustic impedance; specific acoustic resistance
r_t	transition range
r_s	skip distance
R	resistance (acoustic, electrical, mechanical); reflection coefficient; radius of curvature
Re	Reynolds number
$\dot{R}$	range rate
R_m	mechanical resistance
R_r	radiation resistance
R_I	intensity reflection coefficient
R_Π	power reflection coefficient
RL	reverberation level
ROC	receiver operating characteristic
$\mathcal{R}$	universal gas constant
s	spring constant; condensation
sL	apparent source level
S	cross-sectional area; surface area; salinity
S_A, S_B	scattering strengths per unit area
$SENEL$	single event noise exposure level L_{ex} (dBA)
SIL	speech interference level
SL	source level
SPL	sound pressure level
SS	sea state

SSL	source spectrum level
STC	sound transmission class
S_V	scattering strength per unit volume
SWR	standing wave ratio
$\mathcal{S}$	transmitter sensitivity
$\mathcal{SL}$	transmitter sensitivity level
$\mathcal{S}_{ref}$	reference transmitter sensitivity
T	period of motion; temperature; tension; transmission coefficient; reverberation time
T_I	intensity transmission coefficient
T_K	temperature in kelvin
T_{em}, T_{me}	transduction coefficients
T_Π	power transmission coefficient
TL	transmission loss
TNI	traffic noise index (dBA)
TS	target strength
TS_R	target strength for reverberation
TTS	temporary threshold shift
$\mathcal{T}$	membrane tension per unit length
$\vec{u}$	particle velocity
u	particle speed
U	peak particle velocity amplitude; volume velocity
U_e	effective particle velocity amplitude